INTERNATIONAL
WILDLIFE
ENCYCLOPEDIA

THIRD EDITION

Volume 22
INDEX

Marshall Cavendish Corporation
99 White Plains Road
Tarrytown, New York 10591–9001

Website: www.marshallcavendish.com

Library of Congress Cataloging-in-Publication Data

Burton, Maurice, 1898-
    International wildlife encyclopedia / [Maurice Burton, Robert Burton] .-- 3rd ed.
        p. cm.
    Includes bibliographical references (p.).
    Contents: v. 1. Aardvark - barnacle goose -- v. 2. Barn owl - brow-antlered deer -- v. 3. Brown bear - cheetah -- v. 4. Chickaree - crabs -- v. 5. Crab spider - ducks and geese -- v. 6. Dugong - flounder -- v. 7. Flowerpecker - golden mole -- v. 8. Golden oriole - hartebeest -- v. 9. Harvesting ant - jackal -- v. 10. Jackdaw - lemur -- v. 11. Leopard - marten -- v. 12. Martial eagle - needlefish -- v. 13. Newt - paradise fish -- v. 14. Paradoxical frog - poorwill -- v. 15. Porbeagle - rice rat -- v. 16. Rifleman - sea slug -- v. 17. Sea snake - sole -- v. 18. Solenodon - swan -- v. 19. Sweetfish - tree snake -- v. 20. Tree squirrel - water spider -- v. 21. Water vole - zorille -- v. 22. Index volume.
    ISBN 0-7614-7266-5 (set) -- ISBN 0-7614-7267-3 (v. 1) -- ISBN 0-7614-7268-1 (v. 2) -- ISBN 0-7614-7269-X (v. 3) -- ISBN 0-7614-7270-3 (v. 4) -- ISBN 0-7614-7271-1 (v. 5) -- ISBN 0-7614-7272-X (v. 6) -- ISBN 0-7614-7273-8 (v. 7) -- ISBN 0-7614-7274-6 (v. 8) -- ISBN 0-7614-7275-4 (v. 9) -- ISBN 0-7614-7276-2 (v. 10) -- ISBN 0-7614-7277-0 (v. 11) -- ISBN 0-7614-7278-9 (v. 12) -- ISBN 0-7614-7279-7 (v. 13) -- ISBN 0-7614-7280-0 (v. 14) -- ISBN 0-7614-7281-9 (v. 15) -- ISBN 0-7614-7282-7 (v. 16) -- ISBN 0-7614-7283-5 (v. 17) -- ISBN 0-7614-7284-3 (v. 18) -- ISBN 0-7614-7285-1 (v. 19) -- ISBN 0-7614-7286-X (v. 20) -- ISBN 0-7614-7287-8 (v. 21) -- ISBN 0-7614-7288-6 (v. 22)
    1. Zoology -- Dictionaries.   I. Burton, Robert, 1941-   . II. Title.

    QL9 .B796 2002
    590'.3--dc21

                                                        2001017458

Printed in Malaysia
Bound in the United States of America

07 06 05 04 03 02 01   8 7 6 5 4 3 2 1

Brown Partworks
Project editor: Ben Hoare
Associate editors: Lesley Campbell-Wright, Rob Dimery, Robert Houston, Jane Lanigan, Sally McFall, Chris Marshall, Paul Thompson, Matthew D. S. Turner
Managing editor: Tim Cooke
Designer: Paul Griffin
Picture researchers: Brenda Clynch, Becky Cox
Illustrators: Ian Lycett, Catherine Ward
Indexer: Kay Ollerenshaw

Marshall Cavendish Corporation
Editorial director: Paul Bernabeo

Authors and Consultants

Dr. Roger Avery, BSc, PhD (University of Bristol)

Rob Cave, BA (University of Plymouth)

Fergus Collins, BA (University of Liverpool)

Dr. Julia J. Day, BSc (University of Bristol), PhD (University of London)

Tom Day, BA, MA (University of Cambridge), MSc (University of Southampton)

Bridget Giles, BA (University of London)

Leon Gray, BSc (University of London)

Tim Harris, BSc (University of Reading)

Richard Hoey, BSc, MPhil (University of Manchester), MSc (University of London)

Dr. Terry J. Holt, BSc, PhD (University of Liverpool)

Dr. Robert D. Houston, BA, MA (University of Oxford), PhD (University of Bristol)

Steve Hurley, BSc (University of London), MRes (University of York)

Tom Jackson, BSc (University of Bristol)

E. Vicky Jenkins, BSc (University of Edinburgh), MSc (University of Aberdeen)

Dr. Jamie McDonald, BSc (University of York), PhD (University of Birmingham)

Dr. Robbie A. McDonald, BSc (University of St. Andrews), PhD (University of Bristol)

Dr. James W. R. Martin, BSc (University of Leeds), PhD (University of Bristol)

Dr. Tabetha Newman, BSc, PhD (University of Bristol)

Dr. J. Pimenta, BSc (University of London), PhD (University of Bristol)

Dr. Kieren Pitts, BSc, MSc (University of Exeter), PhD (University of Bristol)

Dr. Stephen J. Rossiter, BSc (University of Sussex), PhD (University of Bristol)

Dr. Sugoto Roy, PhD (University of Bristol)

Dr. Adrian Seymour, BSc, PhD (University of Bristol)

Dr. Salma H. A. Shalla, BSc, MSc, PhD (Suez Canal University, Egypt)

Dr. S. Stefanni, PhD (University of Bristol)

Steve Swaby, BA (University of Exeter)

Matthew D. S. Turner, BA (University of Loughborough), FZSL (Fellow of the Zoological Society of London)

Alastair Ward, BSc (University of Glasgow), MRes (University of York)

Dr. Michael J. Weedon, BSc, MSc, PhD (University of Bristol)

Alwyne Wheeler, former Head of the Fish Section, Natural History Museum, London

Contents

Glossary 3028

Bibliography 3043

Internet Resources 3045

Resources for Younger Readers 3048

Wildlife Refuges and Other

    Places of Interest 3050

Wildlife Conservation Agencies

    and Organizations 3055

Index of Common Names 3057

Index of Scientific Names 3081

Index of Places 3099

Index of Animal Behaviors 3107

Comprehensive Index 3113

Glossary

abdomen: in vertebrates, the stomach region; in insects, the third body section

adipose: a term meaning fatty; for example, adipose tissue

age at first breeding: the age at which individuals of an animal species typically breed for the first time, and which in many vertebrates may be after physical sexual maturity has been reached; *see also* sexual maturity

alien species: a plant or animal species that is not native to the habitat in which it is currently found

alpha: a term used to describe a mammal (often a carnivore or primate) that is socially dominant within its group; for example, alpha male

altruistic: a term used to describe behavior by an organism that, while not beneficial to itself, benefits another organism; *see also* symbiotic

Amphibia: the scientific name for the class of cold-blooded vertebrates, distinguished by an aquatic gill-breathing larval stage that is almost always followed by a terrestrial lung-breathing adult stage

amphibian: a member of the class Amphibia; amphibians include the caecilians, frogs, newts, salamanders and toads

amphipod: a member of the subclass Malacostraca, comprising approximately 3,500 known species of minute, laterally flattened freshwater and marine crustaceans; many species of amphipods are important detritus feeders and scavengers

amplexus: the mating embrace of a frog or toad, during which the female's eggs are shed into the water and then fertilized by the male's sperm

anal fin: a fin located near the anus of a fish

antler: one of the pair of deciduous bone appendages that projects from the head of a deer, or member of the deer family

ape: a primate, distinguished from monkeys by having a large brain and body, lengthy forelimbs and no tail; *see also* great ape, lesser ape

appendage: a body part; for example, a limb

Arabia: a geographical region comprising much of the Middle East, including Saudi Arabia, the UAE (United Arab Emirates), Oman, North Yemen and South Yemen

arachnid: a member of the class Arachnida; arachnids include the daddy longlegs (harvestmen or harvest spiders), false scorpions (pseudoscorpions), mites, scorpions, solifugids (sun spiders), spiders, ticks and whip scorpions (Uropygi)

Arachnida: the scientific name for the class of usually terrestrial invertebrates, having an external skeleton and a segmented body divided into two distinct areas, the hind area bearing four pairs of jointed limbs but no antennae

aquatic: a term meaning water living

arboreal: a term meaning tree dwelling

arthropod: a member of the phylum Arthropoda; arthropods include the arachnids, crustaceans and insects

Arthropoda: the scientific name for the phylum containing many terrestrial and marine invertebrates, having an external skeleton, a segmented body and jointed limbs; the largest phylum in the animal kingdom in terms of number of species and biomass

artiodactyl: a member of the order Artiodactyla, or even-toed ungulates; artiodactyls include the antelopes, camels, cattle, deer, gazelles, giraffe, goats, hippos, pigs and sheep

Artiodactyla: the scientific name for the even-toed ungulates, the order of hoofed mammals in which there is an even number of functional toes on each foot; *see also* Perissodactyla, ungulate

Australasia: a geographical region comprising Australia, New Zealand, Tasmania and Melanesia (a group of islands in the southwestern Pacific, including the Solomon Islands, Vanuatu, New Caledonia and Fiji)

Aves: the scientific name for birds, the class of warm-blooded vertebrates in which much of the body is covered with feathers, the forelimbs are modified as wings and the adult female lays eggs

bacterium (*plural* bacteria): a unicellular microorganism that has a cell wall but lacks an organized nucleus; some bacteria can cause disease

baleen: whalebone plates—the thin sheets of keratin that hang down from the roof of a whale's mouth, resembling curtains

or an internal mustache, and which are used in feeding; found only in the large whales of the suborder Mysticeti

barbel: a slender growth on the mouth or nostril of certain fish, used as a sensory organ for touch

bathypelagic zone: the region of the ocean at a depth of approximately 2,000–12,000 feet (600–3,600 m)

benthic: a term that refers to the ocean bottom

binocular vision: a form of vision in which the eyes face forward and can focus simultaneously on an object; each eye provides a slightly different image of the object, but the brain coordinates the two images to produce a single, three-dimensional image; also known as stereoscopic vision, it is characteristic of predatory vertebrates, including humans

biodiversity: biological diversity, or the variety of life

bioluminescence: the production of light by living organisms, such as fireflies, glowworms and some crustaceans and fish

biomass: the total quantity of living matter in a given area

biome: a major ecological community type; for example, coral reef, desert or rain forest

biped: an organism that moves on two feet; *see also* quadruped

bird: a member of the class Aves

bird of prey: a member of the mainly predatory bird order Falconiformes, such as a buzzard, caracara, eagle, falcon or hawk, which has a hooked bill, powerful legs and feet and keen vision; contrary to popular opinion, the owls (order Strigiformes) are not generally considered to be birds of prey

bivalve: a member of the class Bivalvia; the bivalve mollusks include the clams, cockles, mussels, oysters, razor shells, scallops and shipworms

Bivalvia: the scientific name for the bivalve mollusks, or lamellibranchs, the class of marine mollusks that possess a shell composed of calcium carbonate (chalk) and divided into two valves, joined at a hinge

blood pressure: the force created by blood moving through the body of an animal

bloom: a term used to describe a dense growth of planktonic algae, which may give a color to the water in which it occurs

blubber: a thick layer of fatty tissue found below the skin of many large aquatic mammals, including seals and whales, and which acts as a means of thermal insulation

bony fish: a member of the class Osteichthyes; bony fish include most freshwater and many marine fish species, such as the angelfish, bass, carp, catfish, eels, flatfish, groupers, herrings, perch, salmon, snappers, tuna and trout

boreal: a term referring to the northern regions of the Northern Hemisphere, especially those areas dominated by coniferous forest

bovine: a member of the subfamily Bovinae; bovines include the bison, buffalo, oxen and their close relatives

brackish: a term used to describe slightly salty water, such as that found in many coastal lagoons and marshes

breach: to leap out of the water—a habit characteristic of several dolphin, porpoise and whale species; the purpose of breaching may be to dislodge parasites attached to the animal's skin, or to act as a means of communication

bromeliad: a usually tropical herbaceous plant that grows on other plants

brood (*noun*): a collective term applied to a group of young siblings (usually birds) of the same age

brood (*verb*): to sit on or incubate eggs or young

browse: to feed on the buds, shoots, leaves or twigs of trees, bushes or shrubs

brow tines: the projected parts of a deer's antlers

buck: the male of certain species of mammals, especially antelopes, deer, gazelles and rabbits

budding: a form of asexual reproduction in which part of an organism grows and breaks off to become a new individual

buff: a pale, yellow-white color used to describe the appearance of certain animals, especially birds

bull: the male of certain species of mammals, especially bovines, elephants, seals and whales

cache: a food store

calf: the young of certain species of mammals, especially bovines, elephants and whales

canid: a member of the family Canidae; canids include the coyote, dingo, domestic dog, foxes, jackals and wolves

Canidae: the scientific name for dogs, or the family Canidae, a group of medium-sized carnivores distinguished by large canine teeth, long, strong muzzles and large external ears

canopy: the uppermost layer of forest or woodland, formed by the interlocking leaves and branches of trees and often creating

a more or less continuous covering; the canopy of a rain forest intercepts 75 percent of the sunlight that the forest receives

captive breeding: a term applied to any method of bringing animals of the same species into a zoo or similar closed environment in order for them to mate

carapace: a bony casing or shield that covers the back, part of the back, or all of an animal such as a tortoise or turtle

Carnivora: the scientific name for the carnivores, the order of mammals that have powerful jaws and teeth adapted for tearing and eating flesh; most species in the order are mainly predatory, but many also eat vegetable matter and a few (for example, the giant panda) are entirely vegetarian

carnivore: any member of the order Carnivora, including badgers, bears, cats, civets, dogs, ermines, ferrets, hyenas, martens, minks, mongooses, otters, polecats, raccoons and weasels; more generally, the term may be applied to any flesh-eating animal

carnivorous: a term meaning flesh eating; frequently applied to a wide variety of predatory animals, including those that do not belong to the mammalian order Carnivora; *see also* herbivorous, insectivorous, omnivorous

carrion: the decaying flesh of a dead organism

cartilage: an elastic, somewhat translucent connective tissue found in vertebrates

cartilaginous fish: a member of the class Chondrichthyes; cartilaginous fish include the rays, sharks and skates

caste: a specialized form of social insect that carries out specific tasks within its colony; for example, a soldier ant or worker honeybee

caterpillar: the elongated larval stage found in some insects, including butterflies and moths

caudal fin: the tail fin of a fish

cephalopod: a member of the order Cephalopoda; cephalopods comprise the cuttlefish, nautiluses, octopuses and squid

Cephalopoda: the scientific name for the order of marine mollusks that move by expelling water from a tubular siphon under the head, and which possess a group of muscular and usually sucker-bearing arms around the front of the head, a pair of well-developed eyes and (in most species) an ink-containing sac that may be ejected as a means of defense or concealment

Cetacea: the scientific name for the cetaceans, the order of aquatic mammals that possess a nearly hairless body, paddle-shaped forelimbs but no hind limbs, one or two nostrils

opening on the top of the head and a horizontally flattened tail used for locomotion; most cetacean species are marine, but some, particularly the river dolphins, occur in fresh water

cetacean: a member of the order Cetacea; cetaceans include the dolphins, porpoises, river dolphins and whales

chaparral: a biome primarily composed of small-leaved evergreen shrubs, bushes and dwarf trees that form a dense layer of vegetation 3–13 feet (1–4 m) high; found in California, northwestern Arizona, northern Baja California, central Chile, the Mediterranean, the Cape of Africa and southern Australia

chelicera (*plural* chelicerae): one of the front pair of appendages in an arachnid, often specialized as fangs

chemoreceptor: a sense organ that responds to chemical stimuli; for example, taste buds

chlorophyll: a substance that plants and some microorganisms use in order to make food from sunlight

Chondrichthyes: the scientific name for the cartilaginous fish, the class of fish in which the skeleton is wholly or partly composed of cartilage rather than bone; *see also* Osteichthyes

chromatophore: a pigment-bearing cell

chrysalis: *see* pupa

ciguatera: a form of poisoning contracted by eating certain normally edible fish, especially tropical species such as barracuda and snappers, in whose flesh high levels of toxins have accumulated

cilium (*plural* cilia): a tiny, hairlike extension of a cell membrane, often used to move a cell through fluid or to move fluid (such as food or waste) past a cell

circumpolar: a term applied to the region surrounding either the North or South Pole

class: a zoological ranking of animals or plants that share a common set of traits, occurring below the rank of phylum and above that of order

classification: the way in which a species is ranked zoologically in relation to other species

climate: average weather conditions (such as temperature, wind and rain) over a period of years

cloaca: in most vertebrates, a chamber into which the urinary, intestinal and generative (reproductive) canals feed

clutch: a collective term applied to the eggs laid during one nesting cycle; some bird species may produce two or more clutches during a single breeding season

cold-blooded: a term applied to amphibians, fish and reptiles, in which the body temperature is not internally regulated but approximates that of the environment; *see also* warm-blooded

colony: a group of plants or animals, the members of which live in close proximity and interact for their mutual benefit

compound eye: an eye made of many separate visual units

conifer: a cone-bearing tree

coniferous: a term applied to cone-bearing trees and to areas of forest or woodland in which they are the dominant tree type

convergent evolution: the process by which two or more organisms evolve to resemble each other more closely than their ancestors did; *see also* parallel evolution

continental shelf: a shallow underwater plain that borders a continent and which usually ends in a steep slope extending downward to the bottom of the ocean

copepod: a member of the subclass Copepoda, comprising 8,500 to 10,000 known species of minute freshwater and marine crustaceans; copepods often occur in enormous numbers and are of great ecological importance, providing food for many species

coral: the horny skeletal deposit produced by marine polyps; also applied to a single polyp or a colony of polyps

cosmopolitan: a term used to describe species found in most parts of the world and under a wide variety of ecological conditions

cow: the female of certain species of mammals, especially bovines, elephants, seals and whales

crepuscular: a term meaning active at twilight (dawn and dusk)

crocodilian: a member of the order Crocodylia; crocodilians include the alligators, caimans, crocodiles and gharials

Crocodylia: the scientific name for the order of largely aquatic, carnivorous reptiles that have a long, thick-skinned body, internal nostrils that open far back inside the mouth and powerful jaws packed with sharp teeth

crop: in vertebrates (especially birds), an expandable part of the esophagus, in which food may be stored; in invertebrates, an expandable part of the gut, in which food may be digested or stored

crown: in birds, the top of the head

Crustacea: the scientific name for the large class of usually aquatic arthropods that have an external skeleton; a body divided into many segments, each with a pair of often much modified appendages; two pairs of antennae; and gills for breathing

crustacean: a member of the class Crustacea; crustaceans include the amphipods, barnacles, copepods, crabs, crayfish, krill, lobsters, prawns, shrimps, water fleas and wood lice

cud: the food brought up from the stomach for a second chewing, especially in the antelopes, cattle, deer, gazelles and other members of the order Artiodactyla

cuticle: a hard, waxy waterproof covering on the outer surface of many insects and plants

cryptic: a term used to describe coloration designed to camouflage or conceal

cygnet: a young swan

deciduous: a term that refers to the shedding of part of an organism during a certain season or at a regular stage of growth, for example, deciduous trees drop their leaves annually; *see also* evergreen

decomposer: a bacterium or fungus that breaks down, or rots, dead organisms into basic nutrients

decurved: a term meaning curving downward; for example, decurved bill

deforestation: the process by which trees are removed from a particular area, usually to enable agriculture or some other form of land development to take place

delayed implantation: in placental mammals, the process by which the embedding of an embryo in the wall of the mother's uterus is delayed for a period

delta: a wetland biome, comprising a fertile area of alluvial deposits at the mouth of a river

dermis: the sensitive, underlying layer of a vertebrate's skin

desert: a major biome, covering approximately one-seventh of Earth's surface, characterized by lack of plant cover and sometimes defined as any region that has less than 10 inches (25 mm) of rainfall per year, although most deserts receive far less than this; the world's major desert areas are located in the southwestern United States, northern Mexico, coastal Peru and Chile, northern and southwestern Africa, the Middle East, Central Asia and inland Australia

detritivore: a bacterium or other minute organism that lives on the seabed and feeds on dead animals and plant matter

detritus: the organic debris left by living organisms

dew claw: a vestigial digit that does not reach the ground, found on the foot of some mammals, including cats

dew hoof: a vestigial hoof that does not reach the ground, found on the foot of some mammals, including certain species of antelopes, deer and gazelles

dewlap: a flap of loose skin that hangs from the throat of some mammals, including certain species of antelopes and oxen

diatom: a type of microscopic, unicellular alga

digit: the finger or toe of a vertebrate

digitigrade: a term used to describe a manner of movement in which an animal walks on the digits of its feet, the hind part of the foot being raised from the ground

Diptera: the scientific name for the order of flies

dipteran: a member of the insect order Diptera, including the blowflies, crane flies, gnats, horseflies, houseflies, hoverflies, midges and mosquitoes

diurnal: a term meaning active during the day

doe: the adult female of several types of mammals, including deer, kangaroos and rabbits

dorsal: a term relating to the upper surface, or back, of an animal; the dorsal fin of a cetacean or fish is situated on its back; *see also* ventral

DNA: deoxyribonucleic acid—the acid located in the cell nucleus of an organism that is the molecular basis of heredity

drake: a male duck

drone: a male honeybee that has no stinger and collects no honey, having the sole responsibility of mating with the queen

eardrum: a layer of skin in a vertebrate's ear that vibrates when hit by sound waves

echinoderm: a member of the marine phylum Echinodermata, including the brittle stars, featherstars, sea cucumbers, sea lilies, sea urchins and starfish

Echinodermata: the scientific name for a large phylum of marine invertebrates in which the body is radially symmetrical

echolocation: the process by which an animal uses sound to construct a picture of its surroundings

eclipse plumage: the dull plumage worn by male wildfowl during the flightless part of their molt, in order to provide them with camouflage

ecology: the scientific study of the relationship between a living organism and its environment

ecosystem: a community of animals, plants and bacteria and its interrelated physical and chemical environment

egg tooth: an outgrowth at the tip of a bird's upper beak, which is shed soon after hatching

elytron (*plural* elytra): one of the forewings of a beetle or other insect that folds back to protect the functional hind wings

embryo: in animals, the next stage of life after an egg and before birth

endangered species: any species that is on the verge of becoming extinct in the wild

endemic: a term applied to any species that is native to a particular geographic region and that does not occur anywhere else, usually refers to animals with restricted ranges

endoskeleton: a skeleton that is found within the body

epidemic: a situation in which one disease temporarily affects a large proportion of a species' population

epipelagic zone: any part of the ocean into which enough light penetrates to allow photosynthesis

epiphyte: a so-called air plant that grows perched on another plant, nonparasitically; examples of epiphytes include bromeliads and many ferns and orchids

esophagus: a tube that joins the throat to the stomach

estivate: in animals, a form of torpor—to become dormant for an extended period in order to avoid harsh conditions such as heat or drought; *see also* hibernate, torpor

estrous: the period during which female mammals can become pregnant

Eurasia: a term that refers to the landmasses of Europe and Asia, considered as a whole

evaporation: the process by which water turns from a liquid to a gas

even-toed ungulate: *see* Artiodactyla

evergreen: in plants, a term that refers to the retention of leaves all year round; evergreen trees include the cedars, cypresses, firs, pines and spruces; *see also* deciduous

ewe: a female sheep

exoskeleton: an external skeleton—a hard casing that covers the outside of the body, found in a range of animals, including armadillos, crustaceans, insects, mollusks and tortoises

exotic species: *see* alien species

extinct: a term applied to any species that has not been found in the wild for a prolonged period and which is presumed to have disappeared from the wild forever

eye ring: a ring-shaped area of often brightly colored bare skin surrounding a bird's eye

eye stripe: a roughly horizontal, linear plumage feature that arises in front of a bird's eye, continuing behind it; *see also* supercilium

family: a zoological ranking of species that share certain traits, occurring below the rank of order and above that of genus

fauna: animal life—a collective term that describes the animals characteristic of a region, environment type or period; *see also* flora

fawn: a young deer, especially one that has not yet been weaned

feces: the expelled waste products of digestion

feral: a term applied to a wild animal, or population of animals, that has descended from a tame or domesticated ancestor or population

fertile: to be able to support life or produce offspring

filter feeding: a form of feeding whereby minute water-living particles are strained before being consumed; this process is common among aquatic invertebrates and is also found in fish such as basking and whale sharks and mammals such as baleen whales

flagellum (*plural* flagella): a long, whiplike extension of a cell membrane that beats in wavelike undulations; in microorganisms, a group is often used for locomotion

fledging period: the period immediately after a bird's hatching, during which it acquires the feathers required for flight but before it becomes fully independent from the nest

fledgling: a young bird that has recently acquired its first feathers

flora: plant and bacterial life—a collective term that describes the plants and bacteria characteristic of a region, environment type or period; *see also* fauna

foliage: a general term that refers to the assembled mass of leaves, branches and twigs on a tree, bush or shrub

foot: in mollusks, the ventral (lower) surface that is muscular and flattened, forming a sole on which the animal creeps forward

fossil: the remains of an animal, or evidence of its existence, often preserved in ice, tar, peat, rock or volcanic ash

forest: a complex plant community in which trees, bushes and undergrowth grow closely together, sometimes forming a continuous canopy overhead

fry: a term for a young fish or a number of young fish

gander: a male goose

ganoid: a term describing fish scales that consist of bone and a shiny, outer layer resembling enamel

gastropod: a member of the class Gastropoda; gastropods include limpets, sea hares, sea slugs (nudibranchs), terrestrial slugs, snails and whelks

Gastropoda: the scientific name for the class of marine, freshwater or terrestrial mollusks that have a distinct head bearing eyes and tentacles; a well-developed radula (rasping tongue); a large, muscular foot used in locomotion; and a soft body, often covered with a shell

genus (*plural* genera): a zoological ranking of species that share many specific traits, placed below the rank of family and above that of species; the scientific name of a genus is always written in italics

gestation: the period of active embryonic growth inside a mammal's body between the time the embryo attaches itself to the uterus and the time of birth; the females of some mammal species may carry dormant embryos for a period of several months before the embryos become attached to the uterus

gestation period: the period of pregnancy in viviparous animals

gill (*often* gills): the paired respiratory organ of fish and some amphibians, by which oxygen is extracted from water flowing over surfaces within or attached to the walls of the pharynx

gizzard: the toughened part of a bird's digestive system used to grind up food prior to digestion

gonopodium: a modified anal fin found in the male of some fish, with which sperm is inserted into the female's oviduct

graze: to feed on herbage such as grasses; also, to feed on algae or phytoplankton

great ape: any primate belonging to the family Pongidae, including the bonobo (pygmy chimpanzee), chimpanzee, gorilla and orangutan

greenhouse effect: the warming of Earth's atmosphere by gases (especially carbon dioxide) that trap heat

grub: the soft-bodied larva of a beetle

guard hair: one of the long, outer hairs that together make up the coat of a mammal, beneath which lies a layer of shorter

hairs, or underfur; the combination of these two hair types provides efficient thermal insulation

habitat: an environment type in which a particular range of animals and plants is naturally or normally found; for example, a lowland oak wood or a garden pond

haltere: one of a pair of clublike or drumstick-shaped organs present in the true flies (family Tipulidae, order Diptera) that are the modified second pair of wings, and which serve as flight stabilizers

harem: a group of female animals that is attended to by only one male of the species

hemimetabolous: a term meaning incomplete metamorphosis; for example, hemimetabolous insects

hemoglobin: the red, iron-containing substance in blood that carries oxygen

herbivore: an animal that feeds on plants

herbivorous: a term meaning plant eating; *see also* carnivorous, insectivorous, omnivorous

hermaphrodite: an animal that bears both male and female sexual organs

hibernaculum: a den or shelter occupied by a hibernating animal or group of animals (especially certain species of insects and snakes)

hibernate: in warm-blooded animals, to spend the winter in an inactive or dormant state; *see also* estivate, torpor

hierarchy: the relationship between individuals of the same species, or between different species, that determines the order in which animals have access to food, water, mates, nesting or denning sites and similar vital resources

hind: a female deer; *see also* doe

holometabolous: a term meaning complete metamorphosis; for example, holometabolous insects

home range: the area usually covered by an individual animal during a particular period of its life; *see also* range

honeydew: a sweet liquid excreted by some insects, including aphids, psyllids (jumping plant lice), scale insects and the larvae of some species of butterflies and moths; pastoral ants "milk" honeydew-producing insects for this liquid by stimulating them with their antennae

host: an organism that sustains another organism

hybrid: the offspring that results from mating between two different species

hybridization: the mating of two separate species to produce offspring

ichthyologist: a scientist who studies fish

immature: a term used to describe a young animal that has not yet reached sexual maturity

in captivity: a term applied to any animal living in a zoo, captive-breeding program or private collection; animals may be bred in captivity because no specimens are left in the wild

incubation period: the period during which an egg is kept warm by one or more parents or adults until the embryo develops and hatches

indigenous species: any species that is native to the habitat in which it is currently found

insect: a member of the class Insecta

Insecta: the scientific name for insects, the very large class of arthropods that have an external skeleton and a segmented body divided into three distinct areas: the head, thorax and abdomen; the head of an insect bears mandibles and one pair of antennae, the thorax bears three pairs of jointed limbs and frequently one or two pairs of wings, and the abdomen bears no legs but may have other appendages, such as an ovipositor

insecticide: an insect-killing chemical, used to control unwanted populations of insect pests on agricultural land and in gardens

insectivore: an animal that feeds mainly on insects, and also worms or other invertebrates

insectivorous: a term meaning insect-eating; *see also* carnivorous, herbivorous, omnivorous

instar: an immature stage between successive molts in most arthropods

interbreed: the mating of two separate species to produce offspring

intertidal zone: *see* littoral zone

intestine: in vertebrates, the tubular part of the digestive system, extending from the stomach to the anus; in invertebrates, the whole of the digestive system, from the mouth to the anus

invertebrate: any animal without a spinal column, or backbone

Jacobson's organ: either one of a pair of small pits or sacs, situated in the roof of the mouth, and developed as chemoreceptors in amphibians, reptiles and some mammals

joey: a young kangaroo or wallaby

juvenile: a young animal that is not yet sexually mature

kelp: any of the large, brownish species of seaweeds that belong to the order Laminariales

keratin: a fibrous substance formed from hardened skin; keratin is the main structural component of hair, feathers, claws, hoofs, horns, baleen (whalebone plates) and human fingernails

kid: a young goat

kingdom: the largest taxonomic category, above phylum

kleptoparasitism: a term meaning food stealing

krill: any of the approximately 85 known species of shrimplike marine crustaceans belonging to the order Euphausiacea; often refers exclusively to the Antarctic krill, *Euphausia superba*, which is the main food of many whales, seals, penguins, squid and fish

lagomorph: a member of the order Lagomorpha; lagomorphs include hares, jack rabbits, pikas and rabbits

Lagomorpha: the scientific name for an order of small terrestrial mammals in which the ears are large and the eyes are set wide apart on the head; members of the order typically occur in arid steppes, open grasslands, rocky terrain and tundra

larva: an immature, nonreproductive stage in an invertebrate's development

lateral line: a canal that runs along the side of a fish's body, containing pores opening into tubes that are supplied with organs sensitive to low-level vibrations

Laufschlag: in some antelope species, the mating kick—a courtship maneuver in which the male touches the female's underside with a stiff foreleg, placing it either under her flank from the side or between her hind legs from behind; the Laufschlag is performed by blackbucks, dibatags, gazelles, kobs and oryxes, but not by bushbucks, eland, hartebeest, impala, kudu or wildebeest

legume: a plant belonging to the legume, or pea, family

lek: a clearly defined area where certain animals regularly assemble for display or courtship behavior

lesser ape: any of the 11 species of gibbons, belonging to the family Hylobatidae

lichen: a fungus and an algae living closely together in a symbiotic association, developing into a unique form of life that is distinct from either partner

ligament: a tough strip of tissue that joins bones together and holds organs in place

litter: a collective term applied to the siblings produced by a single pregnancy, especially in cartilaginous fish and mammals

littoral zone: the section of the coast that is covered and uncovered by the tides every day; also known as the intertidal zone

longline fishing: a form of large-scale commercial fishing in which boats trail lines up to 80 miles (130 km) long, with up to 12,000 baited hooks on them; longline fishing is a controversial technique because it causes the accidental death of tens of thousands of seabirds each year

maggot: the wormlike soft-bodied larva of a fly

malaria: a blood disease carried by mosquitoes

mamma (*plural* mammae): a mammary gland—a gland in the skin of a mammal, which secretes milk that nourishes the young; mammae are present in both sexes, but are functional only in the adult female

mammal: a member of the class Mammalia

Mammalia: the scientific name for the mammals, the class of warm-blooded vertebrates that have a body covering of hair or fur, prominent external ears and a mouth armed with teeth, and which suckle their young on milk

mammary gland: *see* mamma

mandible: in vertebrates, the lower jaw; in crustaceans, insects and myriapods (for example, centipedes and millipedes), one of the first pair of mouthparts used to bite and crush food

mangrove: any of the trees that belong to the tropical genus *Rhizophora* of the family Rhizophoraceae, in which many prop roots grow around the main trunk to form a dense, tangled mass; mangroves grow along coastlines, playing an important part in coastal land building

mantle: in bivalve mollusks, the flap of muscular tissue located immediately beneath both of the valves

marine: a term meaning relating to the ocean

marsupial: any member of the orders Didelphimorpha, Marsupialia or Paucituberculata; marsupials include the bandicoots, kangaroos, koala, opossums, possums, wallabies and wombats

Marsupialia: the scientific name for the marsupials, the three orders of mammals in which the female has a pouch on her abdomen wherein she carries the young; with the exception of the 70 species of American opossums of the order Didelphimorpha and the seven species of South American caenolestids or rat opossums of the order Paucituberculata, marsupials are confined to Australasia

matriarchal: a term meaning female dominated; for example, matriarchal society

membranous: a term meaning thin and pliable

mesopelagic zone: the region of the ocean at a depth of approximately 600–3,000 feet (200–1,000 m)

metamorphosis: a dramatic change in an animal's form and structure subsequent to birth; for example, the change from a caterpillar into a butterfly, or from a tadpole into a frog

migrate: to move regularly from one clearly defined range to another, particularly at the turn of a season

milt: a fluid containing sperm, released by many amphibians and fish during mating

mimic: in animals, a species that practices mimicry

mimicry: the practice whereby one organism adopts a superficial resemblance to another, or to natural objects (such as tree bark or leaves) among which it lives, in order to improve its chances of survival (for example, by providing protection from predators)

Mollusca: the scientific name for the mollusks, the phylum of usually aquatic, unsegmented invertebrates in which the soft body characteristically is enclosed in a hard shell

mollusk: a member of the phylum Mollusca; mollusks include the bivalves (the clams, cockles, mussels, oysters, razor shells, scallops, shipworms and their relatives), the cephalopods (the cuttlefish, nautiluses, octopuses and squid) and the gastropods (the limpets, sea hares, slugs, snails, whelks and their relatives)

monogamy: the practice of remaining loyal to one mate for a given period; *see also* polygamy

montane: a term used to describe a mountain-dwelling species or the mountainous environment, habitat or region in which it is found; for example, a montane forest

morph: a collective term applied to some individuals of a species, which are distinguishable from other members of their species due to genetically determined differences in form (especially color), although they are capable of interbreeding; for example, light and dark morph birds may breed successfully together, producing both light and dark morph young in the same brood

morphology: the form and structure of a plant or animal; also, the branch of biology that examines the form and structure of plants and animals

mouth brooding: in some fish species, the practice of carrying eggs and fry (young) in the mouth of the adult to improve the survival rate of the offspring, for example by providing protection from predators; mouth-brooding fish include many species in the cichlid family, Cichlidae

mucus: a viscous, slimy liquid that is produced by, and protects, mucous membranes

musk: a substance characterized by its strong odor, produced by a sac in male musk deer, skunks and some other mammals

mustelid: a member of the family Mustelidae of the order Carnivora; mustelids include badgers, ferrets, martens, minks, otters, polecats, skunks, stoats, weasels and wolverines

Mustelidae: the scientific name for the family of small and medium-sized carnivores that have a scent gland located around the anus, which can secrete a pungent, foul-smelling odor; the largest of all the carnivore families

myriapod: a general term for a centipede or millipede

nectar: a sweet liquid made by glands inside flowers in order to encourage animals to visit and pollinate them

nekton: any of the free-swimming animals that live in the pelagic zone of lakes, seas or oceans, such as amphibians, dolphins, fish, squid, turtles and whales

nematocyst: a stinging organ used by aquatic invertebrates such as anemones, corals, hydras and jellyfish to overpower or subdue prey

neoteny: the retention of certain immature or larval characteristics in adulthood; also refers to the attainment of sexual maturity during the larval stage; examples of animals in which neoteny occurs include the axolotl and mudpuppy

neritic zone: the area of the sea over the continental shelf; generally regarded as referring to water of less than 600 feet (200 m) in depth

nestling: a young bird that has not yet left the nest

New World: a geographical term referring to the Western Hemisphere, in particular North, Central and South America; *see also* Old World

niche: a specific position occupied by one species in an ecosystem

nictitating membrane: a third eyelid, found in birds and crocodilians

nocturnal: a term meaning night active

nomadic: a term applied to a species with no permanent territory or range that wanders in search of food and water; *see also* sedentary

nudibranch: any species of sea snail or sea slug belonging to the order Nudibranchia; there are about 45 families of nudibranchs, including some of the most colorful of all marine animals

nutrient: a substance necessary for an organism's life processes

nymph: a sexually immature larval stage of some arthropods

odd-toed ungulate: *see* Perissodactyla

Old World: a geographical term referring to the Eastern Hemisphere excluding Australasia, in particular Europe; *see also* New World

olfactory: a term meaning relating to the sense of smell

omnivore: an animal that includes all food types in its diet

omnivorous: a term applied to animals that feed on all food types, including flesh and vegetable matter; *see also* carnivorous, herbivorous, insectivorous

opercular cavity: the cavity behind the gill coverings of a fish

opposable: a term meaning capable of being placed opposite and against another digit; for example, a thumb

order: a zoological ranking of species that share a range of characteristics, occurring below the rank of class and above that of family

organ: a group of body tissues working together to perform vital functions; for example, the heart

osmosis: the process by which water molecules, or those of other solvents, flow across a permeable membrane from an area of higher concentration to an area of lower concentration

ornithologist: a scientist who studies birds

Osteichthyes: the scientific name for the bony fish, the class of fish in which the skeleton is composed of bone rather than cartilage; the largest class of fish by far in terms of number of species, with more than 24,000 species currently recognized; *see also* Chondrichthyes

ovary: an organ that produces eggs

oviduct: a tube that serves for the passage of eggs from an ovary

oviparous: a term applied to those animals that lay eggs

ovoviviparous: a term applied to those animals that produce eggs which hatch within the body of the mother or immediately after being laid; ovovivipary occurs in many species of fish, insects, lizards, snails and snakes

ovipositor: an organ formed from modified appendages situated at the hind end of female insects and through which they lay eggs

palpus (*plural* palp): a segmented projection on an arachnid's mouthpart; palps are commonly used in taste or touch

pampas: a biome typified by extensive areas of grassland, found in South America, mainly Argentina; receives more annual rainfall than a desert but not enough to support forests; characterized by grass-covered plains, fast-growing vegetation and few trees

parallel evolution: the process whereby two or more taxonomic groups come to share a characteristic, or characteristics, due to the fact that their ecological requirements are similar and their genetic makeup is inherited from a common ancestor; *see also* convergent evolution

parasite: an organism that lives on or in another, known as a host; the parasite benefits from the host, which it usually harms

Passeriformes: the scientific name of the passerines (perching birds), the order of birds in which three toes on each foot point forward and one toe points backward, with the ligaments arranged in such a way that the foot locks onto the branch or other perch when the bird perches or sleeps; passerines also have a well-developed syrinx (voice-producing organ) and typically construct complex, woven nests; about 5,300 species, or 60 percent of all bird species, belong to the Passeriformes

passerine: a perching bird—a member of the Passeriformes, the largest order of birds

pastern: the lower part of a deer's leg

pectoral fin: either of the fins on a fish that correspond to the position in which forelimbs would be found in a quadruped

peep: a colloquial North American term for any of several small species of sandpipers; *see also* shorebird

pelage: the hairy coat, or covering, of a mammal

pelagic: a term meaning relating to the oceans or the open sea

pelvic fin: either of the paired fins situated in a position corresponding to that of hind limbs in a quadruped

perennial: a term applied to a plant that lives and reproduces for a number of years

perissodactyl: a member of the order Perissodactyla, or odd-toed ungulates; perissodactyls include the asses, horses, rhinoceroses, tapirs and zebras

Perissodactyla: the scientific name for the odd-toed ungulates, the order of hoofed mammals that possess either three

functional toes or a single functional toe on each foot; *see also* Artiodactyla, ungulate

permafrost: the permanently frozen ground in and around the North and South Poles

pharyngeal teeth: the throat teeth present in many species of fish

pharynx: the throat—a membrane-lined cavity that forms part of the digestive tract after the mouth

pheromone: an odiferous (smelly) chemical substance secreted and exuded by some animals in order to communicate

phloem: the vascular tissue (tissue composed of channels through which body fluids may flow) in plants that transports sugars and other nutrients

photosynthesis: the biological process by which plants take in energy from light and transform it into carbohydrates via a series of chemical reactions; oxygen is given off as a by-product

phylum (*plural* phyla): one of several broad zoological rankings of the plant and animal kingdoms, occurring above the rank of class and below that of kingdom

phytoplankton: a collective term for plant plankton; *see also* zooplankton

pinna (*plural* pinnae): an ear flap—the largely cartilaginous projecting portion of the external ear of a mammal

placenta: a temporary, hormone-producing organ that allows a mammalian embryo to obtain nourishment from its mother during gestation

placental: refers to mammals with a placenta

plankton: a collective term for minute animal or plant life that floats, or weakly swims, in water; *see also* phytoplankton, zooplankton

plantigrade: a term applied to animals that move with a flat-footed stance, in which the sole and the heel of the foot are placed flush against the ground

plastron: the ventral (lower) part of the shell of a tortoise or turtle, typically composed of nine symmetrically placed bones overlaid with horny plates

plumage: a term used to describe all of the feathers that cover a bird, which grow in distinct tracts on the skin

polar: a term that relates to the North or South Pole or the surrounding regions

pollen: the male sex cells of flowering plants

pollinate: to transfer pollen from the male to the female part of a flowering plant

pollinator: any animal that engages in pollination; important pollinators include insects and bats

polyandry: the practice of having several male mates at one time; *see also* polygyny

polychaete: a marine worm belonging to the class Polychaeta of the phylum Annelida, including the fanworms, lugworms, ragworms and tubeworms

polygamy: the practice of having several mates at one time; *see also* monogamy

polygyny: the practice of having several female mates at one time; *see also* polyandry

polyp: a small, cylindrical, stalklike structure of tissue found in members of the phylum Cnidaria that is attached to a substrate (firm surface) at one end and has a mouth surrounded by tentacles at the other; examples of animals with polyps include anenomes, corals, jellyfish, sea firs and sea pens

powder down: a collective term for a specialized type of feather that continually crumbles or disintegrates at the tip to produce a fine powder, which then is used to help clean the plumage during preening; patches of powder down are found in the bitterns, bowerbirds, herons, parrots and toucans

prairie: a plant community without trees and dominated by grasses; more generally, a large grassland

predator: any species that preys upon another species

prehensile: a term used to describe an animal's limb that is capable of grasping or wrapping round an object; for example, a tail

primary feather: one of the longest, outermost feathers on a bird's wing

primary forest: an undisturbed forest of native trees and other plants that results from natural processes rather than from human action, often known as virgin forest; once deforestation has taken place, primary forest is removed from an area permanently; *see also* secondary forest

primate: a member of the mammalian order Primates, comprising the apes, bushbabies, humans, lemurs, lorises, monkeys, pottos and tarsiers

proboscis: a feeding tube present in butterflies, flies (including the blowflies, crane flies, gnats, horseflies, houseflies, hoverflies, midges and mosquitoes) and moths

proleg: a fleshy abdominal limb of a caterpillar or similar insect larva

protractile: a term used to describe a body part that can be thrust out

protrusible: *see* protractile

pup: the young of certain species of mammals, especially dogs, rats and seals

pupa: the dormant stage of an insect during which the complete metamorphosis from larva to adult form takes place; also known as the chrysalis in butterflies

quadruped: an organism that moves on four feet; *see also* biped

queen: in social ants, termites, bees or wasps, a fully developed female that has the sole responsibility of laying eggs in the colony

race: a colloquial term for subspecies

radula: in mollusks, a rasping tongue—a rasp-like structure of tiny teeth used for scraping food particles off a surface and drawing them into the mouth

rain forest: a major biome characterized by high annual rainfall and tall, fast-growing trees that usually form a continuous overhead canopy; the type of rain forest that occurs varies according to the geographical location and includes equatorial, subtropical, monsoon, montane, mangrove and temperate rain forest; the greatest rain forest expanses are found at tropical latitudes, where the daytime temperature is typically 90° F (32° C) and the annual rainfall is more than 70 inches (1,800 mm)

ram: a male sheep

range: the geographic area within which members of a species typically occur; *see also* home range

raptor: *see* bird of prey

receptor: a cell, or group of cells, that detects specific stimuli, such as heat or pressure

regurgitate: to vomit up incompletely digested food; many bird species (especially birds of prey, seabirds and other waterbirds) regurgitate food with which they feed their young

reptile: a member of the class Reptilia; reptiles include the crocodilians, geckos, lizards, snakes, tortoises and turtles

Reptilia: the scientific name for reptiles, the class of cold-blooded vertebrates that breathe air, have horny or scaly skin and are fertilized internally; except for snakes, reptiles have four limbs, which project to the side of the body

riparian: a term applied to plants and animals that live close to rivers and that are influenced by them

river basin: the area of land drained by a river and its tributaries

rodent: any member of the order Rodentia, including agoutis, beavers, capybaras, chinchillas, coypus (nutria), gerbils, guinea pigs (cavies), hamsters, jerboas, kangaroo mice, kangaroo rats, lemmings, mice, pocket gophers, porcupines, rats, squirrels and voles; sometimes also used to refer to any small mammal, such as a shrew, other than a true rodent

Rodentia: the scientific name for the rodents, the large order of relatively small gnawing mammals that possess a pair of incisors with a chisel-shaped edge in each jaw; just under 40 percent of all mammal species belong to the Rodentia

roe: a collective term for fish eggs, especially when still inside the female

rookery: a collective term for a breeding colony of penguins; also used to describe a colony of rooks, *Corvus frugilegus*, the Eurasian species of crow to which the term was first applied

rostrum: a piercing beak in various insects or arachnids, used to suck juices from plants or animals

rufous: an orange-brown or reddish brown color

rumen: the large first compartment of a ruminant's stomach, containing microorganisms that break down cellulose

ruminant: a mammal that chews cud

rut (*noun*): a term used to describe the mating season in deer and some other related mammals

rutting season: another term for the mating season in deer and some of their relatives

savanna: a tropical grassland biome found in Africa, South America and Australia; savanna receives more rainfall than other types of grassland—perhaps 50 inches (1,250 mm)—but the rainy season is short and is followed by a long dry season, during which the land dries out and becomes semidesert; grasses are the dominant vegetation, as most trees are unable to take root, although acacias (which have specialized leaves) are found in African savannas

scrub: a term applied to vegetation consisting mainly of stunted trees and shrubs

secondary forest: a type of forest that has grown back after forest fire or deforestation; secondary forests may or may not contain exotic tree species but almost always differ in character from primary forests, and are characterized by a lower biodiversity than that of primary forests

sedentary: a term applied to an animal that does not migrate or perform nomadic movements, staying within the same home range all year round; *see also* nomadic

seine net: a type of large fishing net with sinkers on one edge and floats on the other that hangs vertically in the water and is used to enclose fish when its ends are drawn together; purse seines are operated from boats in deep water, whereas drag seines are used off beaches

sexual dimorphism: a term used to describe a situation in which the male and the female of a particular species differ markedly in appearance, for example because of contrasting coloration or differing size

sexual maturity: the age at which an individual of an animal species is physically capable of producing fertile offspring for the first time; *see also* age at first breeding

shorebird: any of a large group of wading birds in the suborder Charadrii of the order Charadriiformes, including the avocets, curlews, dowitchers, godwits, oystercatchers, phalaropes, plovers, sandpipers, snipes and stilts; generally referred to as a wader in Britain

soldier: a type of worker ant or termite distinguished by its larger body, head and jaws; responsible for defending the colony against predators

songbird: a member of the suborder Oscines of the order Passeriformes, in which the syrinx (voice-producing organ) is extremely well developed, enabling the production of sophisticated calls and songs; there are more than 4,000 species of songbirds, including the blackbirds, crows, finches, martins, orioles, sparrows, swallows, thrushes and warblers; *see also* Passeriformes

spawn (*noun*): a collective term for the mass of eggs produced by amphibians, fish and some aquatic invertebrates; for example, frog spawn

spawn (*verb*): in amphibians, fish and some aquatic invertebrates (for example, oysters), to produce or deposit eggs

species: the basic taxonomic unit, used for a distinct kind of plant or animal—the biological ranking below genus; the scientific name of a species is always written in italics

sphincter: a muscle surrounding and controlling a body opening

spinneret: an organ found in caterpillars and spiders that produces threads of silk from the secretions made by silk glands

spiracle: an opening on an animal's body surface that controls the entry and exit of air

stag: a male deer

steppe: an extensive grassland biome found in Europe and Asia, mainly in southern Russia, Siberia and Mongolia; receives more annual rainfall than a desert but not enough to support forests; characterized by grass-covered plains, fast-growing vegetation and few trees

stereoscopic vision: *see* binocular vision

sterile: incapable of supporting life or producing offspring

stotting: a stiff-legged jumping display carried out by some antelope species, for example Thomson's gazelle, *Gazella thomsoni*, in order to advertise their fitness

stridulate: to create a high-pitched sound; a characteristic of male insects, including cicadas, crickets, grasshoppers and katydids (long-horned grasshoppers)

subclass: a taxonomic subdivision containing some of the orders belonging to the same class

subfamily: a taxonomic subdivision containing some of the genera belonging to the same family

suborder: a taxonomic subdivision containing some of the families belonging to the same order

sub-Saharan Africa: a geographical region comprising all of Africa south of the Sahara Desert, excluding Madagascar

subspecies: a taxonomic subdivision comprising one or more populations of a species, the members of which differ in certain respects from other populations of the same species

substrate: a firm substance or layer on which an organism (especially a marine invertebrate or plant) lives or moves; common marine substrates include sand, mud, gravel, rock, corals, piers and shipwrecks

subtropical: a term relating to the regions that border the Tropics

succulent: a plant with fleshy tissues in the leaves, roots or stems that conserve water as an adaptation to life in an arid environment

supercilium: in birds, an eyebrow—a roughly horizontal, linear plumage feature situated directly above a bird's eye; *see also* eye stripe

superfamily: a zoological ranking containing some of the families that belong to the same order

superspecies: a zoological ranking usually applied to groups of species that are geographically separated but which appear to have a common ancestor

swim bladder: a gas-filled balloon inside some fish that helps them to maintain their position in the water

symbiotic: a term used to describe a mutually beneficial

relationship between two organisms, usually belonging to different species; the two organisms generally live in fairly close proximity to each other; *see also* altruistic

syrinx: in birds, the lower larynx, or voice-producing organ; the syrinx is especially well-developed in the songbirds (suborder Oscines)

tadpole: the larva of a frog or toad, characterized by a rounded body, external gills and a long tail bordered by fins

taiga: a major biome found in northern Canada, southern Alaska, Scandinavia, Siberia and Hokkaido, Japan, characterized by a mosaic of habitats, especially conifers (cone-bearing trees), bogs, marshes and small lakes

tapetum lucidum: a reflector situated behind the retina of each eye that reflects any light that has passed through the retina, thereby improving an animal's ability to see in conditions of poor visibility; found in bushbabies and cats among other animals

taxonomy: the scientific classification of plants or animals

temperate: a term denoting moderate climatic conditions; the term temperate zone refers both to the region from the Tropic of Cancer north to the Arctic Circle and to that from the Tropic of Capricorn south to the Antarctic Circle

terrestrial: a term applied to animals that live in or are adapted for living principally on the ground

territory: the area occupied more or less exclusively by an animal or group of animals of the same species

thermal (*noun*): a rising current of hot air occurring over land; used by many large birds, including birds of prey, condors, pelicans, storks and vultures to keep themselves aloft without having to expend energy on flapping their wings

thermoregulation: the way in which an animal regulates its body temperature

thorax: in vertebrates, the part of the body containing the lungs and heart; in insects, the middle body section that bears three pairs of jointed limbs and, often, one or two pairs of wings

threatened species: any species that is at risk of becoming endangered

tissue: a group of body cells that performs the same function; for example, bone or muscle

torpor: in animals, to become dormant for a period in order to survive harsh conditions such as drought or extreme cold or heat; estivation and hibernation are long-term forms of torpor, but certain species may enter torpor regularly on a daily basis; *see also* estivate; hibernate

toxin: a poisonous substance produced by a plant or animal, and which is often employed as a means of defense; a toxin-producing organism secretes toxins in its body, which often displays bold warning colors to potential predators, whereas a venomous animal possesses a specialized body part, such as fangs or a stinger, to deliver its poison into the victim

trachea: the windpipe or the main trunk of the system of tubes by which air passes to and from the lungs in vertebrates

trait: a characteristic that has been inherited

tribe: a more specific taxonomic division within the ranking of family or subfamily

tributary: a river or stream that flows into a larger watercourse

trochophore: a term applied to a free-swimming larval stage found in marine mollusks and polychaete worms, possessing bands of cilia and usually forming part of the plankton; in marine mollusks, a trochophore develops into a veliger before becoming an adult; *see also* veliger

tropical: a term meaning relating to the Tropics

Tropics: the region lying between 23.5° N of the equator (the latitude known as the Tropic of Cancer) and 23.5° S of the equator (the latitude known as the Tropic of Capricorn)

tubercule: a prominent bump or nodule

tundra: a major treeless biome characterized by dark soil with a permanently frozen subsoil and vegetation including mosses, lichens, herbs and dwarf shrubs; tundra stretches as far as the remote northern coasts of Alaska, Canada, Greenland, Scandinavia and Russia, meeting areas of permanent ice cap in Greenland and on Spitsbergen; also found on remote subantarctic islands in the Southern Hemisphere and on those coastal areas of the Ross Peninsula of Antarctica free of permanent ice; equivalent biomes on the upper slopes of mountains, above the tree line and below the region of permanent snow, are often called alpine tundra, as opposed to Arctic tundra

turbid: murky, when referring to water

underfur: the layer of relatively short hairs that grows close to the skin of a mammal, and which lies beneath the layer of longer, outer hairs known as guard hairs; the combination of these two hair types provides efficient thermal insulation

understory: the area immediately below the canopy in forest or woodland, characterized by dense vegetative growth

ungulate: an herbivorous mammal with hooves; ungulates are split into the artiodactyls (even-toed ungulates), including the antelopes, camels, cattle, deer, gazelles, giraffe, goats, hippos, pigs and sheep, and the perissodactyls (odd-toed ungulates),

including the asses, horses, rhinoceroses, tapirs and zebras; *see also* Artiodactyla, Perissodactyla

urine: a liquid containing waste nitrogen excreted by many animals

uterus: the organ within adult female mammals in which unborn mammals develop

valve: the shell of certain mollusks; in a bivalve mollusk, there are two valves joined at a hinge

veliger: a term applied to a free-swimming larval stage found in marine mollusks, possessing bands of cilia, a foot and a shell, and forming part of the plankton; a veliger develops from a trochophore; *see also* trochophore

venom: a poisonous substance secreted by certain animals, for example bees, snakes, spiders or wasps, and used as a weapon with which to attack prey or in defense against a predator; a venomous animal possesses a specialized body part, such as fangs or a stinger, to deliver its poison into the victim, whereas a toxin-producing organism secretes toxins in its body, which often displays bold warning colors to potential predators

ventral: a term relating to the lower surface, or belly, of an animal; *see also* dorsal

vertebra (*plural* **vertebrae**): one of the bony or cartilaginous segments composing the spinal column, or backbone

vertebrate: any animal that has a spinal column, or backbone

vestigial: a term meaning imperfectly developed

vibrissa (*plural* **vibrissae**): one of the modified, bristle-like feathers located around the bill of many birds, especially insectivorous species, which during feeding may serve to prevent prey from escaping; also, one of the specialized, stiff hairs located around the nostrils or elsewhere on the face of many mammal species, and which frequently serves as a tactile organ

viviparous: a term applied to those animals that give birth to live young

wader: *see* shorebird

warm-blooded: a term applied to birds and mammals, in which the body temperature is internally regulated and relatively high and constant, being independent of that of the environment; many mammals are also capable of lowering their body temperature, either on a daily basis or for a prolonged period of hibernation; *see also* cold-blooded

waterfowl: *see* wildfowl

watershed: an area of land that drains ultimately to a particular watercourse or body of water, such as a river, lake or sea

wattle: a colored fleshy lobe hanging from the head or neck of many bird species, especially cassowaries, condors, cotingas, pheasants, plovers, starlings and turkeys

wean: to become accustomed to an alternative source of nourishment other than that provided by a mother's milk

wetland: any area in which water (whether fresh, brackish or salt) is the primary factor controlling the environment and associated plant and animal life; a wetland is located where the water table is at or near the land's surface, or where the land is covered with water to a depth of a few yards; wetlands include marshes, peat bogs, swamps, reed beds, salt marshes, floodplains, deltas and mangroves, and any riverine or coastal bodies of water bordering such wetlands

whalebone plates: *see* baleen

wildfowl: the waterfowl—a collective term for the ducks, geese and swans, which comprise the family Anatidae of the order Anseriformes

windpipe: *see* trachea

wing bar: in birds, a single or double colored bar across the wing

winter deciduous: a term used to describe plants (especially trees) that drop their leaves in the fall

woodland: a complex plant community in which trees grow abundantly but generally far enough apart for their crowns not to intermingle, so that there is no overhead canopy

worker: a sterile or unmated member of an insect colony, responsible for nest building, providing food and tending the young

zoologist: a scientist who studies zoology

zoology: the branch of biology concerned with the classification and study of animals

zoophyte: an invertebrate animal resembling a plant in its appearance or manner of growth, such as a coral or sponge

zooplankton: a collective term for animal plankton; *see also* phytoplankton

Bibliography

BIOMES AND REGIONAL NATURAL HISTORY

Beletsky, L. *Belize and Northern Guatemala: The Ecotravellers' Wildlife Guide*. San Diego, CA: Academic Press, 1998.

Beletsky, L. *Costa Rica: The Ecotravellers' Wildlife Guide*. San Diego, CA: Academic Press, 1998.

Bird, J. *Beneath the North Atlantic*. Hartford, CT: Tidemark Press, 1997.

Bright, M. *Andes to Amazon: A Guide to Wild South America*. London: BBC Worldwide Ltd., 2000.

Crump, D. J., ed. *Hidden Worlds of Wildlife*. Washington, DC: National Geographic Society, 1994.

Ewing, S. *The Great Rocky Mountain Nature Factbook: A Guide to the Region's Remarkable Animals, Plants and Natural Features*. Portland, OR: Westwinds Press, 1999.

Gosner, K. L., and Peterson, R. T. *Peterson Field Guides: A Field Guide to the Atlantic Seashore*. Boston: Houghton Mifflin Co., 1999.

Hosking, E., and Sage, B. *Antarctic Wildlife*. New York: Facts on File Inc., 1983.

Kaplan, E., and Kaplan, S. L. *A Field Guide to the Coral Reefs: Caribbean and Florida*. Boston: Houghton Mifflin Co., 1999.

Kricher, J. *A Neotropical Companion: An Introduction to the Animals, Plants and Ecosystems of the New World Tropics*. Princeton, NJ: Princeton University Press, 1997.

Lambertini, M., and Veneralla, J. *A Naturalist's Guide to the Tropics*. Chicago: University of Chicago Press, 2000.

Paulson, D., and Beletsky, L. *Alaska: The Ecotravellers' Wildlife Guide*. San Diego, CA: Academic Press, 2001.

Pearson, D. L., and Beletsky, L. *Ecuador and the Galapagos Islands: The Ecotravellers' Wildlife Guide*. San Diego, CA: Academic Press, 1999.

Phillips, S. J., et al. *A Natural History of the Sonoran Desert*. Berkeley, CA: University of California Press, 1999.

Rice, T. *Deep Ocean*. London: Natural History Museum, 2001.

Ricciuti, E. R. *Biomes of the World* (series). Tarrytown, NY: Marshall Cavendish Corporation, 1996.

Schafer, K., and Matthews, D. *Beneath the Canopy: Wildlife of the Latin American Rain Forest*. San Francisco: Chronicle Books, 1999.

Steene, R. *Coral Seas*. Willowdale, ON, Canada: Firefly Books, 1998.

Waller, G., ed. *Sealife: A Complete Guide to the Marine Environment*. Robertsbridge, East Sussex, UK: Pica Press, 1996.

Whitmore, T. C. *An Introduction to Tropical Rain Forests*. Oxford, UK: Oxford University Press, 1998.

BIRDS

Able, K. P., ed. *Gatherings of Angels: Migrating Birds and Their Ecology*. Ithaca, NY: Cornell University Press, 1999.

Attenborough, D. *The Life of Birds*. Princeton, NJ: Princeton University Press, 1998.

Brooke, M., and Birkhead, T., eds. *The Cambridge Encyclopedia of Ornithology*. Cambridge, UK: Cambridge University Press, 1991.

Campbell, B., and Lack, E., eds. *A Dictionary of Birds*. Vermillion, SD: Buteo Books, 1985.

Del Hoyo, J., Elliott, A., and Sargatal, J., series eds. *Handbook of the Birds of the World: A 12-Volume Manual of All the Birds of the World*. New York: Buteo Books.
  Vol. 1. *Ostrich to Ducks*: 1992.
  Vol. 2. *New World Vultures to Guineafowl*: 1994.
  Vol. 3. *Hoatzin to Auks*: 1996.
  Vol. 4. *Sandgrouse to Cuckoos*: 1997.
  Vol. 5. *Barn Owls to Hummingbirds*: 1999.
  Vol. 6. *Mousebirds to Hornbills*: 2001.

Dunn, J. L. *National Geographic Field Guide to the Birds of North America*. Washington, DC: National Geographic Society, 1999.

Forshaw, J., ed. *Encyclopedia of Birds*. San Diego, CA: Academic Press, 1999.

Harrison, C., and Greensmith, A. *DK Handbook: Birds of the World*. New York: DK Publishing, 2000.

Jaramillo, A., and Burke, P. *New World Blackbirds: The Icterids*. Princeton, NJ: Princeton University Press, 1999.

Juniper, T., and Parr, M. *Parrots: A Guide to the Parrots of the World*. Yale, CT: Yale University Press, 1998.

Lynch, W. *Penguins of the World*. Willowdale, ON, Canada: Firefly Books, 1997.

Perrins, C. M., and Middleton, A., eds. *The Encyclopedia of Birds*. New York: Facts on File Inc., 1985.

Pizzey, G., and Knight, F. *Field Guide to the Birds of Australia*. London: HarperCollins, 2001.

Sibley, D. A. *National Audubon Society: The Sibley Guide to Birds*. New York: Knopf, 2000.

Sinclair, I., Tarboton, W., and Hockey, P. *The Illustrated Guide to the Birds of Southern Africa*. Princeton, NJ: Princeton University Press, 1995.

Snow, D., and Perrins, C. M., eds. *The Birds of the Western Palearctic*. Concise Edition. Oxford, UK: Oxford University Press, 1998.

Stevenson, T., and Fanshawe, J. *Field Guide to the Birds of East Africa*. San Diego, CA: Academic Press, 2001.

Toops, C. *Owls*. Stillwater, MN: Voyageur Press Inc., 1990.

ENDANGERED SPECIES

Baillie, J., and Groombridge, B., eds. *1996 IUCN Red List of Threatened Animals*. Gland, Switzerland, and Cambridge, UK: I.U.C.N. (World Conservation Union), 1996.

Hildyard, A., ed. *Endangered Wildlife and Plants of the World*. Tarrytown, NY: Marshall Cavendish Corporation, 2001.

FISH

Allen, G., Steene, R., and Allen, M. *A Guide to Angelfishes and Butterflyfishes*. El Cajo, CA: Odyssey Publishing, 1998.

Allen, G. *Marine Fishes of Tropical Australia and South-east Asia*. Perth, Australia: Western Australian Museum, 2000.

Bone, Q., Marshall, N. B., and Blaxter, J. H. *Biology of Fishes*. New York: Blackie Academic & Professional, 1994.

Bright, M. *The Private Life of Sharks: The Truth behind the Myth*. Mechanicsburg, PA: Stackpole Books, 2000.

Ellis, R., and McCosker, J. E. *Great White Shark*. New York: HarperCollins and Stanford University Press, 1995.

Lavett Smith, C. *The National Audubon Society Field Guide to Tropical Marine Fishes: Caribbean, Gulf of Mexico, Florida, Bahamas, Bermuda*. New York: Knopf, 1997.

Lieske, E., and Myers, R. *Coral Reef Fishes*. Princeton, NJ: Princeton University Press, 1998.

Paxton, J. R., and Eschmeyer, W. N., eds. *Encyclopedia of Fishes*. San Diego, CA: Academic Press, 1998.

Pitkin, L. *Coral Fish*. London: Natural History Museum, 2001.

Taylor, L. R., ed. *The Nature Company Guide: Sharks and Rays*. Alexandria, VA: Time-Life Books, 1997.

Tricas, T. C., et al. *The Ultimate Guide: Sharks and Rays*. New York: HarperCollins, 1997.

INVERTEBRATES

Carter, D. *DK Handbook: Butterflies and Moths*. New York: DK Publishing, 2000.

Castner, J. L. *Amazon Insects—A Photo Guide*. Gainesville, FL: Feline Press, 2000.

Dance, S. P. *DK Handbook: Shells*. New York: DK Publishing, 2000.

Discovery Channel. *Explore Your World Handbook: Insects and Spiders*. New York: Discovery Books, 2000.

Evans, A. V., and Bellamy, C. L. *An Inordinate Fondness for Beetles*. Berkeley, CA: University of California Press, 2000.

Foelix, R. F. *Biology of Spiders*. Oxford, UK: Oxford University Press, 1996.

Holldobler, B., and Wilson, E. O. *The Ants*. Cambridge, MA: Harvard University Press, 1990.

Hunt, J. *Octopus and Squid*. Monterey Bay, CA: Monterey Bay Aquarium Press, 1996.

McGavin, G. C. *Insects, Spiders and Other Terrestrial Arthropods*. New York: DK Publishing, 2000.

Michener, C. D. *The Bees of the World*. Baltimore, MD: John Hopkins University Press, 2000.

Moore, J. *An Introduction to the Invertebrates*. Cambridge, UK: Cambridge University Press, 2001.

Norman, M. *Cephalopods: A World Guide*. Hackenheim, Germany: ConchBooks, 2000.

Olsen, L-H., Sunesen, J., and Pedersen, B. V. *Small Woodland Creature*s. Oxford, UK: Oxford University Press, 2001.

O'Toole, C. *Alien Empire : An Exploration of the Life of Insects*. New York: HarperCollins, 1995.

Preston-Mafham, K. *Spiders*. Secaucus, NJ: Chartwell Press, 1998.

Pringle, L. *Scorpion Man: Exploring the World of Scorpions*. New York: Charles Scribner's Sons, 1994.

Schappert, P. *A World for Butterflies: Their Lives, Behavior and Future*. Willowdale, ON, Canada: Firefly Books, 2000.

Scott, J. *Butterflies of North America: A Natural History and Field Guide*. Stanford, CA: Stanford University Press, 1992.

Waldbauer, G. *Millions of Monarchs, Bunches of Beetles: How Bugs Find Strength in Numbers*. Cambridge, MA: Harvard University Press, 2001.

Wilson, E. O. *The Insect Societies*. Cambridge, MA: Harvard University Press, 1974.

Wye, K. R. *The Encyclopedia of Shells*. New York: Knickerbocker Press, 1998.

MAMMALS

Adamson, J. *Born Free: A Lioness of Two Worlds*. New York: Random House, 1987 (1960).

Alderton, D. *Rodents of the World*. New York: Facts on File Inc., 1996.

Balfour, D., et al. *African Elephants: A Celebration of Majesty*. New York: Abbeville Press Inc., 1998.

Barrett, L., and Dunbar, R. *Cousins: Our Primate Relatives*. New York: DK Publishing, 2001.

Brown, A., ed. *Encyclopedia of Mammals*. Tarrytown, NY: Marshall Cavendish Corporation, 1997.

Burton, J. A., and Pearson, B. *Collins Guide to Rare Mammals of the World*. London: Collins, 1987.

Carwardine, M. *Whales, Dolphins and Porpoises*. New York: DK Publishing, 1995.

Clapham, P. *Whales of the World*. Stillwater, MN: Voyageur Press, 2001.

De Waal, F. B. M., et al. *Bonobo: The Forgotten Ape*. Berkeley, CA: University of California Press, 1997.

Emmons, L. H. *Neotropical Rainforest Mammals: A Field Guide*. Chicago: University of Chicago Press, 1997.

Estes, R. D., and Otte, D. *The Safari Companion: A Guide to Watching African Mammals*. White River Junction, VT: Chelsea Green Publishing Co., 1999.

Ewer, R. *The Carnivores*. Ithaca, NY: Comstock Publishing Assoc., 1998.

Fenton, M. B. *Bats*. New York: Facts on File Inc., 1992.

Forsyth, A. *Mammals of North America*. Buffalo, NY: Firefly Books, 1999.

Fossey, D. *Gorillas in the Mist*. Boston: Houghton Mifflin Co., 2000 (1983).

Garbutt, N. *Mammals of Madagascar*. Yale, CT: Yale University Press, 1999.

Geist, V. *Deer of the World: Their Evolution, Behavior, and Ecology*. Mechanicsburg, PA: Stackpole Books, 1998.

Geoffrey, C., ed. *The Year of the Tiger*. Washington, DC: National Geographic Society, 1998.

Goodall, J. *In the Shadow of Man*. Boston: Houghton Mifflin Co., 2000 (1971).

Gould, E., and McKay, G., eds. *Encyclopedia of Mammals*. San Diego, CA: Academic Press, 1998.

Hill, J. E., and Smith, J. P. *Bats: A Natural History*. Austin, TX: University of Texas Press, 1984.

Hoogland, J. L. *The Black-tailed Prairie Dog: Social Life of a Burrowing Mammal*. Chicago: University of Chicago Press, 1995.

Kingdon, J. *The Kingdon Field Guide to African Mammals*. San Diego, CA: Academic Press, 1997.

Kitchener, A. *The Natural History of the Wild Cats*. Ithaca, NY: Cornell University Press, 1998.

Leatherwood, S., and Reeves, R. R. *The Sierra Club Handbook of Whales and Dolphins of the World*. San Francisco: Sierra Club Books, 1983.

Lindsey, S. L., Green, M. N., and Bennett, C. L. *The Okapi*. Austin, TX: Texas University Press, 1999.

Macdonald, D., ed. *Encyclopedia of Mammals*. New York: Facts on File Inc., 2001.

Mech, D. L. *The Wolf: The Ecology and Behavior of an Endangered Species*. Rochester, MN: Minnesota University Press, 1985.

Nowak, R. M., ed. *Walker's Mammals of the World*. Sixth edition. 2 vol. set. Baltimore, MD: John Hopkins University Press, 1999.

Perrin, W., Wursig, B., and Thewissen, J. G. M., eds. *Encyclopedia of Marine Mammals*. San Diego, CA: Academic Press, 2001.

Reynolds, J. E., Wells, R. S., and Eide, S. D. *The Bottlenose Dolphin*. Gainesville, FD: University of Florida Press, 2000.

Rowe, N. *Pictorial Guide to Living Primates*. New York: Pogonias Press, 1997.

Rowland, P., and Strahan, R., eds. *The Mammals of Australia*. Washington, DC: Smithsonian Institution Press, 1995.

Ward, P., and Kynaston, S. *Bears of the World*. New York: Facts on File Inc., 1999.

Whitaker Jr., John O. *The National Audubon Society Field Guide to North American Mammals*. New York: Knopf, 1996.

Wilson, D. E., and Ruff, S., eds. *The Smithsonian Book of North American Mammals*. Washington, DC: Smithsonian Institution Press, 1999.

REPTILES AND AMPHIBIANS

Behler, J. L., and King, F. W. *The National Audubon Society Field Guide to North American Reptiles and Amphibians*. New York: Knopf, 1979.

Bjorndal, K. A., ed. *Biology and Conservation of Sea Turtles*. Washington, DC: Smithsonian Institution Press, 1999.

Cogger, H. G., and Zweifel, R. G. *Encyclopedia of Reptiles and Amphibians*. San Diego, CA: Academic Press, 1998.

Cogger, H. G., ed. *The Little Guides: Reptiles and Amphibians*. Darien, CT: Federal Street Press, 1999.

Conant, R., et al. *Peterson Field Guides: A Field Guide to Reptiles and Amphibians of Eastern and Central North America*. Boston: Houghton Mifflin Co., 1998.

Duellman, W. E., and Trueb, L. *Biology of Amphibians*. Baltimore, MD: John Hopkins University Press, 1994.

Ernst, C. H., and Barbour, R. W. *Turtles of the World*. Washington, DC: Smithsonian Institution Press, 1992.

Greene, H. W. *Snakes: The Evolution of Mystery in Nature*. Berkeley, CA: University of California Press, 2000.

Mattison, C. *Frogs and Toads of the World*. New York: Facts on File Inc., 1987.

Mattison, C. *Lizards of the World*. New York: Facts on File Inc., 1989.

Mattison, C. *Snakes of the World*. New York: Facts on File Inc., 1987.

Mehrtens, J. M. *Living Snakes of the World*. New York: Sterling Publishing Co., 1987.

Ross, C. A., and Garnett, S., eds. *Crocodiles and Alligators*. New York: Facts on File Inc., 1989.

Stafford, P. *Snakes*. Washington, DC: Smithsonian Institution Press, 2000.

Stebbins, R. C. *Peterson Field Guides: A Field Guide to Western Reptiles and Amphibians*. Boston: Houghton Mifflin Co., 1998.

Internet Resources

GENERAL (INCLUDING TAXONOMY)

Biosonar—Seeing with Sound
Interactive web site featuring information, games, exercises and activities related to the use of sonar by bats, dolphins, birds and other animals.
www.biosonar.bris.ac.uk

Earth Life Web
On-line guide to animal and plant life.
www.earthlife.net/insects/thysanop.html

Fire Effects Information System
General information on North American animals, including ways in which fire affects each species.
www.fs.fed.us/database/feis/

Flora and Fauna Database
Cooperative project of the U.S. National Park Service and the University of California, a searchable database of all plants and animals known in the U.S. park system.
www.ice.ucdavis.edu/nps/

Life
Finland-based web site devoted to taxonomy, providing comprehensive lists of animals organized in correct scientific order.
www.funet.fi/pub/sci/bio/life/intro.html

Los Angeles Zoo
Provides on-line fact sheets on a wide variety of animals kept at the zoo.
www.lazoo.org/anicllt.html

Natural History Museum, London
U.K.-based web site providing information on the museum's extensive collections and research, as well as educational and interactive features.
www.nhm.ac.uk/

Nearctica
Extremely large on-line directory to natural history web sites, searchable by the scientific name of the animal or plant required.
www.nearctica.com/nathist/nathist.htm

Perth Zoo, Australia
Australia-based web site offering on-line fact sheets about a range of wildlife, particularly marsupials and other Australasian species.
www.perthzoo.wa.gov.au/wildlife.html

Tree of Life
General information about life on Earth.
phylogeny.arizona.edu/tree/phylogeny.html

Understanding the Plants and Animals of Western Riverside County
Highly detailed species accounts for a selection of North American animals and plants, focusing on Western Riverside County, California.
ecoregion.ucr.edu/mshcp/

University of Michigan Animal Diversity Web
Comprehensive on-line animal encyclopedia, offering many images, sound clips, bibliographies and links to other web sites.
animaldiversity.ummz.umich.edu/

Web of Life
Outlines of animal life, listed by phylum.
curator.org/WebOfLife/Kingdom/P_Mollusca/chitons.htm

Wildlife in the State of Wisconsin
On-line resource of the University of Wisconsin-Richland, providing fact sheets about animals found in Wisconsin and further afield in the U.S.
www.richland.uwc.edu/Depts/Biology/

WNC Nature Center
On-line fact sheets about a range of North American animals, provided by the Western North Carolina Nature Center.
wildwnc.org/af/

Zoo in the Wild
Large on-line encyclopedia giving detailed information on a wide range of animals. Includes distribution maps for many species. Also available in Italian.
www.naturalia.org/ZOO/indexing.html

AMPHIBIANS

Amphiaweb
Information about amphibian biology and conservation.
elib.cs.berkeley.edu/aw/index.html

Frog Calls
A sample of calls from various frog species at the University of Michigan Animal Diversity Web.
animaldiversity.ummz.umich.edu/chordata/lissamphibia/frog_calls.html

Frog Page
Basic information about the life cycle of frogs, plus links to related web sites.
www.geocities.com/TheTropics/1337/

FrogWeb
Information about the decline of frog species.
www.frogweb.gov/

Library of frog and toad photographs
On-line library of images, plus web site links.
gto.ncsa.uiuc.edu/pingleto/herps/frogpix.html

North American Amphibian Monitoring Program
Up-to-date research and surveys of amphibian populations.
www.im.nbs.gov/amphibs.html

Texas Memorial Museum
Guide to the salamanders, frogs and toads of Texas; also covers various reptiles.
www.zo.utexas.edu/research/txherps/

Yahoo! Reptiles and Amphibians
This page offers a wide range of links to web sites about reptiles and amphibians.
dir.yahoo.com/Science/Biology/Zoology/Animals__Insects__and_Pets/Reptiles_and_Amphibians/

Zoological Record: Amphibians
Alphabetical compilation of amphibian web sites.
www.biosis.org/zrdocs/zoolinfo/grp_amph.htm

BIRDS

Birding in Canada
General information about birds and birdwatching with a particular focus on Canada.
www.web-nat.com/bic/

Birds of Eastern North America
Comprehensive descriptions of North American birds and their life cycles.
www.floridaecosystems.org/bird_families_in_eastern_north_a.htm

Creagrus
Private web site with information on a range of birds, plus dolphins and other animals. Special emphasis on Monterey Bay.
www.montereybay.com/creagrus/

Ducks Unlimited
On-line identification guide to North American ducks, including sound clips and details of species' life cycles.
www.ducks.org/waterfowling/gallery/index.asp

International Crane Foundation
Information about cranes and their wetland and grassland habitats throughout the world.
www.savingcranes.org

International Osprey Foundation
Includes a regularly updated database of osprey migration and population data.
www.sancap.com/osprey/tiof.htm

Kakapo Net
New Zealand Government's web site devoted to the kakapo, one of the world's rarest parrots.
www.kakapo.net/en/

Patuxent Bird Identification Center
Photographs, songs, videos, identification tips, maps and life histories of North American birds.
www.mbr-pwrc.usgs.gov/id/framlst/

Patuxent Bird Population Studies
Results of in-depth research into populations and migration habits of North American birds.
www.mbr-pwrc.usgs.gov/

Percy FitzPatrick Institute of African Ornithology
Links to bird web sites, focusing on South Africa.
www.uct.ac.za/depts/fitzpatrick/docs/web sites.html

Peregrine Fund
Detailed information about conservation of the peregrine falcon and many other birds of prey, together with links to related web sites.
www.peregrinefund.org/

Purple Martin Conservation
Information about this North American species, including how to atttact them to your garden.
www.purplemartin.org/

Sampler of the World's Birds
Large on-line catalogue of bird images, organized in scientific order.
www.camacdonald.com/birding/Sampler.htm

World Owl Trust
Provides information about the world's owls.
www.owls.org/Information/info.htm

Zoological Record: Birds
Alphabetical compilation of bird web sites.
www.biosis.org/zrdocs/zoolinfo/grp_bird.htm

BIOMES AND HABITATS

African Wildlife Organization
Provides information about a variety of African biomes and the wildlife that lives there.
www.awf.org/

Aquascope
Sweden-based web site covering wildlife that inhabits beaches, coasts and the open sea.
www.vattenkikaren.gu.se/defaulte.html

BBC Education: World Environmental Changes
Detailed guide to global environments, such as wetlands, forests and oceans. Includes maps, animations and case studies of specific areas.
www.bbc.co.uk/education/landmarks/index.shtml

CoralRealm
Information on coral reef animals, including
fish and marine invertebrates.
www.coralrealm.com/homepage.html

Florida Smart Sea Life and Ocean Studies
Directory
Provides many links to web sites on Florida's
habitats and marine wildlife.
www.floridasmart.com/subjects/ocean/

Northeast Fisheries Science Center
Answers to common questions about a wide
range of marine life.
www.wh.whoi.edu/faq/index.html

Northern Prairie Wildlife Research Center
Information about prairies and their wildlife.
www.npsc.nbs.gov/

Parknet
On-line resource from the U.S. National Park
Service. Provides a wealth of information about
the wildlife and habitats of U.S. national parks.
www.nps.gov/

Rainforestweb
General information about rain forests.
www.rainforestweb.org/

Seaworld Animal Database
Large on-line database, primarily focused on
marine wildlife but also covering other animals
and conservation issues. Includes bibliographies
and many links to useful web sites.
www.seaworld.org/infobook.html

Wild Habitat
Guide to bears, cats, dogs, herbivores and
primates, according to their habitat.
library.thinkquest.org/11234/index.html

Yahoo! Marine Wildlife
This page offers links to web sites on marine life.
dir.yahoo.com/Science/Biology/Zoology/
Animals__Insects__and_Pets/Marine_Life/

FISH

Sharks
All about Sharks
Detailed information about sharks.
members.ozemail.com.au/~bilsons/SHARKS.htm

Pelagic Shark Research Foundation
Offers a wealth of shark-related information.
www.pelagic.org/

Seaworld: Sharks
Host of information on sharks and their relatives.
www.seaworld.org/Sharks/diet.html

Shark Attack!
General information about sharks, with links to
shark-related web sites.
www.pbs.org/wgbh/nova/sharkattack/

Sharks on the Line
A state-by-state analysis of sharks and their
fisheries, at the National Audubon Society.
www.audubon.org/campaign/lo/sol/

Shark Trust
Provides information about sharks and shark
conservation, including many links to web sites.
ds.dial.pipex.com/sharktrust/

Other fish
Australian Museum
General information about Australian fish, plus
links to related sites on the region's marine life.
www.austmus.gov.au/fishes/

Fascinating Facts about Fish
Useful guide from the American Fisheries Society.
www.nefsc.nmfs.gov/faq.html

Fishbase
Search for fish by common or scientific name.
Also provides general information on a very
wide range of fish species.
www.fishbase.org/search.cfm

Fish Information Service
Information about aquarium fish.
www.actwin.com/fish/index.php

Fish out of Time
Dedicated to the coelacanth of the Indian Ocean,
the last survivor of an ancient group of fish.
www.dinofish.com/

Index of Freshwater Fish
Search for freshwater fish by scientific name.
Includes brief summaries on many species.
www.webcityof.com/miffidx.htm

Texas Memorial Museum
The on-line North American freshwater fish
index, including images, maps and information.
www.utexas.edu/depts/tnhc/.www/fish/tnhc/
na/naindex.html

University of Florida: Ichthyology
General information on fish, with useful links.
www.flmnh.ufl.edu/fish/

Virginia Insitute of Marine Science
Variety of information on fish and fisheries.
www.fisheries.vims.edu/

LAND INVERTEBRATES

Butterflies and moths
Butterflies North and South
Collaboration between the Canadian Museum of
Nature and five museums in Peru, a web site
devoted to the biodiversity of butterflies.
www.nature.ca/discover/bfliesns_e.cfm

Captain's European Butterfly Guide
Informative private web site about European
butterflies, including links to related web sites
and a general guide to butterfly photography.
www.butterfly-guide.co.uk

Family Classification of Lepidoptera
Lists of butterfly and moth names, organized in
scientific order. Useful guide to classification.
www.troplep.org/famlist.htm

Lepidoptera Index
This page is a comprehensive alphabetical list of
lepidoptera (butterfly and moth) web sites.
www.chebucto.ns.ca/Environment/NHR/
alphabetical.html

North American Butterflies
On-line guide to North American butterflies.
www.npwrc.usgs.gov/resource/distr/lepid/
BFLYUSA/usa/toc.htm

Zoological Record: Lepidoptera
Alphabetical compilation of lepidoptera
(butterfly and moth) web sites.
www.biosis.org/zrdocs/zoolinfo/grp_lep.htm

Other land invertebrates (including insects and spiders)
Arachnology Home Page
General information on arachnids.
www.ufsia.ac.be/Arachnology/
Arachnology.html

Arthropods
On-line fact sheets on many arthropods,
plus links to other web sites.
www.kheper.auz.com/gaia/biosphere/

California Insects
Exploring California's insects, including a photo
library, on-line slide shows and various
study resources.
www.bugpeople.org/arthropods/

Checklist of Invertebrates
Lists of invertebrate names, organized in
scientific order. Useful guide to classification.
www.interaktv.com/INVERTS/Inverttitl.html

Ohio State University
On-line fact sheets about insect pests and their
control.
www.ag.ohio-state.edu/~ohioline/
hyg-fact/2000/index.html

University of Florida Book of Insect Records
Fascinating collection of miscellaneous
information about insects.
ufbir.ifas.ufl.edu/

Wonderful World of Insects
Extremely useful insect web site, with thousands
of insect-related links.
www.earthlife.net/insects

Yahoo! Arachnids
This page offers links to web sites on spiders,
scorpions, mites, ticks and their relatives.
dir.yahoo.com/Science/Biology/Zoology/
Animals__Insects__and_Pets/Arachnids/

Zoological Record: Arthropods
Alphabetical compilation of arthropod web sites.
www.biosis.org/zrdocs/zoolinfo/gp_index.
htm#arthropods

MAMMALS

Bats
BatAtlas
Information about bats, with a particular focus
on Australian species.
online.anu.edu.au/srmes/wildlife/batatlas.html

Bat Conservation Trust
U.K.-based web site featuring a wide range of
information on bats and their conservation.
Includes links to other web sites about bats.
www.bats.org.uk/

Bears
International Association for Bear Research
and Management
On-line reports on all species of bears, worldwide.
www.bearbiology.com

Cats
BigCats.Com
On-line guide to wild cats, including many
photographs and fact sheets.
bigcats.com/

Cat Specialist Group
Information on wild cats assembled by the Cat
Specialist Group, an international group of
experts working on behalf of the I.U.C.N.
lynx.uio.no/lynx/catfolk/

Cheetah Conservation Fund
Dedicated to cheetah conservation.
www.cheetah.org/home.htm

Cheetah Spot
General on-line guide to cheetahs.
www.cheetahspot.com/

Florida Panther Society
Information about the Florida panther, including
details of conservation attempts.
www.atlantic.net/~oldfla/panther/panther.html

Tiger Information Center
Information on the world's five remaining tiger
subspecies.
www.tigers.org/

World Lynx
Wide variety of information about the several
species and subspecies of lynx.
lynx.uio.no/jon/lynx/lynxhome.htm

Dogs
Canid Specialist Group
Information on wild canids (members of the dog family) assembled by the Canid Specialist Group, an international team of I.U.C.N. experts.
www.canids.org/

North American Wolf Association
Focusing on gray wolves in North America.
www.nawa.org/about_wolves.html

Wolverine Foundation
Dedicated to the protection of the wolverine.
www.wolverinefoundation.org/mainpage.htm

Dolphins, whales and other marine mammals
Cetacea
Very detailed on-line guide to the world's cetaceans (dolphins, porpoises and whales), including a guide to whale-watching and links to other web sites.
www.cetacea.org

Dolphins—The Oracles of the Sea
Useful dolphin web site.
library.thinkquest.org/17963/?tqskip=1

I.F.A.W.: Seals
On-line guide to seals and their conservation, from the International Fund for Animal Welfare.
www.ifawct.org/Seals/seals_default.htm

I.F.A.W.: Whales
On-line guide to whales and their conservation, from the International Fund for Animal Welfare.
www.ifawct.org/Whales/whales_home_page.htm

Save The Manatee Club
Dedicated to promoting the research, rescue and rehabilitation of the manatee.
www.savethemanatee.org

Seal Conservation Society
Information about a range of seal species along with conservation news. Also provides links to related conservation web sites.
www.pinnipeds.fsnet.co.uk/main.htm

Whales of Australia
Information on whales and whale biology, plus a guide to whale-watching in Australian waters.
www.upstarts.net.au/site/non_commercial/whales.html

Whales on the Net: Picture Gallery
On-line library of cetacean photographs.
whales.magna.com.au/DISCOVER/gallery/index.html

Whale-Watching
Information on whales and where to watch them.
www.physics.helsinki.fi/whale/

Marsupials
Gander Academy: The Kangaroo
General information about kangaroos.
www.stemnet.nf.ca/CITE/kangaroo.htm

Marsupial Society of Australia
Includes a checklist of marsupial species.
www.marsupialsociety.org.au/

Monkeys and other primates
Dian Fossey Gorilla Fund International
Information about the endangered mountain gorilla of Central Africa.
www.gorillafund.org/005_gorilla_frmset.html

Duke University Primate Center
Highly informative web site about lemurs, pottos, lorises and their relatives.
www.duke.edu/web/primate/

GeoZoo: Primates
List of primate species organized in scientific order, including links to related web sites.
www.geobop.com/Mammals/Primate/2.htm

Primate Gallery
On-line library of primate photos and artwork, including basic information on some species.
www.selu.com/bio/PrimateGallery/main.html

Other mammals
Foundation for the Preservation and Protection of the Przewalski Horse
Information about Przewalski's horse, the endangered wild horse of Central Asia.
www.treemail.nl/takh/index.htm

I.F.A.W.: Elephants
On-line guide to elephants and their conservation, from the International Fund for Animal Welfare.
www.ifawct.org/Elephants/elephants_home_page.htm

Otternet
Offers a wealth of information about otters.
www.otternet.com/

Ultimate Ungulate Page
On-line guide to the world's hoofed mammals, including information about hyraxes, elephants, dugongs and ungulates.
www.ultimateungulate.com/

General mammal web sites
Animal Info
Searchable database providing information on rare, threatened and endangered mammals.
www.animalinfo.org/

GeoZoo: Mammals
Comprehensive on-line database about mammals.
www.geobop.com/Mammals/

Mammal Society
Although this site focuses on mammals found in the U.K., it is highly informative and includes links to related mammal web sites.
www.abdn.ac.uk/mammal/

Mammal Species of the World
Produced by the Smithsonian Institution, a searchable list of the world's mammal species, including details of distribution and status.
nmnhwww.si.edu/msw/

Yahoo! Mammals
This page offers links to web sites relating to a wide range of mammals.
dir.yahoo.com/Science/Biology/Zoology/Animals__Insects__and_Pets/Mammals

Zoological Record: Mammals
Alphabetical compilation of mammal web sites.
www.biosis.org/zrdocs/zoolinfo/grp_mamm.htm

MARINE INVERTEBRATES

Mollusc Specialist Group
Information from the Mollusc Specialist Group, an international group of I.U.C.N. experts.
bama.ua.edu/~clydeard/IUCN-SSC_html/index.htm

Yahoo! Mollusks
This page offers links to web sites relating to a wide range of mollusks.
dir.yahoo.com/Science/biology/zoology/animals__insects__and_pets/mollusks/

REPTILES

American Alligator
On-line guide from the University of Florida.
www.ifas.ufl.edu/AgriGator/gators/

Axolotls
General information about the axolotl, an endangered species of salamander from Mexico.
www.caudata.org/axolotl/

Belize Zoo: Index of Reptiles and Amphibians
Information on a range of snakes, crocodiles, frogs and iguanas, including videos and sound clips.
www.belizezoo.org/zoo/zoo/herps/

Chelonian Archives
Searchable database of information on the biology and conservation of tortoises and turtles.
www.tortoise.org/cttc1.html

Crocodilian.com
Wide range of information about crocodiles, alligators, caimans and gharials, including links.
www.crocodilian.com/

Dr. Seward's Gila Monster Web Site
General information about gila monsters.
www.drseward.com/

EMBL Reptile Database
Large on-line database devoted to reptiles, searchable by scientific and common names.
www.embl-heidelberg.de/~uetz/LivingReptiles.html

Euro Turtle
General information about sea turtles.
www.ex.ac.uk/telematics/EuroTurtle/

Komodo Dragons
Dedicated to the Komodo dragon, the world's largest living lizard species.
home.rmi.net/~shasta/komodos.htm

Snakes of Florida
Produced by the Florida Museum of Natural History. Although this on-line guide focuses on Florida's snakes, many species occur elsewhere.
www.flmnh.ufl.edu/natsci/herpetology/fl-guide/onlineguide.htm

Turtle Trax
A web site devoted to sea turtles.
www.turtles.org/marines.htm

Yahoo! Reptiles and Amphibians
This page offers a wide range of links to web sites about reptiles and amphibians.
dir.yahoo.com/Science/Biology/Zoology/Animals__Insects__and_Pets/Reptiles_and_Amphibians/

Zoological Record: Reptiles
Alphabetical compilation of reptile web sites.
www.biosis.org/zrdocs/zoolinfo/grp_rept.htm

THREATENED AND ENDANGERED SPECIES

EndangeredSpecie.com
On-line fact sheets about endangered species and the causes of endangerment.
www.endangeredspecie.com

I.U.C.N. Red List 2000
Searchable database of threatened and endangered species, managed by the I.U.C.N., providing up-to-date information about status.
www.redlist.org/

World Conservation Monitoring Centre
On-line fact sheets on 140 endangered species.
endangered.fws.gov/wildlife.html

U.S. Fish and Wildlife Service
Provides information about threatened and endangered animals and plants.
endangered.fws.gov/wildlife.html

World Wildlife Fund
Information about a range of threatened species.
www.panda.org/resources/publications/species/underthreat

Resources for Younger Readers

BOOKS

Animals Animals

Various authors. Series. Tarrytown, NY: Benchmark Books, 2001–.

Animals in Danger

By R. Theodorou. Set of 12 books. Crystal Lake, IL: Heinemann Library, 2000.

Aquatic Life of the World

Edited by B. Giles. 11-volume encyclopedia. Tarrytown, NY: Marshall Cavendish Corporation, 2001.

DK Animal Encyclopedia

Edited by B. Taylor. New York: DK Publishing, 2000.

Earth at Risk: The Living Ocean

By E. Collins. New York: Chelsea House Publications, 1994.

Gray Wolf, Red Wolf

By D. H. Patent. Boston: Houghton Mifflin Co., 1994.

Life in the Deserts

By L. Baker. Princeton, NJ: Two-Can Publishing, 2000.

Nature Unfolds: The Poles

By B. Stonehouse. St. Catharines, ON, Canada: Crabtree Publishing, 2001.

Nature Unfolds: The Tropical Rainforest

By G. Cheshire. St. Catharines, ON, Canada: Crabtree Publishing, 2001.

Rain Forests of the World

11-volume encyclopedia. Tarrytown, NY: Marshall Cavendish Corporation, 2002.

Special Wonders of the Sea World

By D. Buddy and K. Davis. Green Forest, AR: Master Books, 1999.

CD–ROMS

Animals: Their Lives, Habits, and Ecosystems

E.M.M.E., Mac/Windows.

Exploring Land Habitats

Steck-Vaughn, Mac/Windows.

How Animals Move

Focus Multimedia, Windows.

Magic School Bus: Whales and Dolphins

Microsoft, Mac/Windows.

Magic School Bus: World of Animals

Microsoft, Mac/Windows.

Wildlife of the World

Set of 4 CDs. Topics Entertainment. Mac/Windows.

DVDS

Fascinating Nature

Quantum Leap Group, 1999, 93 minutes.

Predators

Telstar Video Entertainment, 2000, 180 minutes.

VIDEOS

The Mammal Story

United Learning, 1998, 13 minutes.

Microcosmos

An amazing chronicle of the insect world filmed with
revolutionary close-up cameras.

Miramax, 2000 (1996), 73 minutes.

National Geographic's Really Wild Animals

Series of 15 videos. National Geographic Society,
1994–1997.

Super Sense—A Voyage in Animal Perception

Set of 2 videos. BBC Worldwide Publishing, 1988.

WEB SITES

Africam

The world's first virtual game reserve. This web site
includes exciting images from more than 20 webcams,
plus a photo gallery and sound links.

www.africam.com

Biosonar—Seeing with Sound

Discover how bats, dolphins and other animals use
sonar. Includes an interactive demonstration of how
sonar works, as well as games and activities.

www.biosonar.bris.ac.uk

Discovery Online Animal Guide

Games, animations, live videos and features on all
sorts of wildlife.

dsc.discovery.com/guides/animals/animals/html

Kids' Castle: Animals

Fun facts, photographs and feature articles about a variety
of animals. Produced by the Smithsonian Institution.

www.kidscastle.si.edu/channels/animals/animals.html

Nature Songs

This web site allows you to listen to the sounds made by all
kinds of wild animals, from monkeys and whales to
rattlesnakes and frogs. You can download the recordings to
your own computer.

www.naturesongs.com/

Natureworks

Brilliant web site linked to a series of videos and TV programs.
Includes nature files on 120 North American animals.

www.nhptv.org/natureworks/

SeaWorld Animal Information Database

Fast facts, games, activities and a library of animal sounds.
Covers land animals as well as ones that live in the sea.

www.seaworld.org/infobook.html

3-D Insects

Interactive web site that features 3-D animations of insects.

www.ento.vt.edu/~sharov/3d/

World Wildlife Fund Kid's Stuff

Educational and fun web site about wildlife and conservation
issues. Includes fact sheets, games and an interactive
funhouse that demonstrates the importance of biodiversity.

www.worldwildlife.org/fun/kids.cfm

Zoo in the Wild

Click on the pictures of animals to find out more about the
life cycle of each species.

www.naturalia.org/ZOO/childing.html

Wildlife Refuges and Other Places of Interest

ZOOS AND AQUARIUMS

American Zoo and Aquarium Association
Home page includes large searchable database of affiliated zoos, aquariums and nature centers.
www.aza.org

Musée
Comprehensive web inventory with up-to-date information about natural history museums, zoos, nature centers and aquariums, worldwide.
www.musee-online.org

WILDLIFE REFUGES IN THE UNITED STATES

Parknet
On-line resource from the National Park Service, offering information about U.S. national parks.
www.nps.gov

U.S. Fish and Wildlife Service
Maintains an extensive system of wildlife refuges.
www.fws.gov

In addition, there are several private web-based organizations providing information on private and public wildlife protection areas. Two are:

Gorp
Quirky private site that tries to maintain current information on all nature preserves, worldwide.
www.gorp.com

Wildernet
Another private site, dedicated to providing details and reviews of wildlife refuges.
www.wildernet.com

ALABAMA
Blowing Wind Cave National Wildlife Refuge
Sanctuary for endangered Indiana and gray bats.
www.fws.gov/~r4eao/nwrbwc.html

Bon Secour National Wildlife Refuge
Saltwater marsh habitat, protecting sea turtles, Alabama beach mice and migratory birds.
www.bonsecour/fws.gov

Key Cave National Wildlife Refuge
Sole remaining breeding site for Alabama cave fish.
http://keycave.fws.gov/

Wheeler National Wildlife Refuge
Waterfowl refuge; feeding site for migrant geese.
www.fws.gov/~r4eao/nwrwlf.html

ALASKA
Alaska Maritime National Wildlife Refuge
Islets and 4,800 miles (7,680 km) of coastline; home to fur seals, sea lions, walruses and polar bears.
www.r7.fws.gov/nwr/akmnwr/akmnwr.html

Izembek National Wildlife Refuge
Lagoon habitat for emperor geese and black brant; also host to salmon, seals, sea lions and whales.
www.r7.fws/nwr/izembek/iznwr.html

Kenai National Wildlife Refuge
Peninsula with a variety of habitat types, protecting moose, caribou and king and red salmon.
www.alaskanet/~northlit/

Kodiak National Wildlife Refuge
Island habitat; home to 400 pairs of bald eagles.
www.r7.fws.gov/

Yukon Flats National Wildlife Refuge
Visited by millions of migratory birds every spring.
www.r7.fws.gov/nwr/yf/yfnwr.html

ARIZONA
Bill Williams National Wildlife Refuge
Includes desert, marsh and forest habitat.
http://southwest.fws.gov/refuges/arizona/billwill.html

Cibola National Wildlife Refuge
Spilling over into California, a site for rare species such as Yuma clapper rail and razorback sucker.
http://southwest.fws.gov/refuges/arizona/cibola.html

Havasu National Wildlife Refuge
Sanctuary for a number of rare species.
http://southwest.fws.gov/refuges/arizona/havasu.html

ARKANSAS
Bald Knob National Wildlife Refuge
Important site for migratory birds.
http://southeast.fws.gov/BaldKnob/Index.html

Big Lake National Wildlife Refuge
Refuge for migrant wildfowl and neotropical birds.
http://biglake.fws.gov/

Felsenthal National Wildlife Refuge
Habitat for endangered red cockaded woodpecker and the bald eagle and American alligator.
http://southeast.fws.gov/felsenthal/index.html

Logan Cave National Wildlife Refuge
Refuge for endangered species such as cave crayfish and gray bats, and the threatened Ozark cavefish.
http://southeast.fws.gov/LoganCave/index.html

CALIFORNIA
Coachella National Wildlife Refuge
One of few remaining sites for fringe-toed lizards.
pacific.fws.gov/coachella/

Don Edwards San Francisco Bay National Wildlife Refuge
Home to many rare species, including harvest mice.
http://desfbay.fws.gov/

Merced National Wildlife Refuge
Hosts a large nesting colony of sandhill cranes.
www.r1.fws.gov/sanluis/merced_info.html

Salinas River Wildlife Refuge
Sonny Bono Salton Sea Wildlife Refuge
Tijuana Slough Wildlife Refuge
All provide important bird sanctuaries, hosting many thousands of visiting migrants annually.
http://pacific.fws.gov/visitor/california.html

COLORADO
Alamosa/Monte Vista National Wildlife Refuge
Supports colonies of whooping and sandhill cranes as well as deer, elk and many other species.
www.r6.fws.gov/ALAMOSAnwr/

Browns Park National Wildlife Refuge
Home to over 300 species of terrestrial wildlife.
www.r6.fws.gov/refuges/Browns/

Rocky Mountain Arsenal
Former military base; now hosts 227 bird species, 32 mammal species and other wildlife.
www.pmrma-www.army.mil/

CONNECTICUT
Stewart B. McKinney National Wildlife Refuge
Sanctuary for roseate terns and many other birds.
http://northeast.fws.gov/ct/sbm.htm

DELAWARE
Bombay Hook National Wildlife Refuge
More than 13,000 acres (5,260 ha) of tidal salt marsh.
http://noretheast.fws.gov/de/bmh.htm

Prime Hook National Wildlife Refuge
Protected expanse of Delaware Bay marshland.
http://northeast.fws.gov/de/pmh.htm

FLORIDA
Arthur R. Marshall Loxahatchee National Wildlife Refuge
Last remaining protected wetland in northern Everglades, including cypress swamp, wet prairie, sawgrass, marsh and tree island habitat.
http://loxahatchee.fws.gov/home/

Caloosahatchee National Wildlife Refuge
Primarily consisting of mangrove islands.
http://caloosahatchee.fws.gov/home/

Cedar Keys National Wildlife Refuge
Numerous islands, providing nesting sites for ibises, cormorants, pelicans and eagles.
http://southeast.fws.gov/refuges/cedarkeys/

Chassahowitzka National Wildlife Refuge
Marsh, swamp and shallow bay habitat supporting manatees, sea turtles and bald eagles.
www.nccentral.com/fcnwr.htm

Crystal River National Wildlife Refuge
Important site for the endangered manatee.
www.nccentral.com/fcnwr/index.html

GEORGIA
Banks Lake National Wildlife Refuge
Wetlands hosting a diversity of fish and reptiles.
http://southeast.fws.gov/BanksLake/

Blackbeard Island National Wildlife Refuge
Mixed habitat including sand dunes, woodlands, palmetto vegetation and marshes.
http://blackbeardisland.fws.gov/

Okefenokee National Wildlife Refuge
The famous swamp hosts cranes, wood storks, alligators, indigo snakes and gopher tortoises.
http://okefenokee.fws.gov/

HAWAII

Hakalau Forest National Wildlife Refuge
Protected rain forest located on the Big Island; supports 8 endangered native species of birds.
http://pacificislands.fws.gov/wnwr/ bhakalaunwr.html

Kakahaia National Wildlife Refuge
Located on the island of Molokai, this refuge hosts large populations of Hawaiian stilts and coots.
http://pacificislands.fws.gov/wnwr/ kakahaianwrindex.html

Kaua'i National Wildlife Refuge
Various island sites host endangered species such as Hawaiian stilts, moorhens and coots.
http://pacificislands.fws.gov/wnwr/ kauainwrindex.html

Oahu National Wildlife Complex
Including the James Campbell and Pearl Harbor Refuges, this complex includes major wetlands.
http://pacificislands.fws.gov/wnwr/ oahunwrindex.html

IDAHO

Bear Lake National Wildlife Refuge
Supports mule deer, moose and nesting colonies of grebes, gulls, terns and herons.
www.ohwy.com/id/b/bearlakenwr.htm

Camas National Wildlife Refuge
Home to ring-necked pheasants and sage grouse.
www.ohwy.com/id/c/camasnwr.htm

Kootenai National Wildlife Refuge
Important site for bald eagles and peregrine falcons.
www.r1.fws.gov/kootenai/kootenai.htm

ILLINOIS

Chautauqua National Wildlife Refuge
Best known for its birds, Chautauqua also protects coyotes, badgers and southern flying squirrels.
http://midwest.fws.gov/IllinoisRiver/ Index2.html

Crab Orchard National Wildlife Refuge
A mixed habitat refuge featuring hardwood forest.
http://midwest.fws.gov/CrabOrchard/

Mark Twain National Wildlife Refuge
Includes some of the last remaining bottom land stands of hardwoods in the U.S.
http://midwest.fws.gov/marktwain/

INDIANA

Muscatatuck National Wildlife Refuge
Hardwood forest and wetland habitats.
www.ohwy.com/in/m/muscanwr.htm

Patoka River National Wildlife Refuge
Hosting a variety of very rare species, including the northern copperbelly water snake.
http://midwest.fws.gov/patokariver/

IOWA

DeSoto National Wildlife Refuge
Extending across the Iowa border with Nebraska; noted for its great diversity of migratory birds.
www.omaha.org/oma/desoto.htm

Neal Smith National Wildlife Refuge
Dedicated to restoring a portion of the nation's once extensive tall-grass prairie oak savanna.
www.tallgrass.org/

KANSAS

Kirwin National Wildlife Refuge
Famous for its many bald and golden eagles.
www.r6.fws.gov/refuges/ks

Quivira National Wildlife Refuge
Diverse refuge supporting whooping cranes, white pelicans, avocets, prairie dogs and coyotes.
www.r6.fws.gov/quivira/

KENTUCKY

Reelfoot National Wildlife Refuge
Host to 75 species of reptiles and amphibians and a large concentration of wintering bald eagles.
http://reelfoot.fws.gov/

LOUISIANA

Atchafalaya National Wildlife Refuge
Home to some of the last Louisiana black bears.
http://southeasternlouisiana.fws.gov/ atchafalaya.html

Catahoula National Wildlife Refuge
Best known for reptiles and amphibians ranging from American alligators to cricket frogs.
http://southeastlouisiana.fws.gov/catahoula.html

Bogue Chitto National Wildlife Refuge
Rare residents include the ringed sawback turtle, inflated heelsplitter mussel and gulf sturgeon.
http://southeasternlouisiana.fws.gov/ boguechitto.html

MAINE

Carlton Pond Waterfowl Production Area
Hosts breeding colonies of endangered black terns.
http://northeast.fws.gov/me/cpw.htm

Moosehorn National Wildlife Refuge
Protected complex of 55 marshes and lakes.
http://moosehorn.fws.gov/

Rachel Carson National Wildlife Refuge
Supports moose, deer, martens, river otters, black bears, gray foxes and more than 250 bird species.
www.wellschamber.org/

MARYLAND

Blackwater National Wildlife Refuge
Contains 40 types of rare wetland communities.
http://friendsofblackwater.org

Eastern Neck National Wildlife Refuge
Home to the endangered Delmarva fox squirrel.
http://northeast.fws.gov/me/esn.htm

Patuxent Research Refuge
Sanctuary for migratory waterfowl.
http://patuxent.fws.gov/

MASSACHUSETTS

Great Meadows National Wildlife Refuge
Important site for many migratory birds.
http://northeast.fws.gov/ma/grm.htm

Monomoy National Wildlife Refuge
Three islands linked by causeways, this primarily beach habitat hosts the threatened piping plover.
www.capecodconnection.com/monomoy/ monomoy.htm

Parker River National Wildlife Refuge
Critical nesting habitat for the piping plover.
www.parkerriver.org/

MICHIGAN

Huron National Wildlife Refuge
Comprises 8 islands in Lake Huron.
www.r6.fws.gov/refuges/huron/

Seney National Wildlife Refuge
25,000 acres (10,120 ha) of wilderness; home to bald eagles, ospreys, loons, otters, beavers, black bears, gray wolves and white-tailed deer.
www.r6.fws.gov/refuges/seney/

MINNESOTA

Agassiz National Wildlife Refuge
Important refuge for more than 200 bird species and 49 mammal species; hosts one of the few resident eastern gray wolf packs in the U.S.
http://midwest.fws.gov/agassiz/

Crane Meadows National Wildlife Refuge
Named for its population of greater sandhill cranes; habitat includes tallgrass prairie and wild rice.
http://midwest.fws.gov/CraneMeadows/

Rice Lake National Wildlife Refuge
Protected area with extensive stands of wild rice.
www.ohwy.com/mn/n/riclknwr.htm

MISSISSIPPI

Grand Bay National Wildlife Refuge
Hosts waterfowl and migrant tropical songbirds.
http://southeast.fws.gov/refuges/GrandBay/

Noxubwee National Wildlife Refuge
Supports quail, deer, wild turkeys, wood storks, bald eagles and American alligators.
http://noxubee.fws.gov/

St. Catherine Creek National Wildlife Refuge
Habitats include bottom- and upland hardwood stands, cypress and swamps.
http://Natchez/ StCatherineCreekNationalWildlifeRefuge.htm

MISSOURI

Clarence Cannon National Wildlife Refuge
Provides nesting sites for the endangered king rail.
midwest.fws.gov/ClarenceCannon/

Mingo National Wildlife Refuge
Important refuge for migratory waterfowl.
midwest.fws.gov/Mingo/

Squaw Creek National Wildlife Refuge
Sanctuary for perhaps the last viable breeding population of the massasauga rattlesnake.
www.ohwy.com/sqcrknwf.htm

MONTANA

Charles M. Russell National Wildlife Refuge
Host to bighorn sheep, elk, bobcats and coyotes.
www.r6.fws.gov/cmr/

Ennis National Fish Hatchery
Major broodstock hatchery, raising fish to adult size, removing the eggs, then dispatching the eggs to production hatcheries to raise.
www.r6.fws.gov/Hatchery/ennis/ennis.htm

National Bison Range
Federally managed lands supporting several bison herds as well as elk and bighorn sheep.
www.ohwy.com/mt/m/natbison.htm

NEBRASKA

Fort Niobrara National Wildlife Refuge
Contains a variety of prairie habitats.
www.r6.fws.gov/refuges/niobrara/

Rainwater Basin Wetland Management District
Hosts migrant wildfowl and shorebirds in spring.
www.r6.fws.gov/refuges/rainwater/index.html

Valentine National Wildlife Refuge
Home to prairie chickens and sharp-tailed grouse.
www.r6.fws.gov/refuges/valentine/

NEVADA

Ash Meadows National Wildlife Refuge
A desert oasis habitat supporting 26 plants and animals found nowhere else in the world.
www.r1.fws.gov/desert/ashframe.html

3051

NEVADA (*continued*)
Desert National Wildlife Range
More than 1.5 million acres (607,000 ha) of desert providing habitat for bighorn sheep along with some 320 bird species and many other animals.
www.r1.fws.gov/desert/desframe.htm

Ruby Lake National Wildlife Refuge
Important nesting area for sandhill cranes.
http://gosouthwest.about.com/cs/rubylake1/

NEW HAMPSHIRE
Lake Umbagog National Wildlife Refuge
Breeding and feeding sites for loons, northern harriers, woodcocks and ospreys.
http://northeast.fws.gov/nh/iku.htm

NEW JERSEY
Cape May National Wildlife Refuge
Edwin B. Forsythe Wildlife Refuge
Located on a peninsula, Cape May is used by very large numbers of visiting migratory birds.
http://forsyth.fws.gov/

Great Swamp National Wildlife Refuge
Hosts many fish, reptiles and amphibians and the largest breeding population of eastern bluebirds.
http://northeast.fws.gov/grs.htm

Supawna National Wildlife Refuge
Predominantly brackish tidal marsh.
http://northeast.fws.gov/nj/spmnr.htm

Wallkill National Wildlife Refuge
Riverine floodplain hosting wood turtles.
http://walkillriver.fws.gov

NEW MEXICO
Bitter Lake National Wildlife Refuge
Supporting 350 species of birds, mammals, reptiles and amphibians.
http://ifw2irm.irm1.r2.fws.gov/refuges/newmex/bitter.html

Bosque del Apache National Wildlife Refuge
Extensive sanctuary for migratory birds.
www.friendsofthebosque.org

NEW YORK
Amagansett National Wildlife Refuge
Preserving fragile shore habitat.
http://northeast.fws.gov/ny/lirc.htm

Montezuma National Wildlife Refuge
Refuge for migratory waterfowl, particularly Canada geese; bald eagles and ospreys nest here.
www.fws.gov/25MNWR/

Wertheim National Wildlife Refuge
Located on the southern shore of Long Island, the refuge is particularly important for migrant birds.
www.friendsofwertheim.org

NORTH CAROLINA
Alligator River National Wildlife Refuge
Swamp, forest and wetland habitat; important as part of a program to reintroduce the critically endangered red wolf to the wild.
http://alligatorriver.fws.gov/

Cedar Island National Wildlife Refuge
Predominantly brackish marshland.
www.gorp.com/gorp/resource/us_nwr/nc_cedar.htm

Mattamuskeet National Wildlife Refuges
Extending into Virginia, habitats include marshes, woodlands and the state's largest natural lake; winter site for swans and snow and Canada geese.
www.ohwy.com/nc/n/mattanwr.htm

Pea Island National Wildlife Refuge
Wintering grounds for greater snow geese.
www.hatteras_nc.com/peaisland/

Pee Dee National Wildlife Refuge
Important wintering grounds for migratory birds.
www.co.anson.nc.us/peedee

NORTH DAKOTA
Audubon National Wildlife Refuge
Established in order to mitigate the species loss occasioned by habitat destruction caused by damming the Missouri River.
www.rb.fws.gov/refuges/AUDUBON/

Chase Lake Wetland Management District
Includes extensive tracts of protected prairie.
www.rb.fws.gov/REFUGES/chase/clpp.htm

Lostwood Wetlands Management District
Supports endangered species such as the piping plover and westslope cutthroat trout; also hosts Canada lynxes, jumping mice and gray wolves.
www.rb.fws.gov/dslcomplex/lostwoodwmd.htm

Sullys Hill National Game Park
Wildlife management program maintaining bison elk and black-tailed prairie dog populations.
www.r6.fws.gov/refuges/sullys/

OHIO
Ottawa National Wildlife Refuge
Includes Cedar Point and West Sister Island refuges; important habitat for migratory birds.
http://midwest.fws.gov/ottawa/ottawa.html

OKLAHOMA
Optima National Wildlife Refuge
Prairie and cactus lands supporting white-tailed deer and various small mammals.
http://southwest.fws.gov/refuges/oklahoma/optima.html

Sequoyah National Wildlife Refuge
Especially important sanctuary for migratory waterfowl and host to nesting bald eagles.
www.gorp.com/gorp/resource/us_nwr/ok_sequo.htm

Wichita Mountains Wildlife Refuge
Home to herds of American bison, elk, deer and Texas longhorn cattle.
http://southwest.fws.gov/refuges/oklahoma/wichita.html

OREGON
Cape Meares National Wildlife Refuge
Provides nesting sites for various seabirds; offshore are California and Steller's sea lions, harbor seals and gray whales.
www.ohwy.com/or/c/capemnwr.htm

Hart Mountain National Antelope Refuge
Dedicated to protecting the pronghorn antelope.
www.ohwy.com/or/h/hartmnar.htm

Klamath Marsh National Wildlife Refuge
Important for cranes, ducks and birds of prey; also home to spotted frogs and Rocky Mountain elk.
www.klamathnwr.org/klamathmarsh.html

Umatilla National Wildlife Refuge
Waterways provide habitat for fish such as walleyes, steelheads, sturgeons and crappies.
www.ohwy.org/u/umatinwr.htm

PENNSYLVANIA
Erie National Wildlife Refuge
Protects a large number of mussel species.
http://erie.fws.gov/

John Heinz National Wildlife Refuge at Tinicum
Provides feeding and resting sites for migrant birds.
http://heinz.fws.gov/

PUERTO RICO
Cabo Rojo National Wildlife Refuge
Subtropical dry forest; protects many species, such as the endangered yellow-shouldered blackbird.
http://southeast.fws.gov/CaboRojo/

Culebra National Wildlife Refuge
Mixed habitats, including forest, mangroves, grasslands and small islands; important refuge for more than 60,000 nesting sooty terns.
http://southeast.fws.gov/Culebra/

Laguna Cartagena National Wildlife Refuge
Sanctuary for yellow-shouldered blackbirds.
http://woutheast.fws.gov/LagunaCartagena/

RHODE ISLAND
Block Island National Wildlife Refuge
Resting stop for migrant birds of prey.
www.ultranet.com/block-island/

Ninigret National Wildlife Refuge
Home to more than 30 percent of Rhode Island's endangered and threatened species.
www.newfs.org/powerofplace/ninigret.html

Sachuest Point National Wildlife Refuge
Feeding and nesting area for migratory birds.
http://newportvisions.com/home/sachuest1.html

Trustom Pond National Wildlife Refuge
Last remaining undeveloped coastal salt pond in Rhode Island, covering 640 acres (260 ha).
www.gorp.com/gorp/resource/us_nwr/ri_trust.htm

SOUTH CAROLINA
Cape Romain National Wildlife Refuge
Home to the largest American wintering population of American oystercatchers.
http://caperomain.fws.gov

Carolina Sandhills National Wildlife Refuge
The refuge includes 30 artificial lakes and consists of transitional habitat shifting from Atlantic coastal plain to piedmont.
http://carolinasandhills.fws.gov/

SOUTH DAKOTA
Lacreek National Wildlife Refuge
Sand hills and prairie habitat; home to whooping cranes, burrowing owls and trumpeter swans.
www.birdingamerica.com/SouthDakota/lacreek.htm

Madison Wetland Management District
Lake and marshland habitat.
www.r6.fws.gov/refuges/madison/

Sand Lake National Wildlife Refuge
Prairie and wetlands providing critical feeding grounds for many migratory bird species.
www.r6.fws.gov/sandlake/

TENNESSEE
Cross Creeks National Wildlife Refuge
Varied habitats support 480 species of birds and mammals; particularly important for waterfowl.
http://crosscreeks.fws.gov/

Dale Hollow National Fish Hatchery
Important mitigation breeding program for various trout species: rainbow, brown and lake trout. Fish bred here are distributed throughout Tennessee, Georgia and Alabama.
http://dalehollow.fws.gov/

TEXAS

Anahuak National Wildlife Refuge
Important migratory stop-over point for waterfowl and neotropical birds.
http://southeast.fws.gov/refuges/texas/anahuac.html

Buffalo Lake National Wildlife Refuge
Includes rivers, marshes and prairie.
http://southeast.fws.gov/refuges/texas/buffalo.html

Muleshoe National Wildlife Refuge
Short-grass prairie habitat; wintering grounds for thousands of lesser sandhill cranes.
http://southeast.fws.gov/refuges/texas/mule.html

Santa Ana National Wildlife Refuge
Varied habitat supporting 400 bird species and more than 450 species of plants; also hosts rare mammals such as the ocelot and jaguarundi.
http://southeast.fws.gov/refuges/texas/santana.html

UTAH

Bear River Migratory Bird Refuge
Extensive area of predominantly wetland habitat.
www.northernutah.com/brefuge.htm

Capitol Reef National Park
Home to brown and black bears, western jumping mice, bison and wild sheep and goats.
www.nps.gov/care/

Ouray National Wildlife Refuge
Protected wetland for fish and waterfowl.
www.rmawildlifesociety.org/ouray.htm

VERMONT

Missisquoi River National Wildlife Reserve
Home to a range of migratory and resident birds.
www.recreation.gov/detail.cfm?ID=(1519)

VIRGINIA

Assateague Island National Seashore
Barrier Island off the Atlantic coast; home to wild horses and other mammals.
www.nps.gov/asis/

Back Bay National Wildlife Refuge
Supports loggerhead turtles, bald eagles, piping plovers and peregrine falcons.
http://backbay.fws.gov/

Chincoteague National Wildlife Refuge
Primarily island habitat; home to the wild Chincoteague ponies.
www.recreation.gov/detail.cfm?ID=(1348)

Great Dismal Swamp
Supports black bears, river otters and rare bats.
www.recreation.gov/cfm?ID=(1406)

WASHINGTON

Dungeness National Wildlife Reserve
World's longest naturally occurring sand spit, supporting 250 species of birds, 41 species of land mammals and 8 marine mammals.
www.recreation.gov/detail.dfm?ID=(1382)

North Cascades National Park
Home to bald eagles, black bears, pumas (cougars), gray wolves and mountain goats.
www.nps.gov/noca

Protection Island National Wildlife Reserve
Important for nesting seabirds; protects one of the largest breeding colonies of rhinoceros auklets.
www.recreation.gov/detail.cfm?ID=(2714)

WEST VIRGINIA

Bluestone National Scenic River
Varied habitat hosting wide diversity of wildlife. on the Appalachian Plateau.
www.nps.gov/blue

Canaan Valley National Wildlife Refuge
Sanctuary for populations of the threatened Cheat Mountain salamander and the endangered Virginia flying squirrel.
www.recreation.gov/detail.cfm?ID=(1331)

WISCONSIN

Apostle Islands National Lakeshore
Comprises old growth forest, 12 miles (19 km) of Lake Superior shoreline and 21 islands.
www.nps.gov/apis/

Horican National Wildlife Refuge
Mosaic of wetlands, forest, brushlands and freshwater marsh habitats.
www.fws.gov/3pao/horicon/

Upper Mississippi River Wildlife and Fish Refuge
Incorporates 264 miles (422 km) of river and associated habitats.
www.emtc.nbs.gov/umv_refuge.html

Necedah National Wildlife Refuge
Refuge for sandhill cranes and endangered species such as Karner blue butterflies, massasauga rattlesnakes and Blanding's turtles.
www.midwest.fws.gov/Necedah/

WYOMING

Grand Teton National Park
Rocky Mountain park and refuge, providing habitat for large mammals such as moose and bighorn sheep along with many birds of prey.
www.nps.gov/grte/

National Elk Refuge
Resident elk herd numbering 8,500 along with moose, deer, pronghorn antelope, trumpeter and tundra swans and bald and golden eagles.
http://nationalelkrefuge.fws.gov/

Yellowstone National Park
Extending into Montana as well, the park is host to grizzly (brown) bears, gray wolves, bison, elk and many other species.
www.nps.gov/yell/

WILDLIFE REFUGES IN CANADA

Parks Canada
Searchable on-line directory with links to all of Canada's wildlife refuges and national parks. Information available in English and French.
www.parkscanada.pch.gc.ca

ALBERTA

Banff National Park
Providing protected habitat for black and grizzly (brown) bears, elk, bighorn sheep and caribou.
www.parkscanada-banff/index.htm

Elk Island National Park
A plains habitat for bison, moose, deer, elk and more than 250 species of birds.
www.parcscanada.gc.ca/parks/alberta/elk_island_e.htm

Waterton Lakes National Park
Varied habitat for bison, deer, moose, elk and grizzly (brown) and black bears
www.parcscanada.gc.ca/parks/alberta/waterton_lakes_e.htm

Wood Buffalo National Park
Larger than Switzerland, this park straddles the border with Northwest Territories and protects one of Canada's last major bison herds.
www.parcscanada.gc.ca/parks/nwtw/wood_buffalo_e.htm

BRITISH COLUMBIA

Glacier National Park
Mountain, valley and glacial habitat supporting brown bears, mountain caribou and wolverines.
http://parkscan.harbour.com/glacier/

Gwaii Haanas National Park
Gwaii Haanas National Marine Conservation Area
Habitat includes coastal rain forest, with black bears, pine martens and deer mice as well as a variety of auklets, puffins and storm petrels.
http://parkscan.harbour.com/gwaii/

Kootenay National Park
Habitat varies from glacier to semiarid grasslands; hosts bighorn sheep, caribou, mountain goats, grizzly and black bears, bison, mule deer, gray wolves, pumas (cougars) and moose.
www.worldweb.com/parkscanada-kootenay/index.html

Mount Revelstoke National Park
Rain forest, subalpine forest, tundra and alpine meadow habitats provide a home for mountain caribou, grizzly (brown) and black bears, wolverines and a wide variety of birds.
http://parkscan.harbour.com/mtrev/

Pacific Rim National Park Reserve
Coastal and island habitats contain gray wolves, pumas (cougars), muskrats, river otters, seals, sea lions, leatherback turtles, tree frogs and snakes; gray whales occur in offshore waters.
http://parkscan.harbour.com/pacrim

MANITOBA

Riding Mountain National Park
Situated on an escarpment rising out of the Manitoba prairie.
www.parcscanada.gc.ca/parks/manitoba/riding_mountain_e.html

Wapusk National Park
This refuge is host to polar bear maternity dens.
www.parcscanada.gc.ca/parks/manitoba/wapusk_e.html

NEW BRUNSWICK

Fundy National Park
Includes the Bay of Fundy, where the world's highest tides occur; used by large marine mammals and an array of coastal invertebrates.
www.parcscanada.gc.ca/parks/new_brunswick/fundy/Fundy_e.html

Kouchibouguac National Park
Coastal island and inland habitat.
www.parcscanada.gc.ca/parks/new_brunswick kouchibouguac/kouchibouguac_e.html

NEWFOUNDLAND AND LABRADOR

Gros Morne National Park
A UNESCO World Heritage Site, providing habitat for woodland caribou.
www.parcscanada.gc.ca/parks/newfoundland/gros_morne/gros_morne_e.html

Terra Nova National Park
Comprises sheltered inlets, bogs, hills, lakes, small ponds and pools.
www.parcscanada.gc.ca/parks/newfoundland/terra_nova/terra_nova_e.html

NORTHWEST TERRITORIES

Aulavik National Park
Located on Banks Island, the park is home to large numbers of musk-oxen as well as Arctic foxes, gray wolves, polar bears and two species of lemmings; offshore are a variety of seals along with beluga whales.
www.parcscanada.gc.ca/parks/nwtw/aulavik/aulavik_e.htm

Nahanni National Park Reserve
Habitat includes sulfur springs, tundra and spruce forest; home to 42 species of mammals and 180 bird species as well as many fish.
www.parcscanada.gc.ca/parks/nwtw/nahanni/nahanni_e.htm

Tuktut Nogait National Park
Supports one of the highest densities of birds of prey in North America.
www.parcscanada.gc.ca/parks/nwtw/tuktut_nogait/tuktut_nogait_e.htm

Wood Buffalo National Park
Larger than Switzerland, this park straddles the border with Alberta and protects one of Canada's last remaining major bison herds.
www.parcscanada.gc.ca/parks/nwtw/wood_buffalo_e.htm

NOVA SCOTIA

Cape Breton Highlands National Park
Extensive tracts of highland habitats.
www.parcscanada.gc.ca/parks/nova_scotia/cape_highlands/Cape_highlands_e.htm

Kemjimkujik National Park
Home to 40 species of mammals, 12 species of fish, 205 bird species, 5 snake species and 3 species of amphibians.
www.parcscanada.gc.ca/parks/nova_scotia/kemjimkujik/kemjimkujik_e.htm

ONTARIO

Fathom Five National Marine Park
Bruce Peninsula National Park
Both parks are located at the northern end of the Niagara escarpment. Fathom Five includes 20 islands; Bruce Peninsula provides access to a wide range of wildlife on the mainland.
www.parcscanada.gc.ca/parks/ontario/fathom_five/fathom_five_e.htm

Georgian Bay Islands National Park
Situated in Hudson Bay, the park is accessible only by boat.
www.parcscanada.gc.ca/parks/ontario/georgian_bay/Georgian_bay_e.htm

Point Pelee National Park
Southernmost point of the Canadian mainland, the park is famous for monarch butterflies in the fall and for the number of migratory bird species that stop here.
www.parcscanada.gc.ca/parks/ontario/point_pelee/point_pelee_e.htm

Pukaskwa National Park
On Lake Superior and consisting largely of old-growth forest, Pukaskwa hosts gray wolves, moose and black bears; an important woodland caribou preserve is also located in the park.
www.parcscanada.gc.ca/parks/ontario/pukaskwa/pukaskwa_e.htm

St. Lawrence Islands National Park
Located in the Thousand Island region, the park hosts a variety of rare wildlife.
www.parcscanada.gc.ca/parks/ontario/st_lawrence_islands/st_lawrence_islands_e/htm

PRINCE EDWARD ISLAND

Prince Edward Island National Park
Smallest national park and refuge in Canada, and the country's most important protected nesting site for endangered piping plovers.
www.parcscanada.com.gc.ca/parks/pei/pei_np/pei_e.html

QUEBEC

Forillon National Park
Peninsular habitat consisting of coasts, cliffs and mountains; features a variety of seabird colonies.
www.parcscanada.gc.ca/parks/quebec/ferillon/en/index.html

Saguenay—St. Lawrence Marine Park
Home to a wide variety of marine wildlife.
www.parcscanada.gc.ca/parks/quebec/saguenay_e.html

SASKATCHEWAN

Grasslands National Park
Mixed prairie and grassland habitat hosting prairie dogs, bison and other species; dinosaur fossils are numerous.
www.parcscanada.gc.ca/parks/saskatchewan/grasslands_e.html

Prince Albert National Park
Coniferous forest habitat; site of the only fully protected nesting colony for white pelicans in Canada. Also home to great gray owls.
www.parcscanada.gc.ca/saskatchewan/Prince_Albert_e.html

YUKON TERRITORY

Ivvavik National Park
Includes a portion of the Porcupine caribou herd range.
www.parcscanada.gc.ca/parks/Yukon/ivvavik/ivvavik_e.htm

Kluane National Park and Reserve
Varied habitats including lakes, alpine meadows and tundra, providing a refuge for many fish and bird species as well as moose, elk, bison, bears and caribou.
http://parkscan.harbour.com/kluane/

Vuntut National Park
Includes a portion of the Porcupine caribou herd range; also provides habitat for muskrats, grizzly (brown) and black bears, gray wolves, wolverines, lynxes, musk-oxen and moose.
http://parkscan.harbour.com/vuntut/

WILDLIFE REFUGES IN MEXICO

Mexico maintains a number of biosphere reserves for the protection of endangered plants and animals. An on-line map with links to information on each of the reserves can be found at:
www.unesco.int/mab/br/brdir/latin-am/Mexicomap.htm

BAJA CALIFORNIA (BAJA PENINSULA)

El Vizcaino Biosphere Reserve
Protecting gray whales, bighorn sheep and pronghorn antelope.
www.vizcaino.gob.mx/vizhome.html

CHIAPAS

La Encrucijada Biosphere Reserve
Providing protected habitat for jaguars, raccoons, crocodiles, iguanas and a wide array of shorebird species.
www.tourbymexico.com/chiapas/encruci/encruci.htm

El Triunfo Biosphere Reserve
Rain forest habitat supporting jaguars, tapirs and quetzals as well as many rare plants.
www.tourbymexico.com/chiapas/triunfo/triunfo.htm

QUERETARO

Sierra Gorda Biosphere Reserve
Home to more than 300 bird species as well as endangered black bears, jaguars, spider monkeys and others.
www.woodrising.com/gesg/biospher.htm

QUINTANA ROO

Sian Ka'an Biosphere Reserve
Supports a particularly wide variety of neotropical birds; also home to crocodiles.
www.ine.gob.mx/ucanp/index7.html

SONORA

Pinacate Biosphere Reserve
Refuge for more than 560 plant species, 56 mammal species, 43 reptile species and 232 bird species.
www.puerto-penasco.com/pinacate.htm

TAMAULIPAS

El Cielo Biosphere Reserve
Featuring a variety of macaws, woodpeckers, parrots, hawks and falcons.
www.bafrenz.com/birds/elcielo.com

VERACRUZ

Las Tuxtlas Biosphere Reserve
Providing protected habitat for hundreds of species of birds.
www.earthfoot.org/places/mx011.htm

YUCATAN

Calakmul Biosphere Reserve
Particularly dedicated to maintaining habitat for endangered jaguars
www.calakmul.org/fochome.html

Celestun Biosphere Reserve
Refuge for waterfowl and wading birds, also contains sea turtle nurseries and provides protected habitat for a variety of fish and crocodile species.
www.mexonline.com/celstun.htm

Rio Legartos Biosphere Reserve
Lagoon habitat featuring flamingos.
www.xaac.com/yucatan/no.htm

Wildlife Conservation Agencies and Organizations

GOVERNMENT AGENCIES

Canadian Wildlife Service
Primary web site with links to all Canadian governmental programs dealing with wildlife and environmental conservation.
www.ec.gc.ca

International Association of Fish and Wildlife Agencies
Web site linking governmental programs at the state, provincial and federal levels for the U.S., Canada and Mexico.
www.teaming.com/infwa.htm

National Marine Fisheries Service
Dedicated to the preservation, protection and enhancement of U.S. fisheries.
www.nmfs.gov

National Ocean Service
A division of the National Oceanic and Atmospheric Administration, specifically concerned with preserving the maritime environment along the coasts of the U.S.
www.nos.noaa.gov

National Oceanic and Atmospheric Administration
Coordinating web site with links to U.S. governmental programs dealing with conservation of oceanic wildlife.
www.noaa.gov

National Park Service
Home page for the National Park Service, with links to specialized web sites offering fact sheets, resources and brochures.
www1.nature.nps.gov

National Resources Conservation Service
A division of the U.S. Department of Agriculture, the N.R.C.S. offers information and assistance in developing conservation programs.
www.nrcs.usda.gov

United Nations Environment Programme, World Conservation Monitoring Center
Provides information about the state of endangered species throughout the world.
www.unep_wcmc.org

U.S. Bureau of Land Management
Dedicated to the development of sustainable land use policies and to habitat preservation.
www.blm.gov

U.S. Bureau of Reclamation
Dedicated to the development of water management policy to protect U.S. wetlands.
www.usbr.gov

U.S. Fish and Wildlife Service
Agency responsible for administration of the U.S. Endangered Species Act. Provides links to conservation and preservation web sites maintained by each of the 50 states.
www.fws.gov

U.S. Forest Service
With state-by-state links, provides information on the nation's forests and woodlands.
www.fs.fed.us

White House Council on Environmental Quality
Coordinates federal environmental efforts.
www.whitehouse.gov/CEQ/

PRIVATE AND NOT-FOR-PROFIT ORGANIZATIONS

America the Beautiful Fund
Dedicated to preserving and protecting the wilderness areas of the U.S.
www.america-the-beautiful.org/home.html

American Land Conservancy
Provides educational resources while working to conserve the habitats of American wildlife.
www.alcnet.org

American Public Information for the Environment
Clearinghouse for environmental information, including links to research and conservation programs.
www.americanpie.org

Americans for the Environment
Advocacy and activist group for the protection of wildlife and the environment.
www.afore.org

BirdLife International
A global alliance of national conservation organizations operating in more than 100 countries, with a web site providing links to many of its partner organizations.
www.wing-wbsj.or.jp/birdlife/

Canadian Nature Federation
Providing information on the endangered species of Canada and links to individual organizations dedicated to wildlife conservation and rehabilitation projects.
www.cnf.ca

Center for Environmental Information
On-line access to environmental and wildlife databases.
www.rochesterenvironment.org

Center for Environmental Study
Think tank dedicated to developing and influencing environmental and conservation policy at the federal level.
www.cesmi.org

C.I.T.E.S.—The Convention on International Trade in Endangered Species of Wild Fauna and Flora
An international agreement between governments that aims to ensure that international trade in specimens of wild animals and plants does not threaten their survival.
www.cites.org

Conservation Action Network
Grassroots organizing for the protection of the world's endangered wildlife.
www.takeaction.worldwildlife.org/

Coral Cay Conservation
Provides resources to help sustain livelihoods through the protection and management of coral reefs and tropical forests. Organizes expeditions for paying volunteers.
www.coralcay.org

Coral Reef Alliance
Dedicated to the protection and preservation of coral reefs, worldwide.
www.coral.org/CORAL.html

Dian Fossey Gorilla Fund International
Supporting research and protection programs for the mountain gorillas studied by Dian Fossey. Includes links to a database on the mountain gorilla habitat maintained by Rutgers University.
www.gorillafund.org/

Ducks Unlimited
Ducks Unlimited, Canada
Dedicated to the protection of North American waterfowl and their wetland habitats.
www.ducks.org
www.ducks.ca/

Earth Island Institute
Providing in-depth information and resources on environmental and wildlife conservation issues.
www.earthisland.org

Earth Share
Supporting grassroots organization and public awareness of environmental issues.
www.earthshare.org

Earth Trust
Education, advocacy and support for research on environmental and wildlife conservation issues.
www.earthtrust.org

EarthWatch
Dedicated to increasing public involvement in conservation issues through expeditions in which the public can participate.
www.earthwatch.com

Ecotrust
Supporting programs and research that focus on preserving the North American rain forest, from San Francisco to Anchorage.
www.ecotrust.org/

EndangeredSpecie.com
Providing information on animal protection legislation and fact sheets about endangered species and the causes of endangerment.
www.endangeredspecie.com

Envirolink Network
A links page providing access to environmentalist and conservationist organizations.
www.envirolink.org/

Environmental Council of the States
A cooperative effort to coordinate the environmental and conservation programs across the U.S.
www.sso.org/ecos/

Environmental Defense Fund
Provides news about environmental and conservation law, and supports fundraising programs to enable action in these areas.
www.edf.org

Environmental Investigation Agency
U.K.-based independent campaigning organization committed to investigating and exposing international environmental crime.
www.eia-international.org/

Environmental Law Institute
Provides information about environmental law, both in the U.S. and internationally.
www.eli.org

Fauna and Flora International
Founded in 1903, the world's longest established international conservation body.
www.fauna-flora.org

Forest Action Network
Offers news about activist projects to protect U.S. woodlands and encourages local groups and individuals to take initiatives in the field.
www.fanweb.org/index/shtml

Forest Stewardship Council
Dedicated to the environmentally appropriate and economically viable management of forests.
www.fscoax.org/noframe.htm

Friends of the Earth
Supports environmental and wildlife conservation activities around the world.
www.foe.org

Fund for Animals
Supports educational programs and research, and provides news updates about conservation issues. Members can access a job-search service for careers in wildlife preservation.
www.envirolink.org/arrs/fund

Greenpeace International
High-profile watchdog organization that confronts violators of international conservation and environmental laws.
www.greenpeace.org

International Fund for Animal Welfare
Organizes campaigns for the protection of animals and their habitats, worldwide.
www.ifaw.org

International Wildlife Coalition
Devoted to advocacy, education and activism in support of wildlife conservation.
www.iwc.org

Intersea Foundation
Dedicated to the study of marine mammals throughout the world.
www.intersea.org

I.U.C.N.—The World Conservation Union
International consortium of nations, government agencies, nongovernmental organizations and thousands of scientists dedicated to conserving the diversity of nature and the equitable and sustainable use of natural resources. Provides link to the Red List of endangered species.
www.iucn.org

National Audubon Society
Primarily focused on the preservation and protection of North American bird species, but with some involvement in other wildlife.
www.audubon.org

National Fish and Wildlife Foundation
Dedicated to the conservation of fish and wildlife species, and to the development of sustainable programs of resource use.
www.nfwf.org

National Watchable Wildlife Program
Cooperative program involving state, federal, private and corporate organizations to provide the public with opportunities for viewing wildlife in its natural habitats.
www.gorp.com/wwildlife/wwhome.htm

National Wildlife Federation
Promotes education, activism and advocacy for wildlife preservation and conservation issues.
www.nwf.org

The Nature Conservancy
Dedicated to the protection of habitat for endangered species. Includes links to state-by-state National Heritage Programs.
www.tnc.org/

North American Wolf Association
Focusing on the protection of gray wolves and on educating the public about the importance of wolves in the ecosystem.
www.nawa.org

The Peregrine Fund
Dedicated to the protection of endangered bird species, both in the U.S. and worldwide.
www.peregrinefund.org

Rainforest Action Network
Organizes rain forest protection programs, grassroots projects, education and fundraising.
www.ran.org/ran/

Rainforest Alliance
Dedicated to the conservation of tropical forests, by developing economically viable and socially desirable alternatives to deforestation.
www.rainforest-alliance.org/

Royal Society for The Protection of Birds
U.K.-based organization primarily focused on the preservation of bird species, worldwide. Has one of the largest memberships of any wildlife not-for-profit organization.
www.rspb.org/

Save the Species
Sponsored by the Discovery Channel, a web site offering interactive games to educate the public on endangered species throughout the world.
www.discovery.com/stories/nature/endangered/endangered.html

Sea Shepherd Conservation Society
Dedicated to research and to the documentation of violations of international conservation and species-protection laws and treaties.
www.estreet.com/orgs/sscs/

Sierra Club
Dedicated to grassroots environmental advocacy.
www.sierraclub.org/

Teaming with Wildlife
Raises funds and lobbies for state-level, non-game wildlife conservation programs.
www.teaming.com

Timberwolf Information Network
Devoted to educating the public of the importance and ecological contributions of the gray wolf, particularly in North America. Offers "adopt a wolf pack" programs.
www.timberwolfinformation.org

TRAFFIC
Wildlife trade monitoring program of the World Wildlife Fund and I.U.C.N. (The World Conservation Union).
www.traffic.org/

Trout Unlimited
Dedicated to protecting, preserving and restoring the coldwater fisheries of North America.
www.tu.org

Voyage of the Odyssey
Gathering data to monitor the levels of synthetic contaminants in the world's oceans.
www.pbs.org/odyssey/

Wetlands International
Dedicated to preserving and restoring wetland habitats and their wildlife.
www.wetlands.agro.nl/

Whale and Dolphin Conservation Society
U.K.-based program dedicated to protecting cetaceans (whales, dolphins and porpoises) through education and fundraising projects.
www.wdcs.org

Whale Conservation Institute
Research and educational programs on whales.
www.whale.org

Whaleman Foundation
Providing research and education about whales, dolphins and porpoises.
www.whaleman.com

Wilderness Society
Monitors U.S. federal and state legislation regarding conservation, and lobbies for conservation issues.
www.wilderness.org

Wildlife Conservation Society
Originally founded as the New York Zoological Society in 1895, now sponsors conservation projects in 50 countries.
www.wcs.org

Wildlife Society
Dedicated to developing and influencing sustainable wildlife management policy in the U.S. Also fosters public awareness of preservation issues through expeditions to wilderness areas.
www.wildlife.org

World Center for Birds of Prey
World headquarters of the Peregrine Fund, dedicated to research and conservation.
www.peregrinefund.org/world.html

World Owl Trust
U.K.-based organization dedicated to preserving the world's owl populations, particularly through breeding programs.
www.owls.org

World Wildlife Fund
Devoted to the protection of wildlife and natural habitats, with branches in most countries of the world. Information for Canada available in English and French.
www.wwf.org

Index of Common Names

Numerals in **bold** refer to volume numbers; those in *italics* refer to picture captions. Index entries in **bold** refer to guidepost or biome and habitat articles.

AMPHIBIANS

Arrow-poison frog (dart frog; poison dart frog) **1**: 93–94; **7**: 905; **20**: 2819
  Blue **1**: *93*
  Golden **1**: 93, 94
  Orange and black **1**: 93
  Strawberry **1**: 93
Axolotl (axototl) **1**: 109–110
  Mexican **1**: 109, *109*, 110
  Tiger **1**: 109

Bullfrog (American bullfrog; jug o'rum) **3**: 312–313; **6**: 744; **7**: 903

Caecilian (naked snake) **3**: 356–357
  Sticky **3**: 356

Fire-toad, Variegated *see* Toad, Yellow-bellied
Frog **7**: 903; **20**: 2819
  American tailed (bell toad) **9**: 1200–1201
  Archey's **9**: 1200
  Arum **15**: 2147
  Barking **19**: 2724
  Bird-voiced **19**: 2724
  Bonyheaded **19**: 2725
  Chorus **19**: 2723, 2724
  Clawed (African clawed toad) **4**: 464–465; **7**: 903
  Common (grass frog) **4**: 510–511; **6**: 744
  Cricket **19**: 2723–2724
  Cuban **8**: 1066
  Edible **6**: 743–744
  Flying *see* Gliding frog
  Gaudy leaf **19**: *2723*
  Ghost **7**: 957
  Goeldi's **19**: 2725
  Grass *see* Leopard frog
  Green **19**: 2724
  Greenhouse **8**: 1066–1067; **12**: 1610
  Hamilton's **9**: 1200, *1201*
  Hochstetter's **9**: 1200–1201
  Marsh **6**: 743, 744
  Meadow *see* Leopard frog
  Mouth-brooding (Darwin's toad; Darwin's frog) **12**: 1678–1679
  Paradoxical **14**: 1876–1877
  Poison-dart *see* Arrow-poison frog
  Pool **6**: 743
  Shad *see* Leopard frog
  Smith (blacksmith frog) **19**: 2723
  Spring peeper **7**: 903, *904*
  Stephens Island **20**: 2869
  Tongueless *see* Toad, Surinam
  Water-holding **10**: 1399; **20**: 2868–2869
  Wood **7**: 903
  see also Bullfrog; Frogs and toads
Frogs and toads 7: 902–905

Giant salamander
  Chinese **9**: 1174, *1174*, 1175
  Japanese **9**: 1174, 1175
  see also Hellbender
Gliding frog (flying frog) **7**: 904, 982–983
  Malayan **7**: 982, *982*

Hellbender (hellbender salamander) **9**: 1174–1175
Herring hopper *see* Leopard frog
Horned frog **9**: 1229–1230; **18**: 2465
  Asian **9**: *1229*, 1230
  Bell's **9**: 1229
  Colombian **9**: *1230*
  Ornate **9**: 1229, 1230
  Wied's **9**: 1229–1230

Jug o'rum *see* Bullfrog

Leopard frog (meadow frog; grass frog; shad frog; herring hopper) **11**: 1447–1448
  Northern **11**: 1447
  Southern **11**: 1447

Marsupial frog **11**: 1573–1574
  Pygmy **11**: 1574, *1574*
Midwife toad **7**: 903, 904; **12**: 1609–1610
  Common (bell toad) **12**: 1609–1610
  Iberian **12**: *1609*
Mountain chicken (*frog*) **8**: 1066
Mud eel *see* Siren
Mudpuppy (water dog) **12**: 1682–1683
  Louisiana water dog **12**: 1682

Naked snake *see* Caecilian
Newt **13**: 1732–1734
  California **13**: 1733, 1734; **19**: *2653*
  Crested (warty newt) **13**: 1732, 1733, *1733*, 1734
  Palmate **13**: 1732, *1732*, 1733
  Siberian **19**: 2616
  Smooth **13**: 1732, 1733, 1734, *1734*

Olm **13**: 1788–1789

Peeper, Spring **19**: 2723, 2724

Reed frog (sedge frog) **15**: 2146–2147
  Brown and white spiny **15**: 2146, 2147
  Five-lined **15**: 2146
  Green **15**: 2146
  Painted (marbled reed frog) **15**: 2146, *2146*, 2147, *2147*
  Small golden spiny **15**: 2146, 2147

Salamander
  Arboreal **11**: 1503–1504, 1505
  California slender **11**: 1503
  Dusky **11**: 1503, 1505
  Eastern tiger **11**: *1504*
  Fire (spotted salamander) **6**: 831–832
  Four-toed **11**: 1503
  Jefferson **11**: *1505*
  Long-tailed **11**: 1503, *1503*
  Lungless **11**: 1503–1505
  Ocoee **11**: 1503
  Pygmy **11**: 1503
  Red-backed **11**: 1503
  Shovel-nosed **11**: 1503
  Slimy **11**: 1504, 1505
  Tiger **1**: 109, 110
  Yellow-blotched **11**: 1505

Siren (*amphibian*; mud eel) **17**: 2381–2382
  Greater **17**: 2381, 2382
  Lesser **17**: *2381*, 2382, *2382*
  Northern dwarf **17**: 2381, 2382
  Southern dwarf **17**: 2381, 2382
Snake, Naked *see* Caecilian
Spadefoot toad **18**: 2465–2466
  Common (garlic toad) **18**: 2465, *2466*
  Couch's **18**: 2466
  Eastern (Hunter's spadefoot toad) **18**: 2465
  Hammond's **18**: 2466
  Mexican **18**: 2465
  Plains **18**: 2466
  Western **7**: 904

Toad **7**: 903, *905*; **20**: 2819
  African clawed *see* Frog, Clawed
  Bell *see* Frog, American tailed; Midwife toad, Common
  Burrowing (Mexican burrowing toad; digger toad) **3**: 326
  Common **4**: 510, 512–513
  Darwin's *see* Frog, Mouth-brooding
  Fire-bellied (firebelly) **6**: 825–826; **7**: 903; **20**: 2819
  Garlic *see* Spadefoot toad, Common
  Giant (cane toad; giant toad) **11**: 1557–1558
  Green **12**: 1720
  Midwife *see* Midwife toad
  Natterjack **12**: 1720–1721
  Spadefoot *see* Spadefoot toad
  Surinam (tongueless frog) **18**: 2578–2579
  Yellow-bellied (variegated fire-toad) **6**: 825
  see also Frogs and toads
Tree frog **7**: 903; **19**: 2723–2725
  Brazilian **7**: 983
  Canyon **19**: 2724
  Cuban **19**: 2723
  Giant **19**: 2724
  Golden (bell tree frog) **19**: 2724
  Red-eyed **7**: *903*
  Rohde's **19**: 2725
  Southern **19**: *2725*

Venomous animals 20: 2817–2820

Water dog *see* Mudpuppy

BIRDS

Abu markub *see* Stork, Shoebill
Accentor **1**: 22–23
  Alpine **1**: 22, 23, *23*
  Hedge (dunnock; hedge sparrow; shuffle-wing) **1**: 22, *22*, 23
  Himalayan **1**: 22
  Siberian **1**: 22
Alarmbird *see* Kookaburra
Albatross **1**: 35–37; **16**: 2264, 2267, 2268
  Black-browed **1**: 35, 36, *36*
  Buller's **1**: 36
  Laysan **1**: 37
  Short-tailed **16**: 2268
  Sooty **1**: 35, 36
  Wandering **1**: *35*, 36, 37; **16**: *2267*
  Waved **1**: 35
Amazon **14**: 1883
  Orange-winged **14**: 1884, *1884*, 1885, *1885*

White-fronted **14**: 1883
  Yellow-headed **14**: 1883
Anhinga, American (snakebird) **5**: 645, 646–647, *646*; **21**: 2907
Antbird **1**: 55–56; **21**: 2969
  Black-faced **1**: 56
  Spotted **1**: 55
Antpitta **1**: 55
Antshrike **1**: 56
  Amazonian **1**: *55*
  Barred **1**: 55
  Black-crested **1**: 55
  Great **1**: *56*
Ant-tanager **19**: 2629
Ant-thrush **1**: 55, 56
  Black-faced **1**: 55, 56
Antvireo **1**: 56
Antwren **1**: 56
  Pygmy **1**: 55
  White-flanked **1**: 55
Anvilhead *see* Hamerkop
Apostlebird **11**: 1522, *1523*
Aracari **19**: 2714
  Collared **19**: 2714, 2715
Asity, Sunbird **12**: 1605
Astrapia, Princess Stephanie's **2**: 205
Auk **8**: 1094, 1095, 1096; **16**: 2265, 2267
  Great (*extinct*) **14**: 1911; **16**: 2268
  Little (dovekie) **11**: 1464–1465; **16**: 2265
  Razorbilled *see* Razorbill
Avadavat (red munia; strawberry finch; strawberry waxbill; tiger finch) **1**: 103–104
  Green **1**: 103
  Red **1**: 103, *103*, 104, *104*
Avocet **1**: 105–106
  American **1**: 105, *105*
  Australian (red-necked avocet) **1**: 105
  Chilean **1**: 105
  Eurasian (pied avocet; black-capped avocet) **1**: 105, 106, *106*

Babbler **1**: 113–114; **18**: 2456
  Common **1**: 114
  Jungle **1**: 113
  Lowland mouse **1**: *113*
  Rufous **1**: *114*
  Scimitar **1**: 113
  Song **1**: 113
  Tit **1**: 113
  Wren **1**: 113
Barber *see* Hornero, Rufous
Bean goose **14**: 1955
  Taiga **14**: 1955
  Tundra **14**: 1955
Becard **4**: 543
  Jamaican **4**: 543
  White-winged **4**: 543, 544
Bee-eater **2**: 180–182
  Carmine **2**: 181
  European **2**: 180, 181, *181*
  Little **2**: *182*
  Rainbow (rainbow bird) **2**: 180
Bellbird **2**: 191–192; **4**: 543
  Bare-throated **2**: 191
  Bearded (mossy-throated bellbird; black-winged bellbird) **2**: 191; **4**: 543, 544

Bellbird (*continued*)
  Crested (pan-pan-panella; Dick-Dick, the devil) **2**: 191–192, *191*
  New Zealand **2**: 191
  Three-wattled **2**: 191; **4**: 543
  White (snowy cotinga) **2**: 191; **4**: 543
Besra **18**: 2467
Betcherrygah *see* Budgerigar
Bib-finch *see* Mannikin, Madagascar
Bird of paradise **2**: 205–207
  Crested **2**: *206*
  Emperor **2**: 205, 206, 207, *207*
  Greater **2**: 206
  King of Saxony **2**: 205
  Lesser **2**: *205*
  Magnificent **2**: 206
  Prince Rudolf's blue **2**: 205, 207
  Raggiana **2**: 206–207
Birds of prey 2: 208–210
  see also Buzzard; Eagle; Falcon; Goshawk; Gyrfalcon; Harrier; Hawk; Hobby; Kestrel; Kite; Merlin; Osprey; Secretary bird
Bishop bird **2**: 213–214
  Burton's black-and-yellow (yellow-shouldered widow bird) **2**: 213
  Fire-crowned (black-winged bishop; black-winged red bishop; red-crowned bishop) **2**: 213, 214
  Golden-backed **2**: 213
  Red **2**: *214*
  Yellow-crowned **2**: 213
Bittern **2**: 220–221; **9**: 1179
  American (bog-bull) **2**: *220*, 221
  Australian **2**: 220
  Eurasian **2**: 220, *221*
  Least **2**: 220
  Little **2**: 220
  Pinnated **2**: 220
  Sun **18**: 2567–2568
Blackbird (Eurasian blackbird; European blackbird) **2**: 224–225; **19**: 2678, 2679
  Austral **13**: 1805
  Red-winged **2**: 224; **15**: 2141–2142; **18**: 2458
  Yellow-headed **2**: 224
Blackcap **20**: 2857
Black-eye, Olive **21**: 2930
Bluebird
  Eastern **18**: *2455*, 2457
Bobolink **13**: 1805, 1807; **18**: *2457*; **20**: 2833
Bobwhite, Northern **15**: 2082, 2083
Bog-bull *see* Bittern, American
Bonxie *see* Skua, Great
Booby **2**: 261–263
  Abbott's **2**: 261, 262
  Blue-footed **2**: 261, *263*
  Brown **2**: 261, 262, 263
  Masked (blue-faced booby) **2**: *261*, 262, *262*, 263
  Pervian (piquero) **2**: 261, 262, 263
  Red-footed **2**: 261, 262
Bowerbird (satinbird) **2**: 269–271
  Archbold's **2**: 271
  Fire-maned **2**: 271
  Golden **2**: 271
  Great gray **2**: *270*
  Regent **2**: 271
  Satin **2**: 270, 271, *271*
  Spotted **2**: 271
Broadbill (*duck*) *see* Scaup, Greater
Broadbill (*Old World tropical bird*) **2**: 282–283
  Black-and-red **2**: *282*
  Blue-crested **2**: *283*
  Dusky **2**: 283

Grauer's (green broadbill; African green broadbill) **2**: 283
  Green **2**: 282
  Long-tailed **2**: 282
  Red-sided **2**: 282
Brolga **5**: 584, 587
Bronze-cuckoo, Shining **5**: 617–618, 619; **19**: 2670
Brownbul **3**: 308
Brush turkey **11**: 1528
  Australian **11**: 1527
Budgerigar (budgerygah; betcherrygah; canary parrot; lovebird; flightbird) **3**: 302–304
Buffalo bird *see* Cowbird, Brown-headed
Bulbul **3**: 308–309
  Black **3**: 308
  Brown-eared **3**: 308
  Common (African bulbul) **3**: *308*, 309
  Red-vented **3**: 309
  Red-whiskered **3**: 309, *309*
  White-cheeked **3**: 308
Bullfinch (common bullfinch; Eurasian bullfinch) **3**: 310–311
Bunting **3**: 319–320
  Corn **3**: 320
  House **3**: 319
  Lapland **3**: 319
  Lark **18**: 2458
  McKay's **3**: 319
  Ortolan **3**: 319, *320*
  Painted **3**: *319*
  Reed **3**: 320; **18**: 2457
  Snow **3**: 319–320
Bush-shrike **17**: 2361
Bustard **3**: 337–339
  Australian **3**: 338
  Great **3**: 337, 338
  Great Indian **3**: 338
  Houbara **3**: 339
  Kori **3**: *337*
Bustard quail
  Barred **3**: 350
  Little *see* Button quail, Common
Butcherbird **3**: 340–341; **17**: 2363
  Black **3**: 340
  Black-backed **3**: 340
  Black-headed **3**: 340
  Gray (silver-backed butcherbird; Derwent jackass; Tasmanian jackass; whistling Jack) **3**: 340, *340*, 341
  Pied **3**: 340, *341*
  White-rumped **3**: 340
Button quail **3**: 349–350; **15**: 2082
  Australian black-breasted **3**: 350
  Common (little button quail; little bustard quail; kurrichane button quail; Andalusian hemipode) **3**: 349, 350
  Yellow-legged **3**: 349
Buzzard **2**: 209; **3**: 351–353
  Black-breasted (black-breasted kite) **3**: 352–353
  Common (Eurasian buzzard; steppe buzzard) **3**: 351–352, *351*, *352*
  Jackal (augur) **3**: 351
  Rough-legged **2**: 209; **3**: 351

Cachalote, Brown **13**: 1820
Cahow **16**: 2268
Calfbird (capuchinbird) **4**: 543
Calyptura, Kinglet **4**: 544
Canary **3**: 368–369
Canvasback **5**: 717; **14**: 1993–1994, 1995
Capercaillie (capercailzie; horse of the woods) **3**: 376–378

Black-billed **3**: 376, 377
  Western **3**: 377
Capuchinbird *see* Calfbird
Cardinal (northern cardinal; red bird) **3**: 387–388
Cariama *see* Seriema
Cassowary **3**: 397–398
  Dwarf **3**: 397
  Northern **3**: 397
  Southern (Australian cassowary; double-wattled cassowary; two-wattled cassowary) **3**: 397, 398, *398*
Castlebuilder *see* Spinetail
Catbird **3**: 399–400
  Abyssinian **3**: 399
  Black **3**: 399
  Gray (common catbird; northern catbird) **3**: 399, *399*, 400, *400*; **4**: 555
  Tooth-billed (stagemaker) **3**: 399
Chachalaca, Plains **20**: 2757
Chaffinch **3**: 414–415
Chaparral cock *see* Roadrunner, Greater
Chat **18**: 2458; **19**: 2678
  Yellow-breasted **21**: 2985
Chickadee *see* Tit
Chiffchaff **20**: 2857
  Iberian **20**: 2859
Chlorophonia **19**: 2627
Chough **4**: 451–452
  Alpine (yellow-billed chough) **4**: 451; **12**: 1673
  Red-billed (common chough; Cornish chough) **4**: 451, *451*, 452, *452*; **12**: 1673
  White-winged **4**: 451; **11**: 1522
Chuck-will's-widow **13**: 1744 **14**: 2012
Churn owl *see* Nightjar, Eurasian
Cicada bird, Black-bibbed **5**: 622
Cinclodes **13**: 1819, 1820
Cisticola, Golden-headed **19**: 2618, *2618*
Clock bird *see* Motmot, Blue-crowned
Cockatiel **4**: 483, 485
  Moluccan **4**: *484*
Cockatoo **4**: 483–485
  Gang-gang **4**: 483
  Palm (great palm cockatoo; great black cockatoo; goliath cockatoo; Cape York cockatoo) **4**: 483, *485*
  Philippine **4**: 485
  Pink **4**: *483*
  Rose-breasted *see* Galah
  Sulfur-crested (white cockatoo) **4**: 483, 485
  White-tailed black **4**: 485
  Yellow-crested **4**: 485
  Yellow-tailed black **4**: 485
Cock-of-the-rock **4**: 490–491
  Andean **4**: 490, 491, 543
  Guianan **4**: 490, 491, 543
Condor **4**: 516–518
  Andean **4**: 516, *516*, *517*; **12**: 1673; **20**: 2841
  California **4**: 516, 517, 518, *518*; **20**: 2841
Conure **14**: 1878
Coot **4**: 523–524
  American **4**: 523, *523*, 524
  Crested (knob-billed coot) **4**: 523
  European (common coot) **4**: 523, 524, *524*
  Giant **4**: 523, 524
  Hawaiian **4**: 523
  Horned **4**: 523

Cordon-bleu (blue waxbill) **21**: 2888, *2889*
Corella, Little **4**: 483, 485
Cormorant **4**: 536–538; **16**: 2265, 2268
  Brandt's **4**: 536
  Cape **4**: 538
  Double-crested (white-crested cormorant) **4**: 536, 537
  Flightless **4**: 536, 537, *537*, 538
  Great (common cormorant; black cormorant) **4**: 536, *536*, *537*
  Guanay **4**: 538
  King **4**: *538*
  Neotropic (olivaceous cormorant) **4**: 536
  Pelagic **4**: 536
  Red-faced **4**: 536
Corncrake (land rail) **4**: 541–542
Cotinga **4**: 543–544
  Banded **4**: 544
  Blue **4**: *543*
  Snowy (white bellbird) **4**: 543
  Swallow-tailed **4**: *544*
Coua **4**: 549; **20**: 2780
Coucal **4**: 549–550
  Black-hooded **4**: 550
  Blue-headed **4**: 549
  Common (crow-pheasant) **4**: 549
  Coppery-tailed **4**: *550*
  Greater black **4**: 549
  Green-billed **4**: 549
  Lesser black **4**: 549
  Pheasant (common coucal; swamp cuckoo) **4**: 549–550, 551
  Senegal **4**: 549, *549*
  White-browed **4**: 549, 550
Courser **4**: 551–552
  Burchell's **4**: 551
  Cream-colored **4**: 551, *551*
  Double-banded **4**: *552*
  Indian **4**: 551
  Jerdon's **4**: 551
  Temminck's **4**: 551–552
Cowbird **4**: 553–555; **18**: 2461
  Bay-winged **4**: 553, 555
  Brown-headed (buffalo bird) **3**: 400; **4**: 553, 554, *554*, 555; **15**: 2141, 2143; **18**: 2459
  Screaming **4**: 553
  Shiny **4**: 553, 555, *555*
Crake **5**: 582–583
  Baillon's **5**: 582, 583
  Carolina *see* Sora
  Little **5**: 582, 583, *583*
  Ruddy **5**: 582
  Spotted **5**: 582
Crane **5**: 584–587
  Black-crowned **5**: 584
  Common (Eurasian crane) **5**: 584, 585, *586*
  Red-crowned (Japanese crane) **5**: 584, *584*, 585–586, 587
  Sandhill **5**: 584, 586, *587*
  Siberian **5**: 585
  Whooping **5**: 584, 585; **20**: 2757
Creeper, Spotted *see* Treecreeper
Crest, Canary Islands **7**: 999
Crocodile bird *see* Plover, Egyptian
Crossbill **5**: 607–609; **18**: 2457
  Common (red crossbill) **5**: 607, 608–609, *608*
  Parrot **5**: 607, 608, *609*
  Scottish **5**: 607, 608
  Two-barred **5**: 607, 608
Crow **5**: 610–613; **18**: 2457
  American (Western crow; Florida crow) **5**: 610, 611, 612, 613, *613*
  Carrion **4**: 452; **5**: 610, 611, *612*, 613
  Fish **5**: 610, 611–612
  Hooded **5**: 610, 612

House **5**: *610*; **18**: 2457, 2459
King *see* Drongo, Black
Northwestern **5**: 610
Pied **5**: *611*
Tamaulipas (Mexican crow) **5**: 610
Wattled *see* Kokako
Crow-pheasant *see* Coucal, Common
Crying bird *see* Limpkin
Cuckoo **5**: 617–620; **21**: 2938
Chestnut-winged **5**: 617
Common (European cuckoo;
Eurasian cuckoo) **5**: 617, 618, *618*,
619, 620; **9**: 1224; **21**: 2898
Didric **5**: 617; **21**: 2898
Great spotted **5**: 617, 619
Indian **5**: 713
Striped **5**: 619
Yellow-billed **5**: *617*
Cuckoo-shrike **5**: 621–622; **21**: 2984
Barred **5**: 621
Black-faced (blue jay; shufflewing)
5: 621, *621*
Ground **5**: 621
Red-shouldered (black cuckoo-
shrike) **5**: 621, 622
White-breasted **5**: 622
White-winged **5**: 622
Curassow **5**: 623–625
Black **5**: 623
Blue-billed **5**: 625
Great (Mexican curassow; great
crested curassow; globose
curassow) **5**: 623, 624
Helmeted **5**: 623
Razor-billed (mutum) **5**: 623, *624*,
625, *625*
Red-billed **5**: 625
Trinidad white-headed **5**: 623
Yellow-knobbed **5**: *623*
Curlew **5**: 626–628
Bristle-thighed **5**: 626, 628
Eskimo (doughbird) **5**: 628
Eurasian (common curlew;
western curlew; European
curlew) **5**: 626, 627
Hudsonian (whimbrel) **5**: 626, 627
Long-billed **5**: 626
Madagascar **5**: 626
Slender-billed **5**: 628
White *see* Ibis, White
Cutthroat *see* Whistler, Golden

Dacnis
Blue **9**: 1213
Scarlet-thighed **9**: 1213
Darter (*bird*) **5**: 645–647
African **5**: 645
Australian **5**: 645
Oriental (Indian darter) **5**: 645
Dickcissel **18**: 2458
Dioch, Sudan *see* Quelea, Red-billed
Dipper **5**: 680–681; **18**: 2457
American (American water ouzel)
5: 680, 681
Brown **5**: 680
Rufous-throated **5**: 680
White-capped **5**: 680
White-throated **5**: 680, *680*
Diver
Black-throated *see* Loon, Arctic
Great northern *see* Loon, Common
Red-throated *see* Loon, Red-
throated
White-billed *see* Loon, Yellow-billed
Diving-petrel **5**: 684–685; **16**: 2267
Common (jack diver; kuaka;
flying penguin) **5**: 684, *684*, 685
Magellanic **5**: 684
Peruvian (potoyunco) **5**: 684–685
South Georgia **5**: 684

Dollarbird **16**: 2204
Dorhawk *see* Nightjar, Eurasian
Dotterel **5**: 704–705
Eurasian **5**: 704, 705, *705*
Doughbird *see* Curlew, Eskimo
Dove
Collared (Eurasian collared dove;
collared turtle dove) **4**: 499–500
Fruit *see* Fruit dove
Palm (town dove; village dove;
garden dove; laughing dove)
13: 1857–1858
Ring *see* Pigeon, Wood
Rock (feral pigeon; domestic
pigeon; racing pigeon; fancy
pigeon) **16**: 2194–2196
Dovekie *see* Auk, Little
Drongo **5**: 712–713
Ashy **5**: 713
Black (king crow) **5**: 712, 713, *713*
Greater racket-tailed **5**: 712
Lesser racket-tailed **5**: 712, *712*
Spangled (hair-crested drongo)
5: 712, 713
Duck
Black-headed **18**: 2538, 2539
Blue **19**: 2710
Blue-billed (spiny-tailed duck)
18: 2538, 2539
Cornfield *see* Whistling duck,
Black-bellied
Eider *see* Eider, Common
Ferruginous **14**: 1993
Freckled **5**: 714
Labrador (*extinct*) **5**: 717
Logger *see* Steamer duck,
Flightless
Long-tailed (oldsquaw) **5**: 717
Maccoa **18**: 2538
Mandarin **5**: 716–717; **21**: 2925,
2970, 2971
Masked **18**: 2538
Muscovy (musky duck; musk
duck) **5**: 717; **12**: 1699–1700
Musk (mold goose) **18**: 2538,
2539
Painted *see* Shelduck, Paradise
Paradise *see* Shelduck, Paradise
Perching **21**: 2925
Pink-headed **5**: 717
Ruddy **18**: 2537, *2537*, 2538, *2538*,
2539
Torrent **19**: 2708–2710
Tree *see* Duck, Wood; Whistling
duck
Tufted **14**: 1993, 1995
White-backed **18**: 2538
White-headed **18**: 2537, 2538–2539,
2539, *2539*
Wood (Carolina duck; summer
duck; tree duck) **5**: 716–717;
21: 2925, 2970–2972
see also Ducks, geese and swans;
Eider; Mallard; Merganser;
Pintail; Pochard; Shelduck;
Stifftail; Teal
Ducks, geese and swans 5: 714–717
Dunlin **16**: 2233
Dunnock *see* Accentor, Hedge

Eagle **2**: 209
Bald (American eagle; white-
headed sea eagle) **1**: 82, 123–125;
2: 208; **21**: 2907
Brown harrier (brown snake-
eagle) **12**: 1589
Crowned (Afrian crowned eagle;
crowned hawk-eagle)
5: 614–616; **8**: 1093
Fish *see* Fish eagle

Golden **7**: 1001–1003
Harpy *see* Harpy eagle
Imperial **7**: 1001
Lesser spotted **21**: 2906
Martial **12**: 1588–1589
Monkey-eating **15**: 2111
River *see* Fish eagle
Verreaux's (black eagle) **7**: 1001,
1003; **12**: 1589
White-tailed **6**: 837, 839
Earth-creeper **13**: 1819
Egret **6**: 750–753
Cattle **6**: 750, 752, 753
Eastern reef (blue reef heron)
6: 750
Great (great white egret; great
white heron; great American
heron; white heron; large heron)
6: 750, 751, *753*
Great white **9**: 1179
Little **6**: 750–751, *753*
Snowy **6**: 750, *751*
Eider **6**: 754–756
Common (eider duck) **6**: 754, 755,
756
King **6**: 754, 755, *755*
Spectacled **6**: 754
Steller's **6**: 754, *754*, *756*
Emu **6**: 782–784
Emu-wren **21**: 2993
Euphonia **19**: 2628
Eve-jarr *see* Nightjar, Eurasian

Fairy bluebird **6**: 790–791
Blue-backed (Asian fairy bluebird;
blue-mantled fairy bluebird)
6: 790
Philippine **6**: 790
Fairy-wren **21**: 2993
Falcon
African pygmy **2**: *209*
Aplomado **2**: *208*
Eleanora's **2**: 210
Lanner **8**: 1110
Peregrine **2**: 209, 210; **8**: 1029,
1110, 1112; **14**: 1921–1923;
16: 2196
Saker **8**: 1110
Falconet, Black-thighed **2**: 208
Fantail **20**: 2843
Fern owl *see* Nightjar, Eurasian
Fieldfare **18**: *2459*
Figbird **8**: 1012
Yellow **8**: 1012
Finch
Bengalese **11**: 1544
Cocos Island **5**: 649
Cut-throat **11**: 1544
Gouldian (rainbow finch; painted
finch) **8**: 1030–1031
Ground **5**: 649
Locust **21**: 2888
Mangrove **5**: 649
Melba **21**: 2938
Red-headed **11**: 1544, *1544*
Tree **5**: 649
Vegetarian **5**: 649
Zebra **21**: 3012–3013
Finches, Darwin's (Galapagos
finches) **5**: 648–650
Finfoot **6**: 823–824
African **6**: 823, *823*
American (sungrebe) **6**: 823, 824;
21: 2907
Asian (masked finfoot) **6**: 823
Fireback **14**: 1934
Siamese **14**: *1935*
Firecrest **7**: 999
Taiwan **7**: 999
Fire-finch **21**: 2938

Fish eagle (African fish eagle;
river eagle) **6**: 837–839
African **6**: 837
Madagascar **6**: 837
Pallas' **6**: 837
Fish owl (fishing owl) **6**: 842–843;
13: 1827
Blakiston's **6**: 843
Brown **6**: *842*, 843
Buffy **6**: 842, *843*
Pel's (African fishing owl; African
fish owl) **6**: 842, 843
Rufous **6**: 842
Tawny **6**: 842–843
Vermiculated **6**: 842
Flamingo **6**: 839, 844–847
Andean **6**: 844, 846, *846*
Chilean **6**: 844, 846
Greater (Caribbean flamingo)
6: 844, 845–846, *845*
James **6**: 844, 846
Lesser **6**: *844*, 846
Flicker **6**: 853–855
Andean **6**: 853
Campo **6**: 853
Chilean **6**: 853
Fernandina's **6**: 853
Gilded **6**: 853, *855*
Northern (common flicker; yellow-
shafted flicker; red-shafted
flicker) **6**: 853, *853*, 854, *854*, 855
Flightbird *see* Budgerigar
Florican
Bengal **3**: 338
Lesser **3**: 338
Flowerpecker **7**: 868–869;
18: 2456
Mistletoe *see* Mistletoe bird
Orange-bellied **7**: *869*
Scarlet-backed **7**: 868
Thick-billed **7**: 869
Tickell's **7**: 869
Flycatcher (*New World*)
Acadian **20**: 2802
Boat-billed **20**: 2801
Many-colored tyrant **20**: 2800
Northern royal **20**: 2800
Scissor-tailed **20**: 2800, *2802*
Tyrant **4**: 555; **7**: 872; **18**: 2455;
19: 2616; **20**: 2800–2802
Vermilion **20**: 2800, 2801, *2801*
Flycatcher (*Old World*) **7**: 872–874;
18: 2456, 2458
Collared **7**: 873, 874
Madagascar paradise **7**: *873*
Pied **7**: 872–873, 874; **18**: 2459
Semicollared **7**: 873
Spotted **7**: 872–873, *874*
Foliage-gleaner **13**: 1819
Francolin **7**: 889–890
Black (black partridge) **7**: 889,
890
Chinese **7**: 890
Coqui **7**: 889
Crested **7**: 890
Gray **7**: 890
Gray-winged **7**: 889
Hartlaub's **7**: 889–890
Natal **7**: *889*
Painted **7**: 890
Swamp **7**: 890
Friarbird **9**: 1215
Helmeted **9**: 1215
Noisy **9**: 1215
Frigate bird **7**: 891–893; **16**: 2266–2267
Ascension **7**: 893
Christmas Island **7**: 893
Great **7**: *892*
Magnificent (man-o'-war bird;
hurricane bird) **7**: *891*, 892

Frogmouth 7: 900–901
    Large 7: 901
    Tawny (mopoke; freckled
        frogmouth) 7: 900, *900*, 901
Fruit dove
    Jambu 7: *911*
    Makatea 7: 912
Fruit pigeon 7: 911–912
    Blue 7: 911
    New Zealand (New Zealand
        pigeon; native pigeon) 7: 912
    Wedge-tailed 7: 911
    Yellow-bellied 7: 911
Fulmar (Atlantic fulmar; Pacific
    fulmar; northern fulmar; fulmar
    petrel) 7: 913–915; 16: 2265, 2268
    Antarctic (silver-gray fulmar;
        southern fulmar) 7: 913

Galah (rose-breasted cockatoo)
    4: 483, 485
Gallinule 7: 919–920
    American (American purple
        gallinule) 7: 919, *919*, 920
    Purple (purple swamphen) 7: 919,
        920
Gannet (northern gannet) 7: 923–925;
    16: 2264, 2265, 2268, *2268*
    Australian (Pacific gannet) 7: 923
    Cape 7: 923, *924*; 16: 2264
Ghostbird *see* Shearwater, Sooty
Gnow *see* Mallee fowl
Goatsucker *see* Nightjar, Eurasian
Godwit 7: 996–998
    Bar-tailed 7: 996, 997, *998*
    Black-tailed 7: 996, 997, *997*, 998;
        21: 2906
    Hudsonian 7: 996, 997
    Marbled 7: 996, 998
Goldcrest 7: 999–1000; 18: 2455
Goldeneye 5: 717
Golden plover 14: 1988, 1989
    American 14: 1988, 1989, 1990
    Eurasian 14: 1988, *1990*
    Pacific 14: 1988
Goldfinch 3: 368; 8: 1016–1017
    American 8: 1016, 1017, *1017*
    Dark-backed (lesser goldfinch)
        8: 1016
    European 8: 1016, *1016*
Gonolek 17: 2361
Goosander *see* Merganser, Common
Goose
    Bar-headed 5: 717
    Barnacle (bar goose) 1: 140–141
    Black 5: 717
    Canada 3: 365–367
    Gray 5: 717; 14: 1955
    Greylag 5: 717; 8: 1073–1074
    Hawaiian (nene; lava goose)
        5: 717; 9: 1163–1165
    Magpie 5: 714
    Maned (wood duck) 21: 2970, 2971
    Mold *see* Duck, Musk
    Pink-footed 14: 1955–1956
    Pygmy 5: 714; 21: 2927
    Ragitata *see* Shelduck,Paradise
    Red-breasted 5: *716*
    Ruddy-headed 5: 716
    Snow 5: *715*
    Spur-winged 5: 717
    Swan 5: 717
    White pygmy 5: 714
    see also Ducks, geese and swans
Goshawk (northern goshawk) 2: 208;
    8: 1027–1029
    Black 8: 1027
    Chanting 8: 1027
    Crested 8: 1027
    White 8: 1027

Grackle 8: 1032–1034
    Boat-tailed 8: 1032, 1034
    Common (bronzed grackle)
        8: 1032, 1033, *1033*, *1034*
    Great-tailed 8: 1032, 1033, 1034
    Slender-billed 8: 1032
Grasswren 21: 2993
Grebe 8: 1061–1063
    Atitlan 19: 2649
    Eared 8: 1061, 1063
    Flightless Titicaca (short-winged
        grebe) 8: 1063
    Great crested 8: 1061, *1061*, 1062
    Great pied-billed (Atitlan grebe)
        8: 1063
    Horned 8: 1061, *1062*, 1063
    Little 8: 1062
    Pied-billed 8: 1061, 1063, *1063*
    Puna (Junin) 8: 1063
    Western 8: 1063
Greenbul 3: 308
Greenfinch (green linnet) 3: 368;
    8: 1064–1065; 11: 1457
Greenlet 20: 2831
    Gray-headed 20: 2831
Green pigeon 7: 911
    Timor 7: 912
Greenshank 16: 2233
Ground thrush *see* Pitta
Grouse 8: 1084–1086
    Black 8: 1085, 1086
    Blue 8: 1084, *1084*
    Ruffed 8: 1084, 1085, *1085*, 1086
    Sage 8: 1084, 1086
    Sand *see* Sand grouse
    Sharp-tailed 8: 1084, 1086
    Spruce 8: 1084, 1086, *1086*
    Willow (willow ptarmigan)
        8: 1084–1086; 15: 2055, *2057*
Guácharo *see* Oilbird
Guillemot (murre) 8: 1094–1096
    Black 8: 1094, 1095, 1096, *1096*
    Brünnich's (thick-billed murre)
        8: 1094, 1095–1096
    Common (common murre)
        8: 1094, 1095, *1095*, 1096
Guinea fowl 8: 1097–1098
    Black 8: 1097
    Crested 8: 1097, 1098
    Helmeted (tufted guinea fowl)
        8: 1097, 1098, *1098*
    Vulturine 8: 1097, *1097*
Gull 16: 2265, *2266*
    Black-headed (European gull;
        common gull) 2: 230–232
    Herring 9: 1187–1189
    Laughing 2: 230
    Mediterranean 2: 230
    Ross's 16: *2266*
Gyrfalcon 2: 210; 8: 1110–1112

Hamerkop (hammerhead;
    anvilhead) 8: 1123–1125; 18: 2551
Hammerhead *see* Hamerkop
Hardhead (*bird*) 14: 1993, 1995
Harpy eagle 2: 208, 209; 8: 1142–1144
    New Guinea 8: 1142
Harrier 8: 1145–1146
    African marsh 8: 1145
    Black 8: 1145
    European marsh 8: 1145, 1146
    Marsh (swamp harrier) 8: 1145
    Montagu's 8: 1145, *1146*
    Northern (marsh hawk; hen
        harrier) 8: 1145, *1145*, 1146
    Pallid 8: 1145
    Pied 8: 1145
Hawk
    Bat 2: 210
    Cooper's 18: 2467

Fish *see* Osprey
    Galapagos 5: 650
    Harrier 17: 2373
    Harris 3: 426
    Marsh *see* Harrier, Northern
    Pigeon *see* Merlin
    Red-tailed 3: *353*, 426
    Savanna 2: 210
    Sharp-shinned 18: 2467
Hawk-cuckoo, Indian 5: 620
Hawk-eagle
    African 12: 1589
    Ayres' 12: 1589
Heath hen (*extinct*) 15: 2036
Hedge sparrow *see* Accentor, Hedge
Helmetbird *see* Vanga, Helmet
Helmet-shrike 17: 2361
Hemipode, Andalusian *see* Button
    quail, Common
Heron 9: 1179–1181
    Black 9: 1181
    Boat-billed 2: 252; 8: 1123
    Gray 9: 1179, *1180*, 1181
    Great blue 9: 1179, 1180, 1181;
        21: *2906*
    Green-backed (green heron)
        9: 1179, *1179*, 1180–1181
    Purple 9: 1179
    Squacco 9: 1179, 1181
Hoatzin 9: 1196–1197
Hobby 9: 1198–1199
    African 9: 1198
Honey buzzard 2: 209; 9: 1211–1212
    Barred 9: 1211
    Crested 9: 1211
    Eurasian 9: 1211
    Long-tailed 9: 1211
Honeycreeper 7: 868; 9: 1213–1214;
    21: 2985
    Green 9: 1213, *1213*, 1214
    Hawaiian 9: 1213–1214
    Purple 9: *1214*
    Red-legged (blue honeycreeper)
        9: 1213
Honeyeater 7: 868; 9: 1215–1216;
    18: 2558
    Blue-faced 9: 1216
    Brown 9: 1215
    Cardinal 9: *1216*
    Pygmy 9: 1215
    Rufous-banded 9: 1216
    Rufous-throated 9: 1216
    Singing 9: 1215
    Strong-billed 9: 1215
    White-cheeked 9: *1215*
    White-gaped 9: 1215
    Yellow-tufted 9: 1215, 1216
Honeyguide 9: 1217–1218
    Greater 9: 1206, 1217–1218
    Lesser 9: *1218*
    Lyre-tailed 9: 1217, 1218
    Malaysian 9: 1217
    Orange-rumped 9: 1217
Hoopoe 9: 1222–1223
    Madagascar 9: 1222
Hornbill 9: 1226–1228
    Abyssinian ground 9: 1226, *1226*
    African red-billed 9: *1228*
    Malabar pied 9: 1227
    Malayan helmeted 9: 1226
    Red-billed dwarf 9: 1226
    Rhinoceros 9: *1227*
    Rufous-necked 9: 1226
    Silvery-cheeked 9: 1226–1227
    Southern ground 9: 1226
    White-crested 9: 1227
Horned pheasant *see* Tragopan
Hornero (*ovenbird*), Rufous (barber)
    13: 1819, 1820, *1820*
Huia (*extinct*) 21: *2886*, 2887

Hummingbird 9: 1255–1257; 12: 1673
    Bee 9: 1255
    Giant 9: 1255; 12: *1672*, 1673
    Hillstar 12: 1672, 1673
    Ruby-throated 9: 1255, 1256, *1256*
    Rufous 9: 1255, *1255*
    Sword-billed 9: 1255
    Violet-eared 9: 1257

Ibis 9: 1274–1276
    Black (red-naped ibis) 9: 1276;
        21: *2905*
    Glossy 9: 1274
    Japanese crested 9: 1274, 1276
    Northern bald (hermit ibis;
        waldrapp) 9: 1274, 1275, 1276
    Sacred 9: 1274, 1275, 1276; 18: *2497*
    Scarlet 9: 1274, 1275, *1275*, 1276,
        1276; 21: *2905*
    White 9: 1274, *1274*; 21: *2905*
Ibisbill 18: 2540
Indigobird 21: 2938
    Village 21: 2936

Jacana (lily-trotter; lotus-bird; water-
    walker) 9: 1289–1290
    African 10: *1398*
    Bronze-winged 9: *1289*, 1290
    Comb-crested (Australian lotus-
        bird) 9: 1289, *1290*
    Northern (American jacana)
        9: 1289, 1290
    Pheasant-tailed 9: 1289
Jackass (*bird*)
    Derwent *see* Butcherbird, Gray
    Tasmanian *see* Butcherbird, Gray
Jackdaw (Eurasian jackdaw; western
    jackdaw) 4: 452; 5: 610;
    10: 1300–1301; 18: 2457
Jaeger 10: 1307–1309
    Long-tailed 10: 1307, *1307*, 1309
    Parasitic (Arctic skua) 10: 1307,
        1309
    Pomarine 10: 1307, 1309
Jay 5: 610; 10: 1317–1319
    Blue 10: 1317, 1319, *1319*
    Eurasian 10: 1317, *1317*, 1318, *1318*,
        1319; 14: 1884; 18: 2457
    Gray (Canada jay) 10: 1317, 1318,
        1319
    Green 10: 1318
    Scrub 3: 426; 10: 1317, 1319
    Siberian 10: 1318, 1319; 19: 2617
    Steller's 10: 1317, 1319
Jay thrush 1: 114
Jewel thrush *see* Pitta
Jungle babbler 1: 113
Jungle fowl (wild jungle fowl)
    10: 1329–1330
    Gray 10: 1329
    Green 10: 1329, *1329*
    Red 10: 1329, 1330, *1330*
Junin *see* Grebe, Puna

Kagu 18: 2567
Kaka 10: 1343
Kea 10: 1343–1344
Kestrel (windhover) 1: 135;
    10: 1345–1347
    American 2: 209; 10: 1345, *1346*;
        18: 2467
    Eurasian 10: 1345, *1345*, 1346, *1347*
    Fox 10: 1345
    Greater 10: 1345
    Mauritius 2: 210; 10: 1345
Killdeer 10: 1348–1349
Kingbird 20: 2802
    Eastern 20: *2800*, 2802
Kingfisher 10: 1355–1357
    Amazon 10: 1355, *1356*

Belted **16**: 2175, *2175*
Common **10**: 1355, 1356, *1357*
Great brown *see* Kookaburra
Green **10**: 1355
Laughing *see* Kookaburra
Malachite **10**: *1355*
Pied **10**: 1355
Racquet-tailed **10**: 1356
Shoe-billed **10**: 1357
Stork-billed **10**: 1356–1357
Striped **10**: 1357
Yellow sacred **10**: 1355
Kinglet
Golden-crowned **7**: 999
Ruby-crowned **7**: 999, 1000
Kite **2**: 209; **10**: 1367–1369
Black (black-eared kite; large Indian kite; pariah kite; yellow-billed kite) **10**: 1367, 1368–1369
Black-breasted *see* Buzzard, Black-breasted
Black-shouldered (black-winged kite) **10**: 1367–1368, 1369
Brahminy **10**: 1367, 1368–1369, *1369*
Everglade snail (Everglade kite; snail kite; snail hawk) **2**: 209; **6**: 788–789
Red **10**: 1367, *1367*, 1368–1369
White-tailed **2**: 209; **10**: 1367
White-winged **10**: 1368
Kittiwake **10**: 1370–1371
Black-legged **10**: 1370, 1371
Red-legged **10**: 1370
Kiwi **10**: 1372–1373
Brown (common kiwi) **10**: 1372, 1373, *1373*
Great spotted (large kiwi) **10**: 1372
Little spotted (little gray kiwi) **10**: 1372
Knot **16**: 2233
Koel **5**: 618, 620, 713
Kokako (wattled crow) **21**: 2886, 2887, *2887*
Kookaburra (laughing kingfisher; laughing jackass; jackie; great brown kingfisher; alarmbird) **10**: 1355, 1385–1386
Blue-winged **10**: 1385
Korhaan, Black **3**: *338*
Kuaka *see* Diving-petrel, Common

Lammergeier (bearded vulture) **10**: 1401–1402; **19**: 2713; **20**: 2840, 2842
Lapwing **10**: 1415–1416; **16**: 2234
Eurasian (northern lapwing; peewit; green plover) **10**: *1415*, 1416
Red-wattled **10**: 1415, *1416*
Sociable **10**: 1415
Lark
Crested **5**: 593–594
Desert **5**: 594
Thekla **5**: 593
Leafscraper **13**: 1819
Leiothrix, Red-billed (Pekin robin) **1**: 113; **16**: 2190
Limpkin (crying bird; wailing bird) **11**: 1453–1454
Linnet **11**: 1457–1458
Locust finch **21**: 2888
Longclaw **14**: 1965, 1966
Yellow-throated **14**: 1965
Loon **11**: 1485–1487
Arctic (black-throated diver) **11**: 1485, *1485*; **21**: 2906
Common (great northern diver) **11**: 1485, 1486, *1486*
Red-throated (red-throated diver; rain goose) **11**: 1485, 1486, 1487, *1487*

Yellow-billed (white-billed diver) **11**: 1485, 1487
Lorikeet **11**: 1488–1489
Little **11**: 1488
Olive-headed (perfect lorikeet) **11**: *1489*
Purple-crowned (blue lorikeet) **11**: 1488
Rainbow (rainbow lory; green-naped lorikeet; red-breasted lorikeet; red-collared lorikeet; blue-bellied lorikeet; coconut lorikeet) **11**: 1488, *1488*, 1489
Scaly-breasted **11**: 1488
Lory **11**: 1493–1494
Dusky **11**: 1493, *1493*
Purple-naped (purple-capped lory) **11**: 1493, 1494, *1494*
Rainbow *see* Lorikeet, Rainbow
Red and blue **11**: 1494
Yellow-backed **11**: 1493
Lotus-bird *see* Jacana
Lovebird **11**: 1495–1496
Black-cheeked **11**: 1495
Black-collared (Swindern's lovebird) **11**: 1495
Black-winged (Abyssinian lovebird) **11**: 1495, 1496
Lovebird (budgerigar) *see* Budgerigar
Lovebird
Fischer's **11**: *1495*
Madagascar (gray-headed lovebird) **11**: 1495, 1496
Masked (yellow-collared lovebird) **11**: 1495, 1496, *1496*
Nyasa (Lilian's lovebird) **11**: 1495
Red-headed **11**: 1495, 1496
Rosy-faced (peach-faced lovebird) **11**: 1495, 1496
Lyrebird (lyretail; native pheasant) **11**: 1509–1510
Northern (Albert's lyrebird) **11**: 1509
Superb **11**: 1509, 1510

Macaw **11**: 1514–1516
Blue-throated **11**: 1516
Blue and yellow (blue and gold macaw) **11**: 1514, *1516*
Glaucous **11**: 1516
Hyacinth (hyacinthine macaw) **11**: 1514, *1515*
Lear's **11**: 1516
Military (great green macaw) **11**: 1514
Red-fronted (red-crowned macaw) **11**: 1516
Red and green **11**: 1514
Scarlet (red and yellow macaw) **11**: 1514, *1514*, 1515
Spix's (little blue macaw) **11**: 1516
Magpie **5**: 610; **11**: 1519–1521
Azure-winged **11**: 1520, *1521*
Black-billed **11**: 1519, 1520–1521
Blue **11**: 1520
Green **11**: 1519
Sri Lanka (Ceylon magpie) **11**: 1519–1520
Yellow-billed **11**: 1519, 1520–1521, *1520*
Magpie lark (mudlark; peewit; peewee; pugwall; little magpie; Murray magpie) **11**: 1522–1523
New Guinea (torrentlark) **11**: 1522
Malimbe **21**: 2896
Red-headed **21**: 2896
Mallard **5**: 717, *717*; **11**: 1524–1526; **17**: 2343
Mallee fowl (mallee hen; lowan; gnow; native pheasant) **11**: 1527–1529

Manakin **11**: 1532–1533
Blue-backed **11**: 1532
Golden-headed **11**: 1532
Gould's **11**: 1532, *1533*
Long-tailed **11**: *1532*, 1533
Red-capped **11**: 1532
White-bearded **11**: 1532
Mannikin **11**: 1544–1545; **21**: 2888
Bronze **11**: 1544
Chestnut-breasted (chestnut-breasted munia; chestnut-breasted finch; chestnut finch; barleyfinch; barley sparrow; barley bird) **11**: 1544, 1545, *1545*
Madagascar (bib-finch) **11**: 1544
Magpie **11**: 1544
Spotted (spotted munia; nutmeg finch; spice finch) **11**: 1544, 1545
Manucode, Trumpeter **20**: 2757
Marabou (marabou stork) **11**: 1553–1554; **18**: 2551
Marlin-spike *see* Tropic bird
Martin **12**: 1590–1592
African river **12**: 1590
Brown-chested **12**: 1590
Crag **12**: 1590
House **12**: 1590, *1590*, 1591
Mascarene **12**: 1590
Purple **2**: 151; **12**: 1590, 1591, *1591*
Sand *see* Swallow, Bank
Tree **12**: 1590
Meadowlark **13**: 1807
Western **8**: *1039*
Merganser **12**: 1599–1601
Aukland Island **12**: 1600
Brazilian **5**: 717; **12**: 1600
Common (goosander) **5**: 717; **12**: 1599, 1600, 1601, *1601*
Hooded **12**: 1599, 1600, *1600*; **21**: 2972
Red-breasted **5**: 717; **12**: 1599, *1599*, 1600, 1601
Merlin (pigeon hawk) **12**: 1602–1603
Mesia, Silver-eared **1**: 114
Mesite **12**: 1604–1605
Brown **12**: 1604–1605
Subdesert (Bensch's monia) **12**: 1604, *1604*, 1605
White-breasted **12**: 1604–1605
Miner (*bird*) **13**: 1819
Minivet **12**: 1614–1615
Ashy **12**: 1614–1615
Fiery **12**: 1614, 1615
Mountain **12**: 1614
Scarlet (flamed minivet; orange minivet) **12**: 1614–1615, *1614*
Small **12**: *1615*
Mistletoe bird (mistletoe flowerpecker) **7**: 868, *868*
Mockingbird **12**: 1622–1624
Bahama **12**: 1622
Blue **12**: 1622
Blue and white **12**: 1622
Galapagos **12**: 1622, 1623–1624
Northern **12**: 1622, *1622*, 1623, 1624
Tropical **12**: *1624*
Monklet **15**: 2060
Moorhen **12**: 1649–1651
Common (common gallinule; blue rail; red-billed mud hen; Florida gallinule) **12**: 1649, 1650
Dusky **12**: 1649
Gough's **12**: 1649; **15**: 2108
Lesser **12**: 1649
Samoan **12**: 1649
San Cristobel (makira) **12**: 1649
Morningbird **21**: 2923
Mother Carey's chicken *see* Storm petrel

Motmot **12**: 1665–1666
Blue-crowned (clock bird) **12**: 1665–1666, *1665*
Rufous **12**: 1665
Turquoise-browed **12**: 1665, *1666*
Mousebird **12**: 1676–1677
Blue-naped **12**: 1676
Red-faced **12**: 1676, *1676*
Speckled **12**: 1676
White-backed **12**: 1676, 1677, *1677*
White-headed **12**: 1676
Munia **21**: 2888
White-rumped (striated munia) **11**: 1544
Murrelet, Xantus' **16**: 2264–2265
Musselpecker *see* Oystercatcher
Mynah **12**: 1714–1716
Bank (Bengal mynah) **12**: 1714, 1716
Brahminy **12**: *1716*
Chinese crested **12**: 1714
Common (Indian mynah) **7**: 912; **12**: 1714–1715, 1716
Golden-crested **12**: 1714
Indian hill (greater hill mynah) **12**: 1714
Papuan **12**: 1714
Rothschild's (Bali mynah; Bali starling; Rothschild's starling; Rothschild's grackle) **12**: *1715*

Needletail
Brown-backed **18**: 2484, *2484*
White-throated **19**: 2600
Nighthawk **13**: 1742
Common **13**: *1744*; **14**: 2013
Night heron **13**: 1737–1739
Black-crowned **13**: 1737–1738, *1738*, 1739, *1739*
Japanese **13**: 1737
Malaysian **13**: 1737
Rufous (Nankeen night heron) **13**: 1737, 1739
White-backed **13**: 1737
White-eared **13**: 1737
Yellow-crowned **13**: 1737, *1737*
Nightingale **13**: 1740–1741; **18**: 2459; **19**: 2678
Nightjar **13**: 1742–1744
Egyptian **13**: 1742
Eurasian (fern owl; churn owl; eve-jarr; dorhawk; goatsucker) **13**: 1742, *1742*, 1743–1744; **14**: 2013
Jungle **13**: 1742
Large-tailed **13**: 1742, *1743*
Long-tailed **13**: 1742
Pennant-winged **13**: 1742
Red-necked **13**: 1742
Spotted **13**: 1742; **14**: 2013
Standard-winged **13**: 1742
White-throated **13**: 1742
Wood *see* Potoo
Noddy **13**: 1755–1756
Black **13**: 1755, 1756
Blue **13**: 1755
Brown **13**: 1755, 1756
Gray **13**: 1755
Lesser **13**: 1755, *1756*
Nunbird **15**: 2061
Black **15**: 2061
White-fronted **15**: 2060–2061
Nunlet **15**: 2060
Nutcracker **13**: 1757–1759
Clark's **13**: 1757, *1757*, 1759, *1759*
European (slender-billed nutcracker; thick-billed nutcracker) **13**: 1757, 1758, *1758*, 1759
Nuthatch **13**: 1760–1762
Coral-billed **20**: 2813
Corsican **13**: 1762

Nuthatch (*continued*)
    Eurasian **13**: 1760–1761, *1760*
    Red-breasted **13**: *1761*, 1762
    White-breasted **13**: *1762*

Oilbird (guácharo; diablotin)
    13: 1784–1785
Oo
    Bishop's **9**: 1215
    Great (oo) **9**: 1215
    Kauai (ooaa) **9**: 1215
Openbill **18**: 2553
Oriole
    Baltimore **1**: 126–127; **20**: 2833
    Black **8**: 1012
    Black-headed **8**: 1013
    Golden **8**: 1012–1013
    Maroon **8**: 1012
Oropendula **13**: 1805–1807
    Chestnut-headed **13**: *1805, 1807*
    Montezuma **13**: 1806
    Russet-backed **13**: *1806*
Osprey (fish hawk) **1**: 82, 124;
    2: 209–210; **13**: 1811–1813;
    21: 2906
Ostrich **13**: 1814–1816
Ouzel
    Ring **19**: 2680
Ovenbird **13**: 1819–1820; **18**: 2455
    Rufous hornero **13**: 1819, 1820, *1820*
    True (hornero) **13**: *1819*
Ovenbird (*wood warbler*) **13**: 1819;
    21: 2985, 2986
Owl
    Barn **2**: 148–150; **3**: 323; **13**: 1825,
        1826, 1829
    Barred **13**: 1825
    Boreal **13**: 1829
    Burrowing **3**: 323–325; **13**: 1828
    Eagle **13**: 1825, 1828, *1829*
    Elf **13**: *1826, 1827*, 1828
    Galapagos **5**: 650
    Great gray **13**: *1825*, 1827
    Hume's desert **13**: 1827, 1828
    Least pygmy **13**: 1825
    Little **13**: *1827*
    Long-eared **13**: 1826
    Masked **18**: 2536
    Pygmy **13**: 1829
    Short-eared **7**: 893; **13**: 1828
    Snowy **13**: 1827–1828, 1829;
        17: 2437–2439; **20**: 2778
    Tawny (brown owl) **13**: 1827;
        19: 2644–2646
    see also Owls; Scops-owl
Owlet, Forest **13**: 1826, 1829
Owlet-nightjar **13**: 1823–1824
    Australian **13**: 1823, 1824, *1824*
    Mountain **13**: *1823*
Owls 13: 1825–1829
Oxpecker (tick bird) **13**: 1830–1831;
    18: 2457
    Red-billed **13**: 1830, *1831*
    Yellow-billed **13**: 1830
Oystercatcher (musselpecker)
    13: 1835–1837
    American **13**: 1835, *1835*, 1837
    American black **13**: 1835
    Black **13**: 1835
    Canary Islands **13**: 1835
    Eurasian **13**: 1835, 1836, 1837, *1837*
    Magellanic **13**: 1835, *1836*
    Pied **13**: 1835
    Sooty **13**: 1835

Painted snipe **13**: 1849–1850
    Old World **13**: 1849, *1849*, 1850
    South American **13**: 1849
Palmchat **13**: 1854
Paradise-whydah **21**: 2936, 2937, 2938

Parakeet **14**: 1878–1880
    American (conure) **14**: 1878
    Blossom-headed **14**: 1878
    Carolina (*extinct*) **14**: 1880
    Crested **14**: 1879
    Eastern slaty-headed **14**: *1878*
    Elegant grass **14**: *1880*
    Grass **14**: 1878–1879
    Gray-breasted **14**: 1878, 1880
    Green **14**: 1878
    Ground **14**: 1879
    Lineolated **14**: 1880
    Long-tailed ground (ground
        parrot; swamp parrot)
        14: 1879–1880, 1885
    Rose-ringed (ring-necked
        parakeet) **14**: 1878, 1879, *1879*
    Short-tailed (night parrot) **14**: 1880
    Slender-billed **14**: 1878
    Turquoisine grass **14**: 1878
Parrot **14**: 1883–1885
    African gray **14**: 1883, 1885
    Black-headed **15**: *2112*
    Eclectus **14**: 1883
    Hanging (bat parakeet) **11**: 1495
Parrotbill (crow tit) **1**: 113;
    14: 1886–1887
    Bearded **18**: 2458
    Black-throated **14**: 1886
    Blyth's **14**: 1886
    Gould's (black-breasted
        parrotbill) **14**: 1886
    Reed bed **14**: 1886
Parrotlet, Green-rumped **14**: 1885
Partridge **14**: 1891–1893
    Bamboo **14**: 1891
    Black *see* Francolin, Black
    Gray (Hungarian partridge)
        14: 1891, 1892, *1892*, 1893, *1893*
    Red-legged **14**: 1891, 1892, 1893
    Rock **14**: *1891*
    Tree **14**: 1891
    Wood **14**: 1891
Peafowl **14**: 1898–1900
    Congo **14**: 1898
    Green **14**: 1898
    Indian (common peafowl; blue
        peafowl; peacock (*male*); peahen
        (*female*)) **14**: 1898, *1898*, 1899,
        1899, 1900
Peep **16**: 2233
Pelican **14**: 1908–1910; **16**: 2264, 2265,
    2267, 2268
    American white **14**: 1908
    Australian **14**: *1909*
    Brown (Chilean pelican; Peruvian
        pelican) **14**: 1908, *1908*, 1909;
        16: 2265
    Dalmatian **14**: 1908
    Great white **14**: 1910, *1910*
    Pink-backed **14**: 1908
    White **14**: 1909
Penguin
    Adélie **1**: 28–29, *80*; **14**: 1912, 1914;
        21: 3003, 3004
    Chinstrap **14**: 1912
    Emperor **1**: 82; **6**: 779–781;
        14: 1911, 1912, 1913, *1914*
    Erect-crested **14**: 1914
    Fjordland **14**: 1914
    Galapagos **14**: 1911, *1913*, 1914
    Gentoo **14**: *1911*
    Jackass **14**: 1911, 1913, 1914
    King **10**: 1358–1360; **14**: 1912, 1913,
        1914
    Little blue (fairy penguin)
        14: 1911, 1913
    Rockhopper **14**: 1913
    Snares Island **14**: 1914
    White-flippered **14**: 1911

Yellow-eyed **14**: 1911, 1914;
    21: 3003–3004
    see also Penguins
Penguins 14: 1911–1914
Peregrine *see* Falcon, Peregrine
Petrel **16**: 2268
    Snow (gray-faced petrel; long-
        winged petrel) **17**: 2435–2436
Phainopepla, Crested **21**: 2890
Phalarope **14**: 1931–1933
    Red (gray phalarope) **14**: 1931,
        1931, 1932–1933
    Red-necked (northern phalarope)
        14: 1931, 1932–1933, *1932*
    Wilson's **14**: 1931, 1932–1933, *1933*
Pheasant **14**: 1934–1936
    Eared **14**: 1934–1935
    Golden **14**: 1935
    Green **14**: 1934
    Lady Amherst's **14**: 1935
    Long-tailed **14**: 1935
    Monal **14**: 1934, *1936*
    Ring-necked **14**: 1934, *1934*,
        1935–1936
    Ruffed **14**: 1935
    Silver **14**: 1934
Phoebe
    Black **20**: 2801
    Eastern (water peewee) **20**: 2800
Piculet **14**: 1937–1938
    Antillean **14**: 1937
    Olivaceous **14**: 1937, 1938, *1938*
    Rufous **14**: 1937
    Speckled **14**: *1937*, 1938
Pigeon
    Imperial **7**: 911
    Mauritius blue **7**: 912
    New Zealand (New Zealand fruit
        pigeon; native pigeon) **7**: 912
    Passenger **3**: 328
    Topknot **7**: 911
    Torres Strait (nutmeg pigeon) **7**: 911
    Wood (ring dove) **21**: 2980–2982
Pintail **14**: 1957–1959; **17**: 2343
    Northern **14**: 1957, *1957*, 1958,
        1959, *1959*
    White-cheeked (Bahama pintail)
        14: 1957, *1958*
Pipit **14**: 1965–1966
    American **14**: 1966
    Bogota **14**: 1965
    Golden **14**: 1965
    Meadow **5**: 620; **14**: 1965, 1966
    Red-throated **14**: 1965
    Rock **14**: 1965, *1966*
    Sprague's **14**: 1965
    Tree **14**: 1965, 1966
    Yellow-breasted **14**: 1965
Piquero *see* Booby, Peruvian
Pitohui **20**: 2819; **21**: 2923
Pitta (painted thrush; ground thrush;
    jewel thrush) **14**: 1970–1971
    African **14**: 1970, 1971
    Banded **14**: *1970*
    Gurney's **14**: 1970
    Indian (Bengal pitta; blue-winged
        pitta) **14**: 1970, 1971, *1971*
    Noisy **14**: 1970
    Steere's **14**: 1970
Plains-wanderer (collared plains-
    wanderer; turkey quail)
    14: 1978–1979
Plantain-eater, Violet **20**: 2780
Ploughbird, Wattled **21**: 2923
Plover **14**: 1988–1990
    Black-bellied **14**: 1988, *1989*
    Blacksmith **10**: 1415
    Crab **4**: 568–569
    Egyptian (crocodile bird) **4**: 551, 552
    Little ringed **14**: 1988

Mountain **14**: 1989
    Ringed **14**: 1988, *1988*, 1990
    Sand **14**: 1988–1989
    Semipalmated **14**: 1988
    Snowy (Kentish plover)
        14: 1988–1989
Pochard **14**: 1993–1995
    Common (European pochard)
        14: 1993, 1994, 1995, *1995*
    Ferruginous **14**: 1993
    Madagascar **5**: 717
    Red-crested **14**: *1993*
    Southern **14**: 1995
Poorwill **14**: 2012–2013
    Common **14**: 2012, *2012*, 2013
    Ocellated **14**: *2013*
Potoo (wood nightjar) **15**: 2032–2033
    Gray (common potoo) **15**: 2032
    Great **15**: 2032, *2032*, 2033, *2033*
    Northern **15**: 2032, 2033
Potoyunco *see* Diving-petrel,
Peruvian
Prairie chicken **15**: 2036–2037
    Greater **15**: 2036, 2037, *2037*
    Lesser **15**: 2036, *2036*, 2037
Pratincole (swallow plover)
    15: 2041–2042
    Australian **15**: 2041, *2042*
    Black-winged **15**: 2041
    Common (collared pratincole; red-
        winged pratincole; locust bird)
        15: 2041, *2041*, 2042
Prion **15**: 2045–2046; **16**: 2268
    Antarctic **15**: 2045
    Broad-billed **15**: 2045, 2046, *2046*
    Fairy **15**: 2045, *2045*
    Fulmar **15**: 2045
    Salvin's **15**: 2045, 2046
    Thin-billed **15**: 2045
Ptarmigan **15**: 2055–2057; **20**: 2778
    Rock **15**: 2055, *2055*, 2056, *2056*
    White-tailed **15**: 2056
Puffback **17**: 2361
Puffbird (silly John) **15**: 2060–2061
    Black-breasted **15**: 2061
    Mustached **15**: *2061*
    Spot-backed **15**: *2060*
    White-necked **15**: 2060, 2061
    White-whiskered **15**: 2060, 2061
Puffin **15**: 2065–2067; **16**: 2268
    Atlantic (common puffin) **15**: 2065,
        2066, *2066*, 2067, *2067*; **16**: 2228,
        2264
    Horned **15**: 2065
    Tufted **15**: 2065, *2065*
Pugwall *see* Magpie lark
Pylylia **21**: 2938

Quail **15**: 2082–2083
    Button *see* Button quail
    Common **15**: 2082, 2083, *2083*
    Crested **15**: 2083
    Gambel's **15**: *2082*
    Lark **3**: 349
    Marbled wood **15**: 2082
Quelea **15**: 2084–2086
    Cardinal **15**: 2085
    Red-billed (Sudan dioch) **15**: *2084*,
        2085, *2085*, 2086; **21**: 2897
    Red-headed **15**: 2085
Quetzal **15**: 2087–2088
    Resplendent (resplendent trogon)
        15: 2087, *2088*

Rail **15**: 2106–2108
    African **15**: 2106
    Clapper **15**: 2106, *2107*
    Flightless **15**: 2108
    King **15**: 2106
    Laysan (*extinct*) **15**: 2108

Lewin's **15**: 2106
Sora **15**: *2106*
Virginia **15**: 2106
Wake Island (*extinct*) **15**: 2108
Water **15**: 2106, 2107, *2108*
Raven **5**: 610; **15**: 2122–2124
Australian **15**: 2122
Chihuahua **15**: 2122
Common **15**: 2122, 2123; **18**: 2455, 2457
Fan-tailed **15**: 2122
Forest **15**: 2122
Little **15**: 2122
Thick-billed **15**: 2122
White-necked **15**: 2122
Razorbill (razorbilled auk) **15**: 2125–2126
Redhead **14**: 1993, 1994–1995, *1994*
Redpoll **15**: 2137–2138
Arctic (hoary redpoll) **15**: 2137–2138
Greenland **15**: 2137
Lesser **15**: 2137
Mealy **15**: 2137, *2137*, 2138
Redshank **16**: 2233
Redstart (*Old World*) **15**: 2139–2140
Black **15**: 2139
Common **15**: 2139, 2140, *2140*; **19**: 2678
Moussier's **15**: 2139
Reedling, Bearded (bearded tit) **14**: 1886, *1886*, 1887, *1887*
Reed warbler **20**: 2859
Blyth's **20**: 2859
Great **5**: 620
Rhea **15**: 2150–2151; **19**: 2692
Greater (common rhea) **15**: 2150, 2151
Lesser (Darwin's rhea) **15**: 2150, *2150*, 2151, *2151*
Riflebird **2**: 207
Prince Albert (magnificent riflebird) **2**: 205
Rifleman (titipounamu; thumbie) **16**: 2164–2165
Roadrunner **3**: *424*, 426; **16**: 2184–2185
Greater (chaparral cock) **16**: 2184, *2184*, 2185, *2185*
Lesser **16**: 2184
Robin
American (Canada robin; northern robin; robin redbreast) **4**: 555; **15**: 2141; **16**: 2188–2190; **19**: 2678; **20**: 2833
European (robin redbreast) **16**: 2191–2193; **19**: 2678
Indian **16**: 2190
see also Tody, Jamaican
Robin chat, Red-capped **6**: 776
Rockfowl **1**: 113
Gray-necked **1**: 114
White-necked **1**: 114
Roller **16**: 2204–2205
Abyssinian (Senegal roller) **16**: 2204
Broad-billed **16**: 2204
Eurasian **16**: 2204
Indian **16**: *2205*
Lilac-breasted **16**: *2204*, 2205
Rook **4**: 452; **5**: 610; **16**: 2206–2207
Rosefinch, Scarlet **4**: 499
Ruff (reeve) **16**: 2212–2214

Saddleback (*bird*; tieke) **21**:2886–2887
Saddlebill **18**: 2551
Sanderling **16**: 2233
Sand grouse **5**: 665; **16**: 2229–2230
Double-banded **16**: *2230*

Namaqua **16**: *2229*, 2230
Pallas's **16**: 2229, 2230
Pintailed **16**: 2230
Spotted **16**: 2229–2230
Sandpiper **16**: 2233–2234
Buff-breasted **16**: 2233, 2234
Common **16**: 2233, *2233*, 2234
Curlew **16**: *2234*
Gray-rumped (gray-tailed tattler) **16**: 2233
Green **16**: 2233, 2234
Pectoral **20**: *2779*, 2833
Purple **16**: 2233, 2234
Solitary **16**: 2234
Terek **16**: 2233
Upland **16**: 2234
Wood **16**: 2234; **21**: 2906
Sapsucker **16**: 2235–2237
Red-breasted **16**: 2235
Red-naped **16**: 2235
Williamson's **16**: 2235, *2236*
Yellow-bellied **16**: 2235, *2235*, 2236, *2237*
Scaup **14**: 1993
Greater (bluebill; broadbill) **14**: 1995
New Zealand **14**: 1993, 1995
Scimitar babbler **1**: 113
Scops-owl
Grand Comoro **13**: 1829
Rain forest **13**: 1827
Seychelles **13**: 1829
Sokoke **13**: *1828*
Torotoroka **13**: 1827
Screamer **5**: 716; **16**: 2262–2263
Horned **16**: 2262
Northern **16**: 2262, 2263
Southern **16**: 2262–2263, *2262*
Screech owl **13**: 1827
Eastern **13**: 1828
Western **13**: 1828
Scrub fowl
Orange-footed (Reinwardt's scrub fowl) **11**: 1527
Philippine (tabon) **11**: 1527
Scythebill **21**: 2968
Seabirds 16: 2264–2268
Secretary bird **2**: 210; **17**: 2316–2318
Seed snipe **17**: 2319–2320
Gray-breasted (D'Orbigny's seed snipe) **17**: 2319
Least (pygmy seed snipe; Patagonian seed snipe) **17**: 2319, *2319*, 2320, *2320*
Rufous-bellied (Gay's seed snipe) **17**: 2319
White-bellied **17**: 2319
Seriema (cariama) **17**: 2321–2322
Burmeister's **17**: 2321
Red-legged (crested seriema; red legged cariama; crested cariama) **17**: 2321–2322, *2321*, *2322*
Serin **4**: 499; **8**: 1016
Shag **4**: 537
Spotted **16**: *2265*
Shearwater **17**: 2334–2336
Christmas **17**: 2336
Cory's **17**: 2334
Fluttering **17**: 2334
Great **17**: 2334, *2334*, 2335
Heinroth's **17**: 2334
Little **17**: 2334
Manx **17**: 2335, *2335*
Short-tailed (muttonbird) **17**: 2335
Sooty (somber shearwater; ghostbird; New Zealand muttonbird) **17**: 2334, 2335
Tailed **16**: 2264
Sheathbill **17**: 2337–2338
Lesser **17**: 2337

Snowy (American sheathbill; pale-faced sheathbill; paddy) **17**: 2337, *2337*, 2338
Sheldgoose **5**: 714
Shelduck **17**: 2341–2343
Common (shelduck) **17**: *2341*, 2342, *2342*
Crested **5**: 717
Paradise (paradise duck; painted duck; rangitata goose) **17**: 2341
Ruddy **17**: 2341–2342, 2343, *2343*
Shikra **18**: 2467
Shrike **17**: 2361–2363; **18**: 2456, 2457, 2458
Loggerhead **17**: 2361, 2362
Northern (great gray shrike) **17**: 2361, *2361*
Pied *see* Triller, Pygmy
Red-backed **17**: 2361, *2363*
Woodchat **17**: *2362*
Shrike-thrush **21**: 2923
Whistling **21**: 2923
Sicklebill **9**: 1255
Silverbill **11**: 1544
Silvereye (*a waxbill species*) **21**: 2888
Silver-eye (*a white-eye species*) **21**: 2930
Siskin **3**: 368; **8**: 1016
Pine (pine linnet) **11**: 1457
Skimmer **16**: 2266; **17**: 2386–2388
African **17**: 2386
Black **17**: 2386, 2387–2388, *2387*, *2388*
Indian **17**: 2386
Skua **1**: 29; **10**: 1307, 1308; **21**: 3004
Arctic *see* Jaeger, Parasitic
Chilean **10**: 1307, 1309
Great (bonxie; Falkland skua) **10**: 1307, *1308*, *1309*
South Polar **10**: 1307–1308, 1309
Skylark (northern skylark) **17**: 2396–2397; **18**: 2459
Small (oriental skylark) **17**: 2396
Smew **12**: 1599, 1600, 1601
Snipe **17**: 2427–2429
Chatham Island **17**: 2429
Common (Wilson's snipe; fantail snipe) **17**: 2427, 2428, *2428*, 2429
Great **17**: 2427, 2429
Jack **17**: 2427
Magellan **17**: *2427*
Noble **17**: 2429
Pintail **17**: 2429
Snow-finch **9**: 1247
Song babbler **1**: 113
Songbirds 18: 2455–2459
Sora (sora rail; Carolina crake) **5**: 583
Sparrow
Chestnut **9**: 1249
Chipping **18**: 2460
Desert **9**: 1247
Grasshopper **18**: 2460
House **7**: 999; **9**: 1247–1249; **18**: 2457, 2459
Ipswich **19**: 2649
Java (Java finch; Java temple bird; ricebird) **10**: 1315–1316; **11**: 1544
Lark **18**: 2460
Song **4**: 555; **18**: 2460–2461
Spanish **9**: 1247, 1249
Tree **9**: 1247
Sparrow hawk **18**: 2467–2469
Levant **18**: 2467
Spinebill **9**: 1215
Spinetail (*an ovenbird species*; castlebuilder) **13**: 1819
Pale-breasted **13**: 1819
Spinetail swift **18**: 2484–2485
Batlike spinetail **18**: 2484
Boehm's spinetail **18**: 2484
Spoonbill **18**: 2496–2498
African **18**: 2497, *2498*

Eurasian **18**: 2496, *2497*, 2498
Lesser **18**: 2496–2497
Roseate **18**: 2496, *2496*, 2497; **21**: 2907
Royal (black-billed spoonbill) **18**: 2496–2497
Yellow-billed (yellow-legged spoonbill) **18**: 2496–2497
Starling **18**: 2521–2523
Bali *see* Mynah, Rothschild's
Common (European starling) **6**: 855; **18**: 2521–2523, *2521*, *2522*, *2523*
Glossy **18**: 2521
Rose-colored **18**: 2521
Rothschild's *see* Mynah, Rothschild's
Wattled **18**: 2521
Steamer duck **5**: 716; **18**: 2524–2525
Flightless (Falkland Island flightless steamer duck; logger duck) **18**: 2524, *2524*, 2525, *2525*
Flying **18**: 2524–2525
Magellanic flightless **18**: 2524, 2525
Stifftail **18**: 2537–2539
Stilt **18**: 2540–2542
Banded **18**: 2540, 2541, 2542
Black **18**: 2540, 2542
Black-necked **18**: 2540, 2541, *2542*
Black-winged (pied stilt; white-headed stilt; longshanks; stilt-bird) **18**: 2540, *2540*, 2541, *2541*
Stint **16**: 2233
Stitchbird **9**: 1215
Stone curlew (thick-knee) **18**: 2545–2546
Beach **18**: 2545
Bush thick-knee **18**: 2545, 2546
Double-striped thick-knee **18**: 2545
Eurasian **18**: 2545, *2545*, 2546
Great thick-knee **18**: 2545
South American thick-knee **18**: 2545–2546
Water thick-knee (water dikkop) **18**: 2545, 2546
Stork **18**: 2551–2553
Adjutant **11**: 1553; **18**: 2551
African wood **18**: 2552
Black **18**: 2551, 2552, *2552*, 2553
Black-necked **18**: 2551
Jabiru **18**: 2551
Lesser adjutant **11**: 1553
Maguari **18**: 2551, 2553
Marabou *see* Marabou
Openbill **18**: 2551
Painted **18**: 2552, *2553*
Saddlebill **18**: 2551
Shoebill (whale-headed stork; abu markub) **8**: 1123; **17**: 2351–2352; **18**: 2551
White **18**: 2551, *2551*, 2552, *2552*, 2553
Wood (wood ibis) **18**: 2551, 2552
Storm petrel (Mother Carey's chicken) **16**: 2266; **18**: 2554–2555
European **18**: 2554, 2555
Galapagos (wedge-rumped storm petrel) **18**: 2554, 2555
Hornby's **18**: 2554
Leach's **18**: *2554*
Wilson's **18**: 2554, 2555, *2555*
Stream-creeper **13**: 1819
Sugarbird **18**: 2558–2559
Blue **18**: 2558
Cape **18**: 2558, 2559
Gurney's **18**: 2558
Sunbird **7**: 868; **18**: 2456, *2458*, 2565–2566
Black **18**: *2565*
Collared **18**: *2566*

Sunbird (*continued*)
Purple **18**: *2458*, 2566
Pygmy **18**: 2455, 2565, 2566
Surfbird **20**: 2790, 2791
Swallow
Bank (sand martin) **2**: 151;
12: 1590, *1592*
Barn (swallow; chimney swallow;
house swallow) **2**: 151–153
Cliff **2**: 151
Golden **2**: 151
Mosque **2**: 151
Red-rumped **2**: 151
Rufous-chested **2**: 151
Tree **2**: 151
Swallow-tanager **18**: 2583–2584
Swallow-wing **15**: 2060
Swan **18**: 2586–2589
Black **5**: *714*; **18**: 2587, *2588*
Black-necked **18**: 2587
Coscoroba **18**: 2586; **21**: 2925, 2927
Mute **5**: 715; **18**: 2586, *2586*, 2588,
2589
Trumpeter **5**: 714, 715, 717;
18: 2587, *2587*, 2588; **20**: 2757
Tundra (Bewick's swan; whistling
swan) **5**: 717; **18**: 2586–2587;
20: 2779
Whooper **5**: 714; **18**: 2586, 2588,
2589
see also Ducks, geese and swans
Swift **19**: 2598–2600
African white-rumped **19**: 2598
Alpine **19**: 2598
Asian white-rumped **19**: 2598
Batlike (Boehm's spinetail)
18: 2484
Black **19**: 2598
Chimney **18**: 2484, 2485, *2485*;
19: 2598
Common **19**: 2598, *2598*, 2599,
2600
Crested **19**: 2598
Edible-nest **19**: 2601
Great dusky **19**: *2599*
Pallid **19**: 2598
Palm (*New World*) **19**: 2598, 2600
Palm (*Old World*) **19**: 2598, 2600
Scissor-tailed **19**: 2600
True **19**: 2598
Vaux's **18**: 2484, 2485; **19**: 2598
White-throated **19**: 2598, 2600
Swiftlet **19**: 2601–2602
Black-nest **19**: 2601
Cave **19**: *2601*
Glossy (white-bellied swiftlet)
19: 2601–2602
White-nest **19**: 2601

Tailorbird **19**: 2618–2619
Ashy **19**: 2618
Ashy-naped **19**: 2618, 2619
Australian **19**: 2618
Black-necked **19**: 2618
Common **19**: 2618, 2619, *2619*
Takahe **19**: 2620–2621
Tanager **18**: 2456, 2458; **19**: 2627–2629
Azure-winged **19**: 2627
Blue-gray **19**: 2627, *2627*, 2628,
2629
Blue-winged mountain **19**: *2628*
Gray-headed **19**: 2629
Hepatic **19**: 2627
Magpie **19**: 2627
Masked **19**: 2627, 2629
Red-rumped paradise **19**: 2627
Scarlet **18**: *2456*, 2457; **19**: 2627,
2628, 2629
Scarlet-rumped black **19**: 2629
Speckled **19**: *2629*

Summer **19**: 2627
Western **19**: 2627
Tattler
Gray-tailed *see* Sandpiper, Gray-
rumped
Wandering **16**: 2233
Tchagra **17**: 2361
Teal **5**: 715; **19**: 2647–2649
Andaman **19**: 2648
Aukland Islands **19**: 2647
Baikal **19**: 2647, 2649
Blue-winged **19**: *2647*
Brown (Auckland teal) **19**: 2649
Campbell Island **19**: 2647, 2649
Cape **19**: 2647
Chestnut **19**: 2647
Gray **19**: 2647, 2648
Green-winged (teal) **19**: 2647,
2648, *2648*
Hottentot **19**: 2647
Laysan **19**: 2648, 2649
Madagascar **19**: 2649
Marbled **19**: 2649
Speckled (sharp-winged teal)
19: 2647, *2649*
Tern (sea swallow) **16**: 2265;
19: 2661–2663
Arctic **16**: 2265; **19**: 2661–2662,
2662, 2663
Black **19**: 2661
Caspian **19**: 2661
Common **19**: 2661, *2661*, 2663
Forster's **16**: 2264; **19**: 2661,
2662–2663
Gull-billed **19**: 2661
Inca **19**: 2661
Roseate **16**: 2268; **19**: 2663
Royal **19**: *2663*
Sooty **19**: 2661, 2663
Whiskered **19**: 2661
White-winged black **19**: 2661
Thornbill (tit; Australian warbler;
wren-warbler) **19**: 2669–2670;
20: 2857
Brown **19**: 2669
Slaty-backed **19**: 2669
Tasmanian **19**: 2669
Western **19**: 2669, 2670
Yellow-rumped **19**: 2669, 2670,
2670
Thrasher **19**: 2671–2673
Bendire's **19**: 2672, 2673
Brown **18**: 2458; **19**: 2671, 2672–2673
California **3**: 426; **19**: 2671, 2673
Crissal **19**: 2673
Curve-billed **19**: *2671*, *2672*
Le Conte's **19**: 2673
Long-billed **19**: 2671
Pearly-eyed **19**: 2671, 2672
Sage **19**: 2671
Scaly-breasted **19**: 2671, 2672
White-breasted **19**: 2671
Thrush **18**: 2458; **19**: 2617, 2678–2680
Gray-cheeked **19**: 2678
Hermit **19**: 2678
Mistle **19**: 2678–2680
Song **19**: 2678, 2679, *2679*, 2680
Varied **19**: *2678*
Wood **19**: 2678
Tinamou **8**: 1040; **19**: 2691–2692
Crested (martineta) **19**: 2691
Great **19**: 2691
Little **19**: 2691
Ornate **19**: 2691–2692
Variegated **19**: 2691, 2692
Tit (titmouse; chickadee) **19**: 2653,
2693–2695
Black-capped chickadee **19**: 2653,
2693
Blue **19**: 2693, *2693*, 2694, 2695

Bush **19**: 2693
Chesnut-backed chickadee
19: *2695*
Great **19**: *2654*, 2693, *2694*, 2695
Long-tailed **19**: 2693, 2694, 2695
Penduline **19**: 2693, 2695
Siberian **19**: 2617
True **19**: 2695
Willow **19**: 2695
Tit babbler **1**: 113
Titipou *see* Rifleman
Titipounamu *see* Rifleman
Titmouse *see* Tit
Tityra **4**: 543–544
Masked **4**: 544
Tody **19**: 2700–2701
Broad-billed **19**: 2700
Cuban (pedorrera) **19**: 2700, *2700*,
2701
Jamaican (robin) **16**: 2190; **19**: 2700
Narrow-billed **19**: 2700
Puerto Rican **19**: 2700, 2701, *2701*
Toucan **19**: 2714–2716
Channel-billed **19**: *2714*
Rainbow-billed **19**: 2714, 2715
Red-breasted **19**: *2716*
Toco **19**: *2715*
Toucanet **19**: 2714
Tragopan (horned pheasant)
19: 2717–2718
Blyth's **19**: 2717, 2718
Cabot's **19**: 2717, 2718
Satyr **19**: 2717
Temminck's **19**: 2717, *2717*, *2718*
Western **19**: 2717, 2718
Treecreeper (spotted creeper)
19: 2721–2722
Australian **19**: 2721
Common **19**: 2721
Eurasian **19**: 2721, *2721*, 2722, *2722*
Short-toed **19**: 2721
Spotted **19**: 2721, 2722
Treerunner **13**: 1819
Trembler **19**: 2671
Brown **19**: 2671
Gray **19**: 2671
Triller
Black-and-white **5**: 621
Black-breasted **5**: 621
Pied **5**: 621
Pygmy (pied shrike) **5**: 621
White-winged **5**: 621
Trogon **20**: 2745–2747
Blue-throated **20**: 2747
Coppery-tailed **20**: *2746*
Elegant **20**: 2745, 2746
Gartered **20**: 2747
Mountain **20**: 2746–2747
Narina **20**: 2745
Resplendent *see* Quetzal,
Resplendent
Slaty-tailed **20**: *2745*
White-tailed **20**: *2747*
Tropic bird (paille-en-queue; marlin-
spike; bo'sun bird; boatswain
bird) **16**: 2265, 2268;
20: 2748–2750
Red-billed **20**: 2748, *2749*, *2750*
Red-tailed **20**: 2748, 2749
White-tailed **20**: 2748, *2748*
Trumpeter (trumpet bird)
20: 2756–2757
Gray-winged **20**: 2756, *2756*, 2757
Green-winged (dark-winged
trumpeter) **20**: 2756, 2757, *2757*
Pale-winged **20**: 2756, 2757
Tui **9**: 1215
Turaco (touraco) **20**: 2780–2781
Black-billed (Congo turaco)
20: 2781

Giant (great blue turaco) **20**: 2780
Hartlaub's **20**: 2780, 2781
Knysna **20**: *2781*
Prince Ruspoli's **20**: 2780
Purple-crested **20**: 2780, *2780*
Ross's **20**: 2780, 2781
Ruwenzori **20**: 2780
Yellow-billed **20**: 2780
Turkey **20**: 2784–2786
Ocellated **20**: 2784, 2786, *2786*
Wild (Merriam's wild turkey; Rio
Grande turkey; Gould's wild
turkey) **20**: 2784, 2785, *2785*
Turnstone (ruddy turnstone)
20: 2790–2792
Black **20**: 2790, 2791, 2792
Turquoisine **14**: 1878
Turtledove **20**: 2793–2794
Twinspot **21**: 2938
Twite **8**: 1016; **11**: 1457
Tyrant, Many-colored **20**: 2800

Umbrellabird **4**: 543

Vanga (vanga-shrike) **12**: 1605;
20: 2813–2814
Bernier's **20**: 2813
Blue **20**: 2813
Chabert's **20**: 2813, 2814, *2814*
Coral-billed **20**: 2813
Helmet (helmetbird) **20**: 2813
Hook-billed **20**: 2813
Pollen's **20**: 2813
Red-shouldered **20**: 2813, 2814
Red-tailed **20**: 2813
Rufous **20**: 2813
Sickle-billed **20**: 2813, *2813*, 2814
Van Dam's **20**: 2813
Veery **19**: 2678
Venomous animals 20: 2817–2820
Vireo **20**: 2831–2833
Blue-headed **20**: 2832
Brown-capped **20**: 2831
Noronha **20**: 2831
Red-eyed **20**: 2831, 2832, 2833, *2833*
Warbling **20**: 2831
White-eyed **20**: 2831–2832
Yellow-throated **20**: 2831
Vulture
Black **20**: 2787, 2840, *2840*, 2841
Cinereous (Eurasian black
vulture) **20**: 2840
Egyptian **20**: 2840–2841, 2842,
2842
Greater yellow-headed **4**: 518;
20: 2787
Griffon **10**: 1401; **20**: 2841, 2842
Hooded **20**: 2841, 2842
Indian black **20**: 2841
King **20**: 2787, 2841, *2841*
Lappet-faced **20**: 2842
Lesser yellow-headed (yellow
headed vulture) **4**: 518; **20**: 2787
Palm nut **20**: 2841, 2842
Turkey **4**: 518; **20**: 2787–2789, 2840,
2841
White-backed **10**: 1401; **20**: *2839*,
2841, 2842
Yellow-headed **20**: 2841
see also Condor; Lammergeier;
Vultures
Vultures 20: 2839–2842

Wagtail **20**: 2843–2845
African pied **20**: 2844
Blue-headed **20**: *2843*, 2844
Cape **20**: 2843
Forest **20**: 2843
Gray **4**: 499; **20**: 2843, 2844, 2845
Madagascar **20**: 2843

Mountain **20**: 2843
 Pied (white wagtail) **20**: 2843,
  2844, 2845, *2845*
 Willie **20**: 2843
 Yellow **20**: 2843, *2843*, 2844, 2845
Wallcreeper **19**: 2721
Warbler, Australian *see* Thornbill
Warbler (*New World*) *see* Wood
 warbler
Warbler (*Old World*) **18**: 2456;
  20: 2857–2859
 Arctic **20**: 2857
 Cetti's **20**: 2857
 Dartford **20**: 2857, 2858
 Grasshopper **20**: 2859
 Marsh **18**: 2458
 Olivaceous **20**: *2859*
 Reed *see* Reed warbler
 Rüppell's **20**: 2857
 Sedge **18**: 2459; **20**: *2858*, 2859
 Willow **20**: 2857, 2859
Warbler-finch **5**: 648, 649, *649*
Water dikkop **18**: 2545, 2546
Waterthrush **21**: 2985
Wattlebird **21**: 2886–2887
Waxbill **21**: 2888–2889, 2938
 Common (St. Helena waxbill)
  21: 2888
 Gray (red-eared waxbill; black-
  rumped waxbill; common
  waxbill) **21**: 2888
 Orange-cheeked **21**: *2888*
 Zebra (orange-breasted waxbill)
  21: 2888, 2889
Waxwing **21**: 2890–2892
 Bohemian **21**: 2890, *2890*, 2891,
  2892, *2892*
 Cedar **21**: 2890, *2891*
 Japanese **21**: 2890
Weaver **21**: 2896–2898
 Baya **21**: 2896
 Cape **21**: 2896
 Masked **21**: 2896, 2898
 Southern masked **21**: *2897*
 Spectacled **21**: 2897
 Village (black-headed weaver)
  21: 2896, 2898
Weka **15**: 2108
Whalebird **15**: 2045
Whippoorwill **14**: 2012, 2013
Whistler (thickhead) **21**: 2923–2924
 Bare-throated **21**: *2923*, *2924*
 Golden (cutthroat; thunderbird;
  whipbird) **21**: 2923–2924
 Olive **21**: 2923
 Rufous **21**: 2923, 2924
 White-breasted **21**: 2923
Whistling duck (tree duck) **5**: 714,
  716; **21**: 2925–2927
 Black-bellied (red-billed whistling
  duck; red-billed tree duck;
  cornfield duck) **21**: 2926, 2927,
  2927
 Black-billed **21**: 2926, 2927
 Fulvous **21**: 2925, *2926*
 Lesser **21**: 2925
 Plumed **21**: 2925
 Wandering **21**: 2925
 White-faced **21**: 2925, *2925*, 2927
White-eye **14**: 1993; **21**: 2930–2931
 African yellow (green white-eye;
  yellow white-eye) **21**: 2931
 Javan **21**: *2930*
 Oriental **21**: *2931*
 Yellow-spectacled (Wallace's
  white-eye) **21**: 2930
Whitethroat, Common **18**: 2459
Whydah **21**: 2936–2938
 Pin-tailed (pin-tailed widowbird)
  21: 2889, 2936, *2936*, 2937, *2937*

Queen (shaft-tailed whydah)
  21: *2938*
Widowbird, Pin-tailed *see* Whydah,
 Pin-tailed
Widow bird, Yellow-shouldered *see*
 Bishop Bird, Burton's black-
 and-yellow
Wiretail, Des Mur's **13**: 1819
Woodcock **17**: 2427; **21**: 2966–2967
 American **21**: 2966, *2966*
 Eurasian **21**: 2966, 2967
Woodcreeper **19**: 2721; **21**: 2968–2969
 Barred **21**: 2968
 Buff-throated **21**: *2968*, 2969
 Great rufous **21**: 2968
 Narrow-billed **21**: 2968
 Spotted-crowned **21**: 2968
 Streak-headed **21**: 2969
 Tawny-winged **21**: 2969
 Wedge-billed **21**: *2969*
Woodlark **17**: 2396
Woodpecker **19**: 2653–2654;
  21: 2977–2979
 Acorn **19**: 2654; **21**: *2978*, 2979
 Downy **19**: 2654
 Great spotted **19**: 2654; **21**: *2979*
 Green **21**: 2965, 2977, *2977*,
  2978–2979
 Ground **21**: 2979
 Hairy **21**: 2978
 Imperial (*extinct*) **21**: 2977
 Ivory-billed (*extinct*) **21**: 2977
 Lesser spotted **19**: 2617
 Pied (spotted woodpecker) **21**: 2977
 Pileated **21**: 2979
 Red-headed **21**: 2978, 2979
 Rufous **21**: 2979
Woodpecker-finch **5**: 648, 649–650,
  650; **16**: 2296
Wood swallow (swallow-shrike)
  21: 2983–2984
 Ashy **21**: 2983, 2984
 Black-faced **21**: *2983*
 Dusky **21**: 2983
 Little **21**: 2983, 2984
 Masked **21**: 2983
 Papuan **21**: 2984
 White-backed **21**: 2984
 White-breasted **21**: 2983, *2984*
 White-browed **21**: 2984
Wood warbler (*New World warbler*)
  18: 2458; **19**: 2617; **21**: 2985–2986
 Black-and-white warbler **21**: 2986
 Black-poll warbler **21**: 2985
 Black-throated blue warbler
  21: 2985
 Blue-winged warbler **21**: 2985
 Chestnut-sided warbler **21**: 2985
 Golden-winged warbler **21**: 2985
 Kirtland's warbler **21**: 2985
 Myrtle warbler **21**: 2986
 Parula warbler **21**: 2986
 Prothonotary warbler **21**: 2986
 Yellow warbler **4**: 553; **21**: 2985, *2986*
Wren **21**: 2993–2995
 Bewick's **21**: 2993
 Bush (South Island bush wren)
  16: 2164
 Cactus **21**: 2993–2994, *2995*
 Common European **7**: 999
 Giant **21**: 2993
 House **21**: 2993, 2995
 Marsh (long-billed marsh wren)
  21: *2994*, 2995
 Rock (*true wren*) **21**: 2994
 Short-billed marsh wren (sedge
  wren) **21**: 2994
 Winter (wren; European wren;
  Jenny wren) **21**: 2993, 2994, 2995
 Zapata **21**: 2994

Wren babbler **1**: 113
Wrentit **1**: 113; **3**: 426
Wrybill **14**: 1989, 1990
Wryneck (snakebird) **21**: 2996–2997
 Eurasian **21**: 2996, *2996*, 2997
 Red-breasted **21**: 2996

Xenops, Plain **13**: 1819

Yellowlegs
 Greater **16**: 2233
 Lesser **16**: 2233
Yuhina **1**: 114

FISH
Acara, Platinum **10**: 1366
Alaska blackfish (Alaska fish;
  devilfish) **1**: 34
Albacore **20**: 2773, 2774, 2775
Alewife **17**: 2327
Alfonsino **18**: 2512
Amberjack **16**: 2242
 Greater **16**: 2242
 Yellowtail **16**: 2242
Anchovy **1**: 45–46
 California (northern anchovy)
  1: 45, 46
 European **1**: 45, 46
Angel, Black **16**: 2245
Angelfish (*freshwater bony fish*) *see*
 Scalare
Angelfish (*marine bony fish*) **1**: 49–50;
  4: 532
 Emperor **1**: *50*
 French **1**: 49, 50
 French old woman **1**: 49
 King **6**: *833*
 Queen **1**: 50
Angelfish (*shark species*) *see* Monkfish
 (*shark species*)
Angelshark *see* Monkfish (*shark species*)
Angler
 Sargassum **1**: 51
 Warty (handfish) **1**: 51
Anglerfish **1**: 51–52; **6**: 833
 Deep-sea **1**: 51; **5**: 653–654
Archerfish **1**: 77–78
 Banded **1**: 78
 Spotted **1**: *77*

Barb, Zebra (striped barb) **21**: 3015
Barbel **8**: 1089, 1090
Barbfish **16**: 2260
Barracuda **2**: 154–155; **4**: 532
 California (Pacific barracuda) **2**: 155
 European **2**: 154, 155
 Great (picuda; becuna) **2**: *154*, 155
 Northern **2**: 154, 155
Bass **2**: 163–164
 Common (European bass) **2**: 163,
  164
 Largemouth **2**: 164, *164*; **8**: 1063;
  18: 2569
 Smallmouth **2**: 164
 Stone (wreckfish) **8**: 1082
 Striped **8**: 1082
 White **8**: 1082
 Yellow **8**: 1082
Batfish (*anglerfish*) **1**: 51, 52
Betta, Penang **6**: 822
Birchir **2**: 195–196
 Ornate **2**: 195
Bitterling **2**: 218–219
Blackfish (*freshwater fish*), Alaska
  1: 34; **12**: 1681
Bleak **12**: 1619
Blenny **2**: 237–238
 Black-headed **2**: 238
 Butterfly **2**: 238

Fanged **2**: 237
Four-eyed **7**: 888
 Viviparous (eelpout) **2**: 238
Blindfish, Kentucky **3**: 409
Bluefish **2**: 245–246
Bluehead **21**: 2991, 2992
Bombay duck (bummalo;
  bummalaw) **2**: 257
Bonito **2**: 258
Bowfin (American mudfish;
  spotfin; freshwater dogfish;
  freshwater wolf) **2**: 272–273;
  5: 686
Bream (common bream; bream flat)
  2: 274–275; **16**: 2182
 Freshwater *see* Tilapia
 Sea **2**: 274
 White (silver bream) **2**: 274
Brill **20**: 2782
Bristlemouth (lightfish) **2**: 278
Broadbill (*fish*) *see* Swordfish
Bugfish *see* Menhaden, Atlantic
Bullhead **3**: 314–315
 European (sculpin; miller's-
  thumb) **3**: 315
Bullhead shark
 Crested **15**: 2027
 Japanese **15**: 2027
Burbot (eel pout) **4**: 495;
  11: 1455
Butterfish (gunnell; rock gunnell;
  nine-eyes) **3**: 342
Butterfly fish (*freshwater fish*)
  3: 347–348
Butterflyfish (*marine fish*) **1**: 49, *49*
 Copperband **1**: 49
Butter hamlet **8**: 1081, 1083

Candirú **3**: 402
Cardinal fish **3**: 389–390; **4**: 532
 Brownspot **3**: 390
 Coral **3**: *390*
 Orange-lined **3**: *389*
Carp (minnow) **3**: 395–396
 Common **3**: 396
 Crucian (Prussian carp) **3**: *396*;
  8: 1018
 see also Minnow
Catfish **3**: 401–402
 Armored **1**: 88–89; **20**: 2755
 Banjo **3**: 401
 Blue **3**: 402
 Channel **3**: 401–402
 Electric *see* Electric catfish
 European (sheatfish; wels; silurus)
  3: 401–402
 Flathead **3**: 401
 Loricariid **1**: 88
 Mailed **1**: 88–89, *88*, *89*
 Sharptooth **3**: *401*
 Talking **1**: 88
 Thorny **1**: 88
Cat shark **3**: 407–408; **17**: 2333
 Variegated **3**: *408*
Cave eel, Blind **18**: 2585
Cave fish **3**: 409–410
 African **3**: 410
 Iraqi **3**: 410
 Mexican (Mexican cave characin)
  3: 409, 410
Chimaera, Plow-nosed **17**: *2330*
Chromis **17**: *2379*
 Blue **5**: 639
Chub **12**: 1618
 Common (European chub)
  12: 1618; **16**: 2182
 Flathead **12**: 1618
Cichlid **3**: 458–459
 Jewel *see* Jewel fish
 Pike **4**: 459

Cichlid (*continued*)
  Red devil 4: *458*
  Rio Grande 4: *459*
  Zebra (convict cichlid) 21: 3015
Clownfish (anemonefish) 4: 533;
  5: 638, 639, *640*
Coalfish (saithe; coley; billet; glassan;
  black pollack) 4: 472–473, 495
Cod (Atlantic cod; codling; codfish)
  4: 472, 495–496; 6: 836
  Arctic (polar cod) 4: 495
Coelacanth (kombessa) 4: 497–498
Convictfish 18: 2576
Corkwing (sea pheasant) 21: 2991
Cornetfish 20: 2766
Cowfish 20: 2758
  Scrawled 20: 2759
Croaker (sea trout) 5: 601–602
  Atlantic 5: 601, 602
  Meager (redfish) 5: 601
  Spotfin 5: 602
  Yellowfin 5: *601*, 602
  see also Drum
Cuchia 18: 2585
Cutlassfish 17: 2415, *2415*

Dab 5: 634–635
Dace 12: 1618–1619; 17: 2411
  Common 12: 1618
  Southern redbelly 12: 1618–1619
Damselfish (demoiselle) 4: 532;
  5: 638–640; 6: 836
  Domino 5: 638
Darter (*fish*) 14: 1917
Dealfish 13: 1768
Devilfish (manta ray; devil ray)
  5: 668–670; 6: 729
  Greater (Atlantic manta ray)
    5: 668, *670*
  Japanese manta ray 5: 669
  Pacific manta ray 5: 668
  Prince Alfred's manta ray 5: 668, 669
  Pygmy Australian devil ray 5: 668
  Southern blue 5: 668
Discus 4: 459
  Blue 14: 2006–2007, *2007*
  Red 14: 2006, *2006*, 2007
Dogfish (grayfish; rock salmon; rock
  eel; fluke) 5: 686–688
  Alligator (bramble shark) 5: 686
  Freshwater *see* Bowfin
  Greater spotted (nurse hound)
    5: 686
  Lesser spotted 5: 686, *686*, 687
  Pacific 5: 686
  Spiny (spur dogfish) 5: 686,
    687–688; 17: 2333
Dolphinfish 5: 693–694
  Common (Ma-hi Ma-hi; dorado)
    5: 693, 694, *694*
  Pompano 5: 693
Dragonet 5: 706–707
Drum
  Black 5: 601, 602
  Freshwater 5: 602

Eagle ray 6: 729–731
  Common 6: 729, 730
  Spotted 6: 730, *730*
Eel (elver; glass eel; leptocephalus)
  6: 745–747, 836
  American 6: 745, 746, 747
  Conger 4: 521–522
  Electric (electric knifefish)
    6: 762–763
  European 6: 745, *745*, 746, *746*,
    747; 14: 1981
Eel-skipper 12: 1684, 1686
Electric catfish 6: 760–761
  Small 6: 760, 761

Electric ray (torpedo fish; torpedo;
  torpedo ray; numbfish;
  crampfish) 6: 764–766; 17: 2332
  Atlantic 6: 764, 765–766; 17: 2332
  Australian crampfish 6: 764
  Pacific 6: 766
Elephant-nose fish 13: 1749, *1750*
Elephant trunk fish *see* Spiny eel,
  Lesser
Escolar (oilfish; scourfish; castor-oil
  fish) 17: 2414, 2415
Eulachon 17: 2410, 2411

Fallfish 12: 1619
Fathead 12: 1618
Featherback (featherfin; Old World
  knife fish) 6: 803–804
Fighting fish 6: 821–822
  Siamese 6: 821, *821*
Filefish 20: 2743, 2744
Fire eel 18: 2490
Firemouth 6: 829–830
Fish 6: 833–836
Flagfish 19: 2705
Flounder (right-eye flounder) 6: 833,
  860–861
  Kamchatka 8: 1121
Flying fish 6: 836; 7: 875–876; 17: 2414
  Atlantic 7: 876
  California 7: 875
  Caribbean 7: 875
  Four-winged 7: 875, *875*, 876
  Two-flying 7: 875
Flying gurnard 7: 880–881
  Atlantic 7: 880, 881, *881*
  Oriental 7: 880–881
Four-eyed fish 7: 887–888
  Finescale 7: 888
  Pacific 7: 888
Frogfish 1: 51, 52

Gar 7: 928–929
  Alligator 7: 928, 929
  Cuban 7: 929
  Longnose (billfish) 7: *928*, 929
  Shortnose 7: 928, 929
  Spotted 7: *928*
  Tropical 7: 929
Garibaldi 5: 638, 639
Garpike (*needlefish*) 12: *1724*, 1725
Gemfish 17: 2414
  Black 17: 2414
Glassfish 6: *834*; 7: 978–979
  Asiatic 7: 978, 979
  Cuban 17: 2379
  Indian (Indian glassy fish)
    7: 978–979, *978*
Gliding frog (flying frog), Wallace's
  7: 982
Goatfish 7: 989–990; 12: 1692
  Bigscale 7: 989, 990
  Manybar 7: 989, 990
  Red 7: 989, 990
  Spotted 7: 989
  Yellow 7: 989, 990, *990*
Goby 6: *834*; 7: 993–995
  Arrow 7: 995
  Blind 7: 995
  Catalina (blue-banded)
    7: 993–994
  Giant 7: 993
  Luzon 7: 993
  Magnus 7: *994*
  Neon 4: 533; 7: 995
  Red striped 7: *995*
Goldfish (golden carp; gibel carp;
  Prussian carp) 8: 1018–1020
Goosefish 1: 51, 52
Gourami 6: 836
  Kissing 10: 1365–1366

Grayling 8: 1047–1048
  Arctic (American grayling;
    Montana grayling; Amur
    grayling; Baikal grayling;
    Kamchatka grayling; East
    Siberian grayling) 8: 1047
  European 8: 1048
Grouper 4: 532; 8: 1081–1083
  Giant 8: 1082
  Golden-striped 8: 1083
  Nassau 8: 1083
  Potato (potato cod; potato bass)
    8: *1081*
  Spotted 8: *1081*, 1082
Grunion 8: 1087–1088
Guaguanche 2: 155
Guavina 7: 993
Gudgeon 8: 1089–1090
Guitarfish (shovelnose ray; shovelnose
  shark; fiddle ray) 8: 1102–1103
  Atlantic (spotted guitarfish) 8: 1103
  Giant 8: 1102, 1103
Guppy (rainbow fish; millions fish)
  8: 1106–1107; 12: 1664
  Veiltail 8: *1106*
Gurnard (sea robin; armored sea
  robin; gurnet) 8: 1108–1109
  Gray 8: 1108, 1109
  Leopard sea robin 8: 1109
  Northern sea robin 8: 1109
  Piper 8: 1109
  Yellow 8: 1108, *1109*

Haddock 4: 495; 6: 836; 8: 1113–1114
Hagfish 17: 2329
Hake 4: 495; 8: 1119–1120
  European 8: 1119, 1120
  Silver 8: 1119, 1120
Halfbeak 12: 1724, 1726
Halibut 7: 927; 8: 1121–1122
  Atlantic 8: 1121, *1121*, 1122
  Greenland (black halibut) 8: 1121
  Pacific 8: 1121
  Shotted 8: 1122
Hammerhead shark 8: 1126–1128
  Common (smooth hammerhead)
    8: 1127
  Great 8: 1127
  Scalloped 8: 1127, *1128*
  Smalleye 8: 1127
  Whitefin 8: 1127
  Winged (winged shark) 8: 1126
Hardhead (*fish*) 12: 1618
Hatchet fish 9: 1161–1162
Herring 9: 1184–1186; 16: 2270
  Atlantic 9: 1184–1186, *1184*, *1186*
  Pacific 9: 1184, 1185
Hogchoker 17: 2444

Ice fish 9: 1277–1278
  Blackfin 9: 1277, *1277*, 1278
  Pike 9: 1277, 1278
Icefish (*glassfish*) 7: 978
  Japanese 7: 978
Icicle fish 7: 978

Jack 16: 2242
  Bigeye 13: *1769*
Jack Dempsey 10: 1302–1303
Jewel fish (African jewel fish; jewel
  cichlid) 10: 1325–1326
Jewfish 8: 1082–1083
John Dory (dory; doree; St. Peter's
  fish; girty) 10: 1327–1328

Kelpfish 2: 238
Killie, Zebra *see* Killifish, Plains
Killifish
  Cuban 19: 2705
  Plains (zebra killie) 21: 3015

King of the herrings 13: 1767, *1767*
King of the salmon 13: 1768
Knife fish 10: 1376–1377
  Banded 10: 1377
  Black ghost 10: *1376*
  Ghost 10: 1376, 1377
  Glass 10: 1376, 1377
  Naked-back 10: 1376, 1377
  Sand 10: 1377

Labyrinth fish 13: 1868
  see also Perch, Climbing
Lamprey (nannie nine eyes; stone
  sucker) 2: 161; 10: 1403–1404;
  17: 2329
  Brook 10: 1403, *1403*
  River 10: 1403, 1404, *1404*
  Sea 10: 1403, 1404; 17: *2332*
Lancet fish (sea wolf; wolf fish)
  10: 1409
  Longnose 10: 1409
  Shortnose 10: 1409
Lantern-eyed fish 18: 2512
Lantern fish 6: 836; 10: 1413–1414;
  13: 1772
Leaf fish (Amazon leaf fish; barbeled
  leaf fish; South American leaf
  fish) 10: 1421–1422
  Malaysian 10: 1421
Ling 11: 1455–1456
  Blue (lesser ling) 11: 1455, 1456
  Common 11: 1455–1456, *1455*
  Spanish (Mediterranean ling)
    11: 1455
Lingcod 20: 2766
Lionfish, Red (turkeyfish; lionfish;
  zebrafish; dragonfish; butterfly
  cod) 6: *835*, 836; 16: 2260–2261
Loach 11: 1473–1474
  Clown 11: *1473*
  Coolie 11: 1473
  Spiny 11: 1474
  Stone 11: 1473, 1474
  Tiger (green tiger botia) 11: 1473,
    1474
Lookdown 16: 2242
Lumpsucker (lumpfish; sea hen; hen
  fish) 11: 1499–1500
Lungfish 11: 1501–1502
  African 11: 1501, 1502, *1502*
  Australian (Burnett salmon)
    11: 1501
  South American 11: 1501, *1501*, 1502
  West African 11: 1501

Maasbanker (South African
  mackerel) 17: 2430
Mackerel 5: 694; 11: 1517–1518
  Atlantic 11: 1517, 1518, *1518*
  Chub 11: 1517
  Horse 11: 1517
  Horse *see* Scad
  Indian 11: 1517
  Island 11: 1517
  Japanese Spanish 11: 1517
  Monterey Spanish 11: 1517
  Pacific Sierra 11: 1517
  Pygmy 11: 1517
  Short 11: 1517
  Snake *see* Snake mackerel
  South African *see* Maasbanker
  Spanish 11: 1517
Madtom 3: 402
Mako, Shortfin 15: 2020–2021,
  2021
Marlin 11: 1561–1562; 19: 2605
  Atlantic blue 11: 1562
  Atlantic white 11: 1561
  Black 11: 1561
  Blue 11: 1561, *1562*

Menhaden **2**: 245; **12**: 1598
    Atlantic (bonyfish; bugfish;
        bunker; fatback; mossbunder;
        pokey; pogy; whitefish) **5**: 694;
        9: 1184; **12**: 1598, *1598*
    Pacific **12**: 1598
Midshipman **19**: 2698
Minnow **12**: 1618–1619
    Eurasian **12**: 1618, *1618*, 1619
    Sheepshead **19**: 2705
    see also Carp
Molly
    Amazon **2**: 234, *234*
    Black **2**: 233–234
    Sail-bearer **2**: 233, *233*, 234
    Sail-fin **2**: 233, 234
    Wedge-face **2**: 233, 234
Monkfish (angler; common
        goosefish) **1**: 51, 52, *52*
Monkfish (*shark species*); angelshark;
        fiddle fish; angelfish) **12**: 1646;
        17: 2329–2330, 2385
Moorish idol **12**: 1652–1653
Moray eel **12**: 1657–1659
    Fine-spotted **12**: *1659*
    Green **12**: 1657–1658, *1658*, 1659
    Leaf-nosed (ribbon eel) **12**: 1658
    Leopard **12**: 1657, 1658
    Zebra **12**: *1657*, 1658
Mudminnow **12**: 1680–1681
    Central **12**: 1680, 1681
    Eastern **12**: 1680, 1681
    European **12**: 1680, 1681
    Olympic **12**: 1680, 1681
Mudskipper **7**: 993; **12**: 1684–1686
Mullet **5**: 694; **12**: 1692–1693
    Flathead **12**: 1692, 1693
    Golden gray **12**: 1692, 1693
    Striped **12**: 1692, 1693
    Striped red (surmullet) **7**: 989, 990
    Thicklip gray **12**: 1692, *1692*, 1693
    Thinlip gray **12**: 1692, 1693
Muskellunge **14**: 1945, 1946

Needlefish (garfish) **12**: 1724–1725
    Atlantic **12**: 1725
    California **12**: *1725*
Nile fish (aba) **10**: 1377; **13**: 1749–1750
Noodlefish **7**: 978
    Chinese **7**: 978

Oarfish **13**: 1767–1768
    King of the herrings **13**: 1767, *1767*
Ocean perch **16**: 2260, 2261
Opah **13**: 1790–1791
    Common (spotted moonfish)
        13: 1790, *1790*, 1791
    Southern **13**: 1790, 1791

Paddlefish **13**: 1845–1846
    Mississippi (spoonbill sturgeon)
        13: 1845, *1845*, 1846, *1846*
Paradise fish **13**: 1868–1869
    Black **13**: 1868
    Round-tailed **13**: 1868
    Spike-tailed **13**: 1868
Parrotfish **4**: 532; **6**: 836; **14**: 1888–1890
    Bullethead **14**: *1888*, 1889
    Bumphead **14**: 1889
    Large **14**: 1889
    Princess **14**: 1888, 1889
    Rainbow **14**: 1890
    Redfin **14**: 1890
    Redlip **14**: *1889*
    Rusty **14**: *1890*
    Striped **14**: 1888, 1889
Peacock eel **18**: 2490
Pearlfish **14**: 1901–1902; **16**: 2272
Perch **6**: 834–835; **14**: 1917–1918
    Blue **10**: 1421

Climbing (climbing fish; climbing
        gourami; labyrinth fish)
        4: 468–469
    European **14**: 1917, *1917*, 1918,
        1918; **16**: 2175
    Nile **13**: 1751–1752
    Ocean **16**: 2260, 2261
    Sacramento **18**: 2569
    Shiner (yellow-banded seaperch)
        18: 2574
    Tule **18**: 2573
    Yellow **14**: 1917
Pickerel
    Chain **14**: 1945, 1946
    Redfin (grass pickerel) **14**: 1945,
        1946
Pike **14**: 1945–1947
    Black-spotted **14**: 1945, 1946
    Northern **14**: 1945, 1946, *1946*
Pilchard (sardine) **14**: 1948–1949
    Australian **14**: 1948
    European **14**: 1948, 1949
    Japanese **14**: 1948
    South American **14**: 1948, *1948*, 1949
    Southern African **14**: 1948;
        17: 2430
Pilot fish **14**: 1950–1951; **16**: 2242
Pineapplefish **18**: 2512
Pinecone fish **18**: 2512
Pipefish **14**: 1960–1961; **20**: 2766
    Bay **14**: 1961
    Broadnosed **14**: 1961
    Dusky **14**: 1960, 1961
    Ghost **20**: 2766
    Greater **14**: *1961*
    Multibar **14**: *1960*
    Network **14**: 1960
    Ringed **14**: 1960
Piranha (piraya) **14**: 1967–1969
    Blackspot **14**: 1969
    Red (common piranha; Natterer's
        piranha) **14**: 1968, *1968*, 1969
    San Francisco **14**: 1967, 1969
    Speckled **14**: 1969
Pirate-perch **20**: 2755
Plaice **14**: 1975–1977
Platy **14**: 1983–1984
    Black **14**: 1983
    Blue (blue moon platy; blue coral
        platy) **14**: 1983
    Golden **14**: 1983
    Hi-fin **14**: *1983*
    Red **14**: 1983
    Southern **14**: 1983, 1984
    Spotted **14**: 1984
    Variable (variegated platy) **14**: 1983
    Wagtail **14**: 1983
Pollack (lythe; pollock; podlok;
        podley) **14**: 2004–2005
Pompadour fish **14**: 2006–2007
Pompano, Florida **16**: 2242
Porbeagle **15**: 2020–2021; **17**: 2331
Porcupine fish **15**: 2022–2023
    Long-spine (spiny puffer) **15**: *2022*
Pufferfish (puffer; balloonfish;
        swellfish; globefish; blowfish)
        4: 532; **6**: 833; **15**: 2062–2064
    Freshwater **15**: 2063, *2063*
    Oceanic **15**: 2063
    Spotted **15**: *2062*
    Star **15**: 2063
Pupfish, Desert **19**: *2705*

Rabbitfish **4**: 532; **15**: 2094–2095
    Masked **15**: *2095*
Rainbowfish (*silverside*) **17**: 2379
    Banded **17**: *2380*
    MacCulloch's **17**: *2380*
Ray **17**: 2329–2330, 2332
    Bullnose **6**: 729, 730

California bat **6**: 729, 730, 731
Cownose **6**: 729
    Round **17**: 2333
    Thornback **17**: 2330
Razorfish **17**: 2367, *2367*, *2368*
    Pearly **21**: 2991
Redcoat **18**: 2511, *2511*, 2512
Redfish (*scorpionfish*), Norway
        16: 2260, 2261
Redfish (*sea trout*) *see* Croaker, Meager
Remora **15**: 2148–2149
    Common **15**: *2149*
    Spearfish **15**: 2148
Ribbonfish **13**: 1768
    Whiptail (king of the salmon)
        13: 1768
Roach (*New World*) **16**: 2182
Roach (*Old World*) **16**: 2182–2183
Rock beauty **1**: 49
Roguefish, Bearded **16**: 2260
Rosefish, Blackbelly (bluemouth;
        bluethroat) **16**: 2260, 2261
Rudd **16**: 2182
Ruffe **14**: 1917, 1918
Sailfish **16**: 2219–2220; **19**: 2605
    Atlantic **16**: 2219
    Indian and Pacific Ocean **16**: 2219,
        2220
Salmon **6**: 836
    Atlantic **1**: 99–101; **16**: 2176
    Cherry (masu) **13**: 1840
    Chinook (tyee salmon; quinnat
        salmon; king salmon; spring
        salmon; Sacramento salmon;
        Columbia River salmon)
        13: 1840, *1840*, 1842
    Chum (keta salmon; dog salmon)
        13: 1840, 1841, 1842
    Coho (silver salmon) **13**: 1840, 1842
    Pacific **13**: 1840–1842
    Pink (humpback salmon) **13**: 1840,
        1841, 1842
    Sockeye (red salmon; blueback
        salmon) **13**: 1840, *1841*, 1842, *1842*
    Zebra **21**: 3015

Sand eel **16**: 2227–2228
    Great **16**: 2228
    Pacific (Pacific sand lance) **16**: 2228
    Small **16**: 2227
    Smooth **16**: 2227
Sand lance **16**: 2227
Saury **12**: 1724
Saury pike (skipper) **17**: 2368
Sawfish **16**: 2238–2239
    Knifetooth **16**: 2238
    Largetooth **16**: 2238
    Smalltooth **16**: 2238, 2239, *2239*
Scad (horse mackerel) **11**: 1517;
        16: 2242–2243
    Atlantic horse mackerel **16**: 2243
    Bigeye (akule) **16**: *2243*
Scalare (angelfish) **4**: 458, 459;
        16: 2244–2245
    Deep angelfish **16**: 2244, 2245
    Lesser angelfish (freshwater
        angelfish) **16**: 2244, 2245
Scat (argusfish) **16**: 2255–2256
    Spotted **16**: 2255, *2255*, 2256
    Tiger **16**: 2255
Scatty **16**: 2255
Scorpionfish **16**: 2260–2261; **20**: 2818,
        2820
    California (sculpin) **16**: 2260, 2261
    Devil **16**: 2260
    Orange **16**: 2260
    Weedy **16**: *2261*
Scrawled fish **20**: 2743
Sculpin (*scorpionfish*) *see*
        Scorpionfish, California

Sea bass **2**: 163, *163*
    Giant (wreckfish) **2**: 163, 164
Sea horse **16**: 2277–2279; **20**: 2766
    Dwarf **16**: 2277
    Longsnout (longnose sea horse)
        16: 2278
    Pygmy **16**: 2277, *2278*
    White's **16**: 2277
    Yellow (Pacific sea horse) **16**: *2277*
Sea moth **16**: 2290–2291
    Dragon (short dragonfish) **16**: 2291,
        Hawaiian **16**: 2290
    Longtail (batfish; sea dragon;
        dragonfish) **16**: 2290, *2290*
Sea pheasant (corkwing) **21**: 2991
Sea snail (*fish*) **11**: 1499
    Common **11**: 1499
Sergeant major **5**: 638, 639
Shad **17**: 2327–2328
    Allis **17**: 2327, *2327*, 2328
    American (common shad; white
        shad) **9**: 1184; **17**: 2327, 2328
    Gizzard **17**: 2328
    Green-tailed (hardhead shad;
        yellow-tailed shad) **17**: 2328
    Hilsa **17**: 2327
    Reeve's **17**: 2327
    Skipjack **17**: 2328
    Twaite (goureen) **17**: 2327
Shanny **2**: 237, 238
Shark **4**: 532; **13**: 1770, 1772
    Basking **2**: 160–162; **17**: 2331
    Blacktip **17**: 2332
    Blue **11**: 1562; **13**: *1770*; **17**: *2329*,
        2331, 2333
    Bonnethead **8**: 1126, 1127
    Bull **17**: 2332
    Caribbean reef **15**: *2148*; **17**: 2329,
        2332
    Cookiecutter **17**: 2332
    Frilled (frill shark) **7**: 896–897;
        17: 2330
    Great white (maneater shark; great
        blue shark; white pointer; white
        shark; white death)
        8: 1058–1060; **17**: 2331, 2332
    Horn **15**: 2027
    Lemon **17**: 2333
    Mako **8**: 1059; **11**: 1562
    Mallethead **8**: 1127
    Megamouth **17**: 2331
    Port Jackson **15**: 2027–2028
    Requiem **17**: 2331, 2332
    Salmon **15**: 2021
    Sand tiger **17**: 2331, 2332
    Saw **16**: 2238
    Scoophead **8**: 1127
    Swell **3**: 407, *407*, 408
    Tiger **17**: 2332
    Whale **6**: 834; **17**: 2331; **21**: 2915–2916
    Whitetip **14**: *1950*
    Zambezi (bull shark; shovelnose
        gray shark; slipway gray shark;
        van Rooyen's shark; cub shark;
        ground shark; whaler shark;
        Lake Nicaragua shark) **17**: 2329;
        21: 3007–3008
    Zebra **21**: 3015
    see also Devilfish; Dogfish; Eagle
        ray; Electric ray; Guitarfish;
        Lamprey; Monkfish; Porbeagle;
        Sawfish; Sharks; Skate; Stingray;
        Wobbegong
Sharks 17: 2329–2333
Sharksucker
    Live (striped remora) **15**: 2148,
        2148, 2149
    Whitefin **15**: 2148
Shiner (*minnow*) **12**: 1618
    Common **12**: 1618

Shiner (*surfperch*) **18**: 2574
    Redfin **18**: 2570
Shrimpfish **17**: 2367–2368; **20**: 2766
    Grooved **17**: 2368
    Smooth **17**: 2368
    Speckled **17**: 2368
Silverside (jacksmelt; sand smelt)
    17: 2379–2380
    Inland (tidewater silverside)
        17: 2379, 2380
    Peruvian **17**: 2380
    Samoan (whitebait) **17**: 2379
Skaamoog (shy eye; skaamhaai;
    izak; lazyshark; leopardshark;
    tigershark) **3**: 407, 408
Skate **17**: 2329–2330, 2333, 2383–2385
    Common (blue skate; grey skate)
        17: 2383, 2384, 2385
    Deep-sea **17**: 2383
    Thornback **17**: *2383*
Skipper (*fish; saury pike*) **17**: 2368
Smelt **17**: 2410–2411
    European **17**: 2410, *2410*
    Longfin **17**: 2410
    Sand *see* Silverside
    Surf **17**: 2410
Snailfish **11**: 1499
Snakehead **17**: 2412–2413
    Giant (Indian snakehead; great
        snakehead) **17**: 2412, 2413
Snake mackerel **17**: 2414–2415
Snapper (*fish*) **4**: 532; **17**: 2423–2424
    Bluestripe **17**: *2424*
    Emperor red **17**: 2423
    Gray **17**: *2423*
    Lane **17**: 2423
    Mangrove red **17**: 2423
    Mutton **17**: 2423
    Yellow **17**: *2423*
    Yellowtail **17**: 2424
Snipefish **20**: 2766
Snoek (barracouta; sierra; couta; sea
    pike; snook) **17**: 2414, 2430–2431
Snook **17**: 2430
Soapfish **8**: 1081
Soldierfish **4**: 532
Sole **6**: 833; **17**: 2444–2445
    Common (Dover sole) **17**: 2444,
        2445, *2445*
    Deepwater **17**: 2444
    Naked **17**: 2444
    Sand **17**: 2444
Spearfish
    Longbill **11**: 1561
    Shortbill **11**: 1561
Spinefoot
    Dusky **15**: 2095
    Little **15**: 2094
    Marbled **15**: 2095
    Orange-spotted **15**: 2094
    Streaked **15**: 2094, 2095
    Vermiculated **15**: 2094, 2095
Spiny eel **18**: 2489–2490
    Barred **18**: 2490
    Blind **18**: 2490
    Lesser (elephant trunk fish) **18**: 2490
    One-striped **18**: 2490
    Tiretrack **18**: 2490
Splittail, Sacramento **12**: 1618
Squawfish **12**: 1618
Squirrelfish **4**: 532; **18**: 2511–2512
    Longspine **18**: *2512*
    Sabre **18**: 2511, 2512
Stargazer **18**: 2519–2520
    Atlantic **18**: 2520
    Mediterranean **18**: 2520
    Northern **18**: 2519
Sterlet **18**: 2556
Stickleback **6**: 836; **18**: 2533–2534
    Brook **18**: 2533

Four-spined **18**: 2533
Nine-spined **18**: 2533
Sea (15-spined stickleback)
    18: 2533–2534
Three-spined **18**: 2533, 2534
Stingray **17**: 2330; **18**: 2543–2544
    Black-spotted (giant reef ray)
        18: *2544*
    Blue-spotted **17**: *2331*; **18**: *2543*
    Southern (clam-cracker; stingaree)
        18: 2544
Stockfish **8**: 1119, 1120
Stonefish **16**: 2260; **18**: 2547–2548;
    20: 2818
    Estuarine **18**: 2547, *2548*
Stoneroller, Central **12**: 1618
Sturgeon **18**: 2556–2557
    Atlantic **18**: 2556, 2557, *2557*
    Russian (beluga) **18**: 2556
    Spoonbill *see* Paddlefish,
        Mississippi
    White **18**: 2556, 2557
Sucker, Marlin **15**: 2148
Suckerfish
    Slender **15**: 2148
    White **15**: 2148
Sunfish (*freshwater*) **2**: 163, 164;
    18: 2569–2570
    Banded **18**: *2570*
    Blue spotted **18**: 2569
    Common (pumpkinseed) **16**: *2174*;
        18: 2569, 2569, 2570
    Long-eared **18**: 2569
    Red-breast **18**: 2569
Sunfish (*marine*)
    Ocean **13**: 1774–1775
    Sharptail **13**: 1774
    Slender **13**: 1774
Surfperch (seaperch; surffish)
    18: 2573–2574
    Midnight **18**: *2573*
    Pink **18**: 2573
    Yellow-banded seaperch (shiner)
        18: 2574
Surgeon, Atlantic Ocean **18**: 2577
Surgeonfish **4**: 532; **18**: 2575–2577
    Five-banded **18**: 2575–2576
    Palette **18**: 2577
Swallower, Black **13**: *1773*
Swamp eel **18**: 2585
Sweeper
    Black-margin **17**: *2441*
    Copper **17**: *2441*
Sweetfish (sweetmouth; ayu; koayu)
    19: 2596–2597
Swordfish (broadbill)
    19: 2605–2607
Swordfish (*paddlefish species*),
    Chinese (swordbill sturgeon)
    13: 1845
Swordtail **14**: 1984; **19**: 2608–2609

Tang
    Blue **18**: 2576, *2576*
    Yellow **18**: 2575, *2577*
Tarpon **19**: 2636–2637
Tench (doctor fish) **12**: 1619
Terapon, Jarbua (zebra fish; tiger
    fish) **21**: 3015
Tetra **19**: 2667–2668
    Black **19**: 2667
    Buenos Aires **19**: 2667
    Cardinal **19**: 2667, *2668*
    Congo **19**: 2667, 2668
    Copeland's **19**: 2667
    Glass **19**: 2667
    Glowlight **19**: 2667, *2667*
    Neon **19**: 2667, 2668
    Rio (flame tetra) **19**: 2667
    Rodway's **19**: 2667

X-ray (water goldfinch; albino
    pristella; pristella tetra)
    21: 2998–2999
Thresher shark (fox shark; sea fox;
    longtail shark) **19**: 2674–2675
    Bigeye thresher **17**: 2333; **19**: 2675
    Common thresher (thintail
        thresher) **19**: 2674, 2675
    Pelagic thresher **19**: 2675
Thryssa, New Guinea **1**: 45
Tilapia (kurper; freshwater bream)
    19: 2689–2690
    Mango **19**: 2690
    Mozambique **19**: 2689
    Red-belly **19**: 2689
Toadfish **19**: 2698–2699
    Oyster **19**: 2698, 2699
Tooth-carp **19**: 2704–2705
Totoaba **5**: 601, 602
Triggerfish **4**: 532; **20**: 2743–2744
    Black **20**: 2744
    Blackbar (humahuma-nuku-nuka-
        a-puaa) **20**: 2744
    Clown **20**: *2743*, 2744
    Gray **20**: 2744
    Orange-lined **20**: 2744
    Yellow-margined **20**: *2744*
Trout **20**: 2751–2753
    Brook **20**: 2751
    Cut-throat **20**: 2751, 2752
    Dolly Varden **20**: 2751
    European (brown trout; Aral
        trout; sea trout; lake trout
        European; Amn-Darya trout)
        20: 2751, 2752, 2753
    Golden **20**: 2751
    Lake (*North American*) **20**: 2751
    Rainbow **20**: 2751, *2751*, *2752*, *2753*
    Sea (*Cynoscion*) *see* Croaker
Trout-perch **20**: 2754–2755
    Blunt-nosed (sandroller) **20**: 2754
Trumpetfish **20**: 2766
Trunkfish (boxfish; cofferfish;
    dingleberry) **20**: 2758–2759
    Four-sided blue **20**: 2758
    Smooth **20**: *2758*
    Spotted **20**: *2759*
Tubenose **20**: 2766
Tubesnout **20**: 2765–2766
Tuna (tunny) **1**: 46; **19**: 2605;
    20: 2773–2775
    Northern bluefin **20**: 2773,
        2774–2775, *2775*
    Skipjack **20**: *2773*, *2774*
    Yellowfin (yellow-finned albacore)
        20: 2773, *2774*
Turbot **20**: 2782–2783
    Mediterranean (Black Sea turbot)
        20: 2782

Unicorn fish, Humpnose **18**: 2576

Venomous animals 20: 2817–2820
Viperfish **10**: 1413; **20**: 2829–2830
    Dana **20**: 2829
    Sloane's **20**: 2829

Walleye (pikeperch) **14**: 1917, 1918
Weakfish **5**: 602
Weatherfish **11**: 1473, 1474
Weever (weeverfish; stingfish; sea
    dragon) **17**: 2445; **20**: 2818;
    21: 2901–2902
    Cape Verde **21**: 2902
    Greater **21**: 2901, 2902
    Guinean **21**: 2902
    Lesser **21**: 2901, *2901*, 2902, *2902*
    Sailfin **21**: 2902
    Spotted **21**: 2902
    Striped **21**: 2902

Whalefish **21**: 2909
    Redmouth **21**: 2909
    Velvet **21**: 2909
Whalesucker **15**: 2148
Whiting **4**: 495; **21**: 2934–2935
Wobbegong (carpet shark) **17**: 2331,
    2333; **21**: 2954–2955
    Cobbler **21**: 2955
    Japanese **21**: 2955
    Northern **21**: 2955
    Ornate **21**: 2955
    Spotted **21**: 2954, 2955
    Tassled **17**: 2331; **21**: *2954*, 2955
Wrasse **4**: 533; **6**: 836; **21**: 2991–2992
    Ballan **21**: 2991
    Broomtail **6**: *836*
    Corkwing **21**: *2992*
    Cuckoo **21**: 2991
    Humphead (Napoleon fish)
        21: *2991*
    Mediterranean rainbow **21**: 2991
Wreckfish *see* Bass, Stone; Sea bass,
    Giant

Zebra fish (zebra danio) **21**: 3014–3015

INVERTEBRATES

Abalone (ormer; sea ear; earshell)
    1: 20–21
    Black **1**: 20, 21
    Green **1**: 21
    Red **1**: 20, 21
Admiral
    Blue **20**: 2810
    European white **12**: 1634
    Italian red **20**: 2810
    Red **6**: 811; **20**: 2810, *2810*, 2811
    White **20**: 2810, 2811, 2812
African devil's flower **11**: 1547, 1548
Agrippa, Giant **13**: 1821, 1822
Amoeba **1**: 41–42
Anemone (sea anemone) **1**: 47–48;
    4: 533
    Beadlet **1**: 48, *48*
    Gem **1**: *47*
    Opelet (snake locks anemone)
        16: 2301
Ant **19**: 2660
    Army (driver ant; foraging ant;
        legionary ant; safari ant; soldier
        ant) **1**: 90–92; **21**: 2969
    Black **14**: *1894*, 1895; **21**: 2964–2965
    Harvesting (harvester ant)
        9: 1156–1157, *1225*
    Honey (honey-pot ant) **1**: 67;
        9: 1204–1205
    Leafcutter (parasol ant) **1**: 68;
        10: 1417–1418
    Meadow *see* Wood ant, Hairy
    Pastoral (*ecological group*) **1**: 68;
        14: 1894–1895
    Slavemaker **21**: 2965
    see also Ants and termites
Ant lion (doodlebug) **1**: 64–65
Ants and termites 1: 66–69
Aphid (plant louse; greenfly; ant cow;
    blackfly) **1**: 74–75; **21**: 2965
    Black bean **1**: 74, *75*
    Peach-potato **1**: 74
    Woolly (American blight) **1**: 74
Appendicularian **17**: 2313, *2313*
Argonaut *see* Nautilus, Paper
Argus, Brown **2**: 243
Army worm **13**: 1822
Backswimmer (boat-fly; water
    boatman; wherry-man)
    1: 118–119; **2**: 188
Barnacle **1**: 138–139; **14**: 1981
    American **1**: 138, 139

Common acorn **1**: 138, 139, *139*
    Goose **1**: 138, 139
Basket star (Gorgon's head) **2**: 158–159
Bee
    Cuckoo **2**: 185; **3**: 318
    Hairy-legged mining **2**: *184*
    Leafcutter (alfalfa bee)
        10: 1419–1420
    Mason **2**: 185
    see also Bees and wasps; Bumblebee
Bees and wasps 2: 183–185
Beetle
    Colorado (Colorado potato beetle; potato bug) **4**: 504–505; **17**: 2344
    Cucujo **6**: 827
    Dung **2**: 187; **16**: 2252–2254
    Fruit **19**: *2611*
    Goliath **2**: 186; **16**: 2252
    Hercules **2**: 186; **16**: 2252
    Oil **2**: 187; **13**: 1782–1783
    Rhinoceros **16**: 2252
    Stag **19**: *2652*
    Striped elaterid **4**: 467
    Tiger **2**: *186*
    Wasp **11**: 1484
    see also Firefly; Glowworm; Ladybug; Weevil
Beetles and bugs 2: 186–188
Bhutan glory **18**: 2582
Birdwing **2**: 211–212; **15**: *2110*; **18**: 2582; **20**: 2806
    Cairns **2**: 211
    Common **2**: 211, 212
    Queen Alexandra's **2**: 211; **3**: 344
    Queen Victoria's **2**: 211, 212
    Rajah Brooke's **2**: 211, *211*, 212
    Tailed **2**: 211
Black widow (hourglass spider; shoe button spider; red mark; red back) **2**: 235–236
Blowfly (bluebottle; greenbottle; cluster fly; attic fly; screwworm fly) **2**: 241–242; **6**: 859
Bluebottle (*Portuguese man-of-war*) **15**: *2029*
Blue butterfly **2**: 243–244; **3**: 344
    Alcon large **2**: *244*
    Chalk-hill **2**: 243
    El Segundo **2**: 244
    Holly **2**: 243
    Large **2**: 243–244
    Long-tailed **2**: 243
    Mission **2**: 244
    Pygmy **2**: 243
    Short-tailed **2**: 243
    Silver-studded **2**: 243, *243*
    Small **2**: 243
    Western pygmy **2**: 244
Bollworm **2**: 255–256; **13**: 1822
    American (New World bollworm; corn earworm; tomato grub tomato fruitworm;) **2**: 255, 256
    Cotton **2**: 255, 256
    Pink **2**: 255, 256
Botfly, Human **6**: 858
Braconid **9**: 1279, 1280
Brandling **6**: 734
Brimstone **21**: 2929
Bristletail **2**: 279
Brittle star **2**: 280–281
Brown butterfly (brown) **3**: 296–297
    Meadow **3**: 296–297
Bug
    Ambush **2**: 188
    Assassin **2**: 188, *188*
    Bed **2**: 188
    Bee assassin **2**: 188
    Forest **17**: 2345
    Green vegetable **17**: 2344
    Parent **17**: 2345

Tortoise **17**: 2344
    see also Aphid; Backswimmer; Beetles and bugs; Cicada; Hopper; Pond skater; Shieldbug; Water scorpion
Bumblebee (humblebee; carderbee) **2**: 185; **3**: 316–318
    White-tailed **3**: *317*
Burnet moth **3**: 321–322, 344
    Doherty's longtail **3**: 321
    Eastern grapeleaf skeletonizer **3**: 321
    Fire grid **3**: 322
    Mountain **3**: 321–322
    Provence **3**: 322
    Six spot **3**: *321*
    Variable **3**: 322
Burying beetle (sexton beetle; gravedigger) **3**: 327–328
    American **4**: 328
Butterflies and moths 3: 343–346
Butterfly
    African leaf **10**: 1332
    Apollo (European Apollo; parnassian) **1**: 76
    Brimstone **21**: 2929
    Cabbage white *see* White butterfly, Small white
    Clouded yellow **21**: 2929
    Comma **3**: 345
    Grayling **3**: 297
    Indian leaf (orange oakleaf butterfly) **10**: 1332
    Kallima **10**: 1331–1332
    Map **20**: 2811
    Question-mark **20**: 2811
    Small white **3**: 416
    Swallowtail **2**: 212
    Thistle (cosmopolitan; painted lady) **20**: 2812
    Vanessa **20**: 2810–2812
    White *see* White butterfly
    Yellow *see* Sulphur butterfly
    see also Birdwing; Fritillary; Monarch; Morpho; Purple emperor; Skipper; Swallowtail
Butterfly bug **9**: 1225
Button shell **19**: 2707

Caddis fly (sedge fly) **3**: 354–355; **16**: 2175
Centipede **3**: 411–413
Chafer **4**: 486, 487, *487*
    Garden **4**: 487
    Rose **4**: 487
Chiggers **12**: 1621
Cicada **4**: 455–457
    Malaysian empress **4**: 455, 456
    Periodic (17-year cicada) **4**: 456, 457
Cirrothauma **13**: 1779
Clam **4**: 462–463
    Giant **4**: 462–463, *462*, *463*
    Hard (quahog) **4**: 462
Click beetle (skipjack; *larva:* wireworm) **4**: 466–467
    Elegant **4**: 466
    Harlequin **4**: 466
Clothes moth **4**: 470–471
    Casemaking **4**: 470, 471
    Common (webbing clothes moth; miller) **4**: 470, 471
Clouded yellow **21**: 2929
Cochineal insect **16**: 2247, *2247*
Cockchafer (May beetle; May bug; June beetle; June bug; common cockchafer; green fig-eater) **4**: 486–487; **16**: 2252
Cockle **4**: 488–489
    Common (edible cockle) **4**: 488, 489
    Common Pacific egg **4**: 488

Fucan **4**: 488
    Giant Pacific **4**: 488
    Jumping **4**: 488
    Prickly **4**: 488
    Smooth giant **4**: 488
    Yellow **4**: 488
Cockroach **4**: 492–494
    American **4**: *492*, 493, 494
    Common (black beetle) **4**: 492–493, *494*
    German (steamfly; shiner) **4**: 493, 494
    Giant Madagascar hissing **4**: 493
Comma **20**: 2811–2812
    Eastern **20**: 2812
    Gray **20**: 2812
    Green **20**: 2812
    Hoary **20**: 2812
    Oreas **20**: 2812
    Satyr **20**: 2811
Conch (fighting conch; trumpet shell; stromb shell) **4**: 514–515; **18**: 2547
    Horse **4**: 514
    Queen **4**: 514, 515, *515*
Cone shell **4**: 519–520; **20**: 2818
    Elephant trunk **4**: 519
    Glory of the sea **4**: 520
    Legate **4**: *520*
    Mediterranean **4**: 520
Copepod **14**: *1981*, 1982
Copper butterfly **4**: 525–526
    Large **4**: *525*, 526
    Purple-edged **4**: 526
    Scarce **4**: 526
    Small **4**: 525–526, *526*
Coquille St. Jacques *see* Scallop
Coral **4**: 527–529
    Black **4**: 527
    Blue **4**: 527
    Elephant's ear **17**: 2441
    Fire **4**: 527
    Horny **4**: 527
    Soft **4**: 527; **17**: 2440–2441
    Stony (true coral) **4**: 527, *528*
    Thorny **4**: 527
Corn borer (European corn borer; maize moth) **4**: 539–540
Cosmopolitan *see* Butterfly, Thistle
Cotton spinner **16**: 2272
Cow-killer **20**: 2815
Cowrie **4**: 556–557
    Chestnut **4**: 556–557
    European **4**: 556, 557
    Mole **4**: *556*
    Money **4**: 557
Crab
    Arrow **18**: *2475*
    Blue edible **4**: 573
    Edible (common crab; great crab) **6**: 741–742
    Ghost **4**: 573
    Harlequin **18**: 2476
    Kelp **18**: 2475
    Land (*crayfish*) **5**: 591
    Mole **18**: 2507
    Pea **4**: 573
    Pigmy **4**: 571
    Porcelain **4**: *571*, 572; **18**: 2507
    Red mangrove **4**: *573*
    River **4**: 571
    Robber (coconut crab) **4**: 571, 572, 573; **9**: 1178; **16**: 2186–2187
    Sally Lightfoot **4**: *570*
    Sand **18**: 2507
    Shore (green crab) **4**: 572; **17**: 2353–2355
    Stone **18**: 2507
    Tasmanian **4**: 571
Crabs and lobsters 4: 570–573

Crane fly (mosquito hawk; bobbing gnat; daddy longlegs. *larva:* leatherjacket) **5**: 588–590; **16**: 2175
Crawdad *see* Crayfish
Crawfish *see* Crayfish; Spiny lobster
Crayfish (crawfish; crawdad; yabbie) **5**: 591–592
    Giant **5**: 592
    Noble **5**: 592
    Signal **5**: 592
    Tasmanian giant **5**: 591
    White-clawed **5**: 592; **16**: 2175
Creeper (*larva*) **18**: 2550
Cricket **5**: 598–600
    Bush *see* Katydid
    Field **5**: 598, 599
    House **5**: 598–599, *599*
    Mole **5**: 598, 599–600
    Stone **5**: *600*
    Wood **5**: 599, 600
Cuttlefish **5**: 631–633
    Common (common cuttle) **5**: 631–632, *631*
    Harvest **5**: *632*
Cutworm **13**: 1822

Daddy longlegs (harvest spider; harvester; harvestman) **5**: 636–637
Damselfly (demoiselle; dancer; damsel; bluet; firetail; sprite; blue arrow) **5**: 641–642
Daphnia (water flea) **5**: 643–644; **10**: 1399
Dead man's fingers (sea fingers; dead man's toes) **5**: 651–652; **17**: 2440
Deerfly **9**: 1235, *1235*, 1236, 1237
Demoiselle (*insect*) *see* Damselfly
Diving beetle (*larva:* water tiger) **5**: 682–683
    Great **5**: 682, *682*
Dogface, California **21**: 2929
Dog-whelk (mud snail) **21**: 2917
    Striped **21**: *2917*
Dor-beetle **16**: 2253
Dragonfly (horse-stinger) **5**: 708–711
    Borneo **5**: 711
    Golden-ringed **5**: 711
    Hawker **5**: 710–711

Earthworm (nightcrawler) **6**: 734–737
    Common **6**: 734
    Giant Australian **6**: 734
Earwig (pincerbug) **6**: 738–740
    Common **6**: 738–739
    Small **6**: 738, 739
Eastern grapeleaf skeletonizer **3**: 321
Elephant tusk shell **6**: 777–778

False scorpion (pseudoscorpion; book scorpion) **6**: 799–800
Fanworm **6**: 801–802
    Red Sea **6**: *802*
Feather duster worm **6**: *801*
Featherstar **6**: 805–806; **16**: 2280
    Rosy **6**: 805–806
Fiddler crab **4**: 573; **6**: 815–817
    European **6**: 817
    Southwestern Atlantic **6**: 816
Fig insect **3**: 417
Firebrat **2**: 279
Firefly (lightning bug; glowworm) **6**: 827–828
    see also Glowworm
Fireworm **16**: 2293
Flatworm **6**: 848–849
Flea **6**: 850–852
    Bird **6**: 852
    Cat **6**: 850, 851, 852
    Gray squirrel **6**: *850*

Flea (continued)
 Hen 6: 850
 Human 6: 850, 851, 852
 Jigger 6: 851
 Mole 6: 852
 Oriental rat 6: 850
 Sticktight 6: 851
Flies 6: 856–859
Fluke (flatworm) 7: 870–871; 14: 2011
 Blood 7: 870, 871
 Chinese liver 7: 871
 Intestinal 7: 870, 871
 Liver 7: 870, 870, 871; 14: 2011
 Lung 7: 871
Fly
 Brine 6: 858
 Drone 9: 1251
 Fruit 7: 909–910
 Hessian 6: 859; 12: 1608
 Narcissus 9: 1251
 Robber 3: 318; 6: 859, 859
 Stable 9: 1242–1243
 Tachina (tachinid fly) 17: 2345;
  19: 2610–2611
 Tsetse 6: 859; 20: 2760–2761
 Window 4: 471
 see also Flies; Horsefly; Housefly;
  Hoverfly
Freshwater shrimp
 European 17: 2364
 Giant 17: 2364
Fritillary 7: 898–899
 Arctic 7: 898
 Dark green 7: 898–899
 Glanville 7: 898, 899
 Gulf 7: 898
 High brown 7: 898–899
 Marsh 7: 898, 899
 Pearl-bordered 7: 898
 Silver-washed 7: 898–899
 Small pearl-bordered 7: 899, 899
Froghopper 9: 1224, 1225
Funnel-web spider 18: 2480; 20: 2820
 Sydney 18: 2482

Gaper 7: 926–927
 Blunt 7: 927
 North American (summer clam;
  rubberneck clam; high-neck
  clam; horse clam; otter-shell clam;
  Washington clam) 7: 926, 927
 Sand (soft-shelled clam; old maid)
  4: 462; 7: 926, 927
 Small 7: 927
Garden spider 18: 2482
 European (diadem spider; cross
  spider) 13: 1800
 Lesser 13: 1800
Geoduck 4: 462
 Pacific 4: 462; 7: 926
Gilt-tail 6: 734
Glory of the sea 4: 520
Glowworm (glowworm beetle)
  7: 987–988
 European 7: 987
 see also Firefly
Grass grub 19: 2603
Grasshopper 8: 1035–1037
 Common field 8: 1037
 Common green 8: 1037
 Large marsh 8: 1036
 Lubber 8: 1035
 Meadow 8: 1036–1037
 Mottled 8: 1037
 Stripe-winged 8: 1037

Hairstreak butterfly 8: 1117–1118
 Brown 8: 1117
 Green 8: 1118
 Hewitson's 8: 1117

Harvestfly, Dogday 4: 456
Hawkmoth 9: 1166–1168
 Bee 9: 1168
 Convolvulus 9: 1166, 1167, 1168
 Death's head 3: 346; 9: 1166–1167,
  1168, 1168, 1210
 Elephant 9: 1166, 1167
 Eyed 9: 1168
 Lime 9: 1166
 Oleander 9: 1167
 Popular 9: 1166
 Privet 9: 1166
 Small hummingbird hawk
  9: 1166
Hermit crab 4: 572; 9: 1176–1178;
  18: 2507
 Common 9: 1176, 1177, 1177, 1178
 Prideaux's 9: 1178
Honeybee 2: 185; 9: 1208–1210
 Eastern 9: 1208, 1209
 European (hive bee; western
  honeybee) 9: 1208–1209
 Giant 9: 1208, 1209
 Small 9: 1208, 1209
Hoofed shell 16: 2248
Hopper 9: 1224–1225
Hornet 9: 1233–1234; 20: 2824
 European (Eurasian hornet)
  9: 1233, 1233, 1234; 20: 2824
 White-faced 9: 1233, 1234
 see also Wasp, Vespula
Hornet moth 9: 1234
 Lunar 9: 1234
Horsefly (stout; cleg) 6: 856;
  9: 1235–1236
Housefly 9: 1241–1243
 Lesser 9: 1241, 1242
Hoverfly 6: 857, 858; 9: 1250–1251;
  20: 2826
 Migrant 9: 1250
Hydra 9: 1265–1266
 Brown 9: 1265, 1266
 Green 9: 1265, 1266

Jellyfish 10: 1320–1322; 13: 1771;
  14: 1980; 20: 2818–2819
 Barrel 16: 2243
 Blue 16: 2243
 Box (sea wasp) 10: 1322;
  20: 2818–2819
 Lion's mane 10: 1320
 Moon (moon jelly; common
  jellyfish) 10: 1321; 16: 2243
Jumping plant louse (lerp) 9: 1224,
  1225

Katydid (long-horned grasshopper;
  bush cricket) 5: 598, 600; 8: 1035,
  1037; 10: 1341–1342
 Texas bush 10: 1341, 1342
King crab 4: 572; 10: 1353–1354
 American 10: 1353–1354, 1354
Krill 4: 564; 10: 1387–1388; 16: 2167

Lacewing 10: 1392–1393
 Brown (larva: aphid lion; aphid
  wolf) 10: 1392–1393
 Dusty-wing 10: 1392, 1393
 Giant 10: 1392
 Green (golden-eye; stinkfly)
  10: 1392, 1393
Ladybug (ladybird; lady beetle)
  2: 187; 10: 1394–1396
 2-spot 10: 1394
 7-spot 10: 1394, 1394
 10-spot 10: 1394
 22-spot 10: 1394
 Eyed 10: 1394–1395
Lampshell (arm-footed lampshell)
  10: 1405–1406

Lancelet (amphioxus) 10: 1407–1408
Lantern fly 9: 1225
 False 9: 1225
Leafhopper (dodger) 9: 1224, 1225
 Rhododendron 9: 1224
Leech 10: 1428–1429
 Horse 10: 1428
 Medicinal 10: 1428, 1428, 1429
Lerp see Hopper
Lightning bug see Firefly
Limpet 11: 1451–1452
 Antarctic 16: 2226
 Blue-rayed (peacock's feathers)
  11: 1451
 Common (European limpet; flitter;
  flither; papshell) 11: 1451, 1452,
  1452
 Keyhole 11: 1451, 1452
 Slipper 11: 1452
 Slit 11: 1451
 Tortoiseshell 11: 1451, 1452
Lion's mane 10: 1320
Lobster 11: 1475–1477
 American 4: 571, 572, 572; 11: 1475,
  1476, 1477
 European (common lobster) 4: 571;
  11: 1475, 1476
 Norway (Dublin prawn) 4: 573;
  11: 1475
 Slipper 4: 570
 see also Crabs and lobsters
Lobworm see lugworm
Locust 11: 1478–1480
 Australian plague 11: 1479
 Desert 11: 1478, 1479, 1479, 1480
 Migratory 11: 1478, 1478, 1479,
  1480, 1480
 Red-legged (red locust) 11: 1478,
  1479, 1480
Long-horned beetle (longicorn beetle;
  wood-boring beetle; timberman.
  larva: round-headed borer)
  2: 188; 11: 1483–1484
 Asian 11: 1484
 House 11: 1484
Longtail, Doherty's 3: 321
Louse
 Graperoot see Vine pest
 Vine see Vine pest
Lousy watchman 16: 2254
Lugworm (lobworm) 11: 1497–1498

Malmignatte 19: 2634–2635
Mantis (mantid; praying mantis;
  praying mantid) 11: 1546–1548
 Chinese 11: 1546
 European 11: 1546, 1547, 1547
 Orchid 11: 1547, 1548
Mantis fly 11: 1549–1550; 21: 2957
 Brazilian 11: 1550
 European 11: 1550
Mayfly 12: 1593–1594
Midge (gnat) 6: 859; 12: 1606–1608
 Biting (no-seeum; sand fly)
  12: 1606, 1607–1608
 Gall 12: 1606, 1608
 Meniscus 6: 859
 Nonbiting (harlequin fly; larva:
  bloodworm) 12: 1606–1607,
  1607, 1608
 Phantom 12: 1606
Millipede 12: 1611–1613
 Pill 12: 1612
 Sierra luminous 12: 1611
 Snake 12: 1612
 Sucking 12: 1612
Mite 3: 318; 12: 1620–1621
 Freshwater 12: 1621, 1621
 Gall 12: 1621
 Harvest (harvest bug) 12: 1620

 Itch 12: 1620, 1621
 Spider 12: 1621
Monarch (milkweed butterfly;
  common tiger butterfly) 3: 345,
  346; 12: 1632–1634; 20: 2810
Morpho 12: 1660–1661
 Common 12: 1661
Mosquito 6: 859; 12: 1662–1664
 Common 12: 1662
Moth
 Angle shades 13: 1821, 1822
 Atlas 1: 102; 3: 343
 Bent-wing 19: 2603–2604
 Boisduval's carpenter 7: 992
 Brown-tail 19: 2611
 Clearwing 3: 343–344
 Common emperor 3: 346
 Cotton 13: 1821, 1822
 Ermine 19: 2686
 Feather-winged 14: 1991
 Footman 19: 2686
 Ghost 19: 2603, 2603, 2604
 Goat 7: 991–992
 Gypsy 3: 346; 19: 2611
 Hercules 3: 344
 Kitten 15: 2076
 Leopard 7: 991
 Lobster 15: 2077
 Madagascan sunset 20: 2805, 2806
 Many-plume 14: 1991, 1991
 Mediterranean brocade 3: 346
 Owlet (Noctuid moth; larva:
  cutworm; army worm) 3: 344;
  13: 1821–1822
 Peppered 14: 1915–1916
 Prominent 15: 2076
 Puss 3: 345; 15: 2076–2077
 Reed leopard 7: 992
 Tapestry 4: 470
 Urania 20: 2805–2806
 Vaporer 3: 346
 Wax 3: 318
 Yucca 21: 3005–3006
 Zodiac 20: 2805, 2806
 see also Bollworm; Butterflies and
  moths; Corn borer
Mother Shipton 13: 1821
Mourning cloak (Camberwell
  beauty) 20: 2810, 2811
Mule-killer 20: 2815
Mussel 12: 1711–1713; 14: 1982
 Australian 12: 1711
 California 12: 1711
 Common (blue mussel; edible
  mussel) 12: 1711, 1712
 Fan 12: 1711
 Green-lipped 12: 1712
 Horse 12: 1711
 Swan 12: 1711
Mussel worm see Ragworm
Mygale see Spider, Bird-eating

Nautilus (pearly nautilus)
  12: 1722–1723
Nautilus, Paper (argonaut) 12: 1722;
  13: 1779
Nutshell 13: 1763–1764
 Brown 13: 1764
 Thin 13: 1763, 1764

Octopus 13: 1779–1781
 Blue-ringed 13: 1781;
  20: 2819–2820, 2819
 Common 13: 1779, 1780, 1781, 1781
 Lesser (curly octopus) 13: 1779
 Pacific 13: 1779
Orange tip, Desert 21: 2928
Orb spider (orb-web spider)
  13: 1800–1802; 18: 2480, 2481, 2482
 Araneus 13: 1801

Otter-shell, Common **7**: 927
Oyster **13**: 1832–1834
 American **13**: 1832, 1834
 European flat **13**: 1832–1833, *1834*
 Japanese **13**: 1832
 Pearl **13**: 1832, 1834
 Portuguese **13**: 1832
 Saddle **13**: 1832
 Thorny **13**: 1832

Palmer worm **19**: 2687
Palolo worm **13**: 1861
Paramecium **14**: 1881–1882
Peacock (*butterfly*) **20**: 2810–2811
Peacock worm **6**: 802
Pedipalpi (whip scorpion; tailless
 whip vinegaroon; scorpion;
 whip spider; short-tailed whip
 spider) **14**: 1906–1907
Peripatus **14**: 1924–1925
Periwinkle **14**: 1926–1928
 Common (edible periwinkle)
  14: 1926, *1926*, 1927
 Flat **14**: 1926, 1927, *1927*, 1928
 Rough **14**: 1926, 1927, 1928, *1928*
 Small **14**: 1926–1927, 1928
Piddock **14**: 1939–1940
 Common **14**: 1939, *1939*, 1940, *1940*
 Great **14**: 1939, 1940
 Little **14**: 1939, 1940
 Paper (American rock borer)
  14: 1939, 1940
 White **14**: 1939–1940
Plankton **13**: 1770; **14**: 1980–1982
Plume moth **14**: 1991–1992
 Artichoke **14**: 1991
 Grape **14**: 1991, 1992
 Triangle **14**: 1991, 1992
 White **14**: *1992*
Pond skater (water strider) **2**: 188;
 14: 2008–2009
Pond snail **14**: 2010–2011
 Dwarf **14**: 2010, 2011
 Ear **14**: 2010, 2011
 Great **14**: 2010, *2010*, 2011
 Marsh **14**: 2011
 North American **14**: 2011
 Wandering **14**: 2010
Portuguese man-of-war
 15: 2029–2031; **16**: 2300; **20**: 2819
Prawn **15**: 2043–2044
 African **15**: 2043
 Burrowing **18**: 2507
 Giant freshwater **15**: 2043
 Malaysian **15**: 2043
 Painted harlequin **15**: *2044*
Processionary moth (procession
 moth; processional moth)
 15: 2047–2048
 Australian **15**: 2048
 Oak **15**: 2047, 2048
 Pine **15**: 2047, *2047*, 2048, *2048*
 Pistachio **15**: 2048
Purple emperor **15**: 2071–2072
 Lesser **15**: *2071*

Rag, King **15**: 2104
Ragworm (mussel worm; clam
 worm; pile worm)
 15: 2104–2105
Ramshorn (trumpet snail; wheel
 snail) **14**: 2010
Razor
 Eggshell **15**: 2127
 Transparent **15**: 2127, *2128*
Razor shell (spoutfish) **15**: 2127–2128
 Pod (jackknife clam) **15**: 2127,
  2127, 2128
 see also Razor
Ringed worm **6**: 805

Salp **16**: 2223–2224
 Giant **16**: 2224
Sand dollar (cake urchin)
 16: 2225–2226
 Common **16**: 2225, 2226
Sawfly (wood wasp; horntail)
 9: 1279, 1280; **16**: 2240–2241
 Black-and-yellow horntail **16**: 2240
 European spruce **16**: 2241
 Jumping-disc **16**: 2240–2241
 Palisade **16**: 2240
 Pear (*larva*: slugworm) **16**: 2240
Scale insect **16**: 2246–2247
 Cottony-cushion **10**: 1396;
  16: 2246, 2247
 Elm scale **16**: 2246, 2247
 Heath scale **16**: 2246, 2247
 Mussel scale **16**: 2246, 2247
 Pine leaf scale **16**: 2246, 2247
Scale worm **16**: 2292
Scallop **16**: 2248–2249
 Doughboy **16**: *2249*
 Great (St. James' shell; coquille
  St. Jacques) **16**: 2248, 2249
 Purple-hinged **16**: 2248
 Queen **16**: *2248*, 2249
 Small (southern scallop) **16**: 2248
Scarab beetle **16**: 2252–2254
 Green **16**: *2252*
 Silver striped **16**: *2253*
 see also Cockchafer
Scorpion **16**: 2257–2259; **20**: 2818, 2820
 Asian forest **16**: 2258
 Emperor **16**: 2257
Sea biscuit **16**: 2226
Sea blubber **10**: 1320
Sea butterfly (pteropod)
 16: 2269–2270
Sea cucumber **14**: 1901–1902, *1901,
 1981*; **16**: 2271–2272
Sea fir (hydroid) **16**: 2273–2274
 Naked **16**: 2273
Sea hare **16**: 2275–2276
 California **16**: *2276*
Sea lemon **16**: 2299, 2300
Sea lily (stone lily) **16**: 2280–2281
Sea mouse **16**: 2292–2293
Sea pansy **16**: 2298
Sea peach **17**: 2313
Sea pen **16**: 2297–2298
Sea-slater (shore-roach; quay louse;
 quay lowder) **17**: 2398–2399
Sea slug (nudibranch) **16**: 2299–2301
 Rose nudibranch **16**: *2299*
Sea snail (*mollusk*) **16**: 2272
 Purple (violet sea snail; violet
  snail; floating shell of the high
  seas) **15**: 2073
Sea spider **17**: 2310–2311
 Giant deep-sea **17**: *2311*
Sea squirt (tunicate) **4**: 556; **14**: *1980*;
 15: 2078; **17**: 2312–2313
 Lightbulb **17**: 2313
Sea urchin (sea hedgehog)
 17: 2314–2315
 Black-spined **17**: 2315
 Edible **17**: *2314*, 2315
 Long-spined **17**: 2315
 Pencil-spined **17**: *2315*
Setaceous Hebrew character **13**: 1821
Shieldbug (stinkbug) **17**: 2344–2345
Shipworm **13**: 1772; **17**: 2349–2350
Shrimp **14**: *1982*; **17**: 2364–2366
 Brine **2**: 276–277
 Cleaner **17**: 2364
 Common European (brown
  shrimp; edible shrimp)
  17: 2364–2365
 Emperor **17**: *2366*
 Fairy **6**: 792–793

Mantis (split-thumb; squilla;
 prawn killer; shrimp mammy;
 nurse shrimp) **11**: 1550, 1551–1552
Opossum **13**: 1795–1796
Oriental **15**: 2043
Pistol **17**: 2364, 2366
Red **17**: *2364*
Red-handed **15**: 2043
Sewing **17**: 2364
Silk moth (*larva*: silkworm) **3**: *345*;
 17: 2376–2378
 Eri **17**: 2377
 Indian **17**: 2377
 Mulberry **17**: *2376*
Silverfish **2**: 279
Silver Y **13**: 1821
Skipper (*butterfly*) **17**: 2391–2392
 Dingy **17**: 2391
 Essex **17**: *2391*, 2392
 Giant **17**: 2392
 Great yucca **17**: 2392
 Lulworth **17**: *2392*
 Regent **17**: 2392
Slater (shore-roach; quay louse; quay
 lowder) **17**: 2398–2399
Slug **17**: 2408–2409
 Banana **17**: 2408
 Gray field **17**: 2408
 Great gray **17**: 2409
 Keelback **17**: 2408, 2409
 Roundback **17**: 2408, 2409
 Shelled **17**: 2408
 Tree **17**: 2408
Slugworm **16**: 2240
Snail
 Bladder **14**: 2010
 Garden (brown garden snail;
  common snail) **7**: 930–932
 Giant **7**: 966–967
 Glutinous **14**: 2010, 2011
 River **14**: 2010
 Valve **14**: 2010
Solifugid (false spider; sun spider;
 wind scorpion; camel spider;
 solipugid) **18**: 2453–2454
 Giant **18**: *2453*
Spider
 Bird-eating (goliath tarantula;
  mygale) **2**: 203–204; **19**: *2634*
 Bolas **13**: 1801–1802; **18**: 2482
 Burrowing **18**: 2480
 Cellar (daddy longlegs) **5**: 636;
  18: 2482
 Crab **5**: 580–581; **18**: 2480, 2482
 Fishing **18**: 2482
 Jumping **18**: 2480, *2480*, 2481, 2482
 Lynx **18**: 2481
 Mothercare **18**: 2483
 Net-casting **18**: 2482
 Nursery-web **18**: 2483; **21**: 2957
 Purse-web **19**: 2719, 2720
 Sheetweb (tunnel-web spider;
  funnel-web spider) **17**: 2339–2340
 Water (European water spider)
  18: 2481–2482; **20**: 2876–2877
 Whip *see* Pedipalpi
 White-tailed **20**: 2818
 Wolf **18**: *2482*, 2483; **21**: 2956–2957
 see also False scorpion; Scorpion;
  Spiders; Tarantula
Spider crab (slender-legged spider
 crab; long-legged spider crab;
 decorator crab) **4**: 572–573;
 17: 2355; **18**: 2475–2476
 Giant **4**: 571; **18**: 2475
 Long-beaked **18**: 2475, 2476
 Spiny (thornback crab) **18**: 2475
Spiders 18: 2480–2483
Spiny lobster (crawfish; langoustine)
 4: 571–572, 573; **11**: 1475, *1476*

California **11**: 1475
Common **4**: 573
West Indian **11**: *1475*
Sponge **18**: 2493–2495
 Barrel **18**: *2495*
 Bath **18**: 2493, 2494, 2495
 Horny **18**: 2493
 Purse **15**: 2074–2075
 Red Sea **18**: *2493*
Sponge fly (spongilla fly) **10**: 1392,
 1393
Springtail **18**: 2504–2505
Squat lobster **4**: 572; **18**: 2506–2507
 Common **18**: 2506
Squid **18**: 2508–2510
 European **18**: 2509
 Flying **18**: 2510
 Giant **13**: 1772; **18**: 2473–2474,
  2509–2510
 Longfin **18**: 2509
 Pop-eyed **18**: *2509*
Squilla *see* Shrimp, Mantis
Star *see* Starfish
Starfish **18**: 2516–2518
 Blood sea star (bloody Henry)
  18: *2517*, 2518
 Burrowing star (sand star)
  18: *2517*, 2518
 Crown-of-thorns **18**: 2517, 2518,
  2518
 Cushion star (starlet) **18**: 2516, 2518
 Scarlet nubbly **18**: *2516*
Stick insect **18**: 2530–2532
 Giant prickly **18**: 2531, *2531*
 Laboratory **18**: 2530, 2531–2532
 Madagascar **18**: 2532
 Prickly **18**: 2530, 2532
 Smooth **18**: 2530, 2531, 2532
Stonefly **18**: 2549–2550
Sulphur butterfly (yellow sulphur
 butterfly) **21**: 2929
 Orange sulphur **21**: 2929
Swallowtail **18**: 2580–2582
 African **18**: 2582
 Asian **18**: 2582
 Citrus (orchard swallowtail)
  18: 2581
 Common **18**: 2580, 2581
 European **3**: *344*
 Five-bar **18**: *2582*
 Giant **18**: 2581
 Giant African **18**: 2581, 2582
 Kite **18**: 2582
 Oriental **18**: 2582
 Royal **18**: 2582
 Ruby-spotted **18**: 2581
 Scarce **18**: 2580
 Tiger **18**: *2580*, 2581
Swift moth **19**: 2603–2604
 Common **19**: 2603, *2604*
 Gold **19**: 2603
 Map-winged **19**: 2604

Tapeworm, Common **6**: 852
Tarantula **2**: 203; **19**: 2633–2635;
 21: 2956
 Mexican superb **19**: *2633*
Termite (white ant) **19**: 2658–2660
 Black-mound **19**: 2658–2659
 Compass (magnetic termite)
  19: 2660
 see also Ants and termites
Thornbug **9**: 1225
Thrips (thunderbug; stormfly;
 bladderfoot; fringewing)
 19: 2676–2677
 Grain (corn thrips) **19**: 2676
 Greenhouse **19**: 2676
 Honeysuckle **19**: 2676
 Lily **19**: *2677*

Thrips (*continued*)
  Onion **19**: 2676
  Western flower **19**: 2677
Tick **19**: 2681–2682
  Common dog **19**: 2682
  Deer **19**: *2681*
  Lone-star **19**: 2682
Tiger moth **19**: 2686–2688
  Cream-spot **19**: 2686, *2686*, 2687
  Garden **19**: 2686, 2687, *2687*
  Jersey **19**: 2687
  Scarlet **19**: 2687
Topshell **19**: 2706–2707
  Brick-red **19**: 2706
  Gray **19**: 2706
  Painted (painted topshell; silver Tommy) **19**: 2706, *2706*, 2707, *2707*
Tortoiseshell, Small **20**: 2812, *2812*
Trapdoor spider **2**: 203; **18**: 2480; **19**: 2719–2720
  Adelaide **19**: 2720
  Brown **19**: 2720
  California **19**: 2720
  Malaysian **19**: *2719*
Treehopper **9**: 1225, *1225*
  Buffalo **9**: 1225
Triton shell, Giant **18**: 2518
True lover's knot **13**: 1821
Tubeworm (parchment-tube worm; innkeeper worm) **13**: 1773; **20**: 2767–2768
Tubifex worm (bloodworm) **20**: 2769–2770
Tumblebug **2**: 188

Underwing
  Beautiful yellow **13**: 1821, 1822, *1822*
  Large yellow **13**: 1822
Urania, Sloane's **20**: 2806
Urchin (*echinoderm*)
  Common heart **17**: 2314
  Keyhole **16**: *2226*
  Pea **16**: 2226
  see also Sea urchin

Velvet ant **20**: 2815–2816
Venomous animals 20: 2817–2820
Viceroy **12**: 1634; **20**: 2810, 2811, 2812
Vine pest (vine mite; vine louse; graperoot louse; grape phylloxera) **20**: 2827–2828

Wasp **18**: 2482–2483
  Chalcid **3**: 416–417; **4**: 471
  Common **20**: 2824, 2826
  Cuckoo **2**: 185; **13**: 1865; **20**: 2824
  Fig **3**: 417
  Gall **2**: 183; **7**: 921–922
  German **20**: 2824, *2825*, *2826*
  Hunting (digger wasp) **9**: 1261–1262; **21**: 2957
  Ichneumon (ichneumon fly) **2**: 183; **9**: 1279–1280
  Mason **9**: 1262
  Mud-dauber **9**: 1262
  Norwegian **20**: 2824
  Paper **2**: 185; **13**: 1864–1866; **20**: 2824
  Potter **9**: 1262, *1262*
  Spider-hunting **2**: 185; **9**: *1261*, 1262
  Tree **20**: 2824
  Vespula (yellowjacket; hornet—*incorrectly*) **20**: 2824–2826
  see also Bees and wasps; Velvet ant
Water beetle (water scavenger beetle) **20**: 2863–2864
  Giant water scavenger beetle **20**: 2863
  Silver **20**: 2863, *2863*, 2864
  see also Whirligig beetle

Water scorpion (toe-biter; needle bug) **20**: 2870–2871
Water stick insect **20**: 2870, *2870*, 2871
Wax insect, Chinese **16**: 2247
Weevil **21**: 2903–2904
  Cotton boll **21**: 2904
  Figwort **21**: 2903, 2904
  Gorse **21**: 2903, 2904
  Grain **21**: 2904
  Large pine **21**: 2903, 2904
  Nut **21**: 2903, *2903*, 2904, *2904*
Wheat-stem borer **16**: 2241
Whelk (drill) **7**: 927; **21**: 2917–2918
  Channeled **21**: 2917, 2918
  Common (white whelk; buccinum; buckie) **21**: 2917, 2918, *2918*
  Red (almond whelk) **21**: 2917
  Red **21**: 2917
  Rough (whelk-tingle; oyster drill) **21**: 2917
Whirligig beetle **21**: 2921–2922
White butterfly **21**: 2928–2929
  Green-veined white **21**: *2928*
  Large white **21**: 2928
  Marbled white **21**: 2929
  Margined white **21**: 2928–2929
  Mustard white **21**: 2928
  Pine white **21**: 2928
  Small white (European cabbage butterfly) **9**: 1280; **21**: 2928
  West Virginia white **21**: 2928, 2929
  Wood white **21**: 2929
Whitefly **3**: 416; **21**: 2932–2933
  Cabbage **21**: 2932, 2933
  Citrus **21**: 2932, 2933
  Greenhouse **21**: 2932, *2932*, 2933
Wood ant **1**: 68; **21**: 2964–2965
  Common **21**: 2964–2965
  Hairy (meadow ant) **21**: 2964, 2965
  Red **21**: 2965
Wood louse (pill bug; sow bug; slater; roly-poly) **21**: 2973–2974
  Pill **21**: 2973
Worm, Horsehair (hair worm; Gordian worm) **9**: 1237–1238
Worm shell (tube mollusk; tube snail; vermetid) **21**: 2989–2990
  Scaled **21**: 2990

Yellow jacket (*hornet*) **9**: 1233
Yellowjacket (*wasp*) *see* Wasp, Vespula

MAMMALS

Aardvark (ant bear) **1**: 16–17
Aardwolf (werewolf) **1**: 18–19
Acouchi **1**: 32
Addax (white antelope; screwhorn antelope) **1**: 24–25; **5**: 665
Agouti **1**: 32–33
  Black **1**: *33*
  Central American **1**: *32*, 33
Agouti (genus *Agouti*) *see* Paca
Alpaca **11**: 1470, 1471, 1472; **12**: 1672
Aner *see* Serval
Angwantibo *see* Potto (*African*), Golden
Anoa **9**: 1288
  Lowland **3**: 305, 306, 307; **9**: 1286
  Mountain **3**: 305, 306, 307; **9**: 1286
Anomalure *see* Scaly-tailed squirrel
Ant bear *see* Aardvark; Anteater, Giant
Anteater **1**: 57–59
  Banded (numbat; marsupial anteater) **1**: 128–129
  Giant (ant bear) **1**: 57, *57*, 58, 59
  Scaly *see* Pangolin

Silky (two-toed anteater) **1**: 57, 58, *58*, 59
Antelope
  Beira **1**: 61; **2**: 189–190
  Blue **1**: 63
  Four-horned *see* Chousingha
  Indian *see* Blackbuck
  Pronghorn **1**: 60; **15**: 2049–2051
  Roan **16**: 2215, 2216
  Royal **1**: 60; **18**: 2571, 2572
  Sable **1**: 61; **13**: *1830*; **16**: 2215–2216
  Screwhorn *see* Addax
  Tibetan *see* Chiru
  White *see* Addax
  Zebra *see* Duiker, Zebra
  see also Addax; Blackbuck; Blesbok; Bontebok; Bushbuck; Dibatag; Dik-dik; Duiker; Eland; Gazelle; Gerenuk; Hartebeest; Impala; Klipspringer; Kob; Kudu; Lechwe; Nilgai; Nyala; Oribi; Oryx; Reedbuck; Springbok; Steenbok; Suni; Waterbuck; Wildebeest
Antelopes 1: 60–63
Apara *see* Armadillo, Three-banded
Ape **1**: 70–73
  Barbary (Barbary macaque) **1**: 136–137; **12**: 1643
  see also Apes; Chimpanzee; Gibbon; Gorilla; Orangutan
Apes 1: 70–73
Argali *see* Sheep, Marco Polo
Armadillo **1**: 85–87
  Fairy **1**: 85, 86
  Giant **1**: 85, 86, 87
  Greater fairy **1**: 87
  Hairy (peludo) **1**: 86, *86*
  Lesser fairy (pichiciago) **1**: 85, 86, 87
  Naked-tailed **1**: 85, 86
  Nine-banded **1**: 85, *85*, 86, 87
  Pygmy (pichi) **1**: 85, 86, 87
  Six-banded **1**: 85, 86, 87
  Three-banded (apara) **1**: 85, 86, 87
Ass
  Domestic (donkey) **21**: 2939, *2940*
Awoura **16**: 2251
Aye-aye **1**: 111–112

Baboon **1**: 115–117; **12**: 1644
  Gelada **1**: 115, 117; **12**: 1643, *1644*, 1645
  Guinea **1**: 115, 116
  Hamadryas (sacred baboon) **1**: 115, *115*, 116; **12**: 1644
  Savanna (chacma baboon; olive baboon; yellow baboon) **1**: 115, *116*, 117
Badger **1**: 120–122
  American **1**: 120, *120*, 122
  Eurasian (brock) **1**: 120, 121, *121*, 122, *122*; **19**: 2653
  Ferret (tree badger; pahmi) **1**: 120
  Hog (sand badger; hog-nosed badger) **1**: 120
  Honey (ratel) **1**: 120; **9**: 1206–1207
  Malaysian stink **1**: 120
  Palawan stink (Calamanian stink badger) **1**: 120
  Stink **1**: 120
Baji *see* River dolphin, Chinese
Bandicoot **1**: 130–131
  Desert (*extinct*) **1**: 130
  Eastern barred **1**: *130*
  Giant **1**: 130, 131
  Long-nosed **1**: 131
  Northern brown **1**: *131*
  Pig-footed (*extinct*) **1**: 130
  Rabbit-eared (greater bilby) **1**: 130–131

Short-nosed **1**: 131
  Western barred **1**: 130
Bandicoot rat **1**: 132–133
  Greater (large bandicoot rat; Indian bandicoot rat) **1**: 132, *132*, 133
  Lesser (Indian mole rat) **1**: 132, 133
  Short-tailed (pest rat) **1**: 132
Band-tailed cat *see* Ringtail
Banteng (tsaine) **3**: 306, *306*, 307; **7**: 936, 937; **21**: 3018
Barbary ape **1**: 136–137
Barfani chita *see* Snow leopard
Bat
  Bumblebee **2**: 167
  Desert **20**: 2822, 2823
  False vampire **6**: 840
  Fish-eating (greater bulldog bat) **2**: 167; **6**: 840–841
  Fishing **2**: 167; **20**: 2822, 2823
  Frog-eating **2**: 167
  Fruit *see* Fruit bat
  Fruit-eating **2**: 169
  Ghost **2**: *169*; **19**: 2702
  Hairless (giant naked bat) **8**: 1115–1116
  Hoary **20**: 2821, 2822
  Horseshoe *see* Horseshoe bat
  Long-eared *see* Long-eared bat
  Mexican fishing **6**: 840
  Mouse-tailed (rat-tailed bat; long-tailed bat) **19**: 2702, *2702*, 2703
  Pallid **20**: *2821*
  Pipistrelle *see* Pipistrelle
  Red **20**: 2821, 2822, *2822*
  Sheath-tailed (sac-winged bat) **19**: 2702, 2703
  Spotted (pinto bat) **2**: 168; **20**: 2821
  Tomb **19**: 2702–2703
  Townsend's big-eared **2**: 168; **20**: 2821
  Vampire *see* Vampire bat
  Vesper **20**: 2821–2823
  Whiskered **20**: 2822, 2823
  see also Bats
Bats 2: 167–169
Beaked whale **2**: 170–171; **21**: 2911
  Blainville's **2**: *170*
  Cuvier's **2**: *171*
  Shepherd's (Tasman beaked whale) **2**: 170
Bear
  Asiatic black (Himalayan black bear; moon bear) **1**: 95–96; **2**: 176
  Black (American black bear; Kermode bear; glacier bear; Everglades bear) **2**: 176, *176*, 222–223; **12**: 1672; **19**: 2653
  Brown (grizzly bear; Kodiak bear; European brown bear; Tibetan brown bear) **2**: 174–175; **3**: 292–295; **12**: 1672; **19**: 2616, 2617, 2653
  Cave (*extinct*) **3**: 295
  Honey (kinkajou) *see* Kinkajou
  Polar (ice bear) **1**: 82; **2**: 174, *175*, 176; **14**: 1998–2000
  Sloth (Indian bear; bhalu) **2**: 176; **17**: 2403–2405
  Spectacled (Andean bear) **2**: 174, 176; **12**: 1672; **18**: 2470–2471
  Sun (honey bear; Malaysian bear; bruang) **18**: 2562–2564
  see also Bears; Panda, Giant
Bears 2: 174–176
Beaver **2**: 177–179; **19**: 2616
  Canadian **2**: 177, 178, 179, *179*
  European **2**: 177, *177*, 178, 179; **21**: 2906
  Mountain *see* Mountain beaver

Beira *see* Antelope, Beira
Beishung *see* Panda, Giant
Beluga whale *see* Whale, Beluga
Bettong **15**: 2116
 Brush-tailed (woylie) **15**: 2116, *2116*
Bighorn sheep **2**: 198–200
 Mexican (desert bighorn sheep) **2**: 198, 199
 Nelson's **2**: 199
 Rimrock **2**: 199
 Rocky Mountain **2**: 199
 Weems' **2**: 199
Bilby
 Greater *see* Bandicoot, Rabbit-eared
 Lesser (*extinct*) **1**: 130
Binturong **2**: 201–202
Bison **2**: 215–217
 American (buffalo) **2**: *215*, 216, *217*; **3**: 305, 306, 307; **8**: *1038*, 1040
 European (wisent) **2**: 215, 216; **3**: 305, 306, 307
 Indian *see* Gaur
 Plains **2**: 215, 216
 Wood **2**: 215, 216
Blackbuck (Indian antelope) **1**: 60; **2**: 226–227; **10**: 1382
Blackfish (*whale*) *see* Killer whale
Bladdernose *see* Seal, Hooded
Blesbok **2**: 239–240, 259, 260; **6**: 759; **16**: 2216
Blesmol **12**: 1628, 1629
Bluebuck **1**: 63
Blue buck (*extinct*) **16**: 2216
Blue buck *see* Nilgai
Blue bull *see* Nilgai
Boar, Wild (wild pig) **19**: 2653, *2655*; **21**: 2907, 2942–2944
Bobak (Himalayan marmot; steppe marmot) **11**: 1568, 1569
Bobcat **2**: 253–254
Bongo **1**: 63; **10**: 1389–1390
Bonobo (pygmy chimpanzee) **1**: *70*, 71, 72; **4**: 439–440, 441
Bontebok **1**: 63; **2**: 239, 259–260; **6**: 759; **16**: 2216
Boodie **15**: 2115, 2116
Boto *see* River dolphin, Amazon
Bottlehead *see* Whale, Northern bottlenose
Brock *see* Badger, Eurasian
Brocket **18**: 2463, 2464
Bruang *see* Bear, Sun
Buffalo 3: 305–307
Buffalo
 Cape (African buffalo; forest buffalo; bushcow) **3**: 305, 306, 307, *307*, 370–372
 Indian (Asian water buffalo) **3**: 305, 306, 307; **9**: 1286–1288
 see also Bison, American; Gaur; Yak; Zebu
Burramys (mountain pygmy possum) **5**: 703
Burro *see* Wild ass, African
Bushbaby **3**: 329–330
 Greater **3**: 330
 Needle-clawed **3**: 330
 Senegal **3**: 329, 330
 South African **3**: 329, 330
 Thick-tailed greater **3**: 329, 330
Bushbuck **1**: 63; **3**: 331–332
Bushcow *see* Buffalo, Cape
Bushdog **3**: 333–334; **17**: 2357
Bushpig (African bush pig; Red River hog) **3**: 335–336

Cacomistle *see* Ringtail
Calling hare *see* Pika

Camel **3**: 361–364; **5**: 665
 Arabian (one-humped camel; dromedary) **3**: 361, 362, 364
 Bactrian (two-humped camel) **3**: 361, 362, *363*, 364
Canecutter *see* Rabbit, Swamp
Capuchin monkey (sapajou; ring-tail monkey) **3**: 379–381; **12**: 1645
 Tufted (weeping capuchin) **3**: 379, 380, 381, *381*
 Wedge-capped **3**: 379, 380
 White-faced **3**: 379, *379*, 380
 White-fronted **3**: 379, 380, *380*
Capybara (water pig; water cavy) **3**: 382–384; **8**: 1040
Caracal (desert lynx; rock cat) **3**: 385–386, 404
Caribou (reindeer) **3**: 391–394; **5**: 655, 657; **20**: *2776*, 2778, *2778*; **21**: 2958
Cat
 Band-tailed *see* Ringtail
 Coon *see* Ringtail
 Desert **3**: 404
 Iriomote **3**: 405
 Margay *see* Margay cat
 Marsupial *see* Marsupial cat
 Miner's *see* Ringtail
 Mountain *see* Ringtail
 Native *see* Marsupial cat
 Otter *see* Jaguarundi
 Pallas' **3**: 404
 Ring-tailed *see* Ringtail
 Rock *see* Caracal
 Sand **3**: 403, 404, *406*
 Servaline *see* Serval
 Weasel *see* Jaguarundi
 see also Bobcat; Caracal; Cats; Cheetah; Civet; Genet; Jaguar; Jaguarundi; Leopard; Lion; Lynx; Ocelot; Palm civet; Puma; Serval; Tiger; Wildcat
Cat-bear *see* Panda, Red
Cats 3: 403–406
Cat-squirrel *see* Ringtail
Cattle
 Brahman *see* Zebu
 Western **21**: 3018
Cavy **8**: 1099, 1100
 Common **8**: 1100, *1101*
 Mountain **8**: 1099, 1100
 Rock (moco) **8**: 1099, 1100, 1101
 Spotted *see* Paca
 Yellow-toothed (cui) **8**: 1099, 1100, *1100*
Chamois **3**: 421–422; **19**: 2623
Cheetah **3**: 403, 406, 427–429; **8**: 1040
Chennai *see* Dhole
Cherogil (daman) **9**: 1270
Chevrotain **18**: 2572
Chickaree **4**: 436–438
Chimpanzee **1**: 70, 71, 72, 73; **4**: 439–442
 Common **4**: 439, 440, 442
 Pygmy *see* Bonobo
Chinchilla **4**: 443–445; **12**: 1672
 Long-tailed **4**: 444
 Mountain *see* Viscacha, Mountain
 Short-tailed **4**: 444
Chipmunk **4**: 448–450
 Eastern **4**: 448, *448*, 449, 450; **19**: 2653
 Least **4**: 449
 Siberian **4**: 448, 449
 Townsend's **4**: 449, *449*
 Western **4**: 448, 450
Chiru (Tibetan antelope) **16**: 2217, 2218, *2218*
Chital *see* Deer, Axis
Chousingha (four-horned antelope) **1**: 60; **13**: 1753, 1754

Chuck (*mammal*) *see* Groundhog
Civet **4**: 460–461
 African **4**: 460, *460*, 461
 Indian (oriental civet) **4**: 460, 461
 Malabar large-spotted **4**: 461
 Malagasy (fanaloka) **4**: 460, 461
 Otter **4**: 460, 461
 Palm *see* Palm civet
 Two-spotted *see* Palm civet, African
Coati **4**: 478–479
 Island **4**: 479
 Mountain **4**: 479
 Ring-tailed (coatimundi) **4**: 479
 White-nosed (coatimundi) **4**: 480
Coatimundi *see* Coati, Ring-tailed; Coati, White-nosed
Colobus monkey **4**: 501–503; **12**: 1644
 Angola pied **4**: 501, 502
 Black **4**: 501, 502, 503
 Black and white (pied colobus) **4**: 502, 503
 Central African red **4**: 502
 Guereza **4**: 501, 502, *502*, *503*
 Olive **4**: 501, 502, 503
 Red **4**: 501, *501*, 502, 503
 Tana River red **4**: 503
 Western red **4**: 502
 Zanzibar red **4**: 503
Coney *see* Hyrax
Coon *see* Raccoon
Coon cat *see* Ringtail
Cottontail **4**: 545–548
 Desert **4**: 545, 546, *547*
 Dice's **4**: 546
 Eastern **4**: 545, 546
 New England **4**: 545, 546
 Nuttall's **4**: 545, 546, *546*
 Omilteme **4**: 546
 Tres Marias **4**: 546
Cougar *see* Puma
Coyote **3**: *425*, 426; **4**: 558–560
Coypu **4**: 561–563
Cui *see* Cavy, Yellow toothed
Cuscus **5**: 629–630
 Black-spotted **5**: 630
 Common **14**: *1930*
 Gray **5**: 630
 Obi **5**: 630
 Silky **5**: 630
 Southern common **5**: 630
 Spotted **5**: 629–630, *629*
 Woodlark Island **5**: 630
Cusimanse **12**: 1636

Daman (cherogil) **9**: 1270
Dassie *see* Hyrax, Cape
Deer 5: 655–658
Deer
 Andean *see* Guemal
 Axis (chital; Ceylon deer; spotted deer; Ganges stag) **1**: 107–108; **21**: 2907
 Barking *see* Muntjac
 Brow-antlered (thamin; Eld's deer; Manipur brow-antlered deer) **2**: 284–285
 Chinese water (Korean water deer; water deer; Chinese river deer; Yangtze River deer; river deer) **4**: 446–447; **5**: 655
 Fallow (Persian fallow deer) **5**: 656, 657; **6**: 794–796
 Hog **1**: 107
 Marsh (swamp deer) **18**: *2462*, 2463, 2464
 Mouse **5**: 655
 Mule (black-tailed deer) **3**: 426; **12**: 1689–1691
 Pampas **18**: 2462–2463, *2463*, 2464
 Père David's **14**: 1919–1920
 Prince Alfred's **5**: 658

 Red *see* Wapiti
 Rib-faced *see* Muntjac
 Roe **5**: 657; **16**: 2201–2203; **17**: 2375
 Schomburgk's (*extinct*) **2**: 285; **5**: 658
 South American **18**: 2462–2464
 Swamp **5**: *657*
 Tufted **12**: 1696
 White-tailed **5**: 656, *656*; **12**: 1689, 1690; **19**: 2653, 2682
 see also Brocket; Caribou; Pudu; Sika; Wapiti
Deer mouse (vesper mouse) **5**: 659–661
 Range **5**: 659, 660
Defassa *see* Waterbuck
Desman **5**: 666–667
 Pyrenean **5**: 666, 667
 Russian **5**: 666, 667, *667*
Devil *see* Tasmanian devil
Devilfish (*whale*) *see* Whale, Gray
Dhole (wild dog; Asiatic wild dog; red wolf; red dog; chennai) **5**: 671–672, 677, 690
Dibatag (Clarke's gazelle) **5**: 673–674; **7**: *952*; **10**: 1382
Dik-dik **1**: 61; **5**: 675–676
 Guenther's **5**: 675, 676, *676*
 Kirk's **5**: 676
 Salt's **5**: 675, 676
 Silver **5**: 676
Dingo **5**: 677–679; **10**: 1336; **19**: 2641
Dog
 Asiatic wild *see* Dhole
 Cape hunting (African wild dog; African hunting dog) **3**: 373–375; **5**: 689; **8**: 1041
 Crab-eating *see* Fox, Crab-eating
 Raccoon *see* Raccoon dog
 Red *see* Dhole
 Short-eared (small eared dog; short-eared zorro) **17**: 2356–2357
 Singing **5**: 677
 Wild *see* Dhole
 see also Bushdog; Coyote; Dingo; Dogs; Fox; Jackal; Wolf
Dogs 5: 689–692
Dolphin **5**: 695, *696*; **13**: 1771–1772
 Atlantic spotted **5**: 698
 Bottlenose (bottle-nosed dolphin; common porpoise; gray porpoise; black porpoise) **2**: 266–268; **4**: 507
 Common (white-bellied porpoise; saddleback dolphin; crisscross dolphin; hourglass dolphin) **4**: 506–509, *509*
 Dusky **5**: 698, *698*
 Ganges **5**: 697; **16**: 2177, 2178
 Hector's **5**: 698
 Indus **5**: 697; **16**: 2177, 2178
 Irrawaddy **21**: 2911
 La Plata (Franciscana) **5**: 697; **16**: 2177, 2178
 Pantropical spotted **5**: 698
 Spinner **5**: 698
 Spotted **5**: *697*
 Striped **5**: 698
 see also Dolphins and porpoises; False killer whale; Killer whale; Pilot whale; Porpoise; Whale, Beluga
Dolphins and porpoises 5: 695–698
Donkey (domestic ass) **21**: 2939, *2940*
Dormouse **5**: 699–701
 African **5**: 699
 Desert **5**: 699
 Edible (fat dormouse) **5**: 699, 700–701, *701*
 Forest **5**: 699
 Garden **5**: 699, *699*, 701

Dormouse (*continued*)
    Hazel (common dormouse;
       sleeper) **5**: 699–700, *700*, 701;
      19: 2653
    Japanese **5**: 699
    Mouse-tailed **5**: 699
Douc **10**: 1410, 1411
Douroucouli **19**: 2697
Drill (*monkey*) **1**: 115; **11**: 1536–1537,
    1538
Dromedary *see* Camel, Arabian
Duckbill *see* Platypus
Duckmole *see* Platypus
Dugong (sea cow; siren) **6**: 724–725
Duiker **1**: 61, 63; **6**: 726–728
    Ader's **6**: 726, 727
    Blue **6**: 726, 728; **18**: 2572
    Common (gray duiker) **6**: 726, *726*,
       727, 728
    Jentink's **6**: 726, 727
    Red **6**: 727, 728, *728*
    Yellow-backed **6**: 726
    Zebra (zebra antelope) **6**: 727
Dunnart, Fat-tailed **11**: 1577, *1578*
Dwarf antelope **18**: 2571
    Bates' **18**: 2571, 2572
Dzo *see* Yak, Domestic

Echidna *see* Spiny anteater
Eland **1**: 62; **6**: 757–759
    Common (Cape eland) **6**: 757, 758
    Giant (Derby eland) **1**: 60; **6**: 758
Elephant **6**: 767–770
    African **6**: 767–769, *767*, 770; **15**: 2113
    Indian (Asian elephant) **6**: 767,
       768, *768*, 769
Elephant seal **6**: 771–773
    Northern **6**: 771–772, *772*, 773
    Southern **6**: 771, 772, *773*; **16**: 2285,
       2286, 2288
Elephant shrew **6**: 774–776
    Checkered **6**: 774, 775, 776
    Forest **6**: 775, *776*
    Long-eared (lesser elephant
       shrew) **6**: *774*, 775
    Yellow-rumped (golden-rumped
       elephant shrew) **6**: 774, 775, 776
Elk
    Irish (*extinct*) **5**: 656
    see also Moose; Wapiti
El tigre *see* Jaguar
Ermine (short-tailed weasel; stoat)
    6: 785–787; **14**: 2003
Euro *see* Wallaroo, Common
Eyra *see* Jaguarundi

False killer whale (false pilot whale;
    pseudorca) **6**: 797–798
Fanaloka (Malagasy civet) **4**: 460, 461
Ferret **14**: 2003
    Black-footed **2**: 228–229
Fire-fox *see* Panda, Red
Fisher (pekan) **11**: 1579, 1580–1581,
    1580; **19**: 2653, 2729
Fitchew *see* Polecat
Flying fox **2**: 167; **7**: 877–879; **15**: 2110
    Gray-headed **7**: 879
Flying lemur **7**: 882–883
Flying mouse *see* Scaly-tailed squirrel
Flying squirrel (*gliding possum*) *see*
    Gliding possum, Lesser
Flying squirrel (*squirrel*) **7**: 884–886
    Eurasian (European flying
       squirrel) **7**: 884, 885
    Kashmir **7**: 884
    North American **7**: 884, 885, 886;
      19: 2616
    Pigmy **7**: 884
    Scaly-tailed *see* Scaly-tailed Squirrel
    Southern **7**: *884*

Fossa **17**: 2373
Foul-marten *see* Polecat
Fox
    Arctic **1**: 83–84; **5**: 691
    Bat-eared **2**: 165–166; **5**: 665, 689
    Blue **1**: 83
    Crab-eating (crab-eating dog;
       crab-eating zorro; wood fox)
      4: 566–567
    Culpeo **5**: 689, 691
    Fennec (fennec) **2**: 165; **5**: 665;
      6: 809–811
    Gray (tree fox) **3**: 426; **5**: 689;
      8: 1045–1046
    Indian (Bengal fox) **5**: *690*
    Island (beach fox) **8**: 1045, 1046
    Raccoon *see* Ringtail
    Red **5**: 689; **15**: 2132–2134; **19**: 2653
    Sand (pale fox) **5**: 689
    San Joaquin kit fox **4**: 560
    South American gray (zorro)
      5: *689*
    Swift (kit fox) **4**: 560; **5**: *691*, *692*
Fox-cat *see* Panda, Red
Franciscana *see* Dolphin, La Plata
Fruit bat (*New World*) **7**: 906, *907*
    Short-tailed **7**: *907*
Fruit bat (*Old World*) **7**: 906–907
    Arabian straw-colored **7**: 908
    Epauletted **7**: 907–908
    Franquet's **7**: 908
    Hammerhead **7**: 908
    Pollen-eating (long-tongued bat)
      7: 908
    Short-eared **7**: 907
    Straw-colored **7**: 908
    Tube-nosed **7**: 908
    see also Flying fox
Fur seal **7**: 916–918; **16**: 2285
    Antarctic (Kerguelen fur seal)
      7: *916*, 917
    Northern (Pribilof fur seal; Alaska
      fur seal) **7**: 916, 917, 918; **16**: *2288*
    South African **7**: *917*
    South American **7**: 918
    Southern **7**: 916, 917, 918, *918*

Galago
    Demidoff's dwarf **3**: 329, 330
    Dwarf **3**: 330
    Silver **3**: 329
Ganges stag *see* Deer, Axis
Gato *see* Jaguarundi
Gaur (Indian bison; seladang) **3**: 305,
    306; **7**: 936–938; **21**: 3018
    Domestic (gayal; mithan) **7**: 937, *938*
Gayal *see* Gaur, Domestic
Gazelle **1**: 61; **7**: 939–942; **10**: 1382
    Clarke's *see* Dibatag
    Dama **7**: 939, 940
    Dorcas **7**: 939, 940, 942
    Giraffe-necked *see* Gerenuk
    Goitered **7**: 940, 941
    Grant's **1**: 62; **7**: 939, *939*, 940, 941
    Mongolian **7**: 939, 942
    Przewalski's **7**: 939
    Sand **7**: 942
    Soemmerring's **7**: 939, 940
    Speke's **7**: 939, 940
    Thomson's (red gazelle) **1**: 61, *61*;
      7: 939, 940, *940*, 941, *942*; **8**: 1040
    Tibetan **7**: 939
    Waller's *see* Gerenuk
Gemsbok (beisa oryx; tufted oryx; true
    gemsbok) **13**: 1808, 1809, *1810*
Genet **7**: 946–948
    Aquatic **7**: 946, 947, 948
    Blotched (tigrine genet) **7**: 946, 947
    Common (feline genet; small-
      spotted genet) **7**: 946, 947, 948

    Ethiopian **7**: 946
    Giant **7**: 946
    Rusty-spotted **7**: 946
Gerbil (sand rat) **2**: 166; **5**: 665;
    7: 949–951
    Fat sand rat **7**: 950, 951
    Great **7**: 949, 951
    Greater Egyptian **7**: *949*
    Indian **7**: 950, 951
    Mongolian (clawed jird) **7**: *950*
    Przewalski's **7**: 949
Gerenuk (giraffe-necked gazelle;
    Waller's gazelle) **7**: 952–953
Giant rat **7**: 964–965
    African *see* Pouched rat, African
      giant
Gibbon **1**: 71, 73; **7**: 968–971
    Agile **7**: 968
    Black-capped (pileated gibbon)
      7: 968, 970
    Bornean (Mueller's gibbon) **7**: 968,
      970
    Concolor (crested gibbon) **7**: 968,
      970, *971*
    Hoolock **7**: 968, 970
    Javan (silver gibbon; moloch
      gibbon) **7**: 968, 970
    Kloss's (dwarf gibbon) **7**: 968
    White-handed (lar gibbon) **1**: *71*;
      7: 968, *968*, 969, *969*
Giraffe **7**: 975–977
Glider *see* Gliding possum
Gliding possum **7**: 984–986
    Greater (greater glider) **7**: 984–985,
      986
    Lesser (lesser glider; sugar glider;
      flying squirrel; sugar squirrel;
      lesser flying phalanger)
      7: 984–985, *984*, 986, *986*
    Pygmy glider (feathertail glider)
      7: 984, 986
Glutton *see* Wolverine
Gnu *see* Wildebeest
Goat **2**: 951
    Domestic **21**: 2951, *2951*, 2952
    Rocky Mountain **16**: 2199–2200;
      19: 2623
    Wild (bezoar; Sind ibex; Chiltan
      goat; Cretan goat) **9**: 1272;
      21: 2951–2953
Golden mole **7**: 1004–1005
    Cape **7**: 1005
    Grant's **7**: 1004, *1004*
    Large **7**: 1004
Gopher, Pocket *see* Pocket gopher
Goral **8**: 1021–1022; **19**: 2623
    Brown (Red goral) **8**: 1021, 1022
    Himalayan **8**: 1021, *1021*, 1022
    Long-tailed **8**: 1021, 1022
Gorilla **1**: 72, 73; **8**: 1023–1026
    Lowland (eastern lowland gorilla;
      western lowland gorilla) **1**: 71;
      8: 1023–1024, *1023*
    Mountain **1**: 73; **8**: 1024–1025, *1026*
Grampus *see* Killer whale
Groundhog (woodchuck; chuck;
    whistle pig) **8**: 1075–1077;
    11: 1568; **12**: 1672
Ground squirrel (rock squirrel;
    suslik) **8**: 1040, 1078–1080; **12**: 1595
    Antelope **8**: 1080, *1080*
    Arctic **8**: 1080
    California **8**: 1078, 1079
    Franklin's **8**: 1078
    Golden-mantled **8**: 1079
    Harris's antelope **8**: 1079, *1079*
    Thirteen-striped **8**: 1078–1079
Grysbok (Cape grysbok) **18**: 2526,
    2527
    Sharpe's **18**: 2526, 2527

Guañaco **11**: 1470, 1471, 1472, *1472*;
    12: 1672
Guayonoche *see* Ringtail, Central
    American
Guemal (Andean deer) **18**: 2463–2464
Guenon **8**: 1091–1093; **12**: 1643
    Red *see* Monkey, Patas
Guereza *see* Colobus monkey, Guereza
Guinea pig (common cavy;
    Patagonian hare; moco; cui)
    8: 1099–1101
    Domestic **8**: 1099, *1099*, 1100
Gundi **8**: 1104–1105

Hamster **8**: 1129–1131
    Common (European hamster)
      8: 1129, *1129*, 1130
    Dwarf **8**: 1129, 1130, *1130*
    Golden **8**: 1129, 1130, 1131, *1131*
    Gray **8**: 1129, 1130–1131
    Romanian **8**: 1131
Hare **8**: 1135–1138
    Belgian **8**: 1135
    Blue **8**: 1137
    Brown (common hare)
      8: 1135–1136, *1135*
    Calling *see* Pika
    Jumping *see* Springhare
    Northern (mountain hare; blue
      hare; varying hare) **8**: 1136, 1137
    Piping *see* Pika
    Plains *see* Jack rabbit, White-tailed
    Prairie *see* Jack rabbit, White-tailed
    Snowshoe **8**: 1136, *1137*, 1138,
      1138; **11**: 1508
    Sumatran *see* Rabbit, Sumatran
    Whistling *see* Pika
Hartebeest **1**: 62; **8**: 1147–1149
    Common (bubal hartebeest; red
      hartebeest; North African
      hartebeest) **8**: 1147, *1147*, 1148,
      1148, 1149, *1149*
    Hunter's **1**: 63
    Lichenstein's **8**: 1147, 1148
Harvest mouse **9**: 1158–1160
    American **9**: 1158, 1159–1160
    Eastern **9**: 1158
    Old World **9**: 1158–1159, *1158*,
      1160, *1160*
    Salt marsh **9**: 1158
    Western **9**: 1158
Hedgehog (hedgepig; urchin) **1**: 27;
    2: 232; **9**: 1171–1173
    Eurasian **9**: 1171–1173, *1172*
    Long-eared **9**: *1171*
Hinney **21**: 2941
Hippopotamus (kiboko) **9**: 1192–1195
    Pygmy **9**: 1192, *1193*, 1195
Hirola **1**: 63
Hog
    Giant forest (western forest hog;
      eastern forest hog; giant hog)
      7: 958–959
    Red River *see* Bushpig
Honey bear *see* Kinkajou
Horse **8**: 1039
    Przewalski's **8**: 1040, *1041*;
      15: 2052–2054
Horseshoe bat **2**: 168; **9**: 1239–1240;
    16: 2254
    Greater **2**: *168*; **9**: 1239, *1239*, 1240
    Large-eared **9**: 1240, *1240*
    Lesser **9**: 1240
Howler monkey **9**: 1252–1254; **12**: 1645
    Black **9**: 1252, *1252*, 1253
    Brown **9**: 1252, 1253
    Mantled **9**: 1252, 1253
    Red **9**: 1252, 1253, *1254*
    Red-handed **9**: 1253

Huemul
  North Andean **5**: 658
  South Andean **5**: 658
Hutia **9**: 1263–1264
  Bushy-tailed **9**: 1264
  Cabrera's **9**: 1264
  Cuvier's **9**: 1264
  Dwarf **9**: 1264
  Garrido's **9**: 1264
  Large-eared **9**: 1264
  Venezuelan **9**: 1263
Hyena **8**: 1040–1041; **9**: 1267–1269
  Brown **9**: 1267, 1268, 1269
  Spotted (laughing hyena)
    3: 374–375; **9**: 1267–1269, *1267,*
    1269
  Striped **9**: 1267, *1268*, 1269
Hyrax (coney) **9**: 1270–1271
  Bush (gray hyrax; yellow-spotted
    hyrax) **9**: 1270, 1271
  Cape (dassie) **9**: 1270, *1270*
  Rock **9**: 1270, 1271
  Tree **9**: 1270, 1271

Ibex **9**: 1272–1273; **12**: 1672
  Alpine (Nubian ibex; Siberian
    ibex) **9**: 1272, *1272*, 1273, *1273*;
    21: 2951
  East Caucasus (tur) **9**: 1272, 1273;
    21: 2951
  Spanish **9**: 1272, 1273; **21**: 2951
  Walia **9**: 1272, 1273; **21**: 2951
  West Caucasus (tur) **9**: 1272, 1273;
    21: 2951
Impala (pallah) **1**: 61; **9**: 1284–1285
Ingwenkala *see* Serval
Inyala *see* Nyala

Jackal **8**: 1041; **9**: 1291–1293
  Black-backed **5**: 692; **9**: 1291–1292,
    1292, 1293, *1293*
  Golden (Asiatic jackal) **5**: 689, 691,
    692; **9**: 1291–1292, *1291*, 1293
  Side-striped **9**: 1291, 1292, 1293
Jack rabbit **8**: 1138; **10**: 1304–1306;
    18: 2561
  Antelope **10**: 1304, 1305
  Black-tailed (jackass hare)
    10: 1304, 1305, *1305*
  White-sided **10**: 1304, 1305
  White-tailed (plains hare; prairie
    hare) **10**: 1304, *1304*, 1305, *1306*
Jaguar (onca; el tigre) **1**: 33; **3**: 406;
    10: 1310–1312
Jaguarondi *see* Jaguarundi
Jaguarundi (jaguarundi; otter cat;
    weasel cat; gato; eyra)
    10: 1313–1314
Jerboa (jumping rat; kangaroo rat)
    5: 665; **10**: 1323–1324
  Marsupial *see* Marsupial mouse
Jird
  Clawed *see* Gerbil, Mongolian
  Midday **7**: 951
Kamoshika *see* Serow, Japanese
Kangaroo **10**: 1333–1336
  Eastern gray **10**: 1333, 1334,
    1335–1336, *1336*
  Gray (great gray kangaroo)
    10: 1333, 1334
  Rat *see* Rat kangaroo
  Red **10**: 1333, *1333*, 1334–1336, *1335*
  Tree *see* Tree kangaroo
  Western gray **10**: 1333, 1334
Kangaroo mouse (hopping mouse;
    jumping mouse; jerboa mouse)
    10: 1337–1338
  American **10**: 1338
  Australian **10**: 1338
  New Guinea **10**: 1337, 1338

Kerinci, Kelinci *see* Rabbit, Sumatran
Kiang **21**: 2939, 2940, *2941*
Kiboko *see* Hippopotamus
Killer whale (orca; grampus; great
    killer whale; blackfish; sea wolf)
    1: 82; **2**: 194; **5**: 695; **6**: 773; **7**: 918;
    10: 1350–1352; **16**: 2211; **21**: 3008
  False *see* False killer whale
Kinkajou (honey bear; nightwalker;
    potto) **10**: 1363–1364
Klipspringer (cliff-springer) **1**: 61;
    10: 1374–1375
Koala (koala bear) **10**: 1378–1380
Kob **1**: 60, 62; **10**: 1381–1382
Koodoo *see* Kudu
Kouprey (Cambodian forest ox)
    3: 307; **7**: 936, 938
Kudu (koodoo; grand kudu) **3**: 331;
    10: 1389–1391
  Greater **10**: 1389, *1389*, 1390, *1390*
  Lesser **10**: 1389, 1390
Kulan **21**: 2939, 2940
Kultarr *see* Marsupial mouse

Laba *see* Paca
Langur **10**: 1410–1412; **12**: 1644
  Brow-ridged **10**: 1411
  Delacour's **10**: 1412
  Douc **12**: 1645
  Golden **10**: 1412
  Golden snub-nosed **10**: 1410
  Hanuman **10**: 1410, 1411, *1411*;
    12: 1643
  Maroon **10**: 1412
  Pig-tailed **10**: 1410, 1411
  Purple-faced **10**: 1410, 1412
  Snub-nosed **10**: 1411
  Spectacled (dusty langur) **10**: 1410,
    1412
  White-bellied **10**: *1412*
  White-headed **10**: 1412
Lechwe (red lechwe; Kafue lechwe;
    black lechwe) **1**: 62;
    10: 1425–1427, *1426*; **21**: *2907*
  Nile (Mrs Grey's lechwe) **1**: 62, 63;
    10: 1425, 1426
Lemming **10**: 1432–1434; **17**: 2438;
    20: 2778
  Bog **10**: 1432, 1433
  Brown **10**: 1433
  Collared **10**: 1432, *1432*, 1433, 1434;
    20: *2777*
  Norwegian **10**: 1432, 1433, 1434
  Steppe **18**: 2528–2529
  Wood **10**: 1432, 1433
Lemur **10**: 1435–1437
  Black **10**: 1435, 1437
  Broad-nosed gentle **10**: 1435
  Crowned **10**: *1437*
  Dwarf **10**: 1435, 1436, 1437
  Flying *see* Flying lemur
  Fork-crowned **10**: 1435
  Gentle **10**: 1435
  Hairy-eared dwarf **10**: 1435
  Mouse **10**: 1435, 1436, 1437
  Ring-tailed **10**: 1435, *1435*, 1436
  Ruffed **10**: 1435, *1436*, 1437
  Sportive (weasel lemur) **10**: 1435,
    1436, 1437
Leopard (panther) **3**: 406; **8**: 1040;
    11: 1444–1446
  Clouded **3**: *404*
Linsang (Oriental linsang) **11**: 1459
  Banded **11**: 1459, *1459*
  Spotted **11**: 1459
Lion **3**: 406; **8**: 1040; **11**: 1460–1463;
    21: 2950
  African **3**: 406
  Mountain *see* Puma
Llama **11**: 1470–1472; **12**: 1672

Long-eared bat **20**: 2822
  American **20**: 2821, 2822
  European **2**: *167*; **20**: 2821, 2822
Loris **11**: 1490–1492
  Pygmy slow (lesser slow loris)
    11: 1490, 1491, 1492
  Slender **11**: 1490, 1491, *1491*, 1492
  Slow **11**: 1490, *1490*, 1491, *1492*
Lynx **11**: 1506–1508; **12**: 1672;
    19: 2616
  Canada **3**: 405; **11**: 1506, 1507, 1508
  Desert *see* Caracal
  Eurasian (Siberian lynx) **11**: 1506,
    1506, 1507, *1507*; **21**: 2906
  Spanish **11**: 1506, 1507, 1508
Lynx tachete *see* Serval

Macaque **11**: 1511–1513; **12**: 1644
  Assam **11**: 1511, 1512, 1513
  Barbary *see* Ape, Barbary
  Bonnet **11**: 1511–1512, 1513
  Celebes black **11**: 1511, 1513
  Crab-eating (long-tailed macaque)
    11: 1511, 1512, 1513, *1513*;
    14: 1942
  Japanese **11**: 1511, *1511*, 1512, 1513
  Lion-tailed **11**: *1512*, 1513; **14**: 1941,
    1942
  Moor **11**: 1511, 1513
  Pig-tailed **14**: 1941–1942
  Rhesus **11**: 1511, 1512, 1513
  Stump-tailed **11**: 1511, 1513
  Tibetan **11**: 1511, 1513
  Toque **11**: 1511, 1512
Manatee **11**: 1534–1535; **21**: 2907
  Amazonian **11**: 1534, 1535
  West African **11**: 1534, 1535
  West Indian **11**: 1534, *1534*, 1535
Mandrill **1**: 115; **11**: 1536–1538;
    12: 1644
Mangabey **11**: 1541–1542
  Agile **11**: 1541, 1542, 1543
  Black-crested **11**: 1541, 1542, *1542*,
    1543
  Crested **11**: 1541
  Crestless **11**: 1541, 1542
  Gray-cheeked **11**: 1541, 1542, 1543
  Red-crowned (collared mangabey)
    11: 1541–1542, *1541*, 1543
  Tana River **11**: 1541, 1542, 1543,
    1543
Mara (Patagonian hare) **8**: 1099,
    1100, 1101
Margay cat **3**: 404; **13**: 1776, 1777
Markhor **11**: 1559–1560; **12**: 1672;
    21: 2951
Marmoset **11**: 1563–1565; **12**: 1644,
    1645
  Buffy-headed **11**: 1565
  Buffy tufted-ear **11**: 1565
  Emperor **11**: *1565*
  Geoffroy's **11**: 1563, 1565
  Pygmy **11**: 1563, 1564, *1564*
  Silky **11**: 1563, 1565
  Silvery **11**: *1563*
  White **11**: 1565
  White ear-tufted **11**: 1563
  see also Tamarin
Marmot **11**: 1568–1570; **12**: 1672
  Alpine **11**: 1568, *1569*, 1570
  Himalayan (bobak; steppe
    marmot) **11**: 1568, 1569
  Hoary (whistler) **11**: 1568, 1569, 1570
  Olympic **11**: 1569, *1570*
  Steppe *see* Bobak
  Vancouver Island **11**: 1569
  Yellow-bellied **11**: *1568*
Marsupial cat (quoll; native cat)
    11: 1571–1572
  Bronze native cat **11**: 1571

Eastern native cat (eastern quoll)
    11: 1571, 1572, *1572*
  Geoffroy's native cat **11**: 1571
  Large-spotted native cat **11**: 1571,
    1572
  Little northern native cat (northern
    quoll) **11**: 1571, *1571*, 1572
  New Guinea native cat **11**: 1571
Marsupial jerboa *see* Marsupial
    mouse
Marsupial mole **11**: 1575–1576
Marsupial mouse (marsupial jerboa;
    dunnart; kultarr) **11**: 1577–1578
  Broad-footed **11**: 1577
  Brush-tailed (tuan) **11**: 1577, 1578
  Fat-tailed dunnart **11**: 1577, *1578*
  Flat-skulled *see* Planigale
  Macleay's **11**: 1578
  Short-haired **11**: 1577
Marten **11**: 1579–1581
  American (American pine marten;
    pine marten) **11**: 1579, 1580;
    19: *2615*, 2616
  Beech (stone marten) **11**: 1579
  Eurasian pine marten **11**: 1579,
    1579, 1580; **19**: 2653
  Yellow-throated **11**: 1579
Meerkat (suricate ; slender-tailed
    meerkat) **12**: 1595–1597
Miner's cat *see* Ringtail
Mink **12**: 1616–1617
  American **12**: 1616, *1616*, 1617
  European **12**: 1616, 1617
Mithan *see* Gaur, Domestic
Moco *see* Cavy, Rock
Mole **12**: 1625–1627
  Eastern **12**: 1625
  European (common mole)
    12: 1625, 1626, *1626*, 1627
  Golden *see* Golden mole
  Kurdistan **12**: 1627
  Marsupial *see* Marsupial mole
  Mediterranean **12**: 1625
  Star-nosed **12**: 1625, 1627
Mole rat (bamboo rat; sand puppy)
    12: 1628–1629
  Common **12**: *1629*
  Indian *see* Bandicoot rat, Lesser
  Naked **12**: *1628*, 1629
  Palestine **12**: 1628–1629
  South American *see* Tuco-tuco
Mongoose **3**: 325; **6**: 814; **12**: 1635–1637
  Banded **12**: 1635, *1637*
  Crab-eating **12**: 1635, 1637
  Dwarf **12**: 1635, 1637
  Egyptian **12**: 1635
  Lesser Indian **12**: 1637
  Malagasy broad-striped **12**: 1635
  Malagasy ring-tailed **12**: *1636*, 1637
  Water **12**: 1636–1637
  Yellow **12**: 1595, *1635*, 1636
Monkey
  Allen's swamp **12**: 1643
  Blue (gentle monkey) **8**: 1091,
    1092, 1093; **12**: 1643
  De Brazza **8**: 1091, 1093; **12**: 1643
  Diana **8**: 1091, *1091*, 1092, 1093;
    12: 1643
  Green (callithrix monkey) **8**: 1091,
    1092, 1093; **12**: 1643
  Mustached **8**: 1091, 1093
  Owl *see* Night monkey
  Owl-faced **8**: 1091
  Patas (red guenon; hussar monkey;
    military monkey) **14**: 1896–1897
  Proboscis **10**: 1410, 1411, 1412;
    12: 1643, *1643*, 1644
  Putty-nosed **8**: 1091
  Red-tailed **8**: 1091, 1092, 1093,
    1093; **12**: 1643

Monkey (*continued*)
Ring-tail *see* Capuchin monkey
Vervet **8:** 1091, *1092;* **12:** 1643
Widow **19:** 2696, 2697
see also Baboon; Barbary ape;
Guenon; Langur; Macaque;
Mandrill; Mangabey; Marmoset;
Monkeys; Saki; Tamarin; Titi;
Uakari
Monkeys 12: 1641–1645
Monk seal **12:** 1647–1648
Caribbean **12:** 1647–1648; **16:** 2286, 2289
Hawaiian **12:** 1647–1648, *1647, 1648;* **16:** 2286, 2289
Mediterranean **12:** 1647, 1648; **16:** 2289
Mono *see* Squirrel monkey
Moose (elk) **5:** 655; **12:** 1654–1656; **19:** *2617*
Mouflon **2:** 198; **12:** 1667–1668
Mountain beaver (sewellel; whistler) **12:** 1674–1675
Mountain cat *see* Ringtail
Mountain lion *see* Puma
Mouse
Bank *see* Vole, Bank
Brush **5:** 660
Cactus **5:** 660
California **3:** 426; **5:** 660
Flying *see* Scaly-tailed squirrel
Four-striped grass **21:** *3016*
Gapper's *see* Red-backed vole, Southern
Harvest *see* Harvest mouse
Hopping *see* Kangaroo mouse
House **9:** 1244–1246
Jerboa *see* Kangaroo mouse
Jumping *see* Kangaroo mouse
Kangaroo *see* Kangaroo mouse
Marsupial *see* Marsupial mouse
Multimammate *see* Multimammate rat
Oldfield **5:** 660
Pocket **5:** 665; **10:** 1338
Red *see* Vole, Bank
Red-backed *see* Red-backed vole
Spiny *see* Spiny mouse
Striped grass *see* Zebra mouse
Vesper *see* Deer mouse
White-footed **5:** 660; **19:** 2682
Wood **21:** 2975–2976
Yellow-necked **21:** 2975, 2976, *2976*
Mouse-hare *see* Pika
Mule **21:** 2941
Mulgara **11:** 1577, 1578
Large **11:** 1578
Multimammate rat (multimammate mouse) **12:** 1694–1695
Muntjac (barking deer; rib-faced deer) **5:** 655; **12:** 1696–1698
Borneo **12:** 1696, 1697
Chinese (Reeve's muntjac) **12:** 1696, 1697, 1698
Fea's **5:** 658; **12:** 1696–1697
Giant **12:** 1697
Hairy-fronted **12:** 1696
Indian **12:** 1696, *1696,* 1697, *1697*
Musang *see* Palm civet, Common
Musk deer **5:** 655; **12:** 1701–1702
Dwarf **12:** 1701, 1702
Mountain (forest musk deer) **12:** 1701, 1702
Siberian **12:** 1701, *1701,* 1702, *1702*
Musk-ox (musk sheep) **12:** 1703–1705
Musk rat *see* Shrew, Musk
Muskrat (musquash) **12:** 1706–1708; **19:** 2616
Round-tailed (Florida water rat) **12:** 1706

Mussel-digger *see* Whale, Gray
Myotis, Long-eared **2:** 168

Naked-tailed rat, Giant **7:** 964, *964,* 965
Narwhal (narwhale) **12:** 1717–1719; **21:** 2911
Native cat *see* Marsupial cat
Native porcupine *see* Spiny anteater
Night monkey (douroucouli; owl monkey) **12:** 1642, 1644, 1645; **13:** 1747–1748
Nightwalker *see* Kinkajou
Nilgai (blue bull; blue buck; nilghai; nylgau) **1:** 60; **13:** 1753–1754
Noctule **20:** 2822, 2823, *2823*
Slender-winged **20:** 2822
Numbat *see* Anteater, Banded
Nyala (inyala) **3:** 331; **13:** 1765–1766
Mountain **13:** 1765, 1766

Ocelot **13:** 1776–1778
Okapi **7:** 975; **13:** 1786–1787
Olingo **10:** 1363
Onager **21:** 2939, 2940, *2940*
Onca *see* Jaguar
Opossum **13:** 1792–1794
Common mouse (murine opossum) **13:** 1792, 1793, *1794*
Dormouse *see* Possum, Dormouse
Pygmy *see* Possum, Dormouse
Virginia (common opossum) **13:** 1792, *1792,* 1793, *1793*
Water (yapok) **13:** 1792, 1793
Orangutan **1:** 70, 71, 72, 73, *73;* **13:** 1797–1799
Orca *see* Killer whale
Organ-grinder *see* Wallaby, Nail-tailed
Oribi **13:** 1803–1804
Oryx **1:** 62; **5:** 665; **10:** 1382; **13:** 1808–1810
Arabian **1:** 63; **13:** 1808, 1809, *1809,* 1810
Beisa *see* Gemsbok
Scimitar **1:** 63; **13:** 1808, 1809, 1810
Tufted *see* Gemsbok
Otter
Canadian (river otter) **16:** 2179, *2179,* 2180
Clawless **16:** 2179
European (common otter) **16:** 2179, 2180, 2181
Giant Brazilian **16:** 2179, 2180
River **16:** 2179–2181
Sea **16:** 2179, 2294–2296
Small-clawed **16:** 2179
Smooth-coated **16:** 2179, 2180
Otter cat *see* Jaguarundi
Otter shrew **13:** 1817–1818; **19:** 2656
Giant **13:** 1817–1818, *1817*
Mount Nimba **13:** 1817, 1818
Ruwenzori **13:** 1817, 1818
Ounce *see* Snow leopard
Ox, Cambodian forest *see* Kouprey

Paca (spotted cavy; laba; agouti) **13:** 1838–1839
False (pacarana) **13:** 1838
Lowland **13:** 1838, *1838,* 1839
Mountain **13:** 1838, 1839
Pacarana *see* Paca, False
Pack rat (trade rat; wood rat) **13:** 1843–1844
Bushy-tailed wood rat **13:** 1843, *1843,* 1844
Desert wood rat **13:** 1844, *1844*
Dusky-footed wood rat **13:** 1843, 1844
Eastern (Florida pack rat) **13:** 1843, 1844

Pademelon (Tasmanian pademelon) **13:** 1847–1848
Calaby's (alpine wallaby) **13:** 1847, 1848
Dusky **13:** 1847, 1848
New Guinea **13:** 1847, 1848
Red-bellied (rufous wallaby) **13:** 1847, *1847,* 1848
Red-legged **13:** 1847, 1848
Red-necked **13:** 1847, 1848
Short-tailed *see* Quokka
Pahmi *see* Badger, Ferret
Pallah *see* Impala
Palm civet **13:** 1855–1856
African (two-spotted civet) **13:** 1855, 1856
Banded **13:** 1855
Borneo banded **13:** 1855
Common (musang; toddy cat) **13:** 1855, 1856, *1856*
Malay banded **13:** 1855, *1855,* 1856
Masked **13:** 1855, 1856
Owston's **13:** 1855
Small-toothed **13:** 1855
Sulawesi **13:** 1855
Palm squirrel **13:** 1859–1860
African **13:** 1859, 1860
Asian (striped palm squirrel) **13:** 1859, 1860
Berdmore's **13:** 1859, 1860
Biafran Bight **13:** 1859
Five-striped **13:** *1860*
Three-striped **13:** *1859*
Western **13:** 1859
Panda
Giant (beishung) **2:** 174, 176; **7:** 960–963
Red **10:** 1363
Red (lesser panda; cat-bear; fox-cat; fire-fox) **15:** 2135–2136
Pangolin (scaly anteater) **13:** 1862–1863
Cape **13:** 1862, *1862,* 1863
Chinese **13:** 1862
Giant **13:** 1862
Indian **13:** 1862, 1863
Long-tailed tree **13:** 1862, 1863
Malayan **13:** 1862, 1863, *1863*
Small-scaled tree **13:** 1862
Panther *see* Leopard; Puma
Patagonian hare *see* Mara
Peccary **14:** 1903–1905
Chacoan **14:** 1903, 1905
Collared **14:** 1903, *1903,* 1904, *1904,* 1905, *1905*
White-lipped **14:** 1903, 1904, 1905
Pectinator, Speke's **8:** 1104
Pekan *see* Fisher
Peludo *see* Armadillo, Hairy
Phalanger **14:** 1929–1930
Dormouse *see* Possum, Dormouse
Lesser flying *see* Gliding possum, Lesser
Pichi *see* Armadillo, Pygmy
Pichiciago *see* Armadillo, Lesser fairy
Pig-tail *see* Macaque, Pig-tailed
Pika (rock rabbit; mouse-hare; rock coney; calling hare; piping hare; whistling hare) **14:** 1943–1944
Pallas' **14:** 1944
Rocky Mountain **14:** 1944
Pilot whale **5:** 695
False *see* False killer whale
Long-finned **14:** 1952, 1953, *1954*
Short-finned **14:** 1952, *1952,* 1953, 1954
Piping hare *see* Pika
Pipistrelle **14:** 1962–1964
Common **14:** 1962, *1962,* 1963, 1964, *1964*

Eastern **14:** 1962, 1964
Soprano **14:** 1962
Western **14:** 1962, *1963,* 1964
Planigale (flat-skulled marsupial mouse) **11:** 1577
Ingram's **11:** 1577, 1578
Platypus (watermole; duckmole; duckbill) **14:** 1985–1987
Pocket gopher **8:** 1040; **14:** 1996–1997
Botta's **14:** *1996*
Northern **14:** *1997*
Polecat (fitchew; foumart; foul-marten) **14:** 2001–2003
European **2:** 228; **14:** 2001, 2002, *2003*
Marbled **14:** 2001
Steppe **2:** 228; **14:** 2001
Striped *see* Zorilla
Porcupine (*New World*) *see* Tree porcupine
Porcupine (*Old World*)
Brush-tailed **5:** 595
Crested (African crested porcupine; African porcupine) **5:** 595–597
Long-tailed **5:** 595
Porcupine
Native *see* Spiny anteater
Spiny *see* Spiny anteater
Porpoise **15:** 2024–2026
Black *see* Dolphin, Bottlenose
Burmeister's **15:** 2024, 2025
Common (bottlenose dolphin) **2:** 266–268; **4:** 507; **15:** 2024
Common (harbor porpoise; puffing pig) **15:** 2024, 2025, *2025, 2026*
Dall's (True's porpoise) **5:** 695, 698; **15:** *2024,* 2025, 2026
Finless **15:** 2024, 2025
Gray *see* Dolphin, Bottlenose
Spectacled **15:** 2024, 2025
White-bellied *see* Dolphin, Common
see also Dolphins and porpoises
Possum
Brush tail (vulpine possum; foxlike possum) **3:** 300–301; **14:** 1929
Dormouse (dormouse opossum; dormouse phalanger; pygmy possum; pygmy opossum) **5:** 702–703
Gliding *see* Gliding possum
Mountain pygmy possum (burramys) **5:** 703
Scaly-tailed **14:** 1929–1930, *1929*
Potoroo *see* Rat kangaroo, Long-nosed
Potto (*African;* Bosman's potto) **11:** 1490; **15:** 2034–2035
Golden (angwantibo; golden angwantibo; Calabar angwantibo) **8:** 1014–1015; **11:** 1490
Potto (*American*) *see* Kinkajou
Pouched rat, African giant (Gambian pouched rat; African giant rat; hamster rat) **7:** 964, 965, *965*
Prairie dog **8:** 1040; **15:** 2038–2040
Black-tailed **15:** 2038, 2039, *2039*
Gunnison's **15:** *2038*
Utah **15:** 2040
White-tailed **15:** 2039
Pronghorn *see* Antelope, Pronghorn
Pseudorca *see* False killer whale
Pudu **5:** 655; **18:** 2463
Northern Andean **18:** 2464
Southern Andean **18:** 2464, *2464*
Puffing pig *see* Porpoise, Common
Puku **10:** 1381, 1382
Puma (cougar; mountain lion; panther) **15:** 2068–2070; **19:** 2653, 2729; **21:** 2907

Quagga *see* Zebra, Plains
Quokka (short-tailed pademelon) **15**: 2089–2090
Quoll *see* Marsupial cat

Rabbit (Old World rabbit; domestic rabbit) **15**: 2091–2093
  Brush **3**: 426; **4**: 546
  Jack *see* Jack rabbit
  Rock *see* Pika
  Snowshoe **18**: 2561
  Sumatran (Sumatran hare; short-eared rabbit; Sumatran short-eared rabbit; Kerinci rabbit; kelinci kerinci) **18**: 2560–2561
  Swamp (canecutter) **4**: 545, 546
  see also Cottontail
Rabbit-rat, White-footed **18**: 2536
Raccoon **15**: 2096–2098
  Bahama **15**: 2096, 2097
  Barbados **15**: 2096
  Common (North American raccoon) **15**: 2096, 2097
  Cozumel Island **15**: 2096
  Crab-eating **15**: 2096, 2097
  Guadeloupe **15**: 2096
  Tres Marías **15**: 2096
Raccoon dog (ussuri dog) **5**: 689; **15**: 2099–2100
Raccoon fox *see* Ringtail
Rat
  Bamboo *see* Mole rat
  Bandicoot *see* Bandicoot rat
  Brown (common rat; Norway rat) **3**: 298–299; **20**: 2764
  Canepiece **15**: 2156
  Cloud **7**: 965
  Giant *see* Giant rat
  Indian Mole *see* Bandicoot rat, Lesser
  Jumping *see* Jerboa
  Kangaroo (*American*) **5**: 665; **10**: 1339–1340
  Kangaroo (*Old World*) *see* Jerboa
  Maned **1**: 1838
  Marmoset **11**: 1566–1567
  Musk *see* Shrew, Musk
  Naked-tailed *see* Naked-tailed rat
  Pack *see* Pack rat
  Pest *see* Bandicoot rat, Short-tailed
  Polynesian **20**: 2764
  Sand *see* Gerbil
  Ship (black rat) **3**: 299; **17**: 2346–2348; **20**: 2764
  Wood *see* Pack rat
Ratel *see* Badger, Honey
Rat kangaroo **15**: 2115–2116
  Desert **15**: 2115, 2116
  Long-nosed (potoroo) **15**: 2116
  Musky **15**: 2115, *2115*
  Rufous **15**: 2115, 2116
  Short-nosed **15**: 2115, 2116
Red-backed vole (red vole; red-backed mouse) **1**: 134; **15**: 2129–2131
  Northern **15**: 2129
  Southern (Gapper's mouse) **15**: 2129, 2130
  Western **15**: 2129
Reedbuck **1**: 61; **15**: 2144–2145
  Bohor **15**: 2144, 2145
  Common (southern reedbuck) **15**: 2144, *2144*, 2145
  Mountain **15**: 2144, 2145, *2145*
Reindeer *see* Caribou
Rhinoceros **15**: 2152–2155
  Black (browse rhinoceros) **15**: 2152, *2152*, 2153, 2154, 2155
  Greater Indian (Asian great one-horned rhinoceros) **15**: 2152, 2153, *2153*, 2154, 2155

Javan **15**: 2152, 2153, 2154–2155
Sumatran **15**: 2152, 2153, *2154*, 2155
White (grass rhinoceros) **15**: 2152, 2153, 2154, 2155, *2155*
Rice rat **15**: 2156–2157
  Brazilian **15**: 2157
  Galapagos **15**: 2157
  Giant (*extinct*) **15**: 2156
Right whale **16**: 2166–2168, 2209; **21**: 2913, 2914
  Northern (black right whale) **16**: 2166, 2167; **21**: 2910
  Pygmy **16**: 2166, 2167, 2168; **21**: 2910
  Southern (black right whale) **16**: 2166, 2167, *2167*, 2168; **21**: 2910
Ringtail (band-tailed cat; cacomistle; cat-squirrel; coon cat; miner's cat; mountain cat; raccoon fox; ring-tailed cat; tepemixtle) **16**: 2171–2172
  Central American (guayonoche) **16**: 2171, 2172
Ring-tailed cat *see* Ringtail
River dolphin **16**: 2177–2178
  Amazon (boto) **5**: 697; **16**: 2176, 2177, 2178
  Chinese (baji) **5**: 697; **16**: 2177, *2177*, 2178
Rock cat *see* Caracal
Rock coney *see* Pika
Rock rabbit *see* Pika
Rock wallaby **16**: 2197–2198
  Brush-tailed **16**: *2197*, 2198
  Little **16**: 2198
  Ring-tailed (yellow-footed rock wallaby) **16**: 2197, 2198, *2198*
  Short-eared **16**: 2197
Rorqual **16**: 2167–2168, 2208–2209; **21**: 2910–2911
  Great northern *see* Whale, Blue
  Lesser *see* Whale, Minke

Sable **11**: 1579, 1581
Saddleback (*seal*) *see* Seal, Harp
Saiga **1**: 61; **16**: 2217–2218
Saki **16**: 2221–2222
  Bald-faced **16**: 2222
  Bearded **16**: 2221, 2222
  Black **16**: 2221, 2222
  Equatorial **16**: 2222
  Red-bearded (monk's saki) **16**: 2221, 2222
  White **16**: 2222
  White-faced **16**: 2221, *2221*, 2222
  White-nosed **16**: 2221, 2222
Sand puppy *see* Mole rat
Sapajou *see* Capuchin monkey
Saarken *see* Snow Leopard
Scalytail *see* Scaly-tailed squirrel
Scaly-tailed squirrel (anomalure; scalytail; scaly-tailed flying squirrel; flying mouse) **16**: 2250–2251
  Beecroft's **16**: 2250, 2251
  Flightless (Cameroon scaly-tailed squirrel) **16**: 2250, *2250*, 2251
  Pel's **16**: 2250, *2251*
  Pygmy **16**: 2251
Sea canary *see* Whale, Beluga
Sea cow *see* Dugong
Seal **1**: 82
  Baikal **10**: 1398; **16**: 2169, 2170
  Bearded **1**: *82*; **16**: 2286
  Caspian **16**: 2169, 2170, *2170*
  Crabeater (white seal) **1**: 82; **4**: 564–565; **12**: 1647; **16**: 2286
  Eared **16**: 2285, *2287*
  Elephant *see* Elephant seal

Fur *see* Fur seal
  Gray (horse head seal) **8**: 1049–1051; **16**: 2289
  Harbor (common seal) **4**: *475*; **8**: 1049, 1132–1134; **12**: 1647
  Harp (Greenland seal; saddleback) **8**: 1139–1141; **16**: *2285*
  Hooded (bladdernose) **9**: 1219–1221
  Leopard **1**: 29, 82; **4**: 565; **10**: 1360; **11**: 1449–1450
  Monk *see* Monk seal
  Ringed **1**: 82; **14**: 1998; **16**: 2169–2170, 2285, 2286
  Ross (big-eyed seal) **16**: 2210–2211
  True (earless) **16**: 2285, 2286
  Weddell **1**: 82; **16**: 2288; **21**: 2899–2900
  see also Sea lion; Seals; Walrus
Sea lion **16**: 2282–2284, 2285, 2287, 2289
  Australian **16**: 2282, 2283, *2283*, *2287*, 2288
  California **16**: 2282, *2282*, 2283, 2284
  Hooker's **16**: 2282, 2283
  Southern **16**: 2282, 2283
  Steller's **16**: 2283, 2284, *2284*, 2289
Seals 16: 2285–2289
Sea wolf *see* Killer whale
Serow **17**: 2323–2324; **19**: 2623
  Japanese (kamoshika) **17**: 2323, 2324, *2324*
  Mainland (maned serow) **17**: 2323, 2324
  Taiwanese **17**: 2323, 2324
Serval (servaline cat; lynx tachete; aner; ingwenkala) **3**: 403; **17**: 2325–2326
Servaline cat *see* Serval
Sewellel *see* Mountain beaver
Shapu **2**: 198
Sheep
  Bighorn *see* Bighorn sheep
  Blue **12**: 1672
  Dall **2**: 198; **12**: *1673*
  Marco Polo (argali) **2**: 198
  Musk *see* Musk-ox
Shrew **17**: 2358–2360
  Dwarf (Savi's pygmy shrew) **12**: 1709
  Elephant *see* Elephant shrew
  Eurasian common **17**: *2358*, 2359
  European pygmy **17**: 2358, 2359, *2359*
  Hero (armored shrew) **9**: 1182–1183
  Least short-tailed (lesser-short-tailed shrew; little short-tailed shrew) **17**: 2358
  Masked (common shrew) **17**: 2358, 2359
  Musk (musk rat; money shrew; house shrew) **12**: 1709–1710; **17**: 2358
  North American pygmy **17**: 2359
  Otter *see* Otter shrew
  Pygmy **17**: 2358
  Red-toothed **17**: 2358
  Short-tailed **17**: 2358, 2359, *2360*
  Smoky **17**: 2358, 2359
  Tree *see* Tree shrew
  Water *see* Tenrec, Web-footed; Water shrew
  White-toothed **12**: 1710; **17**: 2358
Siamang **7**: 969, 970, 971
Sifaka **17**: 2371–2373
  Diadem **17**: 2372
  Tattersall's (golden-crowned sifaka) **17**: 2371, 2372

Verreaux's **17**: 2371, *2371*, 2372, *2372*, *2373*
Sika **17**: 2374–2375
Singsing *see* Waterbuck
Siren (*mammal*) *see* Dugong
Sitatunga **1**: *63*; **3**: 331; **20**: 2760
Skunk **3**: 318; **17**: 2393–2395
  Hog-nosed **17**: 2393–2394, 2395
  Hooded **17**: 2393, 2394
  Spotted **17**: 2393, 2394, 2395
  Striped **17**: 2393, 2394, *2394*, 2395, *2395*
Sleeper *see* Dormouse, Hazel
Sloth **17**: 2400–2402
  Three-toed **15**: *2113*; **17**: 2400, 2401, 2402
  Two-toed **17**: 2400, 2401, 2402
Snake-weasel **21**: 3020
Snow leopard (ounce; xue bao; barfani chita; sarken) **3**: 404, 406; **12**: *1670*, 1672; **17**: 2432–2434
Solenodon **18**: 2452
  Haitian **18**: *2452*
Sperm whale **13**: 1772, *1772*; **18**: 2472–2474, 2510; **21**: 2911, 2913–2914
  Dwarf **18**: 2473; **21**: 2911
  Pygmy **18**: 2472, 2473; **21**: 2911
Spider monkey **12**: 1644; **18**: 2477–2479
  Black-handed **12**: *1641*
  Common **18**: 2477, 2478
  Woolly **18**: 2477, 2478
Spiny anteater (long-beaked echidna; short-beaked echidna; long-nosed echidna; short-nosed echidna; native porcupine; spiny porcupine) **18**: 2486–2488
Spiny mouse **18**: 2491–2492
  Cairo **18**: 2491
  Cape **18**: 2491
  Golden **18**: 2491
Spiny porcupine *see* Spiny anteater
Springbock *see* Springbok
Springbok (springbuck; springbock) **18**: 2499–2501
Springbuck *see* Springbok
Springhare (jumping hare) **18**: 2502–2503
Squirrel
  Douglas **4**: 436, 437
  Eurasian red **4**: 436
  Flying *see* Flying squirrel; Gliding possum, Lesser
  Gray **20**: 2740, *2740*, 2741
  Mearn's **4**: 436, 437
  Oil-palm (African giant squirrel) **13**: 1859, 1860
  Palm *see* Palm squirrel
  Red **20**: 2740, 2741, *2742*
  Red (pine squirrel) **4**: 436, 437, *438*
  Rock *see* Ground squirrel
  Scaly-tailed *see* Scaly-tailed squirrel
Squirrel monkey (titi; mono) **12**: 1644; **18**: 2513–2515
  Gothic **18**: 2513, *2513*, 2515
  Roman **18**: 2513, 2515
Steenbok (steinbok; steinbok; steenbuck) **18**: 2526–2527
Steenbuck *see* Steenbok
Steephead *see* Whale, Northern bottlenose
Steinbok *see* Steenbok
Steinbuck *see* Steenbok
Stick-nest rat **18**: 2535–2536
  Greater **18**: *2535*
Stoat *see* Ermine
Striped weasel **21**: *3021*
  Libyan **21**: 3020

Suni **1**: 61; **18**: 2571–2572
Suricate *see* Meerkat
Suslik *see* Ground squirrel

Tahr **19**: 2612–2613
    Arabian **19**: 2612, 2613
    Himalayan **19**: 2612, 2613, *2613*
    Nilgiri **19**: 2612, *2612*, 2613
Takin **19**: 2622–2623
Tamandua
    Northern **1**: 57, 58, 59, *59*
    Southern **1**: 57, 58, 59
Tamarau **3**: 305, 307; **9**: 1286–1287
Tamarin (marmoset—*incorrectly*)
    12: 1644, 1645; **19**: 2624–2626
    Black-and-gold lion **19**: 2625
    Black-chested mustached **19**: 2624
    Black-faced golden lion **19**: 2625
    Black-mantle **19**: 2624
    Brown-headed (saddle-back
        tamarin) **19**: 2624
    Emperor **12**: 1645; **19**: 2624, *2625*
    Geoffroy's **19**: 2625
    Golden lion **19**: *2624*, 2625
    Golden-rumped lion **19**: 2625
    Lion **19**: 2625, 2626
    Long-tusked **19**: 2625
    Mottle-faced **19**: 2625
    Pied **19**: 2625
    Pinché (cotton-top tamarin)
        12: *1642*; **19**: 2625, *2626*
    Red-chested mustached **19**: 2624
    Red-handed (black tamarin;
        midas tamarin) **19**: 2624, 2626
    Saddleback **12**: *1645*
    White-footed (silvery-brown bare-
        face tamarin) **19**: 2625
Tapir **19**: 2630–2632
    Baird's **19**: 2630, 2631, *2631*, 2632
    Malayan **19**: 2630, *2630*, 2631, 2632
    Mountain (Andean tapir) **19**: 2630,
        2631, 2632
    South American (Brazilian tapir;
        lowland tapir) **19**: 2630, 2631,
        2632
Tarpan (*extinct*) **15**: 2052
Tarsier **19**: 2638–2640
    Horsfield's **19**: 2638, 2639, *2639*
    Philippine **19**: 2638, *2638*, 2639,
        2640
    Spectral (Celebes tarsier) **19**: 2638,
        2639–2640
Tasmanian devil (devil) **5**: 677;
    11: 1571; **19**: 2641–2643
Tasmanian wolf (thylacine; *extinct*)
    5: 677
Tenrec **19**: 2656–2657
    Common **19**: 2656–2657, *2657*
    Hedgehog **19**: 2656, 2657
    Large hedgehog **19**: 2657
    Lesser hedgehog (small hedgehog
        tenrec) **19**: 2657, *2657*
    Long-tailed (shrewlike tenrec)
        19: 2656, 2657
    Rice **19**: 2656, 2657
    Streaked **19**: 2656, *2656*, 2657
    Web-footed (water shrew)
        19: 2656, 2657
Tepemixtle *see* Ringtail
Thamin *see* Deer, Brow-antlered
Tiger **3**: 405, *405*, 406; **19**: 2683–2685;
    21: 2907
Titi **19**: 2696–2697
    Dusky **19**: 2696, *2696*
    Gray **19**: 2696, 2697
    Masked **19**: 2696, 2697, *2697*
    see also Squirrel monkey
Toddy cat *see* Palm civet, Common
Topi **21**: 2950
Tratratratra (*extinct*) **17**: 2373

Tree kangaroo **19**: 2726–2727
    Bennett's **19**: 2726, *2726*
    Goodfellow's **19**: 2726
    Lumholtz's **19**: 2726, *2727*
    Matschie's **19**: 2726
    Scott's **19**: 2727
Tree porcupine **19**: 2728–2729
    Amazon basin porcupines **19**: 2729
    Amazonian porcupine **19**: 2729
    Brazilian **19**: 2728, 2729, *2729*
    Canadian (North American
        porcupine) **19**: 2728, *2728*, 2729
    Prehensile-tailed porcupines
        19: 2729
Tree rat
    Giant **7**: 965
    New Guinea giant **7**: 965
Tree rat (*squirrel*) *see* Tree squirrel
Tree shrew **19**: 2730–2731
    Belanger's **19**: *2730*
    Long-nosed **19**: *2731*
    Mindanao **19**: 2730
    Pen-tailed **19**: 2730, 2731
    Smoothtailed **19**: 2730
Tree squirrel (tree rat) **20**: 2740–2742
    South African **20**: *2741*
Tsaine *see* Banteng
Tuan *see* Marsupial mouse, Brush-
        tailed
Tuco-tuco (South American mole rat;
        tucu-tuco) **20**: 2771–2772
Tur *see* Ibex, East Caucasus; Ibex,
        West Caucasus

Uacari *see* Uakari
Uakari (uacari) **12**: 1645;
    20: 2803–2804
    Bald (white uakari) **20**: 2803–2804,
        2803, *2804*
    Black-headed **20**: 2803, 2804
Urchin (*mammal*) *see* Hedgehog
Urial **2**: 198
Ussuri dog *see* Raccoon dog

Vampire bat **2**: 167; **20**: 2807–2809
    Common **20**: 2807, 2808, *2808*,
        2809, *2809*
    False **6**: 840
    Hairy-legged **20**: 2808, 2809
    White-winged **20**: 2807–2808, 2809
Vaquita **15**: 2024, 2025
Venomous animals 20: 2817–2820
Vicuña **11**: 1470, *1471*, 1472; **12**: 1672
Viscacha **20**: 2834–2836
    Mountain (mountain chinchilla)
        12: 1672; **20**: 2834, 2835–2836,
        2835, *2836*
    Plains **20**: 2834–2835, 2836
Vole **10**: 1347
    Bank (wood vole; bank mouse;
        red mouse) **1**: 134–135; **15**: 2129
    Field (short-tailed vole) **6**: 818–820
    Red *see* Red-backed vole
    Red-backed *see* Red-backed vole
    Sagebrush **18**: 2528, 2529, *2529*

Wallaby **20**: 2846–2848
    Alpine *see* Pademelon, Calaby's
    Banded hare **20**: 2846, 2848
    Bennett's **20**: 2848
    Bridled nail-tailed **20**: *2848*
    Hare **20**: 2846, 2847–2848
    Nail-tailed (organ-grinder)
        20: 2846
    New Guinea forest **20**: 2846
    Parma **20**: *2846*
    Rock *see* Rock wallaby
    Rufous *see* Pademelon, Red-
        bellied
    Scrub **20**: 2848

Wallaroo, Common (euro) **10**: 1333
Walrus **16**: 2285, 2286, 2287, 2288,
    2289, *2289*; **20**: 2851–2853
Wapiti (elk—*incorrectly*; red deer)
    5: *655*, 656, 657; **17**: 2375;
    19: 2653; **20**: 2854–2856
Warthog **20**: 2860–2862
Waterbuck (common waterbuck;
    defassa; singsing) **1**: 62;
    20: 2865–2867
Water cavy *see* Capybara
Watermole *see* Platypus
Water pig *see* Capybara
Water rat
    Florida *see* Muskrat, Round-
        tailed
    Giant **7**: 965
Water shrew **20**: 2872–2873
    American **20**: 2872, 2873
    European **20**: 2872, *2873*
    Mediterranean **20**: 2872
    Pacific **20**: 2872
    see also Tenrec, Web-footed
Water vole **21**: 2884–2885
    Southwestern **21**: 2884, 2885
Weasel **21**: 2893–2895
    Least (cane weasel; common
        weasel) **14**: 2003; **21**: 2893–2895
    Short-tailed *see* Ermine
    Striped *see* Striped weasel
    White-naped **21**: 3020
Weasel cat *see* Jaguarundi
Werewolf *see* Aardwolf
Whale
    Baleen **21**: 2910, 2911, 2913, 2914
    Beaked *see* Beaked whale
    Beluga (belukha; white whale; sea
        canary) **2**: 193–194; **21**: 2911,
        2914, *2914*
    Blue (sulfur-bottom whale; great
        northern rorqual) **2**: 246–248;
        13: 1770; **14**: 1954; **16**: 2208, 2209;
        21: 2910–2911, 2913, 2914
    Bowhead (Arctic whale;
        Greenland whale) **16**: 2166, 2167,
        2168; **21**: 2910, 2913
    Bryde's **16**: 2208, 2209; **21**: 2911,
        2913
    Fin **16**: 2208, 2209; **21**: 2914
    Gray (California gray whale; scrag
        whale; mussel-digger; devilfish)
        8: 1052–1053; **21**: 2910, *2911*,
        2913
    Humpback (humpbacked whale)
        9: 1258–1260; **13**: 1770; **16**: 2208,
        2209, *2209*; **21**: *2910*, 2911, *2913*,
        2914
    Killer *see* Killer whale
    Minke (lesser rorqual; pike whale;
        little piked whale) **16**: 2208,
        2208, 2209; **21**: *2912*, 2914
    Northern bottlenose (North
        Atlantic bottlenose;
        bottlehead; steephead) **2**: 171
    Pilot *see* Pilot whale
    Right *see* Right whale
    Sei **16**: 2208, 2209; **21**: 2914
    Southern bottlenose (Antarctic
        bottlenose whale) **2**: 171
    Sperm *see* Sperm whale
    Toothed **21**: 2910, 2911, 2914
    see also Narwhal; Rorqual; Whales
Whales 21: 2910–2914
Whistle pig *see* Groundhog
Whistler *see* Marmot, Hoary;
    Mountain beaver
Whistling hare *see* Pika
Wild ass **21**: 2939–2941
    African (burro) **21**: 2939
    Asiatic **8**: 1040; **21**: 2939, *2939*, *2941*

Wildcat (African wildcat) **3**: *403*;
    21: 2945–2947
Wildebeest (gnu) **1**: 61; **8**: 1040;
    21: 2948–2950
    Black (white-tailed wildebeest;
        white-tailed gnu) **16**: 2216;
        21: 2948, 2949
    Blue (white-bearded wildebeest;
        brindled gnu) **21**: 2948, *2948*,
        2949, *2949*, 2950
Wisent *see* Bison, European
Wolf **3**: 393
    Ethiopian **5**: 692
    Gray (timber wolf; tundra wolf;
        plains wolf) **5**: 689, 691–692;
        8: 1054–1057; **19**: 2653; **21**: 2906
    Maned **11**: 1539–1540
    Red **5**: 692; **8**: 1054, 1057
    Red (*dhole*) *see* Dhole
Wolverine (glutton) **19**: 2729;
    21: 2958–2960
Wombat **21**: 2961–2963
    Common (coarse-haired wombat)
        21: *2961*, 2962, 2963
    Northern hairy-nosed **21**: 2961,
        2962, 2963
    Southern hairy-nosed **21**: 2961,
        2962, *2962*, 2963
Woodchuck *see* Groundhog
Wood rat *see* Pack rat
Woolly monkey **21**: 2987–2988
    Hendee's (yellow-tailed woolly
        monkey) **21**: 2987, 2988
    Humboldt's **21**: *2987*, 2988
Woylie *see* Bettong, Brush-tailed

Yak **3**: 305, *305*; **21**: 3000–3002
    Domestic (dzo) **21**: 3000, 3001,
        3001, 3002
    Northern (*extinct*) **21**: 3002
    Wild **21**: 3000, 3001, *3002*

Zebra **21**: 3009–3011
    Grevy's **21**: 3010, *3010*, *3011*
    Mountain (Cape mountain zebra;
        Hartmann's mountain zebra)
        21: 3009–3010, 3011
    Plains (Chapman's zebra; Grant's
        zebra; Burchell's zebra—*extinct*;
        quagga—*extinct*) **8**: 1040;
        16: 2216; **21**: *3009*, 3011
Zebra mouse (striped grass mouse)
    21: 3016–3017
Zebu (Brahman cattle) **21**: 3018–3019
Zorilla (zorille; striped polecat)
    21: 3020–3021
Zorille *see* Zorilla
Zorro
    Crab-eating *see* Fox, Crab-eating
    Gray *see* Fox, South American gray
    Short-eared *see* Dog, Short-eared

REPTILES

Adder **1**: 26–27
    Cape Mountain **15**: 2059
    Desert mountain **15**: 2058
    European **7**: 935
    Horned **15**: 2059; **17**: 2370
    Many-horned **15**: 2059
    Namaqua dwarf **15**: 2058
    Night *see* Night adder
    Péringuey's **15**: 2058
    Puff (*African*) **15**: 2058–2059
    Puff (*North American*) *see* Hognose
        snake
    Spread *see* Hognose snake
Agama lizard **1**: 30–31
    Common (rainbow lizard) **1**: 30,
        31, *31*

Desert **1**: 30, 31
Flying **11**: *1466*
Ground **1**: *30*
Starred (hardun; dragon lizard)
  1: 30, 31
Alligator **1**: 38–40
  American **1**: 38, *38*, 40; **21**: *2906*
  Chinese **1**: 38, *39*, 40
Anaconda (water boa) **1**: 43–44;
  17: 2419
  Paraguayan (southern anaconda)
    1: 43
Anole lizard **1**: 53–54
  Brown **1**: *54*
  Green (American chameleon)
    1: 53, 54
  Knight **1**: 53
Arrau *see* Turtle, Amazon river
Asp (aspic viper; June viper)
  1: 97–98

Barba amarilla *see* Fer de lance
Basilisk
  Common (Jesus lizard) **2**: 156
  Plumed (leafy plumb-tree
    basilisk) **2**: 157
Beak-head (*extinct*) **20**: 2762
Black snapper *see* Massasauga
Blind worm *see* Slowworm
Bluebelly *see* Fence lizard
Boa **17**: 2418, 2419
  Common *see* Boa constrictor
  Dwarf **17**: 2419
  Emerald tree **17**: *2418*
  Puerto Rican **2**: 251
  Rosy **2**: 249
  Rubber **2**: 249
  Virgin Islands tree **2**: 251
  Water *see* Anaconda
Boa constrictor (common boa)
  2: 249–251; **17**: 2422
Boomslang **2**: 264–265; **6**: 749;
  13: 1736
Bushmaster **14**: 1972, 1973, 1974

Caiman **3**: 358–360
  Black **3**: 358, 359–360, *360*
  Broad-snouted **3**: 358
  Common (spectacled caiman)
    3: 358, 359, *359*, 360
  Cuvier's dwarf **3**: 358
  Paraguayan (yacare caiman)
    3: 358, 359
  Smooth-fronted **3**: 358
Chameleon **3**: 418–420; **11**: 1467
  American *see* Anole lizard, Green
  Common **3**: 418, 419
  Flap-necked **3**: 418, 419
  Jackson's **3**: 418, 419, *419*
  Setaro's dwarf **3**: 419
  Smith's dwarf **3**: 419
  Stump-tailed **3**: 418
  Three-horned **3**: 419
Chuck (*lizard*) *see* Chuckwalla
Chuckwalla (chuck) **4**: 453–454
  San Esteban Island **4**: 454, *454*
Chunkhead *see* Copperhead
Coachwhip (prairie runner; red
  racer) **21**: 2919–2920
Cobra **2**: 265; **4**: 480–482; **17**: 2418;
  20: 2817
  Cape **4**: 480, 481, *481*, 482
  Egyptian **4**: 480, 481
  Forest **4**: 480, 481
  Indian **4**: 480, *480*, 481–482
  King **17**: 2420
  Mozambique spitting **4**: 480, 481,
    482
  Spitting **4**: 480, 481; **20**: 2818
Cooter **13**: 1851; **20**: 2796

Copperbelly *see* Water snake, Red-
  bellied
Copperhead (*cobra*), Australian
  7: 935
Copperhead (*pit viper*), (chunkhead;
  highland moccasin) **14**: 1972,
Coral snake **4**: 534–535
  Arizona (western coral snake;
    Sonoran coral snake) **4**: 534,
    534, 535
  Common (eastern coral snake;
    Texas coral snake) **4**: 534, 535
Cottonmouth *see* Moccasin, Water
Crocodile **5**: 603–606; **18**: 2546
  American **5**: 603, *603*, 604; **21**: 2907
  Cuban **5**: *605*
  Dwarf **5**: 603
  Indo-Pacific (saltwater crocodile;
    estuarine crocodile) **5**: 603, *604*;
    21: 2907
  Marsh (mugger) **5**: 603, 606
  Nile **5**: 603, 604, 606, *606*
  Orinoco **5**: 603

Diamondback *see* Rattlesnake
Dragon
  Bearded *see* Lizard, Bearded
  Frilled *see* Lizard, Frilled
  Komodo *see* Komodo dragon

Egg-eater
  Common *see* Egg-eating snake,
    Common
  Rhombic *see* Egg-eating snake,
    Common
Egg-eating snake **6**: 748–749
  African **6**: 748; **9**: 1203
  Common (common egg-eater;
    rhombic egg-eater) **6**: 748, 749
  East African **6**: 748
  Indian **6**: 748

Fence lizard (swift; bluebelly)
  6: 807–808
  Eastern (plateau lizard; pine
    lizard; prairie lizard; northern
    fence lizard; southern fence
    lizard) **6**: 807–808, *807*, *808*
  Western **6**: 807, 808
Fer de lance (barba amarillo)
  6: 812–814
Flapshelled turtle (flapshell) **17**: 2442
  Indian **17**: 2442
  Senegal **17**: 2443

Gapper *see* Moccasin, Water
Garter snake (ribbon snake)
  7: 933–935
  Checkered **7**: *935*
  Common **7**: 933, 934, *934*
  Mexican **7**: 933, 934
  Mountain **7**: 933–934
  Narrow-headed **7**: 933
  Plains **7**: 933, 934
  San Francisco **17**: 2422
  Short-headed **7**: 933
  Wandering (western terrestrial
    garter snake) **7**: 933, *933*, 934
Gavial *see* Gharial
Gecko **7**: 943–945; **11**: 1466–1467, 1469
  Banded **7**: 943
  Fat-tailed **7**: 943
  Flying **7**: 945
  Knob-tailed **11**: *1468*
  Leopard **7**: 943
  Madagascar day **7**: *943*
  Northern leaf-tailed **7**: *945*
  Texas banded **7**: 943
  Tokay **7**: 943
  Tree **7**: 944

Gharial (gavial) **7**: 954–956; **21**: 2907
  False **5**: 603; **7**: 954, 955–956
Gila monster **7**: 972–974; **20**: 2818
Glass lizard **6**: 733; **7**: 980–981;
  10:1430
  Eastern **7**: 980, 981
  European **7**: *980*, 981
  Island **7**: 980, 981
  Slender **7**: 980, 981
Goanna *see* Monitor lizard

Habu (Japanese pit viper) **14**: 1973
Hardun *see* Agama lizard, Starred
Hognose snake (hissing sandsnake;
  blow viper; puff adder; spread
  adder) **9**: 1202–1203
  Eastern **9**: 1202, 1203, *1203*;
    17: *2421*
  Southern **9**: 1202, 1203
  Western **9**: 1202, *1202*, 1203
Horned lizard (horned toad)
  9: 1231–1232
  Short-horned **9**: *1231*
  Texas **9**: 1231, 1232
Horned toad *see* Horned lizard

Iguana **9**: 1281–1283
  Desert **9**: 1281, 1282
  Green (common iguana) **9**: *1281*,
    1282, 1283; **11**: *1469*
  Land (terrestrial iguana) **9**: 1281,
    1282, 1283, *1283*
  Marine (Galapagos marine
    iguana) **11**: 1555–1556
  Rhinoceros **9**: 1281, 1282
  Tree **9**: 1282

King snake **10**: 1361–1362
  Arizona mountain **10**: 1361, *1361*
  California **10**: 1361
  Common **4**: 535; **10**: 1362
  Desert **10**: 1361
  Eastern (chain snake) **10**: 1362
  Florida (salt-and-pepper snake)
    10: 1361
Komodo dragon (Komodo
  monitor) **10**: 1383–1384;
  11: 1466; **12**: 1638
Krait **2**: 265
  Sea *see* Sea krait

Lancehead, Yellow-jawed **20**: 2817
Legless lizard **10**: 1430–1431;
  11: 1466
  Burton's **10**: 1431
  California (Geronimo legless
    lizard) **10**: 1430, *1430*, *1431*
Lizard
  Alligator **3**: 426
  Beaded **5**: 665; **7**: 972, 974
  Bearded (bearded dragon; Jew
    lizard) **2**: 172–173
  Centralian blue-tongued
    17: *2389*
  Dragon *see* Agama lizard, Starred
  Fence *see* Fence lizard
  Frilled (frillneck lizard; frilled
    dragon) **7**: 894–895
  Green **8**: 1068–1069
  Jesus *see* Basilisk, Common
  Ocellated (eyed lizard) **8**: 1068
  Pine *see* Fence lizard, Eastern
  Plateau *see* Fence lizard, Eastern
  Prairie *see* Fence lizard, Eastern
  Rainbow *see* Agama lizard,
    Common
  Ruin (Italian wall lizard) **20**: 2849,
    2850
  Sand **16**: 2231–2232
  Short-horned **9**: *1231*

Spiny-tailed **1**: 30
Teiid **19**: 2650
Viviparous (common lizard)
  20: 2837–2838
Water (sail fin lizard) **1**: 30; **2**: 157
  see also Basilisk; Chameleon;
    Chuckwalla; Gecko; Gila
    monster; Iguana; Komodo
    dragon; Lizards; Moloch;
    Skink; Slowworm; Tuatara
Lizards 11: 1466–1469
Luth *see* Turtle, Leatherback

Mamba **2**: 265; **11**: 1530–1531;
  20: 2817
  Black (blackmouth mamba)
    11: 1530, 1531, *1531*; **12**: 1597;
    15: 2103
  Green **11**: 1530, 1531
Massasauga (swamp rattler; black
  snapper) **15**: 2120
Milk snake (king snake) **4**: 535
  Common **10**: 1361
  Louisiana (false coral snake)
    10: 1361
Moccasin
  Highland *see* Copperhead
  Water (cottonmouth; gapper;
    trapjaw) **14**: 1972, 1973
Moloch (thorny devil; mountain
  devil) **5**: *664*, 665; **12**: 1630–1631
Monitor lizard (monitor; goanna)
  9: 1281; **12**: 1638–1640
  Asian water **12**: 1638, 1639, 1640
  Bornean rough-necked **12**: 1639
  Desert **12**: 1639
  Earless **6**: 732–733; **12**: 1638
  Indian **12**: *1638*, 1639, 1640
  Komodo monitor *see* Komodo
    dragon
  Lace **12**: 1639
  Nile **12**: 1638, 1639, 1640
  Savanna **12**: *1639*
  Short-tailed **12**: 1638
Mountain devil *see* Moloch
Mud turtle **12**: 1687–1688; **20**: 2795
  Big Bend **12**: 1687
  Common (Florida mud turtle;
    eastern mud turtle) **12**: 1687,
    1687, 1688
Mugger *see* Crocodile, Marsh
Musk turtle **12**: 1687; **20**: 2795
  Common (stinkpot; stinking-jim)
    12: 1688, *1688*
  Keel-backed **12**: 1687
Mussurana **6**: 814

Night adder **6**: 749; **13**: 1735–1736
  Common (rhombic night adder)
    13: 1735, 1736, *1736*
  Lichtenstein's **13**: 1735
  Snouted **13**: 1735
  Velvety (green night adder)
    13: 1735, *1735*
Night lizard **13**: 1745–1746
  Arizona **13**: 1745, 1746
  Cuban **13**: 1745
  Desert **13**: 1745, 1746
  Granite **13**: 1745, *1745*
  Island **13**: 1745, 1746

Perentie **12**: 1638, 1639, *1640*
Pit viper **13**: 1736; **14**: 1972–1974;
  17: 2420
  Eyelash **14**: *1972*, *1973*
  Himalayan **14**: 1973, 1974
  Japanese *see* Habu
  Wagler's **14**: 1973
  White-lipped **14**: *1974*
Prairie runner *see* Coachwhip

Python **15**: 2079–2081; **17**: 2418, 2419, 2420, 2422
   African (African rock python) **15**: 2079, 2080, *2080*
   Amethystine (Australian scrub python) **15**: 2079, 2080, *2114*
   Ball (royal python) **15**: 2079, 2080, 2081
   Black-headed **15**: 2079
   Carpet (diamond python) **15**: 2079
   Children's **15**: 2079
   Dwarf **15**: 2079
   Green tree (green python) **15**: 2079, *2079*, 2080
   Indian **15**: 2079, 2081
   Reticulated **15**: 2079, 2080, *2081*; **17**: 2419

Racer **15**: 2101–2103
   Antiguan **15**: *2103*
   Black **15**: 2101, *2101*, 2102
   Blue **15**: 2102, *2102*
   Cedar *see* Whip snake, Striped
   Eastern yellow-bellied **15**: *2102*
   Red *see* Coachwhip
   Striped **15**: 2101
Racerunner **19**: 2650
   Prairie **19**: 2650
   Six-lined **19**: 2650
Rat snake (black rat snake; mountain black rat snake; yellow rat snake; gray rat snake; Everglades rat snake; Texas rat snake; chicken snake; pilot snake; striped house snake) **15**: 2117–2118
   Great Plains *see* Snake, Corn
   Green **15**: 2117, 2118
   Indian **15**: 2117; **17**: 2422
   Rosy *see* Snake, Corn
Rattler
   Hardwood *see* Snake, Fox
   Swamp *see* Massasauga
   see also Rattlesnake
Rattlesnake (rattler) **3**: 426; **15**: 2119–2121; **17**: 2422
   Aruba Island **15**: 2120
   Eastern diamondback **15**: 2119, 2120
   Horned *see* Sidewinder
   Mohave **15**: *2120*
   New Mexico ridge-nosed **17**: 2422
   Prairie **15**: *2119*, 2120, 2121
   Pygmy (ground rattler) **15**: 2119, 2120
   Red diamond **15**: 2120, 2121
   Tiger **15**: 2120
   Timber (banded rattlesnake) **15**: 2120
   Western diamondback (coontail rattler) **3**: *426*; **15**: 2120, *2121*
Ridley turtle
   Atlantic **20**: 2797; **21**: 2907
   Pacific (olive turtle) **20**: *2796*

Sandfish **17**: 2390
Sandsnake, Hissing *see* Hognose snake
Sea krait **17**: 2308
   Giant banded (wide-faced sea krait; amphibious sea snake) **17**: *2308*
   Rennell Island **17**: 2308
Sea snake **7**: 935; **17**: 2308–2309; **20**:*2818*, 2820
   Amphibious *see* Sea krait, Giant banded
   Olive **17**: 2309
   Yellow-bellied **17**: 2308

Shingleback **17**: 2390
Sidewinder (horned rattlesnake; horned rattler) **17**: 2369–2370
Skink **17**: 2389–2390
   Bermuda rock **17**: 2390
   Burrowing **11**: *1467*
   Five-lined **17**: 2390
   Florida sand **17**: 2390
   Giant **17**: 2389, 2390
   Great Plains **17**: 2389, 2390
   Kalahari burrowing **17**: 2390
   Keel-bearing **17**: 2389
   Many-lined **17**: 2389, 2390
   Sand **17**: 2389, *2390*
   Snake-eyed **17**: 2389
Slider **13**: 1851; **20**: 2796
Slowworm (blind worm) **10**: 1430; **17**: 2406–2407
   Blue-spotted **17**: 2406
Snake
   Aesculapian **15**: 2117, 2118
   Brown vine **19**: 2733
   Chain **10**: 1362
   Chicken *see* Rat snake
   Coral *see* Coral snake
   Corn (Great Plains rat snake; rosy rat snake) **15**: 2117, 2118
   Dice **20**: 2874, 2875
   Egg-eating *see* Egg-eating snake
   Elephant's trunk **20**: 2874, 2875
   False coral *see* Milk snake, Louisiana
   File **17**: 2419; **20**: 2874, 2875
   Four-lined **15**: 2117
   Fox (hardwood rattler) **15**: 2117, 2118
   Garter *see* Garter snake
   Glossy **17**: 2370
   Grass (water snake; ringed snake) **8**: 1042–1044; **9**: 1203; **15**: 2103; **17**: *2420*; **20**: 2874, 2875
   Hognose *see* Hognose snake
   King *see* King snake
   Ladder **15**: 2117
   Leopard **15**: 2117
   Mexican burrowing **17**: 2419
   Milk *see* Milk snake
   Pilot *see* Rat snake
   Rat *see* Rat snake
   Ribbon *see* Garter snake
   Rough green (vine snake) **19**: 2733
   Salt-and-pepper *see* King snake, Florida
   Sea *see* Sea snake
   Spotted bush **19**: 2733
   Striped house *see* Rat snake
   Thread (flowerpot snake) **17**: 2421
   Tree *see* Tree snake
   Viperine **20**: 2874, *2874*, 2875
   Water *see* Snake, Grass; Water snake
   Whip *see* Whip snake
   Yellow-and-green *see* Whip snake, Dark green
   see also Adder; Anaconda; Asp; Boa constrictor; Boomslang; Cobra; Fer-de-lance; Mamba;Python; Racer; Rattlesnake; Sidewinder; Snakes; Viper
Snake-necked turtle **17**: 2416–2417; **20**: 2795
   Common (eastern snake-necked turtle; long-necked tortoise) **17**: 2416, *2416*, 2417
   Giant *see* Turtle, Broad-shelled river
Snakes 17: 2418–2422

Snapper (*snake*), Black *see* Massasauga
Snapper (*turtle*) *see* Snapping turtle
Snapping turtle **9**: 1190; **12**: 1688; **17**: 2425–2426
   Alligator (alligator snapper) **17**: 2425, 2426; **20**: 2797, 2798
   Common (snapping turtle; snapper) **17**: 2425, 2426, *2426*
Softshell *see* Soft-shelled turtle
Soft-shelled turtle (pancake turtle; flapjack turtle; softshell) **17**: 2442–2443; **20**: 2796
   Asian giant **17**: 2442, 2443
   Black **17**: 2443; **20**: 2798
   Burmese peacock softshell **17**: 2443
   Chinese **17**: 2442
   Florida **17**: 2442, 2443, *2443*
   Smooth **17**: 2442, 2443
   Spiny **17**: 2442, *2442*, 2443
Stinking-jim *see* Musk turtle, Common
Stinkpot *see* Musk turtle, Common
Swift (*lizard*) *see* Fence lizard
Taipan **5**: 665; **20**: *2817*
Tegu **19**: 2650, *2650*
Terrapin **19**: 2664–2666; **20**: 2796
   Basking *see* Turtle, Painted
   Cape **20**: *2799*
   Diamondback **19**: 2664, *2664*, 2665–2666, *2665*, *2666*
   Marsh *see* Turtle, African helmeted
   Painted *see* Turtle, Painted
Thorny devil *see* Moloch
Toad, Horned *see* Horned lizard
Tortoise **17**: 2711–2712; **20**: 2796
   Aldabra **20**: 2799
   Aldabra giant **19**: 2712; **20**: 2799
   Desert **19**: 2711, 2712; **20**: 2799
   Galapagos giant (Indefatigable Island tortoise; Santa Cruz tortoise) **19**: *2711*, 2712, 2713; **20**: 2799
   Gopher **19**: 2711, *2713*; **20**: 2795
   Greek (spur-thighed tortoise) **19**: 2712, *2712*
   Hermann's **19**: 2713
   Leopard **19**: 2712; **20**: *2798*
   Long-necked *see* Snake-necked turtle, Common
   Marion's **19**: 2713
   Speckled Cape (speckled tortoise) **19**: 2712; **20**: 2795
   Texas **19**: 2711, 2712
   Water *see* Turtle, African helmeted
Trapjaw *see* Moccasin, Water
Tree snake **19**: 2732–2733
   Common **19**: 2733
   Golden **19**: 2732, 2733
   Paradise **19**: 2732, *2733*
Tuatara **20**: 2762–2764
   Gunther's (Brother's Island tuatara) **20**: 2763
Turtle
   African helmeted (water tortoise; marsh terrapin) **9**: 1190–1191, *1191*
   Amazon river (arrau) **9**: 1190, *1190*, 1191
   Big-headed **2**: 197
   Box **19**: 2711; **20**: 2796
   Broad-shelled river (Murray River turtle; giant snake-necked turtle) **17**: 2417
   Flapjack *see* Soft-shelled turtle
   Flapshelled *see* Flapshelled turtle
   Green (edible turtle) **8**: 1070–1072; **20**: *2797*, 2798

Hawksbill **9**: 1169–1170; **20**: 2798
   Hidden-necked **9**: 1190–1191; **20**: 2795, 2796
   Leatherback (leathery turtle; luth) **10**: 1423–1424; **20**: 2795, 2796, 2798, 2799
   Loggerhead **11**: 1481–1482; **15**: 2030–2031; **20**: 2797, 2798
   Long-headed **17**: 2442
   Mud *see* Mud turtle
   Murray **17**: *2417*
   Musk *see* Musk turtle
   Olive *see* Ridley turtle, Pacific
   Painted (painted terrapin; basking turtle; basking terrapin; eastern painted turtle; western painted turtle; southern painted turtle; Midland painted turtle) **13**: 1851–1853
   Pancake *see* Soft-shelled turtle
   Papuan (pitted-shell turtle; pig-nosed turtle; hog-nosed turtle; Fly River turtle) **13**: 1866–1867; **20**: 2796
   Ridley *see* Ridley turtle
   Side-necked **20**: 2795, 2796, *2799*
   Snake-necked *see* Snake-necked turtle
   Snapping *see* Snapping turtle
   Soft-shelled *see* Soft-shelled turtle
   Western swamp **9**: 1191; **20**: 2797
   see also Terrapin; Tortoise
Turtles 20: 2795–2799
Venomous animals 20: 2817–2820
Viper **13**: 1736; **17**: 2418, 2420; **20**: 2817
   Aspic *see* Asp
   Blow *see* Hognose snake
   Brazilian **17**: 2421
   Brown saw-scaled **6**: 749
   Gaboon **15**: 2058–2059
   Horned **17**: *2419*
   June *see* Asp
   Lataste's **1**: 97; **21**: 2919
   Long-nosed **1**: 97
   Lowland **6**: 749
   Pit *see* Pit viper
   Rhinoceros **15**: 2058; **17**: *2422*
Wall lizard **20**: 2849–2850
   Common **20**: 2849, *2849*, 2850
   Iberian **20**: 2849, 2850
   Italian (ruin lizard) **20**: 2849, 2850
Water snake **20**: 2874–2875
   Banded (Florida water snake; broad-banded water snake; Gulf salt marsh water snake; mangrove water snake) **20**: 2875
   Green **20**: 2875
   Northern (common water snake) **20**: 2874, 2875
   Red-bellied (copperbelly; yellow-bellied water snake; plain bellied water snake; blotched water snake) **20**: *2875*
Whip snake (whipsnake) **17**: 2421; **21**: 2919–2920
   Dark green (western whip snake; yellow-and-green snake) **21**: 2919, *2919*, 2920
   Green **21**: 2920
   Sonora **21**: 2919, 2920
   Striped (ornate whip snake; cedar racer) **21**: 2919, 2920
   Yellow-faced **21**: 2919, 2920
Whiptail **19**: 2650
   New Mexico **19**: 2650
   Western **19**: 2650
Woma **15**: 2079

Index of Scientific Names

The species described in the *International Wildlife Encyclopedia* are indexed here according to their genus and species names.

According to international zoological convention the scientific name of an animal species is always expressed in *italics*.

Numerals in **bold** refer to volume numbers; those in *italics* refer to picture captions.

A

Abarenicola **11**: 1498
Abastor erythrogrammus **17**: 2382
Abramis brama **2**: 275; **16**: 2182
Abudefduf saxatilis **5**: 638
Acanthaster planci **18**: 2518
Acanthisitta chloris **16**: 2165
Acanthiza
  chrysorrhoa **19**: 2669, 2670
  ewingi **19**: 2669
  inorata **19**: 2670
  pusilla **19**: 2669
  robustirostris **19**: 2669
Acanthochromis polyacanthus **5**: 639
Acanthocinus **11**: 1483, 1484
Acanthomyops nigra **14**: *1894*
Acanthorhynchus **9**: 1215
Acanthostracion quadricornis **20**: 2758, 2759
Acanthoxyla prasina **18**: 2530
Acanthurus
  bahianus **18**: 2577
  triostegus **18**: 2575–2576
Accipiter
  badius **18**: 2467
  brevipes **18**: 2467
  cooperii **18**: 2467
  gentilis **2**: 208; **8**: 1028; **19**: 2617
  melanoleucas **8**: 1027
  nisus **5**: 620; **18**: 2467, 2468
  novaehollandiae **8**: 1027
  striatus **18**: 2467
  trivirgatus **8**: 1027
  virgatus **18**: 2467
Acehlia australiensis **17**: 2311
Aceros nipalensis **9**: 1226
Achatina
  fulica **7**: 966, 967
  marginata **7**: 966, 967
Acherontia atropos **3**: 346; **9**: 1166, 1167, 1210
Acheta domesticus **5**: 598–599
Achorutes nivicola **18**: 2504
Acilius **5**: 683
Acinonyx jubatus **3**: 403, 428; **8**: 1040
Acipenser
  ruthenus **18**: 2556
  sturio **18**: 2556
  transmontanus **18**: 2556, 2557
Acmaea tessulata **11**: 1451, 1452
Acomys **18**: 2492
  cahirinus **18**: 2491
  capensis **18**: 2491
  dimidiatus **18**: 2491, 2492
Acridotheres
  cristatellus **12**: 1714
  ginginianus **12**: 1714
  tristis **12**: 1714, 1715
Acris **19**: 2723, 2724

Acrocephalus
  arundinaceus **5**: 620
  palustris **18**: 2458
  schoenobaenus **18**: 2459; **20**: *2858*
Acrochordus javanicus **20**: 2874, 2875
Acropora **4**: 528
Acryllium vulturinum **8**: 1097
Actinia **1**: 48
Actitis hypoleucos **16**: 2233
Actophilornis africanus **10**: *1398*
Acyrthosiphon **1**: 75
Adalia
  bipunctata **10**: 1394
  conglomerata **10**: *1395*
  decempunctata **10**: 1394
Addax nasomaculatus **1**: 25; **5**: 665
Aechmophorus occidentalis **8**: 1063
Aedes aegypti **12**: 1663, 1664
Aegolius funereus **13**: 1829
Aegotheles
  albertisii **13**: *1823*
  cristatus **13**: *1823*, 1824
Aegypius monachus **20**: 2840
Aenictus **1**: 91
Aeolidia papillosa **16**: 2301
Aeoliscus **17**: 2367
  punctulatus **17**: 2368
  strigatus **17**: 2367, *2367*, 2368
Aepyceros melampus **1**: 61; **9**: 1285
Aepyprymnus rufescens **15**: 2115, 2116
Aequidens latifrons **10**: 1366
Aequipecten **16**: 2248
Aeronautes saxatilis **19**: 2598
Aeshna **5**: 710–711
  cyanea **5**: *711*
  mixta **5**: 709
Aethiomastacembelus **18**: 2490
Aetobatus 730, *730*, 731
Afrixalus **15**: 2146
  fornasinii **15**: 2146, 2147
  spinifrons **15**: 2146, 2147
Afronandus **10**: 1422
Afropavo **14**: 1935
  congensis **14**: 1898
Agalychnis **7**: 982
  callidryas **7**: *903*; **19**: 2723
  spurrellii **7**: *983*
Agama **1**: 30, *30*, 31
Aganippe subtristis **19**: 2720
Agapornis
  canus **11**: 1495
  fischeri **11**: 1495
  lilianae **11**: 1495
  nigigenis **11**: 1495
  personatus **11**: 1495, 1496
  pullarius **11**: 1495
  roseicollis **11**: 1495
  swindernianus **11**: 1495
  taranta **11**: 1495
Agdistis **14**: 1991, 1992
Agelaius phoeniceus **2**: 224; **15**: 2142; **18**: 2458
Agelastes niger **8**: 1097
Agelena labyrinthica **17**: *2339*
Agkistrodon
  contortrix **14**: 1972, 1973
  himalayanus **14**: 1973
  piscivorus **14**: 1972, 1973
Aglais urticae **20**: 2812
Agouti
  paca **13**: 1838, 1839
  takzanowski **13**: 1838, 1839
Agriocharis ocellata **20**: 2784
Agrion **5**: 641, 642

Agriotes **4**: 467
  lineatus **4**: 467
  obscurus **4**: 467
Agrius convolvuli **9**: 1166
Agrostichthys parkeri **13**: 1767
Ahaetulla **17**: 2420
  mycterizans **21**: 2920
Ailuroedus dentirostris **3**: 399
Ailuropoda **2**: 174
  melanoleuca **7**: 961
Ailurus fulgens **10**: 1363
Aipysurus laevis **17**: 2309
Aix
  galericulata **5**: 716–717; **21**: 2925, 2971
  sponsa **5**: 716–717; **21**: 2925, 2970, 2971
Ajaya ajaja **21**: 2907
Alabama argillacea **13**: 1822
Alaria mustelae **7**: 871
Alauda
  arvensis **17**: 2396, 2397; **18**: 2459
  gulgula **17**: 2396
Alburnus alburnus **12**: 1619
Alca torda **15**: 2126
Alcedo
  atthis **10**: 1355, 1356
  cristata **10**: *1355*
Alcelaphus **1**: 62
  buselaphus **8**: 1148
Alces
  alces **12**: 1655
  alces gigas **5**: 655
Alcides **20**: 2805
  zodiaca **20**: 2805, 2806
Alcyonium **5**: 651
  digitatum **5**: 652; **17**: 2440
Aldabrachelys elephantina **20**: 2799
Alectoris
  graeca **14**: *1891*
  rufa **14**: 1891
Alectura lathami **11**: 1527
Alepidomus evermanni **17**: 2379
Alepisaurus
  brevirostris **10**: 1409
  ferox **10**: 1409
Aleuterus scriptus **20**: 2743
Aleyrodes proletella **21**: 2932, 2933
Allactaga **10**: 1324
Alle alle **11**: 1465; **16**: 2265
Allenopithecus nigroviridis **12**: 1643
Alligator
  mississippiensis **1**: 38, 39; **21**: 2906
  sinensis **1**: 38
Allocebus **10**: 1436
Allogalathea elegans **18**: *2506*
Allolobophora longa **6**: 737
Allurus fulgens **15**: 2136
Alopex **5**: 690
  lagopus **1**: 84; **5**: 691
Alopias
  pelagicus **19**: 2674, 2675
  superciliosus **17**: 2333; **19**: 2674, 2675
  vulpinus **19**: 2674, 2675
Alosa
  alosa **17**: 2327
  chrysochloris **17**: 2328
  fallax **17**: 2327
  pseudoharenpus **17**: 2327
  sapidissima **9**: 1184; **17**: 2327, 2328
Alouatta
  belzebul **9**: 1253
  belzebul ululata **9**: 1253
  caraya **9**: 1252, 1253

  fusca **9**: 1252, 1253
  palliatta **9**: 1252, 1253
  seniculus **9**: 1252, 1253
Alpheus **17**: 2364, 2366
Alsophis antiguae **15**: *2103*
Alucita hexadactyla **14**: 1991
Alytes **7**: 904
  cisternasii **12**: *1609*
  obstetricans **12**: 1610
Amandava
  amandava **1**: 103, 104
  formosa **1**: 103
  subflava **21**: 2888
Amandina
  erythrocephala **11**: 1544
  fasciata **11**: 1544
Amazona
  albifrons **14**: 1883
  amazonica **14**: 1884, 1885
  ochrocephala **14**: 1883
Amblonyx cinerea **16**: 2179
Amblyeleotris sungami **7**: 994
Amblyomma americanum **19**: 2682
Amblyrhynchus cristatus **11**: 1556
Amblysomus **7**: 1005
Ambystoma **11**: 1504
  dumerilii **1**: 109
  gracile **1**: 109
  jeffersonianum **11**: *1505*
  mexicanum **1**: 109, 110
  tigrinum **1**: 109; **11**: *1504*
Ameles spallanzania **11**: 1547
Amia calva **2**: 273; **5**: 686
Amitermes
  atlanticus **19**: 2658–2659
  meridionalis **19**: 2660
Ammodorcas clarkei **6**: 674; **7**: 952
Ammodramus savannarum **18**: 2460
Ammodytes
  hexapterus **16**: 2228
  tobianus **16**: 2227
Ammomanes deserti **5**: 594
Ammophila **9**: 1261, 1262
Ammospermophilus **8**: 1079, *1079*, 1080
Amoeba
  discoides **1**: *42*
  proteus **1**: 41–42
Ampedus cardinalis **4**: *467*
Ampeliceps coronatus **12**: 1714
Amphiprion **5**: 640
  clarkii **5**: 640
  ocellaris **5**: 640
  percula **5**: 639
  perideraion **5**: 640
Amphiura **2**: 281
Anabas testudineus **4**: 468, 469, *469*
Anableps
  anableps **7**: 888
  dowi **7**: 888
  microlepis **7**: 888
Analcidometra caribbea **16**: *2281*
Anaplectes **21**: 2896
Anaplophora glabripennis **11**: 1484
Anarhynchus frontalis **14**: 1989
Anarta myrtilli **13**: 1821, 1822
Anas **14**: 1959
  acuta **14**: 1957, 1958
  aucklandica **19**: 2647
  bahamensis **14**: 1957
  bernieri **19**: 2649
  capensis **19**: 2647
  castanea **19**: 2647
  crecca **5**: 715; **19**: 2647, 2648
  flavirostris **19**: 2647, 2649

Anas (continued)
  formosa **19**: 2647
  georgica **14**: 1957
  gibberifrons albogularis **19**: 2648
  laysanensis **19**: 2648
  nestiotis **19**: 2647
  platyrhynchos **5**: 717; **11**: 1525
  punctata **19**: 2647
Anastomus **18**: 2551
Anatis ocellata **10**: 1394–1395
Anax junius **5**: 710
Andrena **2**: 183
Andrias
  davidianus **9**: 1175
  japonicus **9**: 1175
Andricus **7**: 921, 922
Aneides **11**: 1503–1504
Anemonia **1**: 48
  mammilifera **9**: 1178
  sulcata **16**: 2301
Anergates atratulus **1**: 69
Anguilla
  anguilla **6**: 745, 746; **14**: 1980–1981
  rostrata **6**: 745, 746
Anguis fragilis **10**: 1430; **17**: 2407
Anhima cornuta **16**: 2262
Anhinga
  anhinga **5**: 645, 646; **21**: 2907
  melanogaster **5**: 645
  novaehollandiae **5**: 645
  rufa **5**: 645
Anisognathus flavinucha **19**: *2628*
Anniella
  geronimensis **10**: 1431
  pulchra **10**: 1431
  pulchra nigra **10**: *1430*
Anodonta **12**: 1711
Anodorhynchus
  glaucus **11**: 1516
  hyacinthinus **11**: 1514
  leari **11**: 1516
Anolis
  barkeri **1**: 53
  carolinensis **1**: 53
  equestris **1**: 53
  lionotus **1**: 53
  poecilopus **1**: 53
  sagrei **1**: *54*
Anomalurus
  beecrofti **16**: 2251
  peli **16**: 2250
Anomia **13**: 1832
Anomma **1**: 91
Anopheles **12**: 1662, 1663
  darlingi **12**: 1663
  gambiae **12**: 1663
Anoptichthys jordani **3**: 409, 410
Anous
  albivittus **13**: 1755
  minutus **13**: 1755, 1756
  stolidus **13**: 1755
  tenuirostris **13**: 1755
Anoxypristis cuspidata **16**: 2238
Anser **5**: 717; **14**: 1955
  anser **5**: 717; **8**: 1074; **14**: 1955
  brachyrhynchus **14**: 1956
  cygnoides **5**: 717
  fabalis **14**: 1955
  indicus **5**: 717
  serrirostris **14**: 1955
Anseranas semipalmata **5**: 714
Antechinomys **11**: 1577
Antechinus **11**: 1577, 1578
Antedon bifida **6**: 806
Antheraea **17**: 2377
  mylitta **17**: 2377
  paphia **17**: 2377
  pernyi **17**: 2377
  yamamai **17**: 2377
Anthias **8**: 1081

Anthocharis cethura **21**: 2928
Anthonomus grandis **21**: 2904
Anthornis melanura **2**: 191
Anthracoceros coronatus **9**: 1227
Anthreptes **18**: 2455, 2565, 2566, *2566*
Anthus
  antarcticus **14**: 1965
  bogotensis **14**: 1965
  chloris **14**: 1965
  pratensis **5**: 620; **14**: 1965
  rubescens **14**: 1966
  spinoletta **14**: 1965
  spraguei **14**: 1965
  trivialis **14**: 1965
  vervinus **14**: 1965
Antidorcas marsupialis **18**: 2500
Antilocapra americana **1**: 60; **15**: 2050
Antilope cervicapra **1**: 60; **2**: 227
Antrozous pallidus **20**: 2822, 2823
Anuraphis maidi-ridicis **14**: 1895
Anurida maritima **18**: 2504
Aonyx capensis **16**: 2179, 2180
Aotus **12**: 1642; **13**: 1748
  azarai **13**: 1748
  brumbacki **13**: 1748
  hershkovitzi **13**: 1748
  infulatus **13**: 1748
  lemurinus **13**: 1748
  miconax **13**: 1748
  nancymaae **13**: 1748
  nigriceps **13**: 1748
  trivirgatus **13**: 1748
  vociferans **13**: 1748
Apaloderma narina **20**: 2745
Apalone
  ferox **17**: 2443
  mutica **17**: 2443
  spinifera **17**: 2443
  spiniferus spiniferus **17**: *2442*
Apalopteron familiaris **9**: 1215
Apanteles glomeratus **9**: 1280; **21**: 2928
Apantesis **19**: 2687
Apatura **15**: 2071, 2072
Apeltes quadracus **18**: 2533
Aphelocoma coerulescens **3**: 426; **10**: 1317
Aphidius **9**: 1279
Aphis **1**: 74, 75; **21**: 2965
Aphomia sociella **20**: 2826
Aphredoderus sayanus **20**: 2755
Aphriza virgata **20**: 2790
Aphrodite aculeata **16**: 2293
Apion ulicis **21**: 2903, 2904
Apis
  dorsata **9**: 1208, 1209
  florea **9**: 1208, 1209
  indica **9**: 1208, 1209
  mellifera **9**: 1208, 1209
Aplochiton zebra **21**: 3015
Aplodinotus **5**: 602
Aplodontia rufa **12**: 1675
Aplysia **20**: 2275, 2276, *2276*
Apocephalus **10**: 1418
Apodemus **21**: 2975, 2976
Apogon **3**: 390
Apristurus **3**: 407
Aptenodytes
  forsteri **1**: 82; **6**: 773, 780; **14**: 1911
  patagonicus **6**: 779; **10**: 1359; **14**: 1912
Apterodromia evansi **11**: 1550
Apteronotus albifrons **10**: *1376*
Apteryx
  australis **10**: 1372, 1373
  haasti **10**: 1372
  oweni **10**: 1372
Aptychotrema **8**: 1103
Apulus muralis **13**: 1783
Apus
  apus **19**: 2598, 2599
  caffer **19**: 2598
  melba **19**: 2598

  pacificus **19**: 2598
  pallidus **19**: 2598
Aquila
  audax **5**: 679; **6**: 784
  chrysaetos **7**: 1002
  heliaca **7**: 1001
  pomarina **21**: 2906
  verreauxii **7**: 1001; **12**: 1589
Ara
  ararauna **11**: 1514
  chloroptera **11**: 1514
  glaucogularis **11**: 1516
  macao **11**: 1514, 1515
  militaris **11**: 1514
  rubrogenys **11**: 1516
Aramus guarauna **6**: 824; **11**: 1454
Araneus **13**: 1801; **18**: 2482
Araschnia levana **20**: 2811
Arbanitis **19**: *2720*
Arborophila **14**: 1891
Arcella **1**: 41
Archaeolemur **17**: 2373
Archamia **3**: *389*, 390
Archboldia papuensis **2**: 271
Archidoris
  flammea **16**: 2300
  pseudargus **16**: 2300
  pseudoargus **16**: 2299
Archilochus colubris **9**: 1255, 1256
Archispirostreptus **12**: *1613*
Archoplites interruptus **18**: 2569
Archteuthis **18**: 2509
  dux **18**: 2509
  longimanus **18**: 2509
  princeps **18**: 2509
Arctia **19**: 2686
  caja **19**: 2686, 2687
  hebe **19**: 2687
  quenseli **19**: 2687
  villica **19**: 2686–2687
Arctictis binturong **2**: 202
Arctocebus calabarensis **8**: 1015
Arctocephalus **7**: 916, *916*, 917, *917*, 918
Arctogalidia trivergata **13**: 1855
Arctonetta fischeri **6**: 754
Arctonyx collaris **1**: 120
Arctosa perita **21**: 2957
Ardea
  cinerea **9**: 1179
  herodias **9**: 1179, 1180; **21**: *2906*
  purpurea **9**: 1179
Ardeola
  ibis **6**: 750
  ralloides **9**: 1179
Ardeotis
  australis **3**: 338
  kori **3**: 337
  nigriceps **3**: 338
Arenaria
  interpres **20**: 2790, 2791
  melanocephala **20**: 2790
Arenicola **11**: 1498
  cristata **11**: 1497
  ecaudata **11**: 1497
  marina **11**: 1497, *1497*, 1498
Arenicolides **11**: 1498
Argiope **13**: 1801, 1802; **18**: 2481
Argonauta argo **12**: 1722
Argopteron **17**: 2391
Argynnis
  aglaia **7**: 898
  cydippe **7**: 898
  paphia **7**: 898
Argyreus hyperbius **7**: 898
Argyroneta aquatica **18**: 2481–2482; **20**: 2877
Argyropelecus **9**: 1162
Arhopala **2**: 244; **8**: 1117, 1118
Aricia agestis **2**: 243
Ariolimax **17**: 2408, 2409

Arion **17**: 2409
Arizona elegans **17**: 2370
Armadillidium **21**: *2973*, 2974
Armandia lidderdalei **18**: 2582
Arniocera erythropyga **3**: 322
Aromia **11**: 1484
Aromobates **1**: 94
Arothron **15**: 2062, 2063
Artactisteus **7**: 929
Artamus
  cinereus **21**: 2983
  cycanopterus **21**: 2983
  fuscus **21**: 2983, 2984
  leucorhynchus **21**: 2983
  minor **21**: 2983, 2984
  monachus **21**: 2984
  personatus **21**: 2983
  superciliosus **21**: 2984
Artemia salina **2**: 276, 277
Arthropleona **18**: 2504
Arvicola **21**: 2884, 2885
Ascaphus truei **9**: 1200
Asilus crabroniformis **6**: *859*
Asio
  flammeus **5**: 650; **7**: 893; **13**: 1828
  otus **13**: 1826
Aspidelaps **4**: 534
Aspidertes nigricans **17**: 2443; **20**: 2798
Aspidites **15**: 2079, 2080
Aspredinichthys tibicen **3**: 401
Astacopis gouldi **5**: 591, 592
Astacus **5**: 592
Astatatilapia **4**: 459
Asterias **18**: 2517, 2518
Asterina gibbosa **18**: 2518
Asthenes **13**: 1819
Astraea inaequalis **19**: 2706
Astrapia stephaniae **2**: 205
Astropecten **18**: 2517, 2518
Astrophytum **2**: 159
Astroscopus **18**: 2519, 2520
Ateles
  geoffroyi **12**: *1641*
  paniscus **18**: 2477, *2477*, 2478
Atelocynus microtis **17**: 2357
Athene
  blewitti **13**: 1826
  cunicularia **3**: 324; **13**: 1828
  noctua **13**: *1827*
Atherinopsis californiensis **17**: 2379
Athous **4**: 467
Atilax paludinosus **12**: 1636
Atlantisia rogersi **15**: 2108
Atrax **18**: 2482; **20**: 2820
Atretochoana **3**: 357
Atta **10**: 1417, 1418
Attacus
  atlas **1**: 102
  ricini **17**: 2377
Attagis **17**: 2319
Aulichthys japonicus **20**: 2766
Aulorhynchus flavidus **20**: 2766
Aurelia aurita **10**: 1320, 1321–1322
Australophocaena dioptrica **15**: 2024, 2025
Austropotamobius pallipes **5**: 592; **16**: 2175
Autographa gamma **13**: 1821
Automolus **13**: 1819
Avicularia **19**: *2634*
Axis
  axis **1**: 107–108
  calamianensis **1**: 107
  kuhlii **1**: 107
  porcinus **1**: 107
Aythya **14**: 1993
  americana **14**: 1993
  australis **14**: 1993
  ferina **14**: 1993, 1994
  fuligula **14**: 1993

innotata **5**: 717
marila **14**: 1995
novaeseelandiae **14**: 1993, 1995
nyroca **14**: 1993
valisineria **5**: 717; **14**: 1993

B

Bacillus
  rossii **18**: 2530
  thuringiensis **4**: 540
Badis **10**: 1421, 1422
Balaena mysticetus **16**: 2166, 2167;
  21: 2910
Balaeniceps rex **8**: 1123; **17**: 2352;
  18: 2551
Balaenoptera
  acutorostrata **16**: 2208, 2209; **21**: 2914
  borealis **16**: 2208; **21**: 2914
  edeni **16**: 2208; **21**: 2911
  musculus **2**: 247; **5**: 695; **14**: 1954;
  16: 2208; **21**: 2911
  physalus **16**: 2208; **21**: 2914
Balanus nubilis **1**: 138, 139
Balearica paronina **5**: 584
Balistapus undulatus **20**: 2744
Balistes carolinensis **20**: 2744
Balistoides conspicillum **20**: *2743*, 2744
Bambusicola **14**: 1891
Bandicota
  bengalensis **1**: 132, 133
  indica **1**: 132, 133
  savilei **1**: 132, 133
Barbatula barbatula **11**: 1473
Barbourisia rufa **21**: 2909
Barbus barbus **8**: 1089
Barnea **14**: 1939–1940
Bartramia longicauda **16**: 2234
Baryphthengus martii **12**: 1665
Basilicus **2**: 156, 157
Bassaricyon **10**: 1363
Bassariscus **16**: 2171, 2172
Bathothauma **18**: 2509
Bathycrinus carpenterii **16**: 2281
Bathylagus callorhinus **7**: 917
Bathylychnops exilis **7**: 888
Bathyraja abyssicola **17**: 2383
Bathysolea profundicola **17**: 2444
Batrachoseps **11**: 1503, 1504
Batrachostomus **7**: 900, 901
Bdelloura **6**: 849
Beatragus hunteri **1**: 63
Belone belone **12**: 1725
Berardius **2**: 170
Betta
  picta pugnax **6**: 821
  pugnax **6**: 821, 822
  splendens **6**: 821, 822
Bettongia **15**: 2116
  lesueuri **15**: 2115, 2116
  pencillata **15**: 2116, *2116*
Binneya **17**: 2408, 2409
Biorhiza **7**: 922
Bipalium kewense **6**: 849
Birgus latro **4**: 571; **9**: 1177; **16**: 2186,
  2187
Bison
  bison **2**: 215, 216; **8**: 1040
  bison bison **2**: 215, 216
  bison bison athabascae **2**: 215, 216
  bonasus **2**: 215, 216
Biston betularia **14**: 1915–1916
Bitis **15**: 2058
  arietans **15**: 2058, 2059
  caudalis **15**: 2059; **17**: 2370
  cornuta **15**: 2059
  gabonica **15**: 2058
  inornata **15**: 2059
  nasicornis **15**: 2058; **17**: *2422*
  peringueyi **15**: 2058

schneideri **15**: 2058
xerophaga **15**: 2058
Biziura lobata **18**: 2538
Blaniulus guttulatus **12**: 1611–1612
Blarina brevicauda **17**: 2359
Blastocerus dichotomus **18**: 2462, 2463
Blatta **4**: 492–493
Blattella **4**: 493
Bleekeria mitonkurii **16**: 2227
Blennius
  nigriceps **2**: 238
  ocellaris **2**: 238
  pholis **2**: 237, 238
Blepharipoda **18**: 2507
Boa constrictor **2**: 250; **17**: 2422
Bolbonota **9**: 1225
Bolborhynchus lineola **14**: 1880
Boleophthalmus **12**: 1684, 1685, 1686
Bombina **20**: 2819
  bombina **6**: 826
  maxima **6**: 825
  orientalis **6**: 825
  variegata **6**: 825
Bombus **3**: 317, *317*, *318*
Bombycilla
  cedrorum **21**: 2890
  garrulus **21**: 2890, 2891
  japonica **21**: 2890
Bombyliopsis abrupta **19**: 2610
Bombyx mori **3**: *345*; **17**: 2376, 2377
Bonasa umbellus **8**: 1084, 1085
Boophilus **19**: 2682
Boreogadus saida **4**: 495
Bos
  banteng **7**: 936; **21**: 3018
  gaurus **7**: 936–938; **21**: 3018
  indicus **21**: 3018, 3019
  mutus **21**: 3000, 3001
  mutus grunniens **21**: 3000, 3001
  sauveli **7**: 936
Boselaphus tragocamelus **1**: 60; **13**: 1754
Botaurus
  lentiginosus **2**: 220
  pinnatus **2**: 220
  poiciloptilus **2**: 220
  stellaris **2**: 220
Bothriocyrtum californicum **19**: 2720
Bothrops
  ammodytoides **14**: 1974
  asper **20**: 2817
  atrox **6**: 813
  bilineata **17**: 2421
Botia **11**: 1473, *1473*
Botryllus **17**: 2312
Boulengerochromis microlepis **4**: 459
Brachiones przewalskii **7**: 949
Brachycope **21**: 2896
Brachypelma boehmei **19**: 2633
Brachyteles arachnoides **18**: 2477, 2478
Bradypus **17**: 2400, 2402
  infuscatus **15**: *2113*
  torquatus **17**: 2401
  tridactylus **17**: 2401
  variegatus **17**: 2401
Branchiocerianthus **16**: 2273
Branchiomaldane **11**: 1498
Branchiostoma **10**: *1407*, *1407*, 1408
Branta **5**: 717
  canadensis **3**: 366
  leucopsis **1**: 141
  ruficollis **5**: 716
  sandvicensis **5**: 717; **9**: 1164
Brephidium **2**: 243, 244
Brevoortia tyrannus **2**: 245; **9**: 1184;
  12: 1598
Brookesia **3**: 418, 419
Brotogeris tirica **14**: 1878
Bubalus
  bubalis **9**: 1286, 1287
  depressicornis **9**: 1286

mindorensis **9**: 1286
quarlesi **9**: 1286
Bubo bubo **13**: 1825
Buccinum undatum **21**: 2917, 2918
Bucephala clangula **5**: 717
Buceros rhinoceros **9**: 1227
Bucorvus **9**: 1226
Budorcas taxicolor **12**: 1703–1705;
  19: 2622, 2623
Bufo
  bufo **4**: 510, 513
  calamita **12**: 1721
  marinus **7**: *905*; **11**: 1558
  viridis **12**: 1720
Bulinus **7**: 870
Bunaka **7**: 993
Bungarus **2**: 265; **17**: 2308
Bunodactis verrucosa **1**: 47
Buphagus **13**: 1830, 1831
Burhinus
  bistriatus **18**: 2545
  grallarius **18**: 2545
  oedicnemus **18**: 2545, 2546
  superciliaris **18**: 2545
  vermiculatus **18**: 2545
Burramys **5**: 702, 703
Busycon canaliculatum **21**: 2917
Buteo **9**: 1211
  buteo **3**: 352
  galapagoensis **5**: 650
  jamaicensis **3**: 353, 426
  lagopus **2**: 209; **3**: 351
  rufofuscus **3**: 351
Buthus occitans **16**: 2257
Bycanistes brevis **9**: 1226–1227

C

Cabassous **1**: 85
Cacajao **20**: 2803, 2804
Cacatua
  galerita **4**: 483
  haematuropygia **4**: 485
  leadbeateri **4**: 483
  moluccensis **4**: *484*
  roseicapilla **4**: 483
  sanguinea **4**: 483
  sulphurea **4**: 485
Cactospiza **5**: 649
Caecilia **3**: 357
Caecobarbus geertsi **3**: 410
Caecomastacembelus **18**: 2490
Caiman
  crocodilus **3**: 358, 359, *359*
  latirostis **3**: 358
  yacare **3**: 358
Cairina moschata **5**: 717; **12**: 1700
Calamospiza melanocorys **18**: 2458
Calcarius lapponicus **3**: 319
Calcochloris **7**: 1005
Calicalicus **20**: 2813, 2814
Calidris
  alba **16**: 2233
  alpina **16**: 2233
  canutus **16**: 2233
  ferruginea **16**: *2234*
  maritima **16**: 2233
  melanotos **20**: *2779*, 2833
Callaeas
  cinerea **21**: 2886, 2887
  cinerea cinerea **21**: 2886
  cinerea wilsoni **21**: 2886
Calliactis parasitica **9**: 1178
Callianassa subterranea **18**: 2507
Callicebus
  gigot **19**: 2696, 2697
  moloch **19**: 2696, 2697
  personatus **19**: 2696, 2697
  torquatus **19**: 2696, 2697
Callichthys callichthys **1**: 89; **20**: 2755

Callinectes sapidus **4**: 573
Callionymus **5**: 706, *706*, 707
Calliostoma zizyphinum **19**: 2706, 2707
Callipepla gambelii **15**: *2082*
Calliteuthis **18**: 2509
Callithrix **11**: 1563, 1564
  argentata **11**: *1563*
  argentata leucippe **11**: 1565
  aurita **11**: 1565
  chrysoleuca **11**: 1563
  flaviceps **11**: 1565
  geoffroyi **11**: 1563
  imperator **11**: *1565*
  jacchus **11**: 1563
Calliurichthys **5**: 707
Callocephalon fimbriatum **4**: 483
Callophrys **8**: 1118, *1118*
Callorhinus **7**: 916, 917
Callorhynchus milii **17**: *2330*
Calma glaucoides **16**: 2300
Calonectris diomedea **17**: 2334
Caloprymnus campestris **15**: 2115, 2116
Calyptorhynchus **4**: 485
Calyptura cristata **4**: 544
Camarhynchus **5**: 648, 649
Camelus **5**: 665
Campephaga phoenicea **5**: 621, 622
Campephilus **21**: 2977
Campodea **2**: 279
Camponotus **9**: 1205
  ligniperda **1**: 66
Camposcia retusa **18**: 2476
Campostoma anomalum **12**: 1618
Camptorhynchus labradorius **5**: 717
Campylorhamphus **21**: 2968
Campylorhynchus **21**: 2993, *2995*
Cancer pagurus **6**: 741, 742
Canis **5**: 690
  adustus **9**: 1291, 1292
  aureus **5**: 689; **9**: 1291, 1292
  familiaris **5**: 678; **10**: 1336; **19**: 2641
  latrans **3**: 426; **4**: 559
  lupus **5**: 689; **8**: 1054–1057; **19**: 2653;
  21: 2906
  mesomelas **5**: 692; **9**: 1291, 1292
  rufus **5**: 692; **8**: 1054
  simensis **5**: 692
Caperea marginata **16**: 2166, 2167;
  21: 2910
Capra
  aegagrus **9**: 1272; **21**: 2951, 2952
  aegagrus blythi **21**: 2952
  aegagrus chiltanensis **21**: 2952
  aegagrus cretica **21**: 2952
  caucasica **9**: 1273; **21**: 2951
  cylindricornis **9**: 1272, 1273; **21**: 2951
  falconeri **11**: 1560; **12**: 1672; **21**: 2951
  hircus **21**: 2951, *2951*, 2952
  ibex **9**: 1273; **12**: 1672; **21**: 2951
  ibex nubiana **9**: 1273
  ibex sibirica **9**: 1273
  pyrenaica **9**: 1272, 1273; **21**: 2951
  walie **9**: 1273; **21**: 2951
Capreolus
  capreolus **5**: 657; **16**: 2201, 2202;
  17: 2375
  pygargus **16**: 2201, 2202
Caprimulgus **13**: 1742
  aegyptius **13**: 1742
  carolinensis **13**: 1744; **14**: 2012
  europaeus **13**: 1742, 1743; **14**: 2013
  indicus **13**: 1742
  macrurus **13**: 1742
  ruficollis **13**: 1742
  vociferus **14**: 2012
Carapus **14**: 1901
  acus **14**: 1901–1902
  bermudensis **14**: 1901–1902
  homei **14**: 1902

Numerals in **bold** refer to volume numbers; those in *italics* to picture captions.

3083

Carassius
  auratus **8**: 1018, 1019
  auratus gibelio **8**: 1018
  carassius **3**: *396*; **8**: 1018
Carausius morosus **18**: 2530, 2531
Carcharhinus
  leucas **17**: 2329; **21**: *3007, 3008*
  limbatus **17**: 2332
  longimanus **14**: *1950*
  perezi **15**: *2148*; **17**: 2329
Carcharias taurus **17**: 2332
Carcharodon carcharias **8**: 1059; **17**: 2331
Carcinoscorpio **10**: 1354
Carcinus maenas **4**: 572; **17**: 2353, 2354
Cardinalis cardinalis **3**: 388
Cardium edule see *Cerastoderma edule*
Carduelis
  cabaret **15**: 2137
  cannabina **11**: 1458
  carduelis **3**: 368; **8**: 1016
  chloris **3**: 368; **8**: 1065; **11**: 1457
  flammea **15**: 2137, 2138
  flavirostris **8**: 1016; **11**: 1457
  hornemanni **15**: 2137
  psaltria **8**: 1016
  rostrata **15**: 2137
  spinus **3**: 368; **8**: 1016
  tristis **8**: 1016, 1017
Caretta caretta **11**: 1482; **20**: 2797
Carettochelys insculpta **13**: 1867; **20**: 2796
Cariama cristata **17**: 2321, 2322
Carnegiella **9**: 1161
Carollia perspicillata **7**: *907*
Carpodacus erythrinus **4**: 499
Carpophilus hemipterus **19**: *2611*
Caspiomyzon **10**: 1404
Cassiopeia **10**: 1321
Castor
  canadensis **2**: 178; **19**: 2616
  fiber **2**: 178; **21**: 2906
Casuarius
  bennetti **3**: 397
  casuarius **3**: 398; **6**: 782
  unappendiculatus **3**: 397
Catagonus wagneri **14**: 1903, 1904
Catharacta
  chilensis **10**: 1307
  maccormicki **10**: 1307–1308
  skua **10**: 1307
  skua antarctica **10**: *1308*
Catharsius **16**: 2254
Cathartes **4**: 518; **20**: 2841
  aura **4**: 518; **20**: 2788, 2840
  burrovianus **4**: 518; **20**: 2787
  melambrotus **4**: 518; **20**: 2787
Catharus
  fruscescens **19**: 2678
  guttatus **19**: 2678
  minimus **19**: 2678
Causus
  defilipii **13**: 1735
  lichtensteinii **13**: 1735
  resimus **13**: 1735
  rhombeatus **13**: 1735, 1736
Cavia **8**: 1099, 1100
Cebuella pygmaea **11**: 1563, 1564
Cebus
  albifrons **3**: 379, 380
  apella **3**: 379, 380
  capucinus **3**: 379, 380, 381
  olivaceous **3**: 379, 380
Celastrina argiolis **2**: 243
Celeus brachyurus **21**: 2979
Centriscus **17**: 2367, 2368
Centrocercus urophasianus **8**: 1084
Centropomus **13**: 1752
  parallelus **17**: 2430
Centropus
  bernsteini **4**: 549
  chlororhynchus **4**: 550

  cupreicaudus **4**: *550*
  menbeki **4**: 549
  monachus **4**: 549
  phasianus **4**: 549, 550
  senegalensis **4**: 549
  sinensis **4**: 549
  steerii **4**: 550
  superciliosus **4**: 549
Centruroides **16**: 2257
Cepaea hortensis **7**: 930
Cephalopholis **8**: *1083*
Cephalophus **1**: 61; **6**: 726, 727
  adersi **6**: 726, 727
  jentinki **6**: 726, 727
  monticola **6**: 726
  natalensis **6**: 727, 728
  sivicultor **6**: 726
  zebra **6**: 726, 727
Cephalopterus **4**: 543
Cephalorhynchus **5**: 698
Cephaloscyllium ventriosum **3**: 407, 408
Cephea cephea **10**: *1321*
Cephus pygmaeus **16**: 2241
Cepphus grylle **8**: 1094
Cerastes cerastes **17**: *2419*
Cerastoderma edule **4**: 488, 489
Ceratias holboelli **5**: 653, 654
Ceratophrys **9**: 1229, 1230, *1230*
Ceratophyllus
  columbae **6**: 850
  gallinae **6**: 850
  styx **6**: 850
Ceratotherium simum **15**: 2152, 2153
Cercartetus
  caudatus **5**: 703
  concinnus **5**: 703
  lepidus **5**: *702, 703*
  nanus **5**: *703*
Cercocebus **11**: 1541
  agilis **11**: 1541, 1542
  galeritus **11**: 1541, 1542
  torquatus **11**: 1541, 1542
Cercopithecus **8**: 1091, 1092
  ascanius **8**: 1091; **12**: 1643
  cephus **8**: 1091
  diana **8**: 1091; **12**: 1643
  hamlyni **8**: 1091
  mitis **8**: 1091; **12**: 1643
  neglectus **8**: 1091; **12**: 1643
  nictitans **8**: 1091
  patas **14**: 1897
Cerdocyon thous **4**: 567; **17**: *2357*
Ceresa bubalis **9**: 1225
Certhia
  americana **19**: 2721
  brachydactyla **19**: 2721
  familiaris **19**: 2721, 2722
Certhidea **5**: 648, 649
Cerura vinula **3**: 345; **15**: 2076, 2077
Cervus
  axis **21**: 2907
  duvauceli **5**: 657
  elaphus **5**: 656; **19**: 2653; **20**: 2854, 2855
  eldi **2**: 284, 285
  nippon **17**: 2375; **20**: 2854–2855
  schomburgki **5**: 658
  unicolor alfredi **5**: 658
Ceryle rudis **10**: 1355
Cetomimus indagator **21**: *2909*
Cetonia **4**: 487
Cetorhinus maximus **2**: 161; **17**: 2331
Chaenocephalus **9**: 1277, 1278
Chaenodraco **9**: 1277, 1278
Chaeradodus rhombicollis **11**: *1546*
Chaetophractus villosus **1**: 86
Chaetopterus **20**: 2767, *2767*, 2768
Chaetura **18**: 2484
  boehmi **18**: 2484
  gigantea **18**: 2484

  pelagica **18**: 2484, 2485; **19**: 2598
  vauxi **18**: 2484; **19**: 2598
Chaimarrornis leucocephalus **15**: 2139
Chalcides ocellatus **17**: 2390
Chalcolepidius zonatus **4**: 466
Chalcopsitta **11**: 1493
Chama **16**: 2248
Chamaea fasciata **1**: 113; **3**: 426
Chamaelo
  chamaeleon **3**: 418, 419
  dilepis **3**: 418, 419
  jacksoni **3**: 418, 419
  oweni **3**: 419
Champsocephalus **9**: 1277, 1278
Chanda **7**: 979
Channa **17**: 2412, 2413
Channichthys **9**: 1278
Chaoborus **12**: 1606
Chaos carolinensis **1**: 41, 42
Charadrius
  alexandrinus **14**: 1989
  dubius **14**: 1988
  hiaticula **14**: 1988
  montanus **14**: 1989
  morinellus **5**: 705
  semipalmatus **14**: 1988
  vociferus **10**: 1348–1349
Charina bottae **2**: 249
Charonia
  rubicunda **21**: 2917
  tritonis **18**: 2518
Chauliodus **10**: 1413; **20**: 2829
  barbatus **20**: 2829, *2829*
  danae **20**: 2829
  macouni **20**: 2829
  minimus **20**: 2829
  pammelas **20**: 2829
  schmidti **20**: 2829
  sloani **20**: 2829, 2830
  vasnetzovi **20**: 2829
Chauna **16**: 2262, 2263
Cheilinus undulatus **21**: *2991*
Cheilopogon heterurus **7**: 876
Cheirogaleus **10**: 1436
Cheiromeles **8**: 1115, 1116
Chelidonichthys lucerna **8**: 1108, 1109
Chelidoptera tenebrosa **15**: 2060
Chelifer **6**: 799, *800*
Chelmon rostratus **1**: 49
Chelodina
  expansa **17**: 2417
  longicollis **17**: 2416, 2417
Chelonia mydas **8**: 1071; **20**: 2798
Chelon labrosus **12**: 1692, 1693
Chelostoma **2**: 183
Chelydra serpentina **12**: 1688; **17**: 2425, 2426
Chenonetta jubata **21**: 2970, 2971
Chiasmodon niger **13**: *1773*
Chinactis **4**: 535
Chinchilla **12**: 1672
  brevicaudata **4**: 443–444
  laniger **4**: 443, 444
Chionaspis pinifoliae **16**: 2246, 2247
Chionis
  alba **17**: 2337, 2338
  minor **17**: 2337
Chionobathyscus **9**: 1278
Chionodraco **9**: 1277, 1278
Chirocephalus diaphanus **6**: 792, 793
Chironectes minimus **13**: 1792, 1793
Chironex fleckeri **10**: 1322; **20**: 2818–2819
Chironomus **12**: 1607; **20**: 2769
Chiroptes **16**: 2221, 2222
Chiroteuthis veranyi **18**: 2509
Chiroxiphia **11**: 1532, 1533
Chitala **6**: 803, 804
  chitala **6**: 803, *803*, 804
  lopis **6**: 803, 804

Chitra indica **17**: 2442
Chlamydera **2**: 269, 271
Chlamydochaera jefferyi **5**: 621
Chlamydosaurus kingii **7**: 895
Chlamydoselachus anguineus **7**: 897; **17**: 2330
Chlamydotis undulata **3**: 339
Chlamyphorus
  retusus **1**: 87
  truncatus **1**: 85
Chlamys **16**: 2248
  asperrimus **16**: *2249*
  spercularis **16**: 2248
Chlidonias
  hybridus **19**: 2661
  leucopterus **19**: 2661
  niger **19**: 2661
Chloebia gouldiae **8**: 1031
Chloephaga **5**: 714, 716
Chlorocebus **8**: 1091, 1092
  pygerythrus **8**: 1091; **12**: 1643
  sabaeus **8**: 1091; **12**: 1643
Chloroceryle **10**: 1355
Chlorocharis emiliae **21**: 2930
Chlorohydra viridissima **9**: 1265, 1266
Chlorophanes spiza **9**: 1213, 1214
Chlorophonia **19**: 2627
Chlorotalpa **7**: 1005
Chlorurus microrhinos **14**: 1889
Choeropsis liberiensis **9**: 1192
Choloepus **17**: 2400, 2401, 2402
Chondestes grammacus **18**: 2460
Chondropython **15**: 2080
  viridis **15**: 2079, 2080
Chordeiles minor **14**: 2013
Chorthippus **8**: 1036
  brunneus **8**: 1037
  parallelus **8**: 1037
Chortoicetes terminifera **11**: 1479
Chromis **17**: *2379*
  cyanea **5**: 639
Chromodoris
  bullocki **16**: *2300*
  quadricolor **16**: *2301*
Chrotogale owstoni **13**: 1855
Chrysaora **10**: 1320, *1320*
Chrysemys
  picta **13**: 1851, 1852
  picta belli **13**: 1851, 1852
  picta dorsalis **13**: 1851, 1852
  picta marginata **13**: 1851, 1852
  picta picta **13**: 1851, 1852
Chrysiridia **20**: 2805, 2806
Chrysobatrachus **15**: 2146
Chrysochloris **7**: 1005
Chrysococcyx
  americanus **5**: *617*
  caprius **5**: 617; **21**: 2898
  lucidus **5**: 617–618
Chrysocyon **5**: 690
  brachyurus **11**: 1540
Chrysolophus **11**: 1935
Chrysopa carnea **10**: 1392
Chrysopalax **7**: 1004, 1005
Chrysopelea **19**: 2732, 2733
Chrysops **9**: 1235, *1235*, 1236
Chrysotoxum **9**: 1251, *1251*
Chuna burmeisteri **17**: 2321
Cicadetta **4**: 456
Cichlasoma
  citrinellum **4**: *458*
  meeki **6**: 830; **10**: 1303
  nigrofasciatum **21**: 3015
  octofasciatum **10**: 1303
Cicindela hybrida **2**: 186
Cicinnarus magnificus **2**: 206
Ciconia **18**: 2551
  ciconia **18**: 2551, 2552
  maguari **18**: 2551
  nigra **18**: 2551

Cinclocerthia
 gutturalis **19**: 2671
 ruficauda **19**: 2671
Cinclodes **13**: 1819
Cinclus
 cinclus **5**: 680
 leucocephalus **5**: 680
 mexicanus **5**: 680, 681
 pallasii **5**: 680
 shulzi **5**: 680
Ciona intestinalis **17**: 2313
Cionus scrophulariae **21**: 2903, 2904
Circaetus cinereus **12**: 1589
Circus
 aeruginosus **8**: 1145
 cyaneus **8**: 1146
 macrourus **8**: 1145
 melanoleucus **8**: 1145
 pygargus **8**: 1145
 ranivorus **8**: 1145
Cirrothauma **13**: 1779
Cissa **11**: 1519–1520
Cissopis leveriana **19**: 2627
Cisticola **19**: 2618
Cistothorus **21**: 2994, 2995
Civettictis civetta **4**: 460, 461
Cladorhynchus leucocephalus **18**: 2540
Clamator **5**: 617
Clangula hyemalis **5**: 717
Clarias gariepinus **3**: *401*
Clavelina lepadiformis **17**: 2313
Cleidopus gloriamaris **18**: 2512
Clelia clelia **6**: 814
Clethrionomys **15**: 2129
 californicus **1**: 134; **15**: 2129
 centralis **15**: 2129
 gapperi **1**: 134; **15**: 2129, 2130
 glareolus **1**: 134–135; **15**: 2129
 rufocanus **15**: 2129
 rutilus **1**: 134; **15**: 2129
Clevelandia ios **7**: 995
Climacteris **19**: 2721
Clinocardium fucanum **4**: 488
Clione **16**: 2269, 2270
 limacina **16**: 2269, 2270, *2270*
Clitarchus hookeri **18**: 2530, 2531
Cloeon **12**: 1593
Clonorchis **7**: 871
Clossiana
 chariclea **7**: 898
 euphrosyne **7**: 898
 selene **7**: 898, *899*
Clupea **9**: 1184, 1185
Clypeaster rosaceus **16**: 2226
Clytoceyx rex **10**: 1357
Clytus **11**: 1484
Cnemidophorus
 neomexicanus **19**: 2650
 sexlineatus sexlineatus **19**: 2650
 sexlineatus viridis **19**: 2650
 tigris **19**: 2650
Cnemophilus macgregorii **2**: *206*
Cobitis taenia **11**: 1473
Coccidula rufa **10**: 1394
Coccinella septempunctata **10**: 1394
Cochlearius cochlearius **2**: 252; **8**: 1123;
 9: 1179
Cochliomyia hominivorax **6**: 859
Coendou **19**: 2729
 prehensilis **19**: 2728, *2729*
Coenobita **9**: 1176
Coenocorypha **17**: 2427, 2429
Colaptes **6**: 853
 auratus **6**: 853, 854
 campestris **6**: 853
 chrysoides **6**: 853
 fernandinae **6**: 853
 pitius **6**: 853
Coleonyx **7**: 943
Colias **21**: 2929

Colibri **9**: *1257, 1257*
Colinus virginianus **15**: 2082
Colius
 colius **12**: 1676, 1677
 indicus **12**: 1676
 leucocephalus **12**: 1676
 macrourus **12**: 1676
 striatus **12**: 1676
Collessendeis colossea **17**: 2311, *2311*
Collocalia **19**: 2601, 2602
Colluricincla **21**: 2923
Colobus **4**: 501, 502
 angolensis **4**: 501, 502
 guereza **4**: 501, 502
 satanus **4**: 501, 502
Colostethus **1**: 94
Coluber **17**: 2421
 constrictor **15**: 2101, 2102
 constrictor flaviventris **15**: 2102
 viridiflavus **21**: 2919, 2920
Columba
 livia **6**: 850; **16**: 2195
 palumbus **21**: 2981
 palumbus palumbus **21**: 2980
Compsilura concinnata **19**: 2611
Condylura cristata **12**: 1625
Conepatus
 chinga **17**: 2393–2394
 humboldti **17**: 2393–2394
 leuconotus **17**: 2393–2394
 mesoleucus **17**: 2393–2394
 mesoleucus telmalestes **17**: 2394
 semistriatus **17**: 2393–2394
Conger
 conger **4**: 521, 522
 verrouxi **4**: *521*
Conilurus albipes **18**: 2536
Connochaetes **1**: 61
 gnou **16**: 2216; **21**: 2948, 2949
 taurinus **8**: 1040; **21**: 2948, 2949
Conolophus **9**: 1281, 1282
Conophilia **9**: 1216
Conuropsis carolinensis **14**: 1880
Conus **20**: 2818
 gloriamaris **4**: *520*
 legatus **4**: *520*
 mediterraneus **4**: 520
 striatus **4**: *519*
Convoluta **6**: 849
Copris hispanus **16**: 2253–2254
Coracias **16**: 2204
 abysinnicus **16**: 2204
 benghalensis **16**: *2205*
 caudata **16**: 2205
 garrulus **16**: 2204
Coracina
 lineata **5**: 621
 mindanensis **5**: 622
 newtoni **5**: 622
 novaehollandiae **5**: 621
 ostenta **5**: 622
 pectoralis **5**: 622
Coragyps atratus **20**: 2787, 2840
Corallus caninus **17**: *2418*
Corcorax melanorhamphos **4**: 451;
 11: 1522
Cordulegaster boltoni **5**: 711
Cordylophora **16**: 2273
Coris julis **21**: 2991
Coronula **1**: 138
Corucea zebrata **17**: 2390
Corvus **5**: 610
 albicollis **15**: 2122
 albus **5**: *611*
 brachyrhynchos **5**: 610, 611
 caurinus **5**: 610
 corax **15**: 2122, 2123; **18**: 2455
 corone **4**: 452; **5**: 610
 coronoides **15**: 2122
 crassirostris **15**: 2122

cryptoleucus **15**: 2122
 dauuricus **10**: 1300
 frugilegus **4**: 452; **5**: 610; **10**: 1300;
 16: 2207
 mellori **15**: 2122
 monedula **10**: 1300, 1301; **18**: 2457
 ossifragus **5**: 610
 rhipidurus **15**: 2122
 splendens **5**: *610*; **18**: 2457
 tasmanicus **15**: 2122
Corydon sumatranus **2**: 283
Corydoras armatus **1**: *88*
Corynorhinus townsendii **20**: 2821
Coryphaena **5**: 693, 694
Coryphaeshna ingens **5**: 710
Corythaeola cristata **20**: 2780
Corythoichthys flavofasciatus **14**: 1960
Coscinoscera hercules **3**: 344
Coscoroba coscoroba **18**: 2586; **21**: 2925
Cossus cossus **7**: 992
Cossypha natalensis **6**: 776
Cotinga **4**: 543, 544
Cotinis **4**: 487
Cottus gobio **3**: 314–315
Coturnix **15**: 2082, 2083
Cracticus
 cassicus **3**: 340
 louisiadensis **3**: 340
 mentalis **3**: 340
 nigrogularis **3**: 340
 quoyi **3**: 340
 torquatus **3**: 340, *340*, 341
Crago franciscorum **17**: 2364
Crangon crangon **17**: 2364, 2365
Crango septemspinosus **17**: 2364
Crania **10**: 1406
Cranioleuca **13**: 1819
Craseonycteris thonglongyai **2**: 167
Crassotrea **13**: 1832, 1834
Craterscelis murina **1**: *113*
Cratogeomys **14**: 1997
Crax
 alberti **5**: 625
 alector **5**: 623
 blumenbachii **5**: 625
 daubentoni **5**: *623*
 rubra **5**: 623, 624
Creatophora cinerea **18**: 2521
Crematogaster **3**: 1894
Crenicichla **4**: 459
Crepidula fornicata **11**: 1452
Crex crex **4**: 542, *542*
Cricetomys gambianus **7**: 964, 965
Cricetus cricetus **8**: 1130
Cricosaura typica **13**: 1745
Crocidura **12**: 1710
Crocodylus
 acutus **5**: 603, 604; **21**: 2907
 intermedius **5**: 603
 niloticus **5**: 603, 604
 palustris **5**: 603
 porosus **5**: 603; **21**: 2907
 rhombifer **5**: *605*
Crocuta **19**: 2610
 crocuta **3**: 374–375; **9**: 1267, 1268
Crossarchus obscurus **12**: 1636
Crossoptilon **14**: 1934–1935
Crotalus **3**: 426; **15**: 2119
 adamanteus **15**: 2119, 2120
 atrox **3**: 426; **15**: 2120
 cerastes **17**: 2370
 horridus **15**: 2120
 ruber **15**: 2120
 scutulatus **15**: *2120*
 tigris **15**: 2120
 unicolor **15**: 2120
 viridis **15**: *2119*, 2120
 viridis viridis **15**: *2119*, 2120
 willardi obscurus **17**: 2422
Cryodraco **9**: 1278

Cryptobranchus alleganiensis **9**: 1175
Cryptochloris **7**: 1005
Cryptolaemus montrouzieri **10**: 1396
Cryptomys hottentotus **12**: *1629*
Cryptoprocta ferox **17**: 2373
Crypturellus soui **19**: 2691
Ctenocephalides **6**: 850, 851, 852
Ctenodactylus **8**: 1104, 1105
Ctenodiscus crispatus **18**: 2518
Ctenolepisma **2**: 279
Ctenomys **20**: 2771, 2772
Ctenopoma **4**: 468, *468*
Cubanichthys cubensis **19**: 2705
Cuculus
 canorus **5**: 617, 618; **9**: 1224;
 21: 2898
 glandarius **5**: 619
 micropterus **5**: 713
 varius **5**: 620
Cucumeria **16**: 2271, 2272, *2272*
Culaea inconstans **18**: 2533
Culex **12**: 1662, *1662, 1663, 1664*
Culicoides **12**: 1607
Cuon alpinus **5**: 672, 677, 690
Cupido minimus **2**: 243
Curaeus curaeus **13**: 1805
Curculio **21**: *2903, 2903,* 2904
Cursorius
 coromandelicus **4**: 551
 cursor **4**: 551
 rufus **4**: 551
 temminckii **4**: 551
Cyanea **10**: 1320; **14**: 1980; **21**: 2935
Cyanerpes **9**: 1213, *1214*
Cyanocitta **10**: 1317
Cyanocorax yncas **10**: 1318
Cyanopica cyanus **11**: 1520
Cyanopsitta spixii **11**: 1516
Cyclanorbis senegalensis **17**: 2443
Cyclocephala **4**: 487
Cyclopes didactylus **1**: 57
Cyclopterus lumpus **11**: 1499, 1500
Cyclorana **19**: 2723
 alboguttatus **5**: 665
 platycephala **10**: 1399; **20**: 2869
Cyclothone **2**: 278
Cyclura **9**: 1281, 1282
Cygnus
 atratus **18**: 2587
 buccinator **5**: 714; **18**: 2587; **20**: 2757
 columbianus **5**: 717; **18**: 2587; **20**: 2779
 columbianus bewickii **18**: 2586
 columbianus columbianus
 18: 2586–2587
 cygnus **5**: 714; **18**: 2586
 melanocoryphus **18**: 2587
 olor **5**: 715; **18**: 2586
Cymatogaster aggregata **18**: 2574
Cymbirhynchus macrorhynchus **2**: 282
Cynictis pencillata **12**: 1595, 1636
Cynocephalus **7**: 882, 883
Cynogale bennetti **4**: 460
Cynomys **8**: 1040
 gunnisoni **15**: *2038*
 leucurus **15**: 2039
 ludovicianus **15**: 2038, 2039
 mexicanus **15**: 2040
 parvidens **15**: 2040
Cynopterus **7**: 907
Cynoscion **5**: 601, 602
Cyphonia clavata **9**: 1225
Cypraea
 chinensis **4**: *557*
 europaea **4**: 556, 557
 moneta **4**: 557
 spadicea **4**: 556, 557
 talpa **4**: *556*
Cyprinodon
 macularius **19**: *2705*
 variegatus variegatus **19**: 2705

Numerals in **bold** refer to volume numbers; those in *italics* to picture captions.

3085

Cyprinus carpio 3: 396
Cypseloides 19: 2598, *2599*
Cypselurus 7: 875, 876
Cypsiurus parvus 19: 2598
Cystophora cristata 9: 1220

D

Dacelo 10: 1355, 1385, 1386
Dacnis 1213; **18**: 2558
Dactylopius coccus 16: 2247
Dactyloptena 7: 880, 881
Dactylopterus 7: 880, 881
Dallia pectoralis 1: 34
Dama dama 5: 656; 6: 795
Damaliscus
   dorcas 2: 260; 16: 2216
   lunatus 21: 2950
   phillipsii 2: 240; 16: 2216
Damaliscus dorcas 1: 63
Damon variegatus 14: *1906*
Danaus plexippus 3: 345; 12: 1633;
   20: 2810
Danio rerio 21: 3015
Daphnia magna 5: 644
Daphnis neri 9: *1167*
Dascyllus albisella 5: 638
Dasyatis americana 18: 2544
Dasycercus cristicauda 11: 1577, 1578
Dasymutilla occidentalis 20: 2815, *2815*
Dasypeltis 6: 748, *748*, 749; 9: 1203
Dasypoda hirtipes 2: *184*
Dasyprocta 1: 33
Dasypus 1: 85, *85*
Dasyuroides byrnei 11: *1577*
Dasyurus
   albopunctatus 11: 1571
   geoffroii 11: 1571
   hallucatus 11: 1571, 1572
   maculatus 11: 1572
   spartacus 11: 1571
   viverrinus 11: 1571, 1572
Daubentonia madagascariensis 1: 112
Deilephila elpenor 9: *1167*
Delichon urbica 12: 1590
Delphinapterus leucas 2: 194; 21: 2911
Delphinus 4: 506, 507
Demansia psammophis 21: 2919, 2920
Dendragapus 8: 1084
Dendraster exentricus 16: 2226
Dendrelaphis punctulata 19: 2733
Dendroanthus indicus 20: 2843
Dendroaspis 2: 265
   angusticeps 11: 1530, *1530*, 1531
   jamesonii 11: 1530, 1531
   polylepis 11: 1530, 1531; 12: 1597;
     15: 2103
   viridis 11: 1530, 1531
Dendrobaena 6: 734, 735
Dendrobates 1: 94; 20: 2819
   auratus 1: 94
   azureus 1: 93
   granuliferus 1: 93
   histrionicus 1: 94
   lehmanni 1: 94
   leucomelas 1: 93
   pumilio 1: 93, 94, *94*
Dendrochirus 6: 836
Dendrocincla anabatina 21: 2969
Dendrocoelum 6: *848*, 849
Dendrocolaptes certhia 21: 2968
Dendrocopus 19: 2654
Dendrocygna 5: 714; 21: 2925
   arborea 21: 2926
   arcuata 21: 2925
   autumnalis 21: 2926
   bicolor 21: 2925
   eytoni 21: 2925
   javanica 21: 2925
   viduata 21: 2925

Dendrogale murina 19: 2730
Dendrohyrax 9: 1270, 1271
Dendroica
   caerulescens 21: 2985
   coronata 21: 2986
   kirtlandii 21: 2985
   pensylvanica 21: 2985
   petechia 4: *553*; 21: 2985
   striata 21: 2985
Dendrolagus
   bennettianus 19: 2726, 2727
   dorianus 19: 2727
   goodfellowi 19: 2726, 2727
   inustus 19: 2727
   lumholtzi 19: 2726, 2727
   matschiei 19: 2726, 2727
   mbaiso 19: 2727
   scottae 19: 2727
   spadix 19: 2727
   ursinus 19: 2727
Dendronotus 16: 2300
Dendropoma petraea 21: 2990
Dendropos minor 19: 2617
Dentalium 777, 778, *778*
Dermacentor 19: 2682
Dermatobia hominis 6: 858
Dermochelys coriacea 10: 1424; 20: 2795
Dermophis mexicanus 3: 356
Deroceras 17: 2409
   reticulatum 17: 2408
Desmana moschata 5: 666, 667
Desmocerus 11: 1484
Desmodema lorum 13: 1768
Desmodus rotundus 20: 2807, 2808
Desmognathus 11: 1503, 1504
Devonia semperi 16: 2272
Dexiosoma caninum 19: 2610
Diadema 17: 2368
Diaemus youngi 20: 2807–2808
Dialeurodes 21: 2932, 2933
Dialommus fuscus 7: 888
Diapheromera femorata 18: 2531
Diaphus ostenfeldi 10: *1413*
Diatryma 17: 2322
Dicaeum 7: 869, *869*
Dicerorhinus sumatrensis 15: 2152, 2153
Diceros bicornis 15: 2152, 2153
Dicrostonyx 10: *1432*, 1433
Dicrurus
   hottentottus 5: 712, 713
   leuciphaeus 5: 713
   macrocercus 5: 712
   paradiseus 5: 712
   remifer 5: 712
Didelphis virginiana 13: 1792, 1793
Dineutes americanus 21: 2922
Dinomys branicki 13: 1838
Diodon holacanthus 15: 2023
Diomedea
   albatrus 16: 2268
   bulleri 1: 36
   exulans 1: 35; 16: *2267*
   innutabilis 1: 37
   irrorata 1: 35
   melanophris 1: 36
Diopsittaca 11: 1514
Diphylla ecaudata 20: 2808
Diplogale hosei 13: 1855
Diplolepis 7: 921, 922
Diploria 4: 528
Dipodomys
   compactus 10: 1340
   elephantinus 10: 1340
   insularis 10: 1340
   margaritae 10: 1340
   microps 10: 1340
   ordii 10: *1339*, 1340
   spectabilis 10: 1340, *1340*
   stephensi 10: 1340
Dipsosaurus 9: 1281, 1282

Dipturus batis 17: 2383, 2384
Dipus 10: 1324
Dipylidium caninum 6: 852
Dispholidus typhus 2: 264–265; 6: 749;
   13: 1736
Dolichonyx oryzivorus 13: 1805;
   18: 2457; **20**: 2833
Dolichotis 8: 1099, 1100
Dolichovespula 9: 1233
Doliolum 16: 2224, *2224*
Dolomedes 18: 2482
Dorcatragus megalotis 1: 61; 2: 190
Dorcopsis veterum 20: 2846
Dorosoma 17: 2328
Dorylus 1: 90, 91, 92
Doryrhompus dactyliophorus 14: 1960
Draco 11: *1466*
Dromas ardeola 4: 569
Drosophila 7: 909, 910
Drurya antimachus 18: 2581, 2582
Drycopus pileatus 21: 2979
Dryomys 5: 699
Dryophius 19: *2732*
Dugesia 6: 849
Dugong dugon 6: 725
Dulus dominicus 13: 1854
Dumatella carolinensis 3: 399, 400;
   4: 555
Dunckerocampus multiannulatus
   14: *1960*
Dynamena 16: 2274
Dynastes neptunus 2: 186
Dysdera crocata 21: 2974
Dytiscus 5: 682, 683

E

Echeneis 15: 2148, 2149
Echidnophaga gallinacea 6: 851
Echiichthys vipera 21: 2901, 2902
Echinarachnius parma 16: 2225, 2226
Echinocardium cordatum 17: 2314
Echinocyamus pusillus 16: 2226
Echinoprocta rufescens 19: 2729
Echinops telfairi 19: 2657, *2657*
Echinorhinus brucus 5: 686
Echinostrephus molaris 17: 2314–2315
Echinus esculentus 17: 2315
Echiodon 14: 1901
   drummondi 14: 1901, 1902
Eciton 1: 90, 91, 92, *92*
Eclectus roratus 14: *1883*
Ectobius 4: 493
Ectopistes migratorius 3: 328
Edovum puttleri 4: 504–505
Egretta
   alba 6: 750, 751; 9: 1179
   ardesiaca 9: 1181
   garzetta 6: 750
   sacra 6: 750
   thula 6: 750
Eidolon 7: 908
Eisenia 6: 734, *734*, 735
Elacatinus oceanops 7: 995
Elachistodon westermanni 6: 748
Elanus
   caeruleus 10: 1367–1368
   leucurus 2: 209; 10: 1368
Elaphe 15: 2117
   guttata 15: 2117, 2118
   guttata emoryi 15: 2117
   guttata rosacea 15: 2117
   longissima 15: 2117
   obsoleta 15: 2117, 2118
   quatuorlineata 15: 2117
   scalaris 15: 2117
   situla 15: 2117
   triaspis 15: 2117, 2118
   vulpina 15: 2117, 2118
Elaphodus cephalophus 12: 1696

Elaphurus davidianus 14: 1920
Elasmucha grisea 17: 2345
Electrophorus electricus 6: 763
Eledone cirrhosa 13: 1779
Elephantulus 6: 775
   myurus 6: *774*
Elephas maximus 6: 768
Fleutherodactylus planirostris 8: 1067;
   12: 1610
Eliomys quercinus 5: 699
Emarginula elongata 11: 1451
Emberiza
   hortulana 3: 319
   schoeniclus 3: 320; 18: 2457
   striolata 3: 319
Emerita 18: 2507
Empidonax virescens 20: 2802
Emydura maquarii 17: *2417*
Encarsia formosa 2: 185; 3: 416, 417;
   21: 2933
Encheliophis 14: 1901
   homei 14: 1901
Enchelycore pardalis 12: 1657, 1658
Endomychura hypoleuca 16: 2264–2265
Engraulis 1: 45, 46
Enhydra lutris 16: 2179, 2295
Enhydrina 20: 2820
Enicognathus 14: 1878, 1879
Enneacanthus 18: 2569, *2570*
Ensatina 11: 1504, 1505
Ensifera ensifera 9: 1255
Ensis 15: 2127, 2128, *2128*
Entamoeba 1: 41
Enterochromis 4: 459
Entomobrya mawsoni 18: 2504
Eochoreutes 10: 1324
Eopsetta grigorjewi 8: 1122
Eos histrio 11: 1494
Eotragus 1: 60
Eozapus 10: 1324
Ephippiorhynchus 18: 2551
Ephydra hians 6: *858*
Epicauta velata 13: 1783
Epicopeia polydora 20: 2806
Epicrates 251
Epinephelus 8: 1081, *1081*
   itajara 8: 1082–1083
   lanceolatus 8: 1082
   striatus 8: 1083
   tukula 8: *1081*
Epipedobates 1: 94
Epixerus 13: 1859, 1860
Epomophorus 7: 907–908
Epomops franqueti 7: 908
Equus 15: 2052
   asinus 8: 1040; 21: 2939, 2940
   asinus asinus 21: 2939
   burchelli 8: 1040; 21: 3009, 3010
   burchelli antiquorum 21: 3009
   burchelli boehmi 21: 3009
   burchelli burchelli 21: 3011
   burchelli quagga 21: 3011
   caballus 8: 1039; 15: 2052; 21: 2941
   ferus 15: 2052
   grevyi 21: 3010
   hemionus 21: 2939
   hemionus kulan 21: 2939
   hemionus onager 21: 2939, 2940
   kiang 21: 2939, 2940
   przewalskii 8: 1040; 15: 2052–2054
   quagga 16: 2216
   zebra 21: 3009, 3010
   zebra hartmannae 21: 3010
   zebra zebra 21: 3009–3010
Eremitalpa 7: 1004, 1005
Erethizon dorsatum 19: 2728, 2729
Eretmochelys imbricata 9: 1170;
   20: 2798
Eriborus terebrans 4: 540
Ericerus pe-la 16: 2247

Erignathus barbatus **1**: 82; **16**: 2286
Erinaceus europaeus **9**: 1170, 1172
Eriocampa limacina **16**: 2240
Eriococcus **16**: 2246, 2247
Erionota **17**: 2391, 2392
Erisoma lanigerum **1**: 74
Eristalis **9**: 1251
Erithacus rubecula **16**: 2192; **19**: 2678
Erynnis toges **17**: 2391
Esacus **18**: 2545
Eschrichtius robustus **5**: 695; **8**: 1053;
   21: 2910
Esox **14**: 1945, 1946
Estrilda **2**: 2888, *2888*, 2889
Ethmidium maculatum **12**: 1598
Eubalaena
   australis **16**: 2166, 2167; **18**: 2510;
      21: 2910
   glacialis **16**: 2166, 2167; **21**: 2910
Eublepharus macularius **7**: 943
Eubranchipus vernalis **6**: 792
Eucidaris thouarsii **17**: 2315
Eucometis penicillata **19**: 2628
Eucrossorhinus dasypogon **17**: 2331;
   21: 2955
Euderma maculatum **2**: 168; **20**: 2821
Eudia pavonia **3**: 346
Eudocimus **9**: 1274, 1275
Eudromia elegans **19**: 2691
Eudynamys scolopacea **5**: 620, 713
Eudyptes **14**: 1912, 1913, 1914
Eudyptula **14**: 1911
Eulacestoma nigropectus **21**: 2923
Eulemur **10**: 1436, *1437*
Euliphera **2**: 244
Eumeces **17**: 2390
Eumenes coarctatus **9**: 1261
Eumetopias jubatus **16**: 2282, 2283, 2289
Eumomota superciliosa **12**: 1665
Eunectes **1**: 43–44; **17**: 2419
Euoticus **3**: 330
Euphausia superba **10**: 1387, *1387*, 1388
Euphilotes bernardino allyni **2**: 244
Eupholus quinitaenia **21**: *2904*
Euphonia **19**: 2627
Euphractus **1**: 85
Euphydryas aurinia **7**: 898
Euplagia quadripunctaria **19**: 2687
Euplectes **2**: 213
   afer **2**: 213
   aurea **2**: 213
   hordeaceus **2**: 213, 214
   orix **2**: 213
Euplectrus comstockii **3**: 417
Eupodotis afra **3**: *338*
Eupoecilia **4**: 487
Euroleon **1**: 64, 65
Eurostopodus
   argus **14**: 2013
   guttatus **13**: 1742
   mystacalis **13**: 1742
Eurycea **11**: 1503, 1504
Euryceros prevostii **20**: 2813
Eurygaster integriceps **17**: 2344
Eurypegasus **16**: 2290, 2291, *2291*
Eurypyga helias **18**: 2568
Eurystomus **16**: 2204
Euschemon rafflesia **17**: 2392
Eusphyra blochii **8**: 1126, 1127
Eutamias **4**: 448, 449; **19**: 2653
   minimus **4**: 449
   sibiricus **4**: 448, 449
   townsendii **4**: 449
Eutermes triodiae **19**: 2660
Eutoxeres **9**: 1255
Eutrigla gurnardus **8**: 1108, 1109
Eviota **7**: *993*, 994
Exelastis atomosa **14**: 1991
Exocoetus **7**: 875, 876
Extatosoma tiaratum **18**: 2531, *2531*

F

Falco
   alopex **10**: 1345
   biarmicus **8**: 1110
   cherrug **8**: 1110
   columbarius **12**: 1603
   cuvieri **8**: 1198
   eleonorae **2**: 210
   femoralis **2**: *208*
   peregrinus **2**: 209; **8**: 1029, 1110;
      14: 1922
   punctatus **2**: 210; **10**: 1345
   rupicoloides **10**: 1345
   rusticolus **2**: 210; **8**: 1111
   sparverius **2**: 209; **10**: 1345
   subbuteo **9**: 1198, 1199
   tinnunculus **10**: 1345, 1346
Falculea palliata **20**: 2813
Fannia canicularis **9**: 1241, 1242
Fasciola **7**: 870, *870*, 871
Felis
   bieti **3**: 404; **21**: *2945*
   canadensis **11**: 1506, 1507; **12**: 1672
   caracal **3**: 386, 404
   concolor **19**: 2653; **21**: 2907
   iromotensis **3**: 405
   lynx **11**: 1506, 1507; **12**: 1672
   lynx wrangeli **11**: *1507*
   manul **3**: 404; **11**: 1559
   margarita **3**: 403
   pardina **11**: 1506, 1507
   serval **3**: 403
   sylvestris **21**: 2945–2947
   sylvestris lybica **3**: *403*; **21**: 2945, *2945*
   wiedii **3**: 404
   yagouaroundi **10**: 1314
Felis rufus **2**: 254
Felovia vae **8**: 1104
Fennecus zerda **2**: 165; **5**: 665; **6**: 810
Ficedula
   albicollis **7**: 873
   hypoleuca **7**: 872–873; **18**: 2459
   parva **7**: 873
   semitorquata **7**: 873
Ficopomatus enigmaticus **6**: 801
Fissurella costaria **11**: 1451, 1452
Flectonotus pygmaeus **11**: 1574, *1574*
Foetorepus goodenbeani **5**: 706
Fondiator **7**: 876
Forcipomyia **12**: 1607
Forficula auricularia **6**: 738, 739, *740*
Formica **21**: 2964
   aquilonia **21**: 2965
   fusca **21**: 2964–2965
   lugubris **21**: 2965
   pratensis **21**: 2964, 2965
   rufa **1**: 68; **21**: 2964
   sanguinea **21**: 2965
Formicarius analis **1**: 55, 56
Forpus passerinus **14**: 1885
Fossa fossa **4**: 460
Francolinus francolinus **7**: 890
Frankliniella occidentalis **19**: 2677
Fratercula
   arctica **15**: 2065, 2066; **16**: *2228*
   cirrhata **15**: 2065
   corniculata **15**: 2065
Fregata
   andrewsi **7**: 893
   aquila **7**: 893
   magnificens **7**: *891*, 892
   minor **7**: *892*
Fringilla coelebs **3**: 415
Fulica
   alai **4**: 523
   americana **4**: 523, 524
   atra **4**: 523
   cornuta **4**: 523
   cristata **4**: 523
   gigantea **4**: 523

Fulmarus
   glacialis **7**: 914; **16**: 2268
   glacialoides **7**: 913; **17**: 2435
Funambulus **13**: 1859, 1860
   pannanti **13**: *1860*
   tristriatus **13**: *1859*
Funarius **13**: 1819
   rufus **13**: 1819, 1820
Fundulus zebrinus **21**: 3015
Fungia **4**: 528
Funiculini quadrangularis **16**: 2298

G

Gadus morhua **4**: 496; **6**: 836
Galago
   moholi **3**: 329, 330
   senegalensis **3**: 329, 330
Galagoides **3**: 329, 330
Galathea
   dispersa **18**: 2506
   intermedia **18**: 2506
   squamifera **18**: 2506
   strigosa **18**: 2506, 2507, *2507*
Galatheathuria aspera **16**: 2271
Galea **8**: 1099, 1100
   musteloides **8**: *1100*
Galenys pyrenaicus **5**: 666, 667
Galeocerdo cuvier **17**: 2332
Galeodes **18**: 2454, *2454*
Galerida **5**: 593, 594
Galidia elegans **12**: 1637
Galidictis striata **12**: 1635
Gallinago **17**: 2427
   gallinago **17**: 2427, 2428
   media **17**: 2427
   paraguaiae **17**: *2427*
   stenura **17**: 2429
Gallinula
   angulata **12**: 1649
   chloropus **12**: 1649, 1650
   comeris **12**: 1649; **15**: 2108
   pacifica **12**: 1649
   silvestris **12**: 1649
   tenebrosa **12**: 1649
Gallirallus **15**: 2108
Gallus
   gallus **10**: 1329, *1330*
   lafayetti **10**: 1329
   sonnerati **10**: 1329
   varius **10**: 1329
Gamasellus racovitzai **1**: 82
Gammarus **17**: 2364
Gangara **17**: 2391
Garrulus glandarius **10**: 1317, 1318;
   14: 1884; **18**: 2457
Gasteropelecus **9**: 1161
Gasterosteus aculeatus aculeatus
   18: 2533, 2534
Gastrosaccus spinifer **13**: 1795
Gastrotheca **11**: 1573; **19**: 2723
   marsupiata **11**: 1573, 1574
   mertensi **11**: 1574
Gavia
   adamsii **11**: 1485
   arctica **11**: 1485; **21**: 2906
   immer **11**: 1485, 1486
   stellata **11**: 1485
Gavialis gangeticus **7**: 955; **21**: 2907
Gazella **1**: 61; **7**: 939, 940
   dama **7**: 940
   dorcas **7**: 940
   granti **7**: 940
   soemmerringi **7**: 940
   spekei **7**: 940
   thomsoni **1**: 61; **7**: 940; **8**: 1040
   walleri **7**: 952
Gekko
   gecko **7**: 943
   vittatus **7**: *944*

Gelochelidon nilotica **19**: 2661
Gempylus serpens **17**: 2414, 2415
Genetta **7**: 947
   abyssinica **7**: 946
   genetta **7**: 946, 947
   tigrina **7**: 946, 947
   victoriae **7**: 946
Geocapromys **9**: 1263, 1264
Geochelone
   elephantopus **20**: 2799
   gigantea **19**: 2712
   hoodensis **19**: *2711*
   nigrita **19**: 2712
   pardalis **19**: 2712; **20**: *2798*
Geococcyx
   californianus **3**: 426; **16**: 2184, 2185
   velox **16**: 2184
Geocolaptes olivaceus **21**: 2979
Geogale aurita **16**: 2657
Geomys **14**: 1997
Geopsittacus occidentalis **14**: 1880
Geosciurus inauris **12**: 1595
Geositta **13**: 1819
Geospiza **5**: 649
Geotria **10**: 1403, 1404
Geospiza **5**: 649
Geotrupes **16**: 2253, 2254
Gerbillus pyramidum **7**: *949*
Geronticus eremita **9**: 1274
Gerrhonotus **3**: 426
Gerris **14**: 2009
   lacustris **14**: 2009, *2009*
   remigis **14**: 2009
   thoracicus **14**: 2009
Gibbula cineraria **19**: 2706
Gilpinia hercyniae **16**: 2241
Giraffa camelopardalis **7**: 976
Glareola **15**: 2041, 2042
Glaucidium
   minutissimum **13**: 1825
   passerinum **13**: 1829
Glaucomys
   sabrinus **7**: 884; **19**: 2616
   volans **7**: 884, 885
Glirulus japonicus **5**: 699
Glis glis **5**: 699
Globicephala **14**: 1952
   macrorhynchus **14**: 1952, 1953
   melas **14**: 1952, 1953
Glossina **6**: 859; **20**: 2760
   morsitans **20**: 2760
   palpalis **20**: 2760
Glossopsitta
   porphyrocephala **11**: 1488
   pusilla **11**: 1488
Glyph **13**: 1736
Glyphorhynchus spirurus **21**: *2969*
Gnathoncus **6**: 852
Gnathonemus petersii **13**: 1749
Gobio gobio **8**: 1090
Gobiomorus dormitor **7**: 993
Gobiosoma **4**: 533; **7**: 995
Gobius cobitis **7**: 993
Goliathus giganteus **2**: 186
Gomys **8**: 1040
Gonepteryx rhamni **21**: 2929
Gonodactylus guerinii **11**: 1551
Gopherus
   agassizii **19**: 2711, 2712; **20**: 2799
   berlandieri **19**: 2711, 2712
   polyphemus **19**: 2711; **20**: *2795*
Goraschius **13**: 1737
Gordius **9**: *1237*, 1238
Gorgonocephalus **2**: 158, 159
Gorilla
   gorilla beringei **1**: 73; **8**: 1024–1025
   gorilla gorilla **8**: 1023–1024
   gorilla graueri **1**: 71; **8**: 1024
Gorytes **9**: 1262
Gracula religiosa **12**: 1714
Grallina **11**: 1522, 1523

Numerals in **bold** refer to volume numbers; those in *italics* to picture captions.

3087

Grammistes sexlineatus **8**: 1083
Graphium **18**: 2582
  antiphates itamputi **18**: *2582*
Graphiurus **5**: 699
Graphocephala coccinea **9**: 1224
Graphosoma lineatum **17**: 2344
Grapsus grapsus **4**: *570*
Gromphadorhina **4**: 493
  portentosa **4**: 493
Grus
  americana **5**: 584; **20**: 2757
  canadensis **5**: 584
  grus **5**: 584, 585
  japonensis **5**: 584
  leucogeranus **5**: 585
  rubicunda **5**: 584
Gryllus **5**: *598*, 599
Gulo gulo **21**: *2959*, *2959*
Guttera pucherani **8**: 1097, 1098
Gymnachirus nudus **17**: 2444
Gymnammodytes semisquamatus **16**: 2227
Gymnarchus niloticus **10**: 1377; **13**: *1749*, 1750
Gymnetrus russelii **13**: 1767
Gymnocephalus cernua **14**: 1917, 1918
Gymnocorymbus ternetzi **19**: 2667
Gymnogyps californianus **4**: 516; **20**: 2841
Gymnomuraena zebra **12**: 1658
Gymnothorax **12**: 1659
  dovii **12**: *1659*
  funebris **12**: 1657–1658
Gymnotus carapo **10**: 1377
Gypaetus barbatus **10**: 1402; **20**: 2840
Gypohierax angolensis **20**: 2841
Gyps **20**: 2841
  africanus **20**: *2839*
  bengalensis **10**: 1401
  fulvus **10**: 1401
Gyrinophilus **11**: 1504
Gyrinus **21**: 2921
  natator **21**: 2922

H

Hadropithecus **17**: 2373
Haemaphysalis **19**: 2682
Haematopota **9**: 1235, 1236
Haematopus
  ater **13**: 1835
  bachmani **13**: 1835
  fuliginosus **13**: 1835
  leucopodus **13**: 1835
  meadewaldoi **13**: 1835
  ostralegus **13**: 1836
  palliatus **13**: 1835
Haemopis sanguisuga **10**: 1428
Halcyon **10**: 1355, 1357
Haliaeetus
  albicilla **6**: 837
  leucocephalus **1**: 124; **2**: 208; **21**: 2907
  leucoryphus **6**: 837
  vocifer **6**: 838
  vociferoides **6**: 837
Haliastur indus **10**: 1367
Halichoerus grypus **8**: 1050; **16**: 2289
Halichondria **16**: 2299
Haliotis
  cracherodii **1**: 20, 21
  fulgens **1**: 21
  pourtalese **1**: 20
  rufescens **1**: 20, 21
  tuberculata **1**: *20*
Halobaena caerulea **15**: 2045
Halobates **14**: 2008, 2009
Hamirostra melanosternon **6**: 784
Hapalemur **10**: 1436
Hapalochlaena maculosa **13**: 1781; **20**: 2819–2820

Hapalomys
  delacouri **11**: 1566, 1567
  longicaudatus **11**: 1566, 1567
Harpedon nehereus **2**: 257
Harpia harpyja **2**: 208; **8**: 1143
Harpyia **15**: 2076
Harpyopsis novaeguineae **8**: 1142
Harrisina americana **3**: 321
Hassar **1**: 88
Hediste diversicolor **15**: 2104, *2104*, 2105, *2105*
Helarctos malayanus **18**: 2563
Heleophryne
  natalensis **7**: 957, *957*
  purcelli **7**: 957
  rosei **7**: 957
Helicoverpa **2**: 256
Heliocolenus dactylopterus **16**: 2260, 2261
Heliocopris **16**: 2254
Heliopais personata **6**: 823
Heliopsyche **3**: 354
Heliornis fulica **6**: 823, 824; **21**: 2907
Heliothrips haemorrhoidalis **19**: 2676
Helix aspersa **7**: 930–932
Heloderma
  horridum **7**: 972, 974
  suspectum **7**: 972, 973, 974; **20**: 2818
Helostoma temminckii **10**: 1366
Hemachatus **2**: 265
Hemaris **9**: 1168
Hemicentetes semispinosus **19**: 2656, 2657
Hemichromis bimaculatus **10**: 1326
Hemidactylium **11**: 1503, 1504
Hemiechinus auritus **9**: *1171*
Hemigaleus derbyanus **13**: 1855, 1856
Hemigrammus **19**: 2667
Hemigrapsus oregonensis **17**: 2353
Hemimerus **6**: 740
Hemiphaga novaeseelandiae **7**: 912
Hemipus picatus **5**: 621
Hemitaurichthys **1**: 49
Hemitheconyx caudicinctus **7**: 943
Hemitragus **19**: 2612, 2613
Henicopernis longicauda **9**: 1211
Henricia sanguinolenta **18**: 2518
Hepialus **19**: 2603, 2604
Herichthys cyanoguttatum **4**: 459
Hermodice carunculata **16**: 2293
Herpestes
  auropunctatus **12**: 1637
  ichneumon **12**: 1635
  urva **12**: 1635
Hesperoleon **1**: 65
Heteractis crispa **5**: 640
Heteralocha acutirostris **21**: 2886
Heterocephalus glaber **12**: 1629
Heterodon **9**: 1203
Heterodontus **15**: 2027, 2028
Heterogale parvula **12**: 1635
Heterohyrax **9**: 1270, 1271
Heterometrus **16**: 2258
Heteronetta atricapilla **18**: 2538
Heteropsammia **4**: 528
Heterospizias meridionalis **2**: *210*
Heteroxenia **17**: 2441
Hieraaetus **12**: 1589
Himantopterus dohertyi **3**: 321
Himantopus
  himantopus **18**: 2540
  mexicanus **18**: 2540, 2541
  novaezelandiae **18**: 2540
Hinnites giganteus **16**: 2248
Hippocamelus
  antisensis **5**: 658; **18**: 2463–2464
  bisulcus **5**: 658
  bisulcus **18**: 2463–2464
Hippocampus
  bargibanti **16**: 2277
  hippocampus **16**: 2279

  ingens **16**: 2277
  reidi **16**: 2278
  whitei **16**: 2277
  zosterae **16**: 2277
Hippodamia **10**: 1396
Hippoglossus **8**: 1121, 1122
Hippolais pallida **20**: *2859*
Hippopotamus amphibius **9**: 1192
Hippotragus
  equinus **16**: 2215, 2216
  leucophaeus **1**: 63; **16**: 2216
  niger **1**: 61; **13**: *1830*; **16**: 2215, 2216
  niger variani **16**: 2216
Hirundapus caudacutus **19**: 2600
Hirundichthys **7**: 876
Hirundo
  daurica **2**: 151
  medicinalis **10**: 1428
  pyrrhonota **2**: 151
  rupestris **12**: 1590
  rustica **2**: 152; **12**: 1590
  semirufa **2**: 151
  senegalensis **2**: 151
Histioteuthis bonelliana **18**: 2509
Histrio histrio **1**: 51
Holacanthus **1**: 49, 50
Holocentrus rufus **18**: 2512
Holohalaelurus regani **3**: 407, 408
Holothuria **16**: 2271, *2271*
  aspus **14**: 1901
  forskali **16**: 2272
  tubulosa **14**: 1902
Homarus
  americanus **4**: 571; **11**: 1475
  gammarus **4**: 571; **11**: 1475, 1476
Homopus signatus **19**: 2712; **20**: 2795
Homo sapiens **1**: 70–71
Hopkinsia rosacea **16**: 2299
Hormogaster **6**: 734, 735
Houbaropsis bengalensis **3**: 338
Hubbsiella sardina **8**: 1087, 1088
Huso huso **18**: 2556
Hyaena hyaena **9**: 1267, 1268
Hyas **18**: 2476
Hybopsys gracilis **12**: 1618
Hydra
  oligactis **9**: 1265, 1266
  vulgaris **9**: 1265, 1266, *1266*
Hydrobates pelagicus **18**: 2554, 2555
Hydrochaeris hydrochaeris **3**: 383; **8**: 1040; **9**: 1270
Hydromyles **16**: 2270
Hydrophasianus chirurgus **9**: 1289
Hydrophilus
  piceus **20**: 2863, *2863*, 2864
  triangularis **20**: 2863
Hydrophis **20**: 2820
  inornatus **17**: 2309
Hydropotes inermis **4**: 447; **5**: 655
Hydroprogne caspia **19**: 2661
Hydrosaurus **1**: 30
Hydrurga leptonyx **1**: 82; **10**: 1360; **11**: 1450
Hyla **7**: 982
  arenicolor **19**: 2724
  cinerea **19**: 2724
  crucifer **7**: 903; **19**: 2724
  faber **19**: 2725
  goeldii **19**: 2725
  graciosa **19**: 2724
  meridionalis **19**: *2725*
  miliaria **19**: 2725
  venulosa **7**: 983
  zeteki **19**: 2725
Hylobates **1**: 70; **7**: 968, 969
  lar **1**: *71*; **7**: 969
  syndactylus **7**: 969
Hylobius abietis **21**: 2903, 2904
Hylochoerus meinertzhageni **7**: 959
Hylocichla mustelina **19**: 2678

Hylophilus **20**: 2831
  decurtatus **20**: 2831
Hylophylax naevioides **1**: 55
Hylotrupes **11**: 1484
  bajulus **11**: 1484
Hymeniacidon sanguinea **16**: 2300
Hymenolaimus malacorhynchos **19**: 2710
Hymenopus coronatus **11**: 1547, 1548
Hyomys goliath **7**: 965
Hyperolius **15**: 2146
  horstockii **15**: 2147
  marmoratus **15**: 2146, 2147
  quinquevittatus **15**: 2146
  tuberilingius **15**: 2146
Hyperoodon
  ampullatus **2**: 170, 171
  planifrons **2**: 170, 171
Hyperoplus lanceolatus **16**: 2228
Hyphessobrycon
  copelandi **19**: 2667
  flammeus **19**: 2667
  rodwayi **19**: 2667
Hypnos monopterygium **6**: 764
Hypoatherina temminckii **17**: 2379
Hypogastrura
  armata **18**: 2504
  viatica **18**: 2504
Hypomesus pretiosus **17**: 2410
Hypoplectrus **8**: 1081
Hyposita corallirostris **20**: 2813
Hypoterus **13**: 1752
Hypsignathus monstrosus **7**: 908
Hypsipetes **3**: 308
Hypsiprymnodon moschatus **15**: 2115
Hypsypops **5**: 639
  rubicunda **5**: 638
Hysterocarpus traski **18**: 2573
Hystrichopsylla talpae **6**: 852
Hystrix **5**: 595, 596

I

Icerya purchasi **16**: 2246, 2247
Ichthyomyzon **10**: 1404
Ichthyophaga **6**: 837
Ichthyophis **3**: 357
  glutinosus **3**: 356
Ictalurus
  furcatus **3**: 402
  punctatus **3**: 401–402
Icteria virens **21**: 2985
Icterus galbula **1**: 127; **20**: 2833
Ictonyx striatus **21**: 3021
Idiurus zenkeri **16**: 2251
Idolum diabolicum **11**: 1547, 1548
Iguana **9**: 1282
  iguana **9**: 1281, 1282; **11**: *1469*
Ilyodrilus **20**: 2769
Inachis io **20**: 2810–2811
Inachus dorsettensis **18**: 2476
Indicator
  archipelagicus **9**: 1217
  indicator **9**: 1206, 1217–1218
  minor **9**: *1218*
  variegatus **9**: 1217–1218
  xanthonotus **9**: 1217
Indostomus paradoxus **20**: 2766
Inia geoffrensis **5**: 697; **16**: 2176, 2177, 2178
Iphiclides podalirius **18**: 2580
Irediparra gallinacea **9**: 1289
Irena
  cyanogaster **6**: 790
  puella **6**: 790, 791
Ischnomyrmex **9**: 1156, 1157
Isistius **17**: 2332
Isoodon
  macrourus **1**: *131*
  obesulus **1**: 131

Issoria lathonia **7**: 898
Istiblennius enosimae **2**: 237
Istiophorus
    albicans **16**: 2219
    platypterus **16**: 2219, 2220; **19**: 2605
Isurus oxyrinchus **15**: 2020–2021
Ixobrychus **2**: 220
Ixodes **19**: 2682
    dammini **19**: 2681, 2682
    pacificus **19**: 2682
    ricinus **19**: 2682
Ixoreus naevius **19**: 2678

J

Jabiru **18**: 2551
Jacana spinosa **9**: 1289, 1290
Jaculus **10**: 1324
Janthina **15**: 2073
Jordanella floridae **19**: 2705
Jordanicus **14**: 1901, 1902
Julbernardia pellegriniana **16**: 2251
Jynx **21**: 2996, 2997

K

Kallima **10**: 1331, 1332
Kassina **15**: 2146
Katsuwonus pelamis **20**: *2773*
Kerodon **8**: 1099, 1100
Ketupa **6**: 842; **13**: 1827
    blakistoni **6**: 843
    flavipes **6**: 842
    ketupa **6**: 842
    zeylonensis **6**: 843
Kheper aegyptiorum **16**: 2252
Kinosternon **20**: 2795
    hirtipes murrayi **12**: 1687
    subrubrum **12**: 1687, 1688
Klauberina riversiana **13**: 1745, 1746
Kobus **10**: 1381
    ellipsiprymnus **1**: 62; **20**: 2866
    kob **1**: 62; **10**: 1381, 1382
    leche **1**: 62; **10**: 1425, 1426; **21**: *2907*
    megaceros **1**: 62; **10**: 1425, 1426
    vardonii **10**: 1381, 1382
Kogia **18**: 2472
    breviceps **18**: 2473; **21**: 2911
    simus **18**: 2473; **21**: 2911
Kogmania **21**: 2974

L

Labia minor **6**: 738, 739
Labrochromis **4**: 459
Labroides phthirophagus **21**: 2992
Labrus bimaculatus **21**: 2991
Lacerta
    agilis **16**: 2232
    viridis **8**: 1068, 1069
    vivipera **20**: 2838
Lachesis muta **14**: 1972, 1973
Lactophrys **20**: *2758*
Laevicardium **4**: 488
Lagenorhynchus obscurus **5**: 698
Lagidium
    crassus **20**: 2834
    peruanum **20**: 2834
    wolffsohni **12**: 1672
Lagocephalus lagocephalus **15**: 2063
Lagopus **20**: 2778
    lagopus **8**: 1084; **15**: 2055
    leucurus **15**: 2055, 2056
    mutus **15**: 2055
Lagostomus maximus **20**: 2834, 2835
Lagostrophus fasciatus **20**: 2846
Lagothrix **18**: 2477
    flavicauda **21**: 2987, 2988
    lagotricha **21**: 2987, 2988
Lagurus lagurus **18**: 2528, 2529

Lalage **5**: 621
Lama
    glama **11**: 1470, 1471
    guanicoe **11**: 1470, 1471
    pacos **11**: 1470, 1471
Lamna
    ditropis **15**: 2021
    nasus **15**: 2021; **17**: 2331
Lampetra **11**: 1403, 1403, 1404
Lampris **13**: 1790, 1791
Lampropeltis
    getulus **4**: 535; **10**: 1361, 1362
    pyromelana **10**: 1361, *1361*
    triangulum **10**: 1361
    zonata **10**: 1361
Lampyris **6**: 828; **7**: 988
    noctiluca **7**: 987
Lanius **17**: 2361
    collurio **17**: 2361
    exbitor **17**: 2361
    ludovicianus **17**: 2361, 2362
    senator **17**: *2362*
Lanthanotus borneensis **6**: 733; **12**: 1638
Laothoe populi **9**: 1166
Larosterna inca **19**: 2661
Larus
    argentatus **9**: 1188
    atricilla **2**: 230
    fuscus **9**: 1189
    melanocephalus **2**: 230
    ridibundus **2**: 231
Lasiorhinus
    krefftii **21**: 2961, 2962
    latifrons **21**: 2961, 2962, *2963*
Lasiurus
    borealis **20**: 2821, 2822, *2822*
    cinereus **20**: 2821, 2822
Lasius **14**: 1895, *1895*
Laterallus ruber **5**: 582
Laterinia laterinia **9**: 1225
Lates niloticus **13**: 1752
Laticauda
    colubrina **17**: 2308, 2309
    crockeri **17**: 2308
    semifasciata **17**: 2309
Latimeria chalumnae **4**: 497–498
Latrodectus **2**: 235
    hesperus **2**: 236, *236*
    mactans **2**: 235, *235*, 236
    variolus **2**: 236
Lehmannia **17**: 2408, 2409
Leicesteria **12**: 1664
Leimadophis epinephelus **20**: 2819
Leiopelma
    archeyi **9**: 1200
    hamiltoni **9**: 1200; **20**: 2869
    hochstetteri **9**: 1201
Leioptilus lienigianus **14**: 1992
Leiothrix lutea **16**: 2190
Leipoa ocellata **11**: 1509, 1527, 1528
Leithrix **1**: 113
Lemmiscus curtatus **18**: 2528, 2529
Lemmus **10**: 1433
Lemniscomys **21**: 3016, 3017
Lemur **10**: 1435, 1436
Leontopithecus **19**: 2625, 2626
Leopardus
    pardalis **13**: 1777
    wiedi **13**: 1776
Lepas anatifera **1**: 138, 139
Lepidocephalus guntea **11**: 1474
Lepidochelys
    kempii **20**: 2797; **21**: *2907*
    olivacea **20**: *2796*
Lepidocolaptes
    affinis **21**: 2968
    angustirostris **21**: 2968
    souleyetii **21**: 2969
Lepidoderma albohirtum **11**: 1557
Lepidosaphes ulmi **16**: 2246, 2247

Lepisma saccharina **2**: 279
Lepisosteus
    oculatus **7**: 928
    osseus **7**: 929
    platostomus **7**: 929
Lepomis
    auritus **18**: 2569
    gibbosus **16**: *2174*; **18**: 2569
    megalotis **18**: 2569
Leporillus **18**: 2535, *2535*, 2536
Leptailurus serval **17**: 2326
Leptasterias groenlandica **18**: 2518
Leptidea sinapis **21**: 2929
Leptinotarsa decemlineata **4**: 505
Leptocircus **18**: 2582
Leptoconops **12**: 1607
Leptodactylus fallax **8**: 1066
Leptomicrurus **4**: 534
Leptomithrax spinulosus **18**: 2475
Leptomyrmex **9**: 1205
Leptonychotes weddelli **1**: 82; **16**: 2210,
    2288; **21**: 2899–2900
Leptopelis **15**: 2146, 2147
Leptopterus
    chabert **20**: 2813
    madagascarinus **20**: 2813
Leptoptilos **18**: 2551
    crumeniferus **11**: 1554; **18**: 2551
    dubius **11**: 1553
    javanicus **11**: 1553
Leptosiren paradoxa **11**: 1501, 1502
Lepus **18**: 2561
    alleni **10**: 1304, 1305
    americanus **8**: 1136, 1138; **11**: 1508;
        18: 2561
    californicus **8**: 1138; **10**: 1304, 1305
    callotis **10**: 1304, 1305
    europaeus **8**: 1136
    timidus **8**: 1136, 1137
    timidus scoticus **8**: 1137
    townsendii **8**: 1138; **10**: 1304, 1305
Leto staceyi **19**: 2603–2604
Leuciscus
    cephalus **12**: 1618; **16**: 2182
    leuciscus **12**: 1618
Leucopsar rothschildi **12**: 1716
Leuresthes tenuis **8**: 1087, 1088, *1088*
Leurognathus marmoratus **11**: 1503
Lialis burtonis **10**: 1431
Liasis **15**: 2080
    childreni **15**: 2079
Libellula quadrimaculata **5**: 710
Lichanura trivirgata **2**: 249
Lichenostomus melanops **9**: 1215, 1216
Lichmera indistincta **9**: 1215
Ligia **2974**
    baudiensis **17**: 2398–2399
    oceanica **17**: 2398
Lima **12**: 1713
Limacina **16**: 2270
    retroversa **16**: 2269–2270
Limanda limanda **5**: 635
Limax **17**: 2409
    maximus **17**: 2409
Limenitis
    archippus **3**: 345; **12**: 1634; **20**: 2810,
        2811
    arthemis arthemis **20**: 2810, 2811
    camilla **12**: 1634
Limnaea **7**: 870
Limnodrilus **20**: 2769
Limnogale mergulus **19**: 2656, 2657
Limnophilus **3**: 354
Limosa
    fedoa **7**: 996
    haemastica **7**: 996
    lapponica **7**: 996, 997
    limosa **7**: 996; **21**: 2906
Limothrips cerealium **19**: 2676
Limulus polyphemus **10**: 1353, 1354

Linckia **18**: 2517
Lingula **10**: 1405–1406
Liothrips vaneeckei **19**: 2677
Liphistius **19**: *2719*
Lipotes vexillifer **5**: 697; **16**: 2177, 2178
Lissemys punctata **17**: 2442
Lithobius forficatus **3**: 412, *412*, 413
Lithodes maia **18**: 2507
Lithrurus umbratilis **18**: 2570
Litocranius walleri **7**: 953
Litoria infrafrenaia **19**: *2724*
Littorina **14**: 1926
    littoralis **14**: 1926, 1927
    littorea **14**: 1926, 1927
    neritoides **14**: 1926, 1927
    saxatilis **14**: 1926, 1927
Liza
    aurata **12**: 1692, 1693
    ramada **12**: 1692, 1693
    tricuspidens **12**: 1692, 1693
Lobodon carcinophagus **1**: 82; **4**: 565;
    12: 1647; **16**: 2286
Lochimas **13**: 1819
Locusta
    migratoria **11**: 1479
    migratoria migratorioides **11**: 1478
Locustella naevia **20**: 2859
Loligo **18**: 2508–2510
Lonchura
    castaneothorax **11**: 1544, 1545
    cucullata **11**: 1544
    fringilloides **11**: 1544
    nana **11**: 1544
    punctulata **11**: 1544
    striata **11**: 1544
Lophiomys imhausi **13**: 1838
Lophius piscatorius **1**: 51, 52, *52*
Lophocebus **11**: 1541
    albigena **11**: 1541, 1542
    albigena aterrimus **11**: 1542
Lophodytes cucullatus **12**: 1599;
    21: 2972
Lopholithodes **18**: 2507
Lophophorus **14**: 1934
    impeyanus **14**: 1936
Lophura **14**: 1934
    diardi **14**: *1935*
Lorentzimys nouhuysi **10**: 1337, 1338
Loriculus **11**: 1495
Loris tardidradus **11**: 1490
Lorius **11**: 1493, 1494
Lota lota **4**: 495; **11**: 1455
Loxia
    curvirostra **5**: 607, 608
    leucoptera **5**: 607
    pytyopsittacus **5**: 607
    scotica **5**: 607
Loxocemus bicolor **17**: 2419
Loxodonta africana **6**: 768; **15**: 2113
Lucanus cervus **19**: 2652
Lucastris **9**: 1250
Lucilia bufonivora **2**: 242
Lullula arborea **17**: 2396
Lumbricus **6**: 734, 735, 737
Luminodesmus sequoiae **12**: 1611
Luscinia megarhynchos **13**: 1741;
    18: 2459; **19**: 2678
Lutjanus
    analis **17**: 2423
    apodus **17**: 2423
    argentimaculatus **17**: 2423
    griseus **17**: 2423
    kasmira **17**: 2424
    sebae **17**: 2423
    synagris **17**: 2423
Lutra
    canadensis **16**: 2179, *2179*, 2180
    lutra **16**: 2179, 2180
Lutraria lutraria **7**: 926, 927
Lutrogale perspicillata **16**: 2180

Numerals in **bold** refer to volume numbers; those in *italics* to picture captions.

3089

Luxilus **12**: 1618
Lycaena **4**: 525–526
Lycalopex **5**: 690
   culpaeus **5**: 689
   griseus **5**: *689*
Lycaon pictus **3**: 374; **5**: 689; **8**: 1041
Lycophotia varia **13**: 1821
Lycosa **11**: 1550; **19**: 2635; **21**: 2956,
   2956, 2957
   barbipes **19**: 2634
   cuneata **19**: 2634
   fabrilis **19**: 2633
   pulverulenta **19**: 2634
   tarentula **2**: 203; **19**: 2633, 2634
Lymnaea **14**: 2010, 2011
Lymnocryptes minimus **17**: 2427
Lynx
   canadensis **3**: 405; **19**: 2616
   lynx **21**: 2906
Lyonogale **19**: 2731
Lyristes **4**: 456
   plebejus **4**: 456
Lysandra coridon **2**: 243
Lysiosquilla **11**: *1552*

M

Macaca
   arctoides **11**: 1511
   assamensis **11**: 1511
   fascicularis **11**: 1511
   fuscata **11**: 1511, 1512
   maura **11**: 1511
   mulatta **11**: 1511, 1512
   nemestrina **14**: 1942
   nigra **11**: 1511
   radiata **11**: 1511
   silenus **14**: 1941
   sinica **11**: 1512
   sylvanus **1**: 137; **12**: 1643
   thibetana **11**: 1511
Machaerhamphus alcinus **2**: 210
Macolor macularis **18**: 2573
Macrobrachium **15**: 2043
   carcinus **15**: 2043; **17**: 2364
   rosenbergii **15**: 2043
Macrocheira kaempferi **4**: 571; **18**: 2475
Macroclemys temmincki **17**: 2425, 2426;
   20: 2797
Macrodactyla doreensis **5**: 640
Macroderma gigas **2**: 169
Macrodipteryx
   longipennis **13**: 1742
   vexillarius **13**: 1742
Macrodontia **11**: 1483, 1484
Macrogalidia musschenbroeki **13**: 1855
Macrogeomys **14**: 1996
Macroglossum stellatarum **9**: 1166
Macrognathus **18**: 2490
   aculeatus **18**: 2490
   aral **18**: 2490
   pancalus **18**: 2490
Macronyx **14**: 1965
   croceus **14**: 1965
Macroperipatus torquatus **14**: 1924
Macropodia rostrata **18**: 2475, 2476
Macropodus **13**: 1868, 1869
Macropus **10**: 1335–1336
   macropus fuliginosus **10**: 1333, 1334
   macropus giganteus **10**: 1333, 1334
   parma **20**: *2846*
   robustus **10**: 1333
   rufogriseus fruticus **20**: 2848
   rufus **10**: 1333, 1334
Macrotermes michaelseni **19**: 2658
Macrotis
   lagotis **1**: 130–131
   leucura **1**: 130
Macrotrema caligans **18**: 2585
Macrovipera schweizeri **17**: 2422

Maculinea **2**: 243–244, *244*
Madoqua **1**: 61; **5**: 675, 676
Magicicada **4**: 456, 457
Maia squinado **17**: 2355; **18**: 2475, 2476
Makaira **11**: 1561, 1562
Malaclemys terrapin **19**: 2664, 2665
Malacoptila
   mystacalis **15**: *2061*
   panamensis **15**: 2060
Malapterurus **6**: 760, 761
Malimbus **21**: 2896
Mallomys rothschildi **7**: 965
Manacus **11**: 1532
Mandrillus
   leucophaeus **11**: 1536, 1537
   sphinx **11**: 1536, 1537; **12**: 1644
Maniola jurtina **3**: 296–297
Manis **13**: 1863
   crassicaudata **13**: 1862, 1863
   gigantea **13**: 1862
   javanica **13**: 1862, 1863
   pentactyla **13**: 1862
   temminckii **13**: 1862, 1863
   tetradactyla **13**: 1862, 1863
   tricuspis **13**: 1862
Mansonia **12**: 1664
Manta **5**: 668, 669
Mantispa **11**: 1549, *1549*, 1550
Mantis religiosa **11**: 1546, 1547
Margaronis **13**: 1819
Margarops
   fuscatus **19**: 2671
   fuscus **19**: 2671
Marmaronetta angustirostris **19**: 2649
Marmessus **8**: 1117, 1118
Marmosa murina **13**: 1792, 1793
Marmota **12**: 1672
   bobak **11**: 1568, 1569
   caligata **11**: 1568, 1569
   flaviventris **11**: *1568*
   marmota **11**: 1568, 1569
   monax **8**: 1076; **11**: 1568
   olympus **11**: 1569
   vancouverensis **11**: 1569
Marsupiobdella **10**: 1429
Martes
   americana **11**: 1579, 1580; **19**: 2616
   flavigula **11**: 1579
   foina **11**: 1579
   martes **11**: 1579; **19**: 2653
   pennanti **11**: 1579; **19**: 2653
   zibellina **11**: 1579
Massoutiera mzabi **8**: 1104
Mastacembelus **18**: 2490
   armatus **18**: 2490
   brichardi **18**: 2490
   circumcinchus **18**: 2490
   erythrotaenia **18**: 2490
Masticophis
   bilineatus **21**: 2919, 2920
   flagellum **21**: 2919, 2920
   lateralis **15**: 2101
   taeniatus **21**: 2919, 2920
Mastigoproctus giganteus **14**: 1906, *1907*
Mastomys **12**: *1694*, 1695
Mastophora **18**: 2482
Mastotermes **18**: 2658, 2660
Masturus lanceolatus **13**: 1774
Maticora **4**: 534
Mayetiola destructor **6**: 859; **12**: 1606
Mazama **18**: 2463, 2464
Megabunus diadema **5**: 637
Megaceryle alcyon **16**: 2175
Megachasma pelagios **17**: 2331
Megachile
   centuncularis **10**: 1419, 1420
   rotundata **10**: 1420
Megaderma lyra **6**: 840
Megadyptes antipodes **14**: 1911; **21**: 3004
Megaladapis **17**: 2373

Megalagrion **5**: 642
Megaloceros **5**: 656
Megalomys **15**: 2156
Megalops **19**: 2636, *2636*, 2637
Megamuntiacus vuquangensis **12**: 1697
Meganeuramonyi **5**: 711
Meganyctiphanes norvegica **16**: 2167
Megaphobema mesomelas **18**: 2483
Megapodius
   cumingii **11**: 1527
   reinwardt **11**: 1527
Megaptera novaeangliae **5**: 695; **9**: 1259;
   16: 2208; **21**: 2911
Megarhynchus pitangua **20**: 2801
Megarhyssa **16**: 2241
Megascolides **6**: 735
   australis **6**: 734
Megathymus **17**: 2391, 2392
   yuccae **17**: 2392
Megophrys **9**: 1229
   nasuta **9**: *1229*, 1230
Meiacanthus atrodorsalis **2**: 237
Melanargia galathea **21**: 2929
Melanerpes
   erythrocephalus **21**: 2978
   formicivorus **19**: 2654; **21**: *2978*,
   2979
Melanogrammus aeglefinus **6**: 836;
   8: 1114
Melanoptila glabrirostris **3**: 399
Melanosuchus niger **3**: 358
Melanotaenia
   maccullochi **17**: 2380
   trifasciata **17**: *2380*
Melanotis
   caerulescens **12**: 1622
   hypoleucus **12**: 1622
Meleagris
   gallopavo **20**: 2784, 2785
   gallopavo gallopavo **20**: 2784
   gallopavo intermedia **20**: 2784
   gallopavo merriami **20**: 2784
   gallopavo mexicana **20**: 2784
   gallopavo osceola **20**: 2784
   gallopavo silvestris **20**: 2784
Melese laodamia **19**: 2688
Meles meles **1**: 120, 121; **19**: 2653
Melibe **16**: 2300
Melichneutes robustus **9**: 1217
Melichthys niger **20**: 2744
Melierax **8**: 1027
Meliphaga unicolor **9**: 1215
Melitaea cynxia **7**: 898, 899
Melithreptus validirostris **9**: 1215
Mellisuga helenae **9**: 1255
Mellita sexies perforata **16**: *2226*
Mellivora capensis **1**: 120; **9**: 1207, 1217
Meloë proscarabaeus **13**: 1782, *1782*,
   1783
Melogale **1**: 120
Melolontha **4**: 486, 487
   melolontha **4**: 486
Melophorus **9**: 1205
   bagoti **9**: 1205, *1205*
Melospiza melodia **4**: 555; **9**: 1247;
   18: 2461
Melursus ursinus **17**: 2404
Menetes berdmorei **13**: 1859, 1860
Menidia beryllina **17**: 2380
Menura
   alberti **11**: 1509
   novaehollandiae **11**: 1509, 1510
Mephitis **17**: 2393, 2394
Merganetta armata **19**: 2708–2710
Mergus
   albellus **12**: 1599
   australis **12**: 1600
   merganser **5**: 717; **12**: 1599, 1600
   octosetaceus **5**: 717; **12**: 1600
   serrator **5**: 717; **12**: 1599

Meriones
   meridianus **7**: 951
   unguiculatus **7**: *950*
Merlangius merlangus **21**: 2935
Merluccius
   aungustianus **8**: 1119, 1120
   australis **8**: 1119, 1120
   bilinearis **8**: 1119, 1120
   capensis **8**: 1119, 1120
   gayi **8**: 1119
   merluccius **8**: 1119, 1120
   productus **8**: 1120
Merodon equestris **9**: 1251
Merops
   apiaster **2**: 180, 181
   ornatus **2**: 180
   pusillus **2**: *182*
Mesocapromys **9**: 1263, 1264
Mesochaetopterus rickettsi **20**: 2768
Mesocricetus
   auratus **8**: 1130
   neutoni **8**: 1131
Mesoenas **12**: 1604–1605
   benschi **12**: 1605
Mesoplodon **2**: 170
   densirostris **2**: *170*
Messor **9**: 1156, 1157, *1157*
Meta **13**: 1801
   segmentata **13**: *1800*
Meterodon platyrhinos **17**: 2421
Metoecus paradoxus **20**: 2826
Metopidius indicus **9**: 1289
Metridium **1**: 47, 48
Micrathena **13**: 1801
Micrathene whitneyi **13**: 1827
Microcavia **8**: 1099, 1100
Microcebus **10**: 1436
Microcentrum retinerve **10**: 1341
Microciona **16**: 2299
Microdipodops **10**: 1338
Microgale **19**: 2656, 2657
Microhierax fringillarius **2**: 208
Microlophium **1**: 75
Micromys minutus **9**: 1158, 1159
Micropogon undulatus **5**: 601, 602
Micropotamogale **19**: 2656
   lamottei **13**: 1817, 1818
   ruwenzorii **13**: 1817, 1818
Micropterus salmoides **8**: 1063;
   18: 2569
Microstomum **6**: 848
Microtus **6**: 818
   agrestis **6**: *818*, 819
Micruroides euryxanthus **4**: 534, 535
Micrurus
   fulvius **4**: 534, 535
   nigrocinctus **4**: 535
Miliaria calandra **3**: 320
Milvus
   migrans **10**: 1367, 1368
   milvus **10**: 1367
Mimas tiliae **9**: 1166
Mimus
   gilvus **12**: *1624*
   gundlachii **12**: 1622
   polyglottos **12**: 1622, 1623
Minyobates **1**: 94
Miracinonyx inexpectatus **3**: 429
Mirounga
   angustirostris **6**: 771, 772; **16**: 2285
   leonina **6**: 771, 772
Misgolas rapax **19**: 2720
Misgurnus **11**: 1473, 1474
Misumena vatia **5**: 580, 581, *581*
Mitu **5**: 623, 625, *625*
Mniotilta varia **21**: 2986
Mobula **5**: 668, 669
Modiolaria **12**: 1713
Modiolus modiolus **12**: 1711
Moenkenhausia oligolepsis **19**: 2667

Moho **9**: 1215
Mola
  mola **13**: 1774, 1775
  ramsayi **13**: 1774
Moloch horridus **5**: 665; **12**: 1631
Molothrus
  ater **3**: 400; **4**: 553, 554; **15**: 2141;
    18: 2459
  badius **4**: 553
  bonariensis **4**: 553
  rufoaxillaris **4**: 553
Molva **11**: 1455, 1456
Momotus momota **12**: 1665, 1666
Monachus
  monachus **12**: 1647, 1648; **16**: 2289
  schauinslandi **12**: 1647, 1648;
    16: 2286
  tropicalis **12**: 1647, 1648; **16**: 2286
Monasa **15**: 2060
  atra **15**: 2061
  morphoeus **15**: 2060–2061
Monocentris japonica **18**: 2512
Monocirrhus **10**: 1421, 1422
Monodon monoceros **12**: 1718; **21**: 2911
Monomorium **9**: 1157
Monopterus albus **18**: 2585
Mordacia **10**: 1403, 1404
Morelia **15**: 2080
  amethistina **15**: 2079, 2080, *2114*
  spilota **15**: 2079
Morone
  dolomieur **2**: 164
  labrax **2**: 164
  mississipiensis **8**: 1082
  salmoides **2**: 164
Morpho
  hecuba **12**: 1660, 1661
  neoptolemus **12**: 1660, 1661
  peleides **12**: 1661
  rhetenor **12**: 1660, 1661
  sulkowskyi **12**: 1660, 1661
Morus
  bassana **7**: 923, 924; **16**: *2268*
  capensis **7**: 923
  serrator **7**: 923
Moschus **5**: 655
  berezovskii **12**: 1701, 1702
  chrysogaster **12**: 1701, 1702
  moschiferus **12**: 1701, 1702
Motacilla
  aguimp **20**: 2844
  alba **20**: 2843
  capensis **20**: 2843
  cinerea **4**: 499; **20**: 2843
  clara **20**: 2843
  flava **20**: 2843, *2843*
  flaviventris **20**: 2843
Mugil cephalus **12**: 1692, 1693
Mulloidichthys **7**: 990
Mullus **7**: 989, 990
Mungos mungo **12**: 1635
Muntiacus
  atherodes **12**: 1696, 1697
  crinifrons **12**: 1696
  feae **5**: 658; **12**: 1696, 1697
  muntjak **12**: 1696, 1697
  reevesi **12**: 1696, 1697
Murexia **11**: 1577
Mus avellanarius **19**: 2653
Musca domestica **9**: 1241, 1242
Muscardinus avellanarius **5**: 699, 700,
    702
Muscicapa striata **7**: 872–873
Muscivora forficata **20**: 2800
Mus musculus **9**: 1245
Musophaga **20**: 2780, 2781
Mustela
  erminea **6**: 786; **14**: 2003
  eversmanni **2**: 228; **14**: 2001
  lutreola **12**: 1616, 1617

  nigripes **2**: 229
  nivalis **14**: 2003; **21**: 2893–2895
  putorius **2**: 228; **14**: 2001, 2002
  putorius furo **14**: 2001
  vison **12**: 1616, 1617
Mutilla
  europaea **20**: 2815
  lilliputiana **20**: 2815
Mya **7**: 926
  arenaria **7**: 927
  binghami **7**: 927
  truncata **7**: 927
Mycteria **18**: 2551
  americana **18**: 2552
  ibis **18**: 2552
  leucocephala **18**: *2553*
Myctophum punctatum **10**: 1414
Mydaus javanensis **1**: 120
Myiopsitta monachus **14**: 1878
Mylagra azureocapilla **2**: 283
Myliobatis **6**: 729, 730
Mylopharodon conocephalus **12**: 1618
Myocastor coypus **4**: 562; **5**: 667
Myomimus **5**: 699
Myoprocta pratti **1**: 32
Myopus **10**: 1433
Myotis
  adversus **6**: 840
  evotis **2**: 168
  mystacinus **20**: 2822, 2823
Myripristis **18**: 2511
Myrmecobius fasciatus **1**: 129
Myrmecocystus **9**: 1205
Myrmeleon **1**: 64
Myrmeleotettix **8**: 1036, 1037
Myrmotherula **1**: 55
Mysateles **9**: 1263, 1264
Mysis relicta **13**: 1796
Mytilus **12**: 1711, 1712
Myxas **14**: 2010, 2011
Myzomela cardinalis **9**: *1216*
Myzostoma cirriferum **6**: 805
Myzus persicae **1**: 74

N

Naemorhedus
  baileyi **8**: 1022
  caudatus **8**: 1022
  goral **8**: 1022
Naja **2**: 265; **4**: 480–482; **20**: 2818
  haje **4**: 481
  melanoleuca **4**: 481
  mossambica **4**: 481
  naja **4**: 481
  nigricollis **4**: 481
  nivea **4**: 481
Nandinia binotata **13**: 1855, 1856
Nandus **10**: 1421, 1422
Nannopterum harrisi **4**: 536
Narceus **12**: 1611
Narcine tasmaniensis **6**: 764
Nasalis **10**: 1411
  larvata **12**: 1643
Naso **18**: 2576
Nassarius glans **21**: 2917
Nasua **4**: 478, *478*, 479
Nasuella olivacea **4**: 478, 479
Nasutitermes **1**: 68, *68*
Natrix **20**: 2874
  maura **20**: 2874, 2875
  natrix **8**: 1043; **9**: 1203; **15**: 2103;
    17: *2420*; **20**: 2874, 2875
  tesselata **20**: 2874, 2875
Naucrates ductor **14**: 1951; **16**: 2242
Nautilus
  belauensis **12**: 1722, 1723
  pompilius **12**: 1722, 1723
  repertus **12**: 1722
  stenomphalus **12**: 1722, 1723

Neanura **18**: 2505
Necrophorus **3**: 327, 328
Necrosyrtes monachus **20**: 2841
Nectarinia
  amethystina **18**: 2565
  asiatica **18**: 2458, 2566
Nectonema **9**: 1237
Necturus **12**: 1682, 1683
Negaprion brevirostris **17**: 2333
Nematus compressicornis **16**: 2240
Nemesia **19**: 2719
Nemorhaedus **17**: 2324
Nemoura **18**: 2549
Neoceratodus forsteri **11**: 1501
Neofelis nebulosa **3**: *404*
Neofiber alleni **12**: 1706
Neomys
  anomalus **20**: 2872
  fodiens **20**: 2872
Neomysis integer **13**: *1795*, 1796
Neopagetopsis **9**: 1278
Neophasia menapia **21**: 2928
Neophema
  elegans **14**: *1880*
  pulchella **14**: 1878–1879
Neophoca cinerea **16**: 2282, 2283, 2288
Neophocaena phocaenides **15**: 2024,
    2025
Neophron percnopterus **20**: 2840–2841
Neoseps reynoldsi **17**: 2390
Neotermis militaris **1**: 68
Neotoma **13**: 1843, 1844, *1844*
Neotragus
  batesi **18**: 2571, 2572
  moschatus **1**: 61; **18**: 2571, 2572
  pygmaeus **1**: 60; **18**: 2571, 2572
Nepa cinerea **20**: 2871, *2871*
Nephila **13**: 1800–1801
Nephrops norvegicus **4**: 573; **11**: 1475
Nephruros **11**: *1468*
Neptunea antiqua **21**: 2917
Nereis **15**: 2104, 2105
Nerodia **20**: 2874
  cyclopion **20**: 2875
  erythrogaster **20**: 2875, *2875*
  fasciata **20**: 2875
  sipedon **20**: 2874, 2875
Nesiarchus nasutus **17**: 2414
Nesoctites micromegas **14**: 1937
Nesokia indica **1**: 132
Nesolagus netscheri **18**: 2561
Nesomimus **12**: 1622
Nesoryzomys **15**: 2157
Nestor **10**: 1343, 1344
Netta **14**: 1993
  erythrophthalma **14**: 1995
  rufina **14**: *1993*
Nettapus **5**: 714; **21**: 2927
Neuroterus **7**: 922
  quercus-baccarum **7**: 921, 922, *922*
Newmaster rubiginosa **16**: 2280
Nezara viridula **17**: 2344
Nilssonia formosa **17**: 2443
Nipponia nippon **9**: 1274
Nithoptera priamus **15**: *2110*
Noctilio leporinus **2**: 167; **6**: 840, *840*, 841
Noctua pronuba **13**: 1822
Nomadacris septemfasciata **11**: 1478,
    1479
Nomeus albula **15**: 2031
Nonnula **15**: 2060
Notemigonus crysoleucas **16**: 2182
Notharchus **15**: 2061
Nothoprocta ornata **19**: 2691
Notiomystis cincta **9**: 1215
Notomys **10**: 1338
  alexis **10**: *1337*
Notonecta
  glauca **1**: 118, 119
  hoffmanni **1**: 118

insulata **1**: 118, 119
undulata **1**: 119
Notopterus **6**: 803, 804
Notornis mantelli **19**: 2621
Notoryctes typhlops **11**: 1576
Novumbra hubbsi **12**: 1680, 1681
Nucifraga
  caryocatactes **13**: 1757, 1758
  columbiana **13**: 1757
Nuculus **13**: 1763, 1764
Numenius **5**: 626–628
Numida meleagris **8**: 1097
Nyctalemon **20**: 2805
  patroclus **20**: 2805, *2806*
Nyctalus noctula **20**: 2822
Nyctea scandiaca **13**: 1827; **17**: 2438;
    20: 2778
Nyctereutes **5**: 689; **15**: 2100
Nyctibius
  grandis **15**: 2033
  griseus **15**: 2032
  jamaicensis **15**: 2032
Nycticebus **11**: 1490, 1491
Nycticorax **13**: 1737, 1738
Nycticryphes semicollaris **13**: 1849
Nyctiphrynus ocellatus **14**: 2013
Nymphalis antiopa **20**: 2810, 2811
Nymphicus hollandicus **4**: 483
Nymphon gracile **17**: 2310, 2311
Nystalus maculatus **15**: *2060*

O

Obelia **16**: 2273
Oceanites
  hornbyi **18**: 2554
  oceanicus **18**: 2554
  tethys **18**: 2554
Oceanodroma leucorhoa **18**: *2554*
Ochotona **14**: 1944
  pallasi **14**: 1944
  princeps **14**: 1944
Ochrogaster lunifer **15**: 2048
Octopus
  apollyon **13**: 1779
  arborescens **13**: 1779
  cyanea **13**: *1779*
  hongkongensis **13**: 1779
  vulgaris **13**: 1779, 1780
Ocypode cursor **4**: 573
Ocyurus chrysurus **17**: 2424
Odobenus rosmarus **7**: 927; **16**: 2285;
    20: 2852
Odocoileus **5**: 656
  hemionus **12**: 1689, 1690
  hemionus columbianus **12**: 1689
  hemonius **3**: 426
  virginianus **12**: 1689, 1690; **19**: 2653,
    2682
Odontesthes regia **17**: 2380
Odontodactylus scyllarus **11**: *1551*
Odontophorus gujanensis **15**: 2082
Oedistoma pygmaeum **9**: 1215
Okapia johnstoni **13**: 1787
Ommatophoca rossi **16**: 2211
Omocestus **8**: 1036, 1037
Oncomeris flavicornis **17**: 2344
Oncorhynchus
  aguabonita **20**: 2751
  clarki **20**: 2751, 2752
  clarki clarki **20**: 2751
  clarki lewisi **20**: 2751
  clarki pleuritians **20**: 2751
  gorbuscha **13**: 1840
  keta **13**: 1840, 1841
  kisutch **13**: 1840
  masou **13**: 1840
  mykiss **20**: 2751, 2752
  nerka **13**: 1840
  tshawytscha **13**: 1840

Numerals in **bold** refer to volume numbers; those in *italics* to picture captions.

3091

Ondatra zibethicus **5**: 667;
  12: 1706–1708; **19**: 2616
Oniscus **21**: 2973, 2974
Onuxodon **14**: 1901
Onychiurus **18**: 2505
Onychogalea fraenata **20**: 2846
Onychorhynchus mexicanus **20**: 2800
Onychoteuthis banksi **18**: 2510
Opheodrys aestivus **19**: 2733
Ophiocoma **2**: 159
Ophiodon elongatus **20**: 2766
Ophion **9**: 1279
Ophiophagus **2**: 265
  hannah **17**: 2420
Ophiothrix **2**: 159, 281
Ophisaurus **6**: 733; **7**: 980; **10**: 1430
  apodus **7**: 981
  attenuatus **7**: 980, 981
  compressus **7**: 980, 981
  ventralis **7**: 980, 981
Ophisternon candidum **18**: 2585
Ophiura **2**: 281
Opifex **12**: 1664
Opisthocomus hoazin **9**: 1197
Opsanus tau **19**: 2698, 2699
Orcaella brevirostris **21**: 2911
Orchopeas howardi **6**: *850*
Orcinus orca **1**: 82; **5**: 695; **10**: 1351;
  16: 2211
Oreamnos americanus **16**: 2200
Oreaster occidentalis **18**: *2516*
Orectochilus **21**: 2921, 2922
Orectolobus **21**: 2954, 2955
Oreochromis **19**: 2689–2690
Oreoica gutturalis **2**: 191, 192
Oreoscoptes montanus **19**: 2671
Oreotragus oreotragus **1**: 61; **10**: 1375
Oreotrochilus **12**: 1672
Oriolia bernieri **20**: 2813
Oriolus **8**: 1012, 1013
Ornithodoros **19**: 2682
Ornithoptera **2**: 212; **20**: 2806
  alexandrae **3**: 344; **18**: 2582
Ornithorhynchus anatinus **14**: 1986
Ortalis vetula **20**: 2757
Orthogeomys **14**: 1997
Orthopsittaca **11**: 1514
Orthotomus **19**: 2618, 2619
Ortygospiza locustella **21**: 2888
Orycteropus afer **1**: 17
Oryctolagus cuniculus **15**: 2092
Oryx **1**: 62; **5**: 665
  dammah **1**: 63; **13**: 1808, 1809
  gazella **13**: 1808, 1809
  gazella beisa **13**: 1808
  gazella callotis **13**: 1808
  gazella gazella **13**: 1808
  leucoryx **1**: 63; **13**: 1808, 1809
Oryzomys **15**: *2157*
  alfaroi **15**: 2156
  antillarum **15**: 2156
  bauri **15**: 2156
  couesi **15**: 2156
  melanotis **15**: 2156
  palustris **15**: 2157
Oryzorictes **19**: 2656, 2657
Osbornictis piscivora **7**: 946, 947
Osmerus eperlanus **17**: 2410
Osmylus fulvicephalus **10**: 1392
Osteolaemus **5**: 603
Ostracion meleagris **20**: 2758
Ostrea edulis **13**: 1832, 1833
Ostrinia nubilalis **4**: 540
Otaria flavescens **16**: 2282, 2283
Otis tarda **3**: 337, 338
Otocyon **5**: 689
  megalotis **2**: 166; **5**: 665
Otolemur **3**: 330
  argentatus **3**: 329
  crassicaudatus **3**: 329, 330

Otus
  asio **13**: 1827, 1828
  insularis **13**: 1829
  ireneae **13**: *1828*
  kennicotti **13**: 1827, 1828
  madagascariensis **13**: 1827
  pauliani **13**: 1829
  rutilus **13**: 1827
Ourebia ourebia **13**: 1804
Ovibos moschatus **12**: 1703–1705
Ovis
  ammon **2**: 198
  canadensis **2**: 198, 199
  dalli **2**: 198
  laristan **2**: 198
  musimon **2**: 198
  orientalis **2**: 198; **12**: 1668
Oxybelis aeneus **19**: 2733
Oxyporhamphus **7**: 876
Oxyura
  australis **18**: 2538
  dominica **18**: 2538
  jamaicensis **18**: 2537, 2538
  leucocephala **18**: 2537
  maccoa **18**: 2538
Oxyuranus
  microlepidotus **5**: 665
  scutellatus **20**: *2817*
Ozotoceros bezoarticus **18**: 2463

P

Pachycephala **21**: 2923, 2924
Pachycheles **18**: 2507
Pachygrapsus crassipes **17**: 2353
Pachyphantes **21**: 2896
Pachyptila
  belcheri **15**: 2045
  salvini **15**: 2046
  turtur **15**: *2045*
  vittata **15**: 2045, 2046
Pachyramphus **4**: 543
Pacifastacus leniusculus **5**: 592
Padda oryzivora **10**: 1316; **11**: 1544
Padochela hemphilli **18**: 2476
Pagetopsis **9**: 1278
Pagodroma nivea **17**: 2436
Pagophilus groenlandica **8**: 1140
Paguma larvata **13**: 1855
Paguropsis typica **9**: 1178
Pagurus
  bernhardus **9**: 1176, 1177
  megistas **9**: *1176*
  prideauxi **9**: 1178
Palaemon **18**: 2043, 2044
Paleosuchus
  palpebrosus **3**: 358
  trigonatus **3**: 358
Palinurus **4**: 573; **11**: 1475
  argus **11**: *1475*
  elephas **11**: 1476, *1476*
  interruptus **11**: 1475
Palophus titan **18**: 2530
Palpares **1**: 64, 65
Pan
  paniscus **1**: *70, 71*; **4**: 439–440
  troglodytes **1**: 71; **4**: 439, 440
Panaxia dominula **19**: 2687
Pandaka pygmaea **7**: 993
Pandinus imperator **16**: 2257
Pandion haliaetus **13**: 1812; **21**: 2906
Pangio kuhlii **11**: 1473
Panthera **3**: 406
  leo **3**: 406; **8**: 1040; **11**: 1461
  onca **3**: 406; **10**: 1311
  pardus **3**: 406; **8**: 1040; **11**: 1445,
    11: 1559
  tigris **3**: 405; **19**: 2683–2685;
    21: 2907
  tigris altaica **19**: 2683

  tigris amoyensis **19**: 2683
  tigris corbetti **19**: 2683
  tigris sumatrae **19**: 2683
  tigris tigris **19**: 2683
  uncia **11**: 1559; **12**: 1672; **17**: 2433
  unis **3**: 404
Pantholops hodgsoni **16**: 2217, *2218*
Pantodon buchholzi **3**: 347–348
Panulirus pencillatus **4**: 573
Panurus biarmicus **14**: 1886, 1887;
  18: 2458; **19**: 2693
Panyptila sancti-hieronymi **19**: 2600
Papagomys armandvillei **7**: 965
Papilio
  aegeus **18**: 2581
  anchisiades **18**: 2581
  cresphontes **18**: 2581
  dardanus **18**: 2582
  glaucus **18**: *2580*, 2581
  hospiton **18**: 2580
  machaon **3**: 344; **18**: 2580, 2581, *2581*
  memnon **18**: 2582
Papio **11**: 1536
  cynocephalus **1**: 115, 116
  hamadryas **1**: 115, 116; **12**: 1644
  leucophaeus **1**: 115
  papio **1**: 115, 116
  sphinx **1**: 115
Pappogeomys **14**: 1997
Papyrocranus **6**: 803, 804
Parabuteo unicinctus **3**: 426
Paracanthurus hepatus **18**: 2577
Parachanna **17**: 2412
Paracheirodon **19**: 2667, 2668
Paradipus **10**: 1324
Paradisaea **2**: 206–207
Paradoxornis **14**: 1886
Paradoxurus **13**: 1856
  hermaphroditus **13**: 1855, 1856
Paraexocoetus **7**: 876
Paragordius **9**: 1237
Parahyaena brunnea **9**: 1267, 1268
Parambassis ranga **7**: 978–979
Paramecium
  bursaria **14**: 1881, 1882
  caudatum **14**: 1881, *1881*, 1882, *1882*
Parapagurus pilosimanus **9**: 1178
Paraplesiops **5**: 668
Parascyllium variolatum **3**: *408*
Parategeticula **21**: 3005
Paraxerus capapi **20**: *2741*
Pardosa **21**: 2957
Parenteroxenos dogieli **16**: 2272
Parioglossus **7**: 995
Parnassius **1**: 76
Parophasma galinieri **3**: 399
Parthenolecanium corni **16**: *2246*
Parthenope investigatoris
  18: 2475–2476
Parula americana **21**: 2986
Parupeneus **7**: 990
Parus **19**: 2653, 2693
  atricapillus **19**: 2653, 2693
  caeruleus **19**: 2693, 2694
  cinctus **19**: 2617
  major **19**: 2654, 2693
  montanus **19**: 2695
  rufescens **19**: *2695*
Passer **9**: 1247
  domesticus **7**: 999; **9**: 1247, 1248;
    18: 2457
  eminebey **9**: 1249
  hispaniolensis **9**: 1247
  montanus **9**: 1247; **10**: 1316
  simplex **9**: 1247
Passerculus princeps **19**: 2649
Passerina ciris **3**: *319*
Pasteurella pestis **6**: 850
Patagona gigas **9**: 1255; **12**: 1673
Patella vulgata **11**: 1451, 1452

Patina pellucida **11**: 1451
Pauxi pauxi **5**: 623
Pavo **14**: 1898, 1899
Pecari tajacu **14**: 1903, 1904
Pecten **16**: 2248, 2249
Pectinator spekei **8**: 1104
Pectinophora gossypiella **2**: 256
Pedetes capensis **18**: 2502–2503
Pedionomus torquatus **14**: 1979
Pegasus volitans **16**: 2290
Pegusa lascaris **17**: 2444
Pelagia noctiluca **10**: 1320
Pelagothuria natatrix **16**: 2271
Pelamis platurus **17**: 2309
Pelargopsis capensis **10**: 1356–1357
Pelecanoides **5**: 684, 685
Pelecanus
  conspicillatus **14**: *1909*
  crispus **14**: 1908
  erythrorhynchos **14**: 1908
  occidentalis **14**: 1908, 1909; **16**: 2265
  onocrotalus **14**: 1910
  rufescens **14**: 1908
Peliolepisma **2**: 279
Pelobates **18**: 2466
Pelochelys bibroni **17**: 2443
Pelomedusa subrufa **9**: 1191; **20**: *2799*
Peloscolex **20**: 2769
Pempheris **17**: 2441
Peneus
  duorarum **17**: *2364*
  setiferus **17**: 2364
Pennatula **16**: *2298*
Pentatoma rufipes **17**: 2345
Peradorcas concinna **16**: 2198
Perameles
  bougainville **1**: 130
  gunnii **1**: *130*
  nasuta **1**: 131
Perca
  flavescens **14**: 1917, 1918
  fluviatilis **14**: 1917, 1918; **16**: 2175
Percopsis **20**: 2754, 2755
Perdix perdix **14**: 1891, 1892
Periclemenes
  amethysteus **17**: 2365
  imperator **17**: 2366
Pericrocotus **5**: 621; **12**: 1614
  brevirostris **12**: 1614
  cinnamomeus **12**: *1615*
  flammeus **12**: 1614, 1615
  igneus **12**: 1614
  solaris **12**: 1614
Perillus bioculatus **4**: 504
Periophthalmodon **12**: 1684
  schlosseri **12**: 1684, 1685
Periophthalmus **12**: 1684–1685
  chrysospilos **12**: 1684, 1685
  kalolo **12**: 1684–1685
  koelreuteri **12**: 1685
Peripatoides novaezealandiae **14**: 1925
Peripatus **14**: 1924, *1925*
Periplanta **4**: 493
Perisoreus
  canadensis **10**: 1317
  infaustus **10**: 1318; **19**: 2617
Perissocephalus tricolor **4**: 543
Perla **18**: 2549
Perna canaliculus **12**: *1712*
Pernis **9**: 1211
  apivorus **2**: 209; **9**: 1211, 1212
  celebensis **9**: 1211
  ptilorhynchus **9**: 1211
Perodicticus potto **15**: 2035
Perognathus **10**: 1338
Perolicula **15**: 2082
Peromyscus **19**: 2682
  boylii **5**: 660
  californicus **3**: 426; **5**: 660
  eremicus **5**: 660

floridanus **5**: 661
leucopus **5**: 660
maniculatus **5**: 659, 660
polionotus **5**: 660
Peroryctes
   broadbeuti **1**: 130, 131
   eremiana **1**: 130
Petaurus **7**: 985
   breviceps **7**: 985
Petrobius maritimus **18**: 2505
Petrochelidon nigricans **12**: 1590
Petrodromus tetradactylus **6**: 775
Petrogale
   brachyotis **16**: 2197
   penicillata **16**: 2197, 2198
   xanthopus **16**: 2197, 2198, *2198*
Petroica **16**: 2190
Petrolisthes **4**: *571*; **18**: 2507
Petromyzon **10**: 1404
   marinus **10**: 1403, 1404; **17**: *2332*
Petrosia **18**: *2495*
Pezoporus wallicus **14**: 1879–1880
Phacochoerus
   aethiopicus **20**: 2861
   africanus **20**: 2861
Phaethon
   aethereus **20**: 2748
   lepturus **20**: 2748
   rubricauda **20**: 2748, 2749
Phagocata **6**: 849
Phainopepla nitens **21**: 2890
Phalacrocorax
   aristotelis **4**: 537
   auritus **4**: 536, 537
   bougainvilli **4**: 538
   capensis **4**: 538
   carbo **4**: 536
   olivaceous **4**: 536
   pelagicus **4**: 536
   penicillatus **4**: 536
   urile **4**: 536
Phalaenoptilus nuttalli **14**: 2013
Phalanger
   intercastellanus **5**: 630
   lullulae **5**: 630
   mimilus **5**: 630
   orientalis **14**: *1930*
   rothschildi **5**: 630
   sericeus **5**: 630
Phalangium opilio **5**: 636
Phalaropus
   fulicarius **14**: 1931, 1932
   lobatus **14**: 1931
Phaner **10**: 1436
Pharomachrus mocinno **15**: 2088
Pharus
   legumen **15**: 2127
   pellucidus **15**: 2127, *2128*
Phascogale **11**: 1577
   calwa **11**: 1578
Phascolarctos cinereus **10**: 1379
Phasianus **14**: 1934
   colchicus **14**: 1934, 1935
   versicolor **14**: 1934
Phedina borbonica **12**: 1590
Pheidole **9**: 1156, 1157, *1225*
Phelsuma madagascariensis **7**: *943*
Phenacogramma interruptus **19**: 2667
Phengodes **7**: 988
Phibalura flavirostris **4**: *544*
Phigys **11**: 1493
Philaenus spumarius **9**: 1224
Philanthus **9**: 1262
Philemon **9**: 1215
   buceroides **9**: 1215
   corniculatus **9**: 1215
Philesturnus
   carunculatus **21**: 2886
   carunculatus carunculatus **21**: 2886
   carunculatus rufusater **21**: 2886

Philohela minor **21**: 2966
Philomachus pugnax **16**: 2213
Philothamnus semivariegatus **19**: 2733
Philydor **13**: 1819
Phlogophora meticulosa **13**: 1822
Phoca
   hispida **1**: 82; **16**: 2285
   sibirica **10**: 1398
   vitulina **8**: 1133; **12**: 1647
Phocaena **15**: 2026
   phocaena **15**: 2024, 2025
   sinus **15**: 2024, 2025
   spinipinnis **15**: 2024, 2025
Phocarctos hookeri **16**: 2282, 2283
Phocoenoides dalli **5**: 695; **15**: 2024, 2025
Phodopus sungorus **8**: 1130
Phoenicopterus
   andinus **6**: 844
   chilensis **6**: 844
   jamesi **6**: 844
   minor **6**: 844
   ruber **6**: 844, 845
   ruber roseus **6**: 844
   ruber ruber **6**: 844
Phoenicurus
   moussieri **15**: 2139
   phoenicurus **15**: 2139, 2140
Pholadidea loscombiana **14**: 1939, 1940
Pholas dactylus **14**: 1939, 1940
Pholcus phalangioides **18**: 2482
Pholis gunnellus **3**: 342
Phoneutria **18**: 2820
Phonygammus keraudreni **20**: 2757
Photinus **6**: 828
   pyralis **6**: 827
   scintillans **6**: 828
Photuris **6**: 828
   pennsylvanica **6**: 828
Phoxinus
   erythrogaster **12**: 1618–1619
   phoxinus **12**: 1618, 1619
Phrixothrix **7**: 988
Phronima **15**: *2078*; **16**: 2224
Phryganea **3**: 354, 355
Phrynocephalus **1**: 30
Phrynosoma
   cornutum **9**: 1231, 1232
   douglassi **9**: *1231*
Phtheirichthys lineatus **15**: 2148
Phylidonris nigra **9**: *1215*
Phyllastrephus **3**: 308
Phyllirhoe **16**: 2300
Phyllobates **1**: 94; **20**: 2819
   terribilis **1**: 93
Phyllomedusa **19**: 2725
   burmeisteri **19**: 2723
Phyllopertha **4**: 487
   horticola **4**: 487
Phyllophryne didactyla **1**: 93
Phylloscopus
   collybita **20**: 2857
   sibilatrix **21**: 2985
   trochilus **20**: 2857
Phyllotoma aceris **16**: 2240
Phylloxera vitifoliae **20**: 2827, 2828
Phyllurus cornutus **7**: *945*
Physalia **20**: 2819
   physalis **14**: 1980–1981; **15**: 2029, 2030; **20**: 2798
   utricularis **15**: *2029*
Physeter
   catodon **18**: 2510
   macrocephalus **18**: 2472; **21**: 2911
Pica
   nuttalli **11**: 1519, 1520
   pica **11**: 1519, 1520
Picoides **21**: 2977
   major **21**: 2979
   villosus **21**: 2978

Picromerus bidens **17**: 2344
Picumnus
   innominatus **14**: 1938
   olivaceus **14**: 1937
Picus **21**: 2977
   viridis **21**: 2965, 2977
Pieris
   brassicae **21**: 2928, *2929*
   rapae **3**: 416; **21**: 2928
Piliocolobus **4**: 501, 502
   badius **4**: 502
   oustaleti **4**: 502
Pimephales promelas **12**: 1618
Pinaroloxias **5**: 649
Pinctada **13**: 1834
Pinguinus impennis **14**: 1911; **16**: 2268
Pinna fragilis **12**: 1711
Pionites melanocephala **15**: 2112
Piophila **18**: 2505
Pipa pipa **18**: 2579
Pipile pipile **5**: 623
Pipistrellus
   hesperus **14**: 1962
   pipistrellus **14**: 1962, 1963
   pygmaeus **14**: 1962
   subflavus **14**: 1962
Pipra
   erythrocephala **11**: 1532
   mentalis **11**: 1532
Piranga
   flava **19**: 2627
   ludoviciana **19**: 2627
   olivacea **2**: 2457; **19**: 2627
   rubra **19**: 2627
Pisaura **21**: 2957
   mirabilis **18**: 2483
Pithecia **16**: 2221
   aequatorialis **16**: 2222
   albicans **16**: 2222
   irrorata **16**: 2222
   monachus **16**: 2221, 2222
   monachus milleri **16**: 2222
   pithecia **16**: 2221, 2222, *2222*
Pithecophaga jeffreyi **15**: 2111
Pitta
   angolensis **14**: 1970, 1971
   brachyura **14**: 1970
   guajana **14**: *1970*
   gurneyi **14**: 1970
   steerei **14**: 1970
   versicolor **14**: 1970
Pizonyx vivesi **2**: 167; **6**: 840; **20**: 2822, 2823
Plagiodontia **9**: 1263, 1264
   aedium **9**: 1264
Plagiolepis **9**: 1205
   trimeni **9**: 1205
Plagionotus **11**: 1484
   detritus **11**: *1484*
Planigale **11**: 1577
   ingrami **11**: 1577, 1578
Planorbis **7**: 870
Plasmodium **12**: 1663
Platalea
   ajaja **18**: 2496
   alba **18**: 2497
   flavipes **18**: 2496
   leucorodia **18**: 2496, 2497
   minor **18**: 2496
   regia **18**: 2496
Platambus **5**: 683
Platanista **5**: 697; **16**: 2177, 2178
Platichthys flesus **6**: 860, 861
Platyarthrus **21**: 2974
Platynereis megalops **15**: 2105
Platyptilia
   calodactyla **14**: 1992
   carduidactyla **14**: 1991
   gonodactyla **14**: 1991
Platyrhina **8**: 1103

Platyrhinoidis **8**: 1103
Platyspiza **5**: 649
Platysternon megacephalum **2**: 197
Plebejus
   argus **2**: 243
   icariodes missionensis **2**: 244
Plecoglossus
   altivelis altivelis **19**: 2597
   altivelis ryukyuensis **19**: 2597
Plecotus
   auritus **2**: *167*; **20**: 2821, 2822
   macrotis **20**: 2821, 2822
   townsendii **2**: 168
Plectronemia **3**: 354
Plectrophenax
   hyperboreus **3**: 319
   nivalis **3**: 319, 320
Plectropterus gambiensis **5**: 717
Plegadis falcinellus **9**: 1274
Plethodon **11**: 1504
   cinereus **11**: 1503
   glutinosus **11**: 1504
Pleuronectes platessus **14**: 1976
Ploceus **21**: 2896
   cucullatus **21**: 2896
   intermedius **21**: 2896
   ocularis **21**: 2897
   philippinus **21**: 2896
   velatus **21**: *2897*
Plumaria setacea **16**: *2273*
Plusiotis gloriosa **16**: *2253*
Pluvialis **14**: 1988, *1989*
   apricaria **14**: 1988
   dominica **14**: 1988, 1989
   obscurus **14**: 1988
   squatarola **14**: 1988
Pluvianus aegyptus **4**: 551, 552
Podarcis
   hispanica **20**: 2849
   muralis **20**: 2849, 2850
   sicula **20**: 2849, 2850
Podargus strigoides **7**: 901
Podica senegalensis **6**: 823
Podiceps
   auritus **8**: 1061
   cristatus **8**: 1061
   nigricollis **8**: 1061
   taczanowskii **8**: 1063
Podilymbus
   gigas **8**: 1063; **19**: 2649
   podiceps **8**: 1061, 1062
Podisus **17**: 2344
   maculiventris **4**: 504
Podocnemis expansa **9**: 1191
Podura **18**: 2505
   aquatica **18**: 2504–2505
Poecilia
   formosa **2**: 234
   latipinna **2**: 234
   reticulata **8**: 1107; **12**: 1664
   sphenops **2**: 234
   velifera **2**: 234
Poecilictis libyca **21**: 3020
Poecilogale albinucha **21**: 3020
Pogona **5**: 665
   barbata **2**: 172
   vitticeps **2**: 173
Pogonias chromis **5**: 601, 602
Pogonichthys macrolepidotus **12**: 1618
Pogonomyrmex **9**: 1156, 1157
Polemaetus bellicosus **12**: 1589
Polihierax semiorquatus **2**: *209*
Polistes **2**: 185; **20**: 2824
   annularis **13**: 1865
Pollachius
   pollachius **14**: 2004–2005
   virens **4**: 473
Pollenia rudis **2**: 242
Pollicepes pollicepes **1**: 139
Polyboroides **17**: 2373

Polycarpa aureta 17: *2312*
Polycentrus 10: 1422
Polyergus rufescens 1: 69
Polygonia
    c-album 3: 345
    comma 20: 2812
    faunus 20: 2812
    gracilis 20: 2812
    interrogationis 20: 2811
    oreas 20: 2812
    progne 20: 2812
    satyrus 20: 2811
Polyipnus 9: 1162
Polyodon spathula 13: 1845, 1846
Polyprion 2: 164
    americanus 8: 1082
Polypterus 2: 196
    birchir 2: 196
    ornatipinnis 2: 195
Polysticta stelleri 6: 754
Pomacanthus
    imperator 1: *50*
    semicirculatus 1: 49
    striatus 1: 49
Pomacentrus coeruleus 5: 638
Pomatoceros 6: 801
    triqueter 6: 801
Pomatomus saltatrix 2: 245
Pomatostomus isidorei 1: *114*
Pomponia imperatoria 4: 455, 456
Pongo
    pygmaeus 1: 70
    pygmaeus abelii 13: 1797, 1798
    pygmaeus pygmaeus 13: 1797, 1798
Pontoporia blainvillei 5: 697; 16: 2177, 2178
Porania pulvillus 18: 2518
Porcellana longicornis 18: 2507
Porcellio 21: 2974
    scaber 21: 2973, 2974
Porichthys 19: 2698
Porina 19: 2603
Porites 4: 528
Porphyrio
    martinica 7: 919, 920
    porphyrio 7: 919
Porpita 16: 2300
Portia 18: 2482
Porzana
    carolina 5: 582, 583; 15: *2106*
    palmeri 15: 2108
    porzana 5: 582
    pusilla 5: 582
Potamochoerus porcus 3: 336
Potamogale 19: 2656
    velox 13: 1817, 1818
Potorous 15: 2116
Potos flavus 10: 1364
Prennas 5: 639, 640
    biaculeatus 5: *640*
Prenolepis imparis 9: 1204
Presbytis 10: 1411
    entellus 12: 1643
    melalophus 10: *1410*
    thomasi 10: *1412*
Priodontes giganteus 1: 85
Prionace glauca 17: 2331
Prionodon
    linsang 11: 1459
    pardicolor 11: 1459
Prionodura newtoniana 2: 271
Prionotus
    carolinus 8: 1109
    scitulus 8: 1109
Prionus 11: 1484
Pristella maxillaris 21: 2998, 2999
Pristis
    pectinata 16: 2238, 2239
    perotteti 16: 2238
Pristolepis 10: 1421, 1422

Probosciger aterrimus 4: 483, 484
Procapra 7: 939
Procapromys geayi 9: 1263
Procavia 9: 1270, 1271
    capensis 9: 1271
Procelsterna cerulea 13: 1755
Procnias
    alba 4: 543
    averano 4: 543, 544
    tricarunculata 2: 191; 4: 543
Procolobus verus 4: 501, 502
Procyon
    cancrivorus 15: 2096, 2097
    gloveranni 15: 2096
    insularis 15: 2096, 2097
    lotor 15: 2096, 2097
    maynardi 15: 2096, 2097
    minor 15: 2096, 2097
    pygmaeus 15: 2096, 2097
Prognathochromis 4: 459
Progne
    subis 2: 151; 12: 1590, 1591
    tapera 12: 1590
Prognichthys 7: 876
Promerops
    cafer 18: 2558, 2559
    gurneyi 18: 2558
Propithecus
    diadema 17: 2372
    diadema diadema 17: 2372
    diadema perrieri 17: 2372
    tattersalli 17: 2372
    verreauxi 17: 2372
Propyrrhura 11: 1514
Prosthemadera novaeseelandiae 9: 1215
Proteles cristatus 1: 19
Proteus anguinus 13: 1789
Protonotaria citrea 21: 2986
Protopterus
    aethiopicus 11: 1501, *1502*
    annectens 11: 1501
Protoxerus 13: 1860
    aubinnii 13: 1859
    stangeri 13: 1859
Prunella
    collaris 1: 22
    himalayana 1: 22
    modularis 1: 22
    montanella 1: 22
Psammomys obesus 7: 950, 951
Psammoperca 13: 1752
Psarisomus dalhousiae 2: 282
Psarocolius
    angustifrons 13: *1806*
    montezuma 13: 1805, 1806
    wagleri 13: *1805*
Psephurus gladius 13: 1845
Pseudacris 19: 2723
    nigrita 19: 2724
Pseudanthias 8: 1081
Pseudemydura umbrina 9: 1191; 20: 2797–2798
Pseudemys 13: 1851
Pseudeos 11: 1493
    fuscata 11: 1493
Pseudibis papillosa 9: 1276
Pseudis paradoxa 14: 1877
Pseudobalistes flavimarginatus 20: *2744*
Pseudobranchus
    axanthus 17: 2382
    striatus 17: 2382
Pseudocalyptomena graueri 2: 283
Pseudocarcinus gigas 4: 571
Pseudoceros 6: 849
    ferrugineus 6: *849*
Pseudochaenichthys 9: 1278
Pseudochilidon 12: 1590
    eurystomina 12: 1590
    sirintarae 12: 1590
Pseudococcus 10: 1396

Pseudocreobotra wahlbergi 11: 1548
Pseudogyps 20: 2841
Pseudois najaur 12: 1672
Pseudopontia paradoxa 21: 2929
Pseudorca crassidens 6: 798
Pseudoseisura lophotes 13: 1820
Pseudosphromenus dayi 13: 1868
Pseudotriakis microdon 3: 407
Pseudupeneus 7: 990
    grandisquamis 7: 990
Psittacula
    cyanocephala 14: 1878
    himalayana finschii 14: *1878*
    krameri 14: 1878, 1879
Psittacus erithacus 14: 1883
Psolus diomediae 16: 2271
Psophia
    crepitans 20: 2756, 2757
    leucoptera 20: 2756, 2757
    viridis 20: 2756, 2757
Pteridophora alberti 2: 205
Pterocles
    alchata 16: 2230
    bicinctus 16: 2230
    burchelli 16: 2230
    namaqua 16: 2230
Pterocnemia pennata 15: 2151
Pterodroma cahow 16: 2268
Pteroglossus torquatus 19: 2714, 2715
Pterognatha gigas 11: 1484
Pterois volitans 16: 2260
Pteromalus
    marginalis 21: 2928–2929
    oleracea 21: 2928
    puparum 3: 416; 21: 2928
    virginiensis 21: 2928
Pteromylaeus 6: 730
Pteronarcys californica 18: 2549
Pteronura brasiliensis 16: 2179, 2180
Pterophorus
    pentadactyla 14: *1992*
    periscelidactylus 14: 1991, 1992
Pterophylla camellifolia 10: 1341
Pterophyllum 4: 459
    altum 16: 2244, 2245
    leopoldi 16: 2244, 2245
    scalare 4: 458; 16: 2244, 2245
Pteroptyx malaccae 6: *828*
Pteropus 7: 907; 15: 2110
    giganteus 2: 167; 7: 877, 878, 879
    poliocephalus 7: 878, 879
    vampyrus 2: 167; 7: 877, 878
Ptilocercus lowi 19: 2730, 2731
Ptilonorhynchus violaceus 2: 270
Ptiloris 2: 207
    magnificus 2: 205
Ptyas mucosus 15: 2117; 17: 2422
Ptychocheilus 12: 1618
Ptychozoon 7: 945
Pudu
    mephistophiles 5: 655; 18: 2463, 2464
    pudu 18: 2463, 2464
Puffinus
    assimilis 17: 2334
    gavia 17: 2334
    gravis 17: 2334
    griseus 17: 2334, 2335
    heinrothi 17: 2334
    mauretanicus 17: 2335
    nativitatis 17: 2336
    tenuirostris 16: 2264; 17: 2335
Pugettia producta 18: 2475
Pulex irritans 6: 850, 851
Puma 15: 2069, 2070
Pungitius pungitius pungitius 18: 2533
Puntius eugrammus 21: 3015
Pusa
    caspica 16: 2169, 2170
    hispida 14: 1998; 16: 2169, 2170
    sibirica 16: 2169, 2170

Pycnodonta 13: 1832
Pycnogonum 17: 2311
    stearnsi 17: 2311
Pycnonotus
    barbatus 3: 309
    cafer 3: 309
    leucogenys 3: 308
Pygathrix 10: 1411; 12: 1645
Pygocentrus 14: 1967
    cariba 14: 1969
    nattereri 14: 1968
    piraya 14: 1967
Pygoplites diacanthus 6: *833*
Pygoscelis 14: 1913
    adeliae 1: 29, *80*; 14: 1912
    antarctica 14: 1912
    papua 14: *1911*
Pylocheles 9: 1176
Pylodictis olivaris 3: 401
Pylopagurus 9: 1176
Pyrocephalus rubinus 20: 2800, 2801
Pyrophorus 6: 828
    noctilucus 6: 827
Pyrosema 15: 2078, *2078*
    atlanticum 15: 2078
Pyrosomella 15: 2078
Pyrostremma 15: 2078
Pyrrhocorax
    graculus 4: 451; 12: 1673
    pyrrhocorax 4: 451, 452; 12: 1673
Pyrrhopyge 17: 2391
Pyrrhula pyrrhula 3: 311
Python 15: 2079, 2080
    anchietae 15: 2079
    molurus 15: 2079; 17: 2422
    regius 15: 2079, 2080
    reticulatus 15: 2079, 2080; 17: 2419, 2422
    sebae 15: 2080
Pytilia melba 21: 2938
Pyxipoma 21: 2989

Q

Quelea 15: 2084
    cardinalis 15: 2085
    erythrops 15: 2085
    quelea 15: 2085; 21: 2897
Quiscalus
    major 8: 1032
    mexicanus 8: 1032
    palustris 8: 1032
    quiscula 8: 1032, 1033

R

Raja
    binoculata 17: 2383
    clavata 17: 2330, 2383
Rallus 15: 2106
    aquaticus 15: 2106, 2107
    caerulescens 15: 2106
    elegans 15: 2106
    longirostris 15: 2106
    pectoralis 15: 2106
Ramphastos 19: 2714, 2716
    dicolorus 19: *2716*
    sulfuratus 19: 2714
    toco 19: *2715*
    vitellinus 19: 2714
Ramphocelus passerinii 19: 2629
Ramphocinclus brachyurus 19: 2671
Ramphotyphlops braminus 17: 2421
Rana
    catesbeiana 3: 313; 6: 744; 7: 903
    esculenta 6: 743
    lessonae 6: 743
    pipiens 11: 1447, 1448
    ridibunda 6: 743
    sylvatica 7: 903

temporaria **4**: 511
utricularia **11**: 1447
Ranatra linearis **20**: 2870, 2871
Rangifer tarandus **3**: 391, 392; **5**: 655; **20**: 2778
Ranzania laevis **13**: 1774
Raphicerus
  campestris **18**: 2527
  melanotis **18**: 2527
  sharpei **18**: 2527
Rastrelliger
  brachyosoma **11**: 1517
  feughni **11**: 1517
  kanapurta **11**: 1517
Rattus
  exulans **20**: 2764
  norvegicus **3**: 298–299; **20**: 2764
  rattus **17**: 2347; **20**: 2764
Recurvirostra
  americana **1**: 105
  andina **1**: 105
  avosetta **1**: 105, 106
  novaehollandiae **1**: 105
Redunca **1**: 61
  arundinum **15**: 2144, 2145
  fulvorufula **15**: 2144, 2145
  redunca **15**: 2144, 2145
Regalecus glesne **13**: 1767, *1767*, 1769
Regulus
  calendula **7**: 999
  goodfellowi **7**: 999
  ignicapillus **7**: 999
  regulus **7**: 999, 1000; **18**: 2455
  satrapa **7**: 999
  teneriffae **7**: 999
Reinhardtius
  evermanni **8**: 1121
  hippoglossoides **8**: 1121
Reithrodontomys **9**: 1159
  humulis **9**: 1158
  megalotis **9**: 1158
  raviventris **9**: 1158
Remora
  australis **15**: 2148
  brachyptera **15**: 2148
  osteochir **15**: 2148
  remora **15**: 2148, *2149*
Remorina albescens **15**: 2148
Renilla **16**: 2298
Repomucenus **5**: 707
Reticulitermes **1**: 68
Rhabdamia **3**: 390
Rhabdomys pumilio **21**: 3016, *3017*
Rhacophorus
  nigropalmatus **7**: 982, 983
  reinwardtii **7**: 982, 983
Rhagium **11**: 1484
Rhagodes nigracinctus **18**: 2454
Rhantus **5**: 683
Rhea americana **15**: 2151
Rhina **8**: 1103
  ancylostoma **8**: *1102*
Rhincodon typhus **6**: 834; **17**: 2331; **21**: *2915*, 2916
Rhinecanthus aculeatus **20**: 2744
Rhinobatos **8**: 1103
  lentiginosus **8**: 1103
  rhinobatos **8**: 1103, *1103*
Rhinoceros
  sondaicus **15**: 2152, 2153
  unicornis **15**: 2152, 2153
Rhinoderma
  darwinii **12**: 1678–1679
  rufum **12**: 1678, 1679
Rhinolophus
  ferrum **2**: 168
  ferrumequinum **9**: 1239, 1240
  hipposideros **9**: 1240
  philippienensis **9**: *1240*
Rhinomuraena quaesita **12**: 1658

Rhinophrynus dorsalis **3**: 326
Rhinopias aphanes **16**: *2261*
Rhinopithecus **10**: 1411
Rhinoplax virgil **9**: 1226
Rhinopoma
  hardwickei **19**: 2703
  microphyllum **19**: 2703
  muscatellum **19**: 2703
Rhinoptilus
  africanus **4**: 552
  bitorquatus **4**: 551
Rhipidura leucophrys **20**: 2843
Rhizostoma **10**: 1320, 1321
Rhodalia cardinalis **10**: 1396
Rhodeus sericeus **2**: 219
Rhodites rosae **7**: 921
Rhodonessa caryophyllacea **5**: 717
Rhodostethia rosea **16**: *2266*
Rhombobomys opimus **7**: 949
Rhopalodina **16**: 2271
Rhyacophila **16**: 2175
Rhynchdermus **6**: 849
  terrestris **6**: 848
Rhynchobatos **8**: 1103
  djiddensis **8**: 1102, 1103
Rhynchocyon
  chrysopygus **6**: 774, 775
  cirnei **6**: 774
Rhynochetus jubatus **18**: 2567
Rhyssa persuasoria **9**: 1279, *1279*, 1280
Riparia riparia **2**: 151; **6**: 850; **12**: 1590
Rissa
  brevirostris **10**: 1370
  tridactyla **10**: 1370, 1371
Rivulus **19**: 2704
Roccus
  chrysops **8**: 1082
  saxatilis **8**: 1082
Rollandia microptera **8**: 1063
Rollulus **14**: 1891
Roncador stearnsi **5**: 602
Rondeletia loricata **21**: 2909
Rostanga rufescens **16**: 2299–2300
Rostratula benghalensis **13**: 1849, *1849*, 1850
Rostrhamus
  hamatus **6**: 788
  sociabilis **2**: 209; **6**: 789
Rousettus **2**: 167
  aegyptiacus **7**: *906*
Rupicapra rupicapra **3**: 422
Rupicola
  peruviana **4**: 491, 543
  peruviana aurea **4**: 491
  peruviana sanguinolenta **4**: 491
  rupicola **4**: 491, 543
Rutilus
  rutilus **16**: 2183
  rutilus aralensis **16**: 2182
  rutilus caspicus **16**: 2182
  rutilus fluviatlis **16**: 2182
  rutilus henckli **16**: 2182
Ruvettus pretiosus **17**: 2415
Ruwenzorornis johnstoni **20**: 2780
Rynchops
  albicollis **17**: 2387
  flavirostris **17**: 2387
  nigra **17**: 2387

S

Sabella **6**: 801
  pavonina **6**: 802
  spallanzanii **6**: 801
  starte indica **6**: *802*
Sagittarius serpentarius **2**: 210; **17**: 2316–2318
Saguinus **19**: 2624, 2625
  bicolor **19**: 2625
  fuscicollis **12**: *1645*; **19**: 2624

  fuscicollis tripartitus **19**: 2624
  geoffroyi **19**: 2625
  imperator **12**: 1645; **19**: 2624
  inustus **19**: 2625
  labiatus **19**: 2624
  leucopus **19**: 2625
  midas **19**: 2624
  mystax **19**: 2624
  nigricollis **19**: 2624
  oedipus **12**: 1642; **19**: 2625
Saiga
  tatarica **1**: 61; **16**: 2217, 2218
  tatarica mongolica **16**: 2217
Saimiri
  boliviensis **18**: 2513, 2514
  oerstedii **18**: 2513
  sciureus **18**: 2513, 2514
  ustus **18**: 2513, 2514
  vanzolinii **18**: 2513
Sakesphorus canadensis **1**: 55
Salamandra salamandra **6**: 832
Salamandrella keyserlingii **19**: 2616
Salangichthys microdon **7**: 978
Salanx
  cavieri **7**: 978
  chinensis **7**: 978
Salmo
  salar **1**: 100; **16**: 2176
  trutta **16**: 2175; **20**: 2751
  trutta aralensis **20**: 2751
  trutta fario **20**: 2751
  trutta lacustris **20**: 2751, 2752
  trutta macrostigma **20**: 2751
  trutta oxianus **20**: 2751
  trutta trutta **20**: 2751, 2752
Salpa **16**: *2223*
  maximus **16**: 2224
Salpinctes obsoletus **21**: 2994
Salpingotus **10**: 1324
Salvelinus
  fontinalis **20**: 2751
  malma **20**: 2751
  namaycush **20**: 2751
Sandelia **4**: 468
Sarcogyps calvus **20**: 2841
Sarcophilus harrisii **5**: 677; **19**: 2642
Sarcophyton trocheilophorum **17**: 2441
Sarcorhamphus papa **20**: 2787, 2841
Sarda sarda **2**: 258
Sardina pilchardus **14**: 1948, 1949
Sardinops
  melanostictus **14**: 1948
  neopilchardus **14**: 1948
  ocellatus **14**: 1948; **17**: 2430
  sagax **14**: 1948, 1949
Sargocentron
  rubrum **18**: 2511, 2512
  spiniferum **18**: 2511, 2512
  violaceum **18**: 2512
Sarotherodon
  galilaeus **19**: 2689, *2689*
  galilaeus galilaeus **19**: 2690
Sasia abnormis **14**: 1937
Sauromalus
  obesus **4**: 454
  varius **4**: 454, *454*
Saxicoloides fulicata **16**: 2190
Sayornis
  fusca **20**: 2800
  nigricans **20**: 2801
Scaeochlamys lividus **16**: 2248
Scaphiopus **18**: 2466
  bombifrons **18**: 2466
  couchii **18**: 2466
  hammondi **18**: 2466
  holbrooki **18**: 2465
  multiplicatus **18**: 2465
Scarabaeus sacer **16**: 2252
Scardinius erythrophthalmus **16**: 2182
Scartelaos **12**: 1684, 1685

Scarus **14**: 1888, 1889
  croicensis **14**: 1888, 1889
  ferrugineus **14**: 1890
  guacamaia **14**: 1889–1890
  iserti **14**: 1888, 1889
  perrico **14**: 1889
  rubripinne **14**: 1890
  rubroviolaceus **14**: *1889*
  sordidus **14**: 1888, 1889
  taeniopterus **14**: 1888, 1889
Scatophagus **16**: 2256
  argus **16**: 2256
  rubifrons **16**: 2255
  tetracanthus **16**: 2255, 2256
Sceliphron **9**: 1262
Sceloporus **6**: 807
  occidentalis **6**: 807, 808
  undulatus **6**: 807, 808
  undulatus elongatus **6**: *807*
Scenopinus fenestralis **4**: 471
Schetba rufa **20**: 2813
Schistocerca gregaria **11**: 1478, 1479
Schistometopum thomense **3**: 356
Schistosoma **7**: 871
Schizothaerus nuttallii **7**: 927
Schoblia **21**: 2974
Sciaena **5**: 602
  aquila **5**: 601
Scincus
  mitranus **17**: 2390
  scincus **17**: *2390*
Sciurus **20**: 2740
  carolinensis **5**: 700; **20**: 2740, 2741
  vulgaris **4**: 436; **20**: 2740, 2741, *2742*
Sclerurus **13**: 1819
Scolopax rusticola **21**: 2966, 2967
Scolopendra **3**: 411
  gigantea **3**: 411
  morsitans **3**: 411, 412
Scomber
  japonicus **11**: 1517
  scombrus **11**: 1517, 1518
Scomberesox **17**: 2368
Scomberomorus
  concolor **11**: 1517
  maculatus **11**: 1517
  niphonius **11**: 1517
  sierra **11**: 1517
Scophthalmus
  maximus **20**: 2782, *2782*, 2783
  maximus maestica **20**: 2782
  rhombus **20**: 2782
Scopus umbretta **8**: 1124; **18**: 2551
Scorpaena **20**: 2818
  brasiliensis **16**: 2260
  guttata **16**: 2260
  scrofa **16**: 2260
Scorpaenopsis diabolus **16**: *2260*
Scotopelia **6**: 842; **13**: 1827
  bouvieri **6**: 842
  peli **6**: 842, 843
  ussheri **6**: 842
Scotornis climacurus **13**: 1742
Scudderia texensis **10**: 1341
Scutigera forceps **3**: 411–412
Scutisorex somereni **9**: 1182
Scyliorhinus
  canicula **5**: 686, 687
  stellaris **5**: 686
Scyllaea **16**: 2300
Scyllarides **4**: 570
Scypha compressa **15**: 2075
Sebastes
  marinus **16**: 2260
  viviparus **16**: 2260
Sehirus dubius **17**: 2344–2345
Seiurus **21**: 2985
  aurocapillus **13**: 1819; **21**: 2985
Selar crumenophthalmus **16**: *2243*
Selasphorus rufus **9**: 1255

Numerals in **bold** refer to volume numbers; those in *italics* to picture captions.

3095

Selene vomer **16**: 2242
Selevina betpakdalaensis **5**: 699
Semibalanus balanoides **1**: 138, *139*
Semiotus affinis **4**: 466
Semotilus corporalis **12**: 1619
Sepia officinalis **5**: 632
Sepioteuthis sepioidea **18**: *2509*
Sericulus
  bakeri **2**: 271
  chrysocephalus **2**: 271
Serinus
  canaria **3**: 369
  dermus **8**: 1016
  serinus **4**: 499
Seriola
  dumerili **16**: 2242
  lalandi **16**: 2242
Serpulorbis
  gigas **21**: 2990
  novaehollandiae **21**: 2990
  squamigerus **21**: 2990
Serrasalmus **14**: 1967, *1967, 1969*
  spilopleura **14**: 1969
Sertularella speciosa **16**: *2274*
Sertularia **16**: 2274
  abietena **16**: 2274
Sesarma meinerti **4**: *573*
Sesia apiformis **9**: 1234
Setifer
  setosus **19**: 2656
  spinosus **19**: 2656, 2657
Setonix brachyurus **15**: 2090
Sialia sialis **18**: 2457
Siganus
  guttatus **15**: 2094
  javus **15**: 2094, 2095
  luridus **15**: 2095
  puellus **15**: 2095
  rivulatus **15**: 2095
  spinus **15**: 2094
  vermiculatus **15**: 2094, 2095
Sigmoceros lichensteini **8**: 1148
Siliquaria **21**: 2989
Silurus glanis **3**: 402; **6**: 760
Singa **13**: 1801
Sinobdella **18**: 2490
Siphamia **3**: 390
  versicolor **3**: 389
Siphonogorgia **17**: 2440
Sipyloidea sipylus **18**: 2532
Siren
  intermedia **17**: 2382
  lacertina **17**: 2382
Sirex **9**: 1279
  gigas **16**: 2240
Sirpus **4**: 571
Sistrurus **3**: 426; **15**: 2119
  catenatus **15**: 2120
  miliarius **15**: 2120
Sitophilus granarius **21**: 2904
Sitta
  canadensis **13**: 1762
  carolinensis **13**: *1762*
  europaea **13**: 1760, 1761
  whiteheadi **13**: 1762
Smerinthus ocellata **9**: 1168
Sminthillus limbatus **8**: 1066
Sminthopsis
  crassicaudata **11**: 1577
  douglasi **11**: 1578
Sminthurides **18**: 2505
Sminthurus viridis **18**: 2504
Smithorius rufolateralis **2**: 282
Solea solea **17**: 2445
Solenodon
  arredondoi **18**: 2452
  cubanus **18**: 2452
  marcanoi **18**: 2452
  pardoxus **18**: 2452, *2452*
Somateria **6**: 754, 755

Sorex
  araneus **17**: 2359
  bendirei **20**: 2872
  cinereus **17**: 2358, 2359
  fumeus **17**: 2358, 2359
  hoyi **17**: 2358, 2359
  minutus **17**: 2359
  palustris **20**: 2872, 2873
Sparisoma **14**: 1888
Spathius exarator **4**: 471
Speleomantes **11**: 1503, 1504
Speothos venaticus **3**: 334; **17**: 2357
Spermophilus **8**: 1040
  beecheyi **8**: 1078, 1079
  citellus **8**: 1080
  franklinii **8**: 1078
  lateralis **8**: 1079
  parryi **8**: 1080
  suslicus **8**: 1080
  tridecemlineatus **8**: 1079
Sphaeramia, nematoptera **3**: *390*
Sphaerodactylus **11**: 1466
Sphecia bembeciformis **9**: 1234
Sphecotheres
  viridis **8**: 1012
  viridis flaviventris **8**: 1012
Spheniscus **14**: 1914
  demesus **14**: 1911
  mendiculus **14**: 1911
Sphenodon
  guntheri **20**: 2763, 2764
  punctatus **20**: 2763, 2764
Sphiggurus **19**: 2729
Sphinx ligustri **9**: 1166
Sphyraena
  argentea **2**: 155
  barracuda **2**: 155
  borealis **2**: 155
  guachancho **2**: 155
  sphyraena **2**: 155
Sphyrapicus
  thyroideus **16**: 2235
  varius **16**: 2235, 2236
Sphyrna
  corona **8**: 1127
  couardi **8**: 1127
  lewini **8**: 1127
  media **8**: 1127
  mokarran **8**: 1127
  tiburo **8**: 1126, 1127
  tudes **8**: 1127
  zygaena **8**: 1127
Spilocuscus
  maculatus **5**: 629–630
  rufoniger **5**: 630
Spilogale
  gracilis **17**: 2393, 2394
  putorius **17**: 2393
  pygmaea **17**: 2393, 2394
Spinachia spinachia **18**: 2533
Spinus pinus **11**: 1457
Spirinchus thaleichthys **17**: 2410
Spirorbis **6**: 801
  borealis **6**: 801
  pagenstecheri **6**: 801
Spiza americana **18**: 2458
Spizella passerina **18**: 2460
Spodoptera littoralis **3**: 346
Spondylus **13**: 1832
Spongia **18**: 2493
  officinalis **18**: 2494
Squalodon **16**: 2178
Squalus
  acanthias **5**: 686, 687; **17**: 2333
  suckleyi **5**: 686
Squatina **12**: 1646; **16**: 2244; **17**: 2385
  argentina **12**: 1646
  californica **12**: 1646
Squilla **11**: 1551, 1552
Stagmomantis carolina **11**: 1546

Stauropus fagi **15**: 2077
Steatornis caripensis **13**: 1785
Steganopus tricolor **14**: 1931
Stegostoma fasciatum **21**: 3015
Steinernema carpocapsae **4**: 540
Stenella
  attenuata **5**: 698
  coeruleoalba **5**: 698
  frontalis **5**: 698
  longirostris **5**: 698
Stenobothrus **8**: 1036
  lineatus **8**: 1037
Stenophylax **3**: 354
Stenoptilia pneumonanthes
  14: 1991–1992
Stenorhynchus seticornis **18**: *2475*
Stephanoaetus coronatus **5**: 615; **8**: 1093
Stephopoma **21**: 2989
Stercorarius
  longicaudus **10**: 1307
  parasiticus **10**: 1307, 1308
  pomarinus **10**: 1307
Stereolepis gigas **2**: 164
Sterna
  dougallii **16**: 2268; **19**: 2663
  forsteri **16**: 2264; **19**: 2661
  fuscata **19**: 2661
  hirundo **19**: 2661
  maxima **19**: *2663*
  paradisaea **16**: 2265; **19**: 2661, 2662
  vittata **19**: 2661
Sternoptyx **9**: 1162
  diaphana **9**: 1162
Sternotherus **20**: 2795
  carinatus **12**: 1687
  odoratus **12**: 1687, 1688
Stethophysma grossum **8**: 1036
Sthenoteuthis **18**: 2508
Stichodactyla haddoni **5**: 640
Stichopus
  chloronotus **14**: 1901
  regalis **14**: 1902
Stictocarbo punctatus **16**: *2265*
Stictonetta naevosa **5**: 714
Stigmella ridiculosa **3**: 344
Stiltia isabella **15**: 2041
Stizostedion vitreum **14**: 1917, 1918
Stoichactis **1**: 47, 48
Stomoxys calcitrans **9**: 1242–1243
Strangalia **11**: 1484
Streptopelia
  decaocto **4**: 500
  senegalensis **13**: 1858
  turtur **4**: 499; **20**: 2794
Strigamia maritima **3**: 411, 412
Strix
  aluco **13**: 1827; **19**: 2645
  butleri **13**: 1827
  nebulosa **13**: 1827
  varia **13**: 1825
Strombus **4**: 514–515
  gigas **4**: 514, 515
  goliath **4**: 514, 515
Strongylocentrotus **17**: 2315
Strongylura
  exilis **12**: *1725*
  marina **12**: 1725
Struthidea cinerea **11**: 1522
Strutio
  camelus **13**: 1815
  camelus syriacus **13**: 1814
Strymonidia **8**: 1118
Sturmia scutellata **19**: 2611
Sturnella neglecta **8**: *1039*
Sturnus
  pagodarum **12**: *1716*
  roseus **18**: 2521
  vulgaris **6**: 855; **18**: 2521, 2522
Stylatula elongata **16**: 2297
Stylodipus **10**: 1324

Subcoccinella vigintiquatuorpunctata
  10: 1396
Suillotazus marchei **1**: 120
Sula
  abotti **2**: 261
  bassana **16**: 2264
  capensis **16**: 2264
  dactylatra **2**: 261
  leucogaster **2**: 261, 262
  nebouxii **2**: 261
  sula **2**: 261
  variegata **2**: 261
Suncus **17**: 2358
  etruscus **12**: 1709
  murinus **12**: 1710; **17**: 2358
Suricata suricatta **12**: 1596
Sus scrofa **19**: 2653; **21**: 2907, 2943
Sutorectus tentaculatus **21**: 2955
Sygnathus
  acus **14**: *1961*
  floridae **14**: 1960
  leptorhynchus **14**: 1961
  typhle **14**: 1961
Sylvia
  atricapilla **20**: 2857
  communis **12**: 2459
  undata **20**: 2857, 2858
Sylvicapra **6**: 727
  grimmia **6**: 726, 727
Sylvilagus
  aquaticus **4**: 546
  audubonii **4**: 546
  bachmani **3**: 426; **4**: 546
  dicei **4**: 546
  floridanus **4**: 546
  graysoni **4**: 546
  insonus **4**: 546
  nuttallii **4**: 546
  transitionalis **4**: 546
Sylviorthorhynchus desmursii **13**: 1819
Sympetrum
  flaveolum **5**: 709
  fonscolombei **5**: 709
  striolatum **5**: 710
Symphasis varia **11**: 1550
Symphodus melops **21**: 2991
Symphysodon **4**: 459
  aequifasciatus **14**: 2006
  discus **14**: 2006, 2007
Synallaxis **13**: 1819
  subpudica **13**: 1819
Synanceia **5**: 668
  horrida **16**: 2260; **18**: 2547
  verrucosa **18**: 2547, 2548
Synaptomys **10**: 1433; **20**: 2778
Syncerus caffer **3**: 371
Sypheotides indica **3**: 338
Syrmaticus **14**: 1935
Syrphus **9**: 1251
  balteus **9**: 1250
  ribesii **6**: *857, 858*
Syrrhaptes paradoxus **16**: 2229

T

Tabanus **9**: 1235, 1236, *1236*
  distinguendus **6**: *856*
  sudeticus **9**: 1235
Tachornis squamata **19**: 2598
Tachpleus **10**: 1354
Tachuris rubigastra **20**: 2800
Tachybaptus ruficollis **8**: 1062
Tachycineta
  bicolor **2**: 151
  euchrysea **2**: 151
Tachyeres
  brachypterus **5**: 716; **18**: 2524, 2525
  leucocephalus **5**: 716; **18**: 2524
  patachonicus **5**: 716; **18**: 2524
  pteneres **5**: 716; **18**: 2524

Tachyglossus **18**: 2486
  aculeatus **18**: 2487
Tadorna
  ferruginea **17**: 2341
  tadorna **17**: 2342
  variegata **17**: 2341
Tadorna cristata **5**: 717
Taeniopygia **21**: 3012
  guttata **21**: 3013
Taeniura
  lymma **18**: 2543
  melanospilos **18**: 2544
Talpa
  caeca **12**: 1625
  europaea **12**: 1625, 1626
  micrura **12**: 1625
  streeti **12**: 1627
Tamandua
  mexicana **1**: 57
  tetradactyla **1**: 57
Tamiasciurus
  douglasii **4**: 436, 437
  hudsonicus **4**: 436, 437
  mearnsi **4**: 436, 437
Tamias striatus **4**: 448, 449;
  19: 2653
Tangara **19**: 2627
  chilensis **19**: 2627
  larvata **19**: 2629
  nigrocincta **19**: 2627
Tanysiptera galatea **10**: 1356
Tanytarsus **12**: 1607
Tapera naevia **5**: 619
Taphozous **19**: 2703
  mauritianus **19**: 2703
Tapirus
  bairdii **19**: 2630, 2631
  indicus **19**: 2630, 2631
  pinchaque **19**: 2630, 2631
  terrestris **19**: 2630, 2631
Taricha **20**: 2819
  torosa **13**: 1733, 1734;
  19: 2653
Tarsius
  bancanus **19**: 2638, 2639
  dianae **19**: 2638, 2639
  pumilus **19**: 2638, 2639
  spectrum **19**: 2638, 2639
  syrichta **19**: 2638, 2639
Tasmacetus sheperdi **2**: 170
Tatera indica **7**: 950
Tauraco
  corythaix **20**: *2781*
  hartlaubi **20**: 2780
  macrorhynchus **20**: 2780
  ruspolii **20**: 2780
  schuttii **20**: 2781
Taurotragus **7**: 942
  derbianus **1**: 60; **6**: 757, 758
  oryx **6**: 757, 758
Taxidea taxus **1**: 120
Tayassu pecari **14**: 1903, 1904
Tegenaria **17**: 2339
  domestica **17**: 2339, 2340
Tegeticula **21**: 3005
  maculata **21**: 3005
  synthetica **21**: 3005
  yuccasella **21**: 3005, *3005,*
  3006
Teinopalpus imperialis **18**: 2582
Tenodera aridifolia sinensis
  11: 1546
Tenrec ecaudatus **19**: 2656, 2657
Tenualosa
  ilisha **17**: 2327
  reevesi **17**: 2327
Teraba major **1**: *56*
Terapon jarbua **21**: 3015
Teratornis incredibilis **4**: 516
Terebratella **10**: 1406

Terebratulina **10**: 1406
Teredo navalis **17**: 2350, *2350*
Terpsiphone mutata **7**: *873*
Terrapene **20**: 2796
Tersina viridis **18**: 2584
Testacella **17**: 2409
Testudo
  graeca **19**: 2712
  hermanni **19**: 2713
Tethys **16**: 2300
  californica **16**: 2276
Tethyum pyriforme **17**: 2313
Tetracanthagyna plagiata **5**: 711
Tetracerus quadricornis **1**: 60;
  13: 1753
Tetragonopterus **19**: 2667
Tetramorium caespitum **1**: 69
Tetrao
  parvirostris **3**: 376, 377
  tetrix **8**: 1084
  urogallus **3**: 377
Tetraodon palembangensis
  15: 2063
Tetrapturus **11**: 1561
  albidus **11**: 1561
  angustirostris **11**: 1561
  audax **11**: 1561
  pfluegeri **11**: 1561
Tetrarope barbata **16**: 2260
Tetrix **8**: 1036
Thalassoica antartica **17**: 2435
Thalassoma bifasciatum
  21: 2991, 2992
Thalassornis leuconotus
  18: 2538
Thaleichthys pacificus **17**: 2410,
  2411
Thamnophilus doliatus **1**: 55
Thamnophis **7**: 933, 934
  elegans **7**: 934
  marcianus **7**: 935
  radix **7**: 934
  sirtalis **7**: 934
  sirtalis tetrataenia **17**: 2422
Thaumetopoea
  pityocampa **15**: 2047, 2048
  processionea **15**: 2047, 2048
  solitaria **15**: 2048
Thea vigintiduopunctata
  10: 1394
Thecla **8**: 1118
  betulae **8**: *1117*
Theophila religiosae **17**: 2377
Theraphosa leblondi **2**: 204
Theridion sisyphium **18**: 2483
Thermobia domestica **2**: 279
Theropithecus gelada **1**: 115,
  116; **12**: 1643
Thinocorus
  orbignyianus **17**: 2319
  rumicivorus **17**: 2319, 2320
Thomisus onustus **5**: 580, 581
Thomomys **8**: 1040; **14**: 1997
  bottae **14**: *1996*
  talpoides **14**: *1997*
Thoracocharax **9**: 1161
Thraupis
  cyanoptera **19**: 2627
  episcopus **19**: 2628
  virens **19**: 2627
Threskiornis aethiopicus **9**: 1274;
  18: *2497*
Thrips
  flavus **19**: 2676
  tabaci **19**: 2676
Thryomanes bewickii **21**: 2993
Thrysites atun **17**: 2431
Thryssa scratchleyi **1**: 45
Thunnus **19**: 2605; **20**: 2773, 2774
Thylogale **13**: 1847, 1848

Thymallus
  arcticus **8**: 1047
  arcticus baicalensis **8**: 1047
  arcticus grubii **8**: 1047
  arcticus pallasi **8**: 1047
  thymallus **8**: 1047, 1048
Thymelicus
  lineola **17**: 2391, 2392
  octeon **17**: 2392
Thysania agrippina **3**: 344;
  13: 1821, 1822
Thysanoessa inermis **16**: 2167
Tibicen **4**: 456
  canicularis **4**: 456
Tichodroma muraria **19**: 2721
Tilapia **4**: 459
  mariae **19**: 2690
  zillii **19**: 2689
Tiliqua multifasciata **17**: 2389
Tinamus major **19**: 2692
Tinca tinca **12**: 1619
Tineola bisselliella **4**: 470, 471
Tipula oleracea **5**: 589
Tityra semifasciata **4**: 544
Tmetothylacus tenellus **14**: 1965
Tockus
  camurus **9**: 1226
  erythrorhynchus **9**: 1228
Todus
  angustirostris **19**: 2700
  mexicanus **19**: 2700
  multicolor **19**: 2700, 2701
  subulatus **19**: 2700
  todus **16**: 2190; **19**: 2700
Tolypeutes **1**: 85
  matacus **1**: 85
Tomistoma schlegelii **5**: 603;
  7: 954
Tomocerus longicornis **18**: 2504
Torgos tracheliotus **20**: 2842
Tornierella **15**: 2146
Torpedo
  fuscomaculata **6**: *764*
  marmorata **6**: 765
  nobiliana **6**: 764, 765; **17**: 2332
  torpedo **6**: 766
Totoaba macdonaldi **5**: 601, 602
Toxostoma
  bendirei **19**: 2672
  dorsale **19**: 2673
  lecontei **19**: 2673
  longirostre **19**: 2671
  redivivum **3**: 426; **19**: 2671,
  2673
  rufum **18**: 2458; **19**: 2671,
  2672
Toxotes
  chatereus **1**: *77*
  jaculator **1**: 78
Trachinotus carolinus **16**: 2242
Trachinus **20**: 2818
  araneus **21**: 2902
  armatus **21**: 2902
  draco **21**: 2901, 2902
  lineolatus **21**: 2902
  pellegrini **21**: 2902
  radiatus **21**: 2902
Trachipterus
  arcticus **13**: 1768
  trachypterus **13**: 1768
Trachops cirrhosus **2**: 167
Trachurus **16**: 2242
  trachurus **16**: 2243; **17**: 2430
Trachycardium
  egmontianum **4**: 488
  muricatum **4**: 488
  quadragenarium **4**: 488
Trachydosaurus rugosus
  17: 2390
Trachypithecus **10**: 1411

Tragelaphus **3**: 331; **13**: 1765
  angasi **3**: 331; **13**: 1765, 1766
  bongo **10**: 1390
  buxtoni **13**: 1765, 1766
  eurycerus **1**: 63;
  10: 1389–1390
  imberbis **3**: 331; **10**: 1390
  scriptus **1**: 63; **3**: 332
  spekei **1**: 63; **3**: 331; **20**: 2760
  strepsiceros **10**: 1390
Tragopan
  blythi **19**: 2717, 2718
  caboti **19**: 2717
  melanocephalus **19**: 2717
  satura **19**: 2717
  temmincki **19**: 2717
Tremarctos **2**: 174
  ornatus **12**: 1672; **18**: 2471
Trialeurodes vaporariorum
  3: 416; **21**: 2932, *2932, 2933*
Trichechus
  inunguis **11**: 1534, 1535
  manatus **11**: 1534, 1535;
  21: 2907
  senegalensis **11**: 1534, 1535
Trichiurus nitens **17**: 2415
Trichoglossus
  chlorolepidotus **11**: 1488
  euteles **11**: *1489*
  haematodus **11**: 1488, 1489
Trichoniscus **21**: 2974
Trichosurus
  arnhemensis **3**: 300, 301
  caninus **3**: 300, 301
  vulpecula **3**: 300, 301
Trichys
  fasciculata **5**: 595
  lipura **19**: 2729
Tridacna gigas **4**: 462, 463
Trigla lyra **8**: 1109
Trigonoceps **20**: 2841
Trimeresurus
  albolabris **14**: *1974*
  flavoviridis **14**: 1973
  wagleri **14**: 1973
Trimma **7**: 994
Trimmatom nanus **6**: 834; **7**: 993
Trinectes maculatus **17**: 2444
Tringa
  brevipes **16**: 2233
  flavipes **16**: 2233
  glareola **16**: 2234; **21**: 2906
  incana **16**: 2233
  melanoleuca **16**: 2233
  nebularia **16**: 2233
  ochropus **16**: 2233
  solitaria **16**: 2234
  totanus **16**: 2233
Trionyx **17**: 2442
Tritonia **5**: 652
  hombergi **5**: 652; **16**: 2300
Triturus **13**: 1732
  cristatus **13**: 1732, 1733
  helveticus **13**: 1732, 1733
  vulgaris **13**: 1732, 1733
Trochus niloticus **19**: 2707
Troglodytes **21**: 2993
  aedon **21**: 2993
  troglodytes **7**: 999; **21**: 2993,
  2994
Trogon
  elegans **20**: 2745, 2746, *2746*
  massena **20**: 2745
  mexicanus **20**: 2746–2747
  rufus **20**: 2747
  violaceus **20**: 2747
  viridis **20**: *2747*
Trogonoptera **2**: 212
Troides **2**: 211, 212
Tropheus duboisi **10**: *1400*

Numerals in **bold** refer to volume numbers; those in *italics* to picture captions.

3097

Tropicranis albocristatus **9**: 1227
Tropidacris latriellei **8**: 1035
Trygonorhina **8**: 1103
Tryngites subruficollis **16**: 2233, 2234
Trypanosoma
 brucei **20**: 2760
 brucei brucei **20**: 2761
 brucei congolense **20**: 2761
 brucei gambiense **20**: 2760
 rhodesiense **20**: 2760
Tubifex
 costatus **20**: 2769
 tubifex **20**: 2769, 2770
Tubularia
 bellis **16**: 2274
 indivisa **16**: 2274
 larynx **16**: 2274
Tuea pellionella **4**: 470
Tunga penetrans **6**: 850, 851
Tupaia
 belangeri **19**: 2730
 glis **19**: 2730
 tana **19**: 2731
Tupinambis nigripunctatus **19**: 2650
Turdoides caudatus **1**: 114
Turdus
 merula **2**: 225; **19**: 2678
 migratorius **4**: 555; **15**: 2141; **16**: 2189; **19**: 2678; **20**: 2833
 philomelos **19**: 2678, 2679
 pilaris **18**: 2459
 torquatus **19**: *2680*
 viscivorus **19**: 2678
Turnix
 melanogaster **3**: 350
 sylvatica **3**: 350
Tursiops truncatus **2**: 267
Tylos **21**: 2974
Tympanuchus
 cupido **15**: 2037
 pallidicinctus **15**: 2037
 phasianellus **8**: 1084
Typhlacontias gracilis **17**: 2390
Typhlogobius californiensis **7**: 995
Typhlonectes **3**: 357
Typhologarra widdowsoni **3**: 410
Tyrannus **20**: 2802
 tyrannus **20**: 2802
Tyto
 alba **2**: 148–150; **13**: 1825
 novaehollandiae **18**: 2536

U

Uca **4**: 573; **6**: 815
 annulipes **6**: 815, 816
 paradussumieri **6**: 817
 pugilator **6**: *817*
 tangeri **6**: 817
 uruguayensis **6**: 816
Umbonia crassicornis **9**: 1225
Umbra
 krameri **12**: 1680, 1681
 limi **12**: 1680, 1681
 pygmaea **12**: 1680, 1681
Umbrina roncador **5**: *601*, 602
Upucerthia **13**: 1819
Upupa
 epops **9**: 1223
 marginata **9**: 1222
Uraeginthus angolensis **21**: *2889*
Uraeotyphlus **3**: 357
Urania **20**: 2805
 leilus **20**: 2805, *2805*
 sloanus **20**: 2806
Uranomys **7**: 964
 sapientis **7**: *964*
Uranoscopus **18**: 2520

Uria aalge **8**: 1095
Urocissa **11**: 1520
Urocyon **5**: 689
 cinereoargenteus **3**: 426; **8**: 1045, 1046
 littoralis **8**: 1045, 1046
Urogale **19**: 2731
 everetti **19**: 2730
Urolophus halleri **17**: 2333
Uromastix **1**: 30
Uromys caudimaculatus **7**: 965
Urosalpinx cinerea **21**: 2917
Ursavus elmensis **2**: 174
Ursus
 americanus **2**: 223; **12**: 1672; **19**: 2653
 arctos **3**: 293, 295; **12**: 1672; **19**: 2616, 2653
 arctos arctos **3**: 293
 arctos horribilis **3**: 293
 arctos middendorffi **3**: 293; **19**: 2617
 arctos nelsonii **3**: 295
 arctos pruinosus **3**: 293, 295
 maritimus **1**: 82; **2**: *175*; **14**: 1999
 spelaeus **3**: 295
 thibetanus **1**: 96; **19**: 2653

V

Vandellia cirrhosa **3**: 402
Vanellus
 armatus **10**: 1415
 gregarius **10**: 1415
 indicus **10**: 1415
 vanellus **10**: 1415, 1416; **16**: 2234
Vanessa
 atalanta **20**: 2810, 2811
 canace **20**: 2810
 cardui **20**: 2812
 indica **20**: 2810
Vanga curvirostris **20**: 2813
Varanus
 albigularis **12**: *1639*
 bengalensis **12**: 1639
 brevicaudata **12**: 1638
 giganteus **12**: 1638, 1639
 griseus **12**: 1639
 komodoensis **10**: 1384; **11**: 1466; **12**: 1638
 niloticus **12**: 1638, 1639
 rudicollis **12**: 1639
 salvator **12**: 1638, 1639
 varius **12**: 1639
Varecia **10**: 1436, *1436*
Velella **15**: 2073; **16**: 2300
Verasper variegatus **8**: 1122
Vermetus cristatus **21**: 2990
Vermicularia knorri **21**: *2990*
Vermivora
 chysoptera **21**: 2985
 pinus **21**: 2985
Vespa crabro **9**: 1234; **20**: 2824
Vespula **20**: 2824
 austriaca **20**: 2824
 germanica **9**: 1233; **20**: 2824
 maculata **9**: 1233, 1234
 norvegica **20**: 2824
 rufa **20**: 2824
 squamosa **20**: 2824
 sylvestris **20**: 2824
 vulgaris **9**: 1233; **20**: 2824
Vicugna vicugna **11**: 1470, 1471; **12**: 1672; **20**: 2772
Vidua
 chalybeata **21**: 2936
 macroura **21**: 2889, 2936, 2937
 paradisaea **21**: 2936
 regia **21**: *2938*

Vipera
 ammodytes **1**: 97
 aspis **1**: 98
 berus **1**: 27; **7**: 935
 latastei **1**: 97
Vireo
 flavifrons **20**: 2831
 gilvus **20**: 2831
 gracilirostris **20**: 2831
 griseus **20**: 2831–2832
 olivaceus **20**: 2831, 2832
 philadelphicus **20**: 2831
 solitaricius **20**: 2832
Vireosa **7**: 995
Viverra
 civettina **4**: 461
 zibetha **4**: 460
Volucella **9**: 1251; **20**: 2826
 bombylans **9**: 858; **9**: 1251
Vombatus ursinus **21**: 2961, 2962
Vormela pergusna **14**: 2001
Vulpes **5**: 690
 bengalensis **5**: *690*
 pallida **5**: 689
 velox **5**: 560; **5**: *692*
 velox macrotis **5**: *691*
 velox mutica **4**: 560
 vulpes **5**: 689; **15**: 2133
 vulpes necator **15**: 2133
Vultur gryphus **4**: 516, 517; **12**: 1673; **20**: 2841

W

Wilsonia pusilla **21**: 2985
Wuchereria **12**: 1664
Wyulda squamicaudata **14**: 1929, 1930

X

Xanthocephalus xanthocephalus **2**: 224
Xanthopan morgani predicta **9**: 1168
Xantusia
 arizonae **13**: 1745, 1746
 henshawi **13**: 1745
 vigilis **13**: 1745, 1746
Xenia **17**: 2440, 2441
Xenicus
 gilviventris **16**: 2164
 longipes **16**: 2164
 lyalli **16**: 2165
Xenomystus **6**: 803, 804
 nigre **6**: 804
Xenopirostris
 damii **20**: 2813
 polleni **20**: 2813
Xenops minutus **13**: 1819
Xenopsylla cheopis **6**: 850
Xenopus laevis **4**: 465; **7**: 903
Xenus cinereus **16**: 2233
Xestia c-nigrum **13**: 1821
Xiphias gladius **19**: 2606
Xiphocolaptes major **21**: 2968
Xiphophorus
 helleri **14**: 1984; **19**: 2609
 maculatus **14**: 1983, 1984
 variatus **14**: 1983
Xiphorhynchus guttatus **21**: *2968*, 2969
Xixuthrus heros **11**: 1483
Xyleutes cinereus **7**: 992
Xylopargus **9**: 1176
Xyrichthys novacula **21**: 2991
Xysticus cristatus **5**: 581

Y

Ypsolophus pometellus **19**: 2687

Z

Zaedyus pichiy **1**: 85
Zaglossus **18**: 2486
 bruijni **18**: 2487
Zalembius rosaceus **18**: 2573
Zalophus californianus **16**: 2282, 2283, 2285
Zanclus cornutus **12**: 1653
Zanobatus **8**: 1103
Zapteryx **8**: 1103
Zebrasoma flavescens **18**: 2575
Zenkerella insignis **16**: 2251
Zeus faber **10**: 1328
Zeuzera pyrina **7**: 991
Ziphius cavirostris **2**: 170, *171*
Zirfaea crispata **14**: 1939, 1940
Zoarces viviparus **2**: 238
Zosterops **21**: 2930
 flavus **21**: *2930*
 lateralis **21**: 2930
 palpebrosa **21**: 2931
 senegalensis **21**: 2930, 2931
 wallacei **21**: 2930
Zygaena
 carniolica **3**: 321
 ephialles **3**: 322
 exulans **3**: 321–322
 filipendulae **3**: *321*
 occitanica **3**: 322
Zygogeomys **14**: 1996, 1997

Index of Places

The species described in the *International Wildlife Encyclopedia* are indexed here according to the continent, island group, country, state or region where they are found.

Numerals in **bold** refer to volume numbers.

A

Abbaya, Lake **13**: 1751–1752
Aden, Gulf of **4**: 568–569
Afghanistan **1**: 95–96; **7**: 889–890; **12**: 1614–1615; **17**: 2432–2434; **21**: 2893–2895, 2930–2931, 2951–2953

AFRICA
Amphibians
  3: 356–357; **4**: 464–465, 512–513; **6**: 831–832; **13**: 1732–1734; **15**: 2146–2147; **19**: 2723–2725
Birds
  1: 22–23, 105–106, 113–114; **2**: 151–153, 180–182, 220–221, 224–225, 282–283; **3**: 308–309, 319–320, 337–339, 349–350, 351–353, 414–415; **4**: 523–525, 541–542, 549–550, 551–552; **5**: 584–587, 593–594, 607–609, 614–616, 617–620, 621–622, 626–628, 645–647, 704–705, 712–713; **6**: 750–753, 823–824, 837–839, 842–843, 844–847; **7**: 872–874, 889–890, 911–912, 923–925, 996–998, 1001–1003; **8**: 1012–1013, 1016–1017, 1027–1029, 1061–1063, 1064–1065, 1097–1098, 1123–1125, 1145–1146; **9**: 1179–1181, 1198–1199, 1211–1212, 1217–1218, 1222–1223, 1226–1228, 1247–1249, 1274–1276, 1289–1290; **10**: 1300–1301, 1315–1316, 1317–1319, 1345–1347, 1355–1357, 1367–1369, 1370–1371, 1401–1402, 1415–1416; **11**: 1457–1458, 1495–1496, 1544–1545, 1553–1554; **12**: 1588–1589, 1590–1592, 1649–1651, 1676–1677; **13**: 1737–1738, 1740–1741, 1742–1744, 1760–1762, 1811–1813, 1814–1816, 1830–1831, 1849–1850, 1857–1858; **14**: 1878–1880, 1883–1885, 1908–1910, 1921–1923, 1931–1933, 1937–1938, 1965–1966, 1970–1971, 1988–1990, 1993–1995; **15**: 2041–2042, 2082–2083, 2084–2085, 2106–2108, 2122–2124, 2139–2140; **16**: 2204–2205, 2212–2214, 2229–2230, 2233–2234; **17**: 2316–2318, 2351–2352, 2361–2363, 2386–2388, 2427–2429; **18**: 2467–2469, 2484–2485, 2496–2498, 2521–2523, 2540–2542, 2545–2546, 2551–2553, 2554–2555, 2558–2559, 2565–2566; **19**: 2598–2600, 2678–2680, 2693–2695, 2721–2722; **20**: 2745–2747, 2780–2781, 2790–2792, 2793–2794, 2839–2842, 2843–2845, 2857–2859; **21**: 2888–2889, 2896–2898, 2925–2927, 2930–2931, 2936–2938, 2977–2979, 2980–2982, 2993–2995, 2996–2997
Fish
  2: 195–196; **3**: 389–390, 395–396, 407–408, 409–410; **4**: 458–459; **6**: 745–747, 760–761, 803–804, 860–861; **7**: 993–995; **8**: 1018–1020, 1058–1060, 1119–1120; **10**: 1325–1326, 1421–1422; **11**: 1501–1502; **12**: 1684–1686; **13**: 1749–1750, 1751–1752; **16**: 2292–2293; **17**: 2412–2413; **18**: 2533–2534, 2547–2548; **19**: 2689–2690, 2704–2705
Invertebrates
  1: 20–21, 90–92; **2**: 235–236; **3**: 296–297, 321–322, 327–328; **4**: 488–489; **6**: 741–742, 777–778, 805–806; **7**: 898–899, 966–967, 991–992; **9**: 1156–1157, 1166–1168, 1204–1205; **10**: 1331–1332; **11**: 1478–1480; **15**: 2047–2048, 2076–2077, 2127–2128; **18**: 2453–2454; **20**: 2760–2761
Mammals
  1: 16–17, 18–19, 60–63, 115–117; **2**: 165–166, 239–240; **3**: 305–307, 329–330, 331–332, 335–336, 361–364, 370–372, 373–375, 385–386, 427–429; **4**: 439–442, 460–461, 501–503; **5**: 595–597, 675–676, 699–701; **6**: 724–725, 726–728, 757–759, 767–770, 774–776; **7**: 906–907, 939–942, 946–948, 949–951, 952–953, 958–959, 964–965, 975–977, 1004–1005; **8**: 1014–1015, 1091–1093, 1147–1149; **9**: 1182–1183, 1192–1195, 1206–1207, 1239–1240, 1267–1269, 1270–1271, 1272–1273, 1284–1285, 1291–1293; **10**: 1323–1324, 1374–1375, 1381–1382, 1389–1391, 1425–1427; **11**: 1444–1446, 1460–1463, 1541–1543; **12**: 1628–1629, 1635–1637, 1647–1648, 1694–1695, 1709–1710; **13**: 1765–1766, 1786–1787, 1803–1804, 1808–1810, 1817–1818, 1855–1856, 1859–1860, 1862–1823; **14**: 1896–1897; **15**: 2034–2035, 2091–2093, 2132–2134, 2144–2145, 2152–2155; **16**: 2215–2216, 2250–2251; **17**: 2325–2326; **18**: 2491–2492, 2499–2501, 2502–2503, 2526–2527, 2571–2572; **19**: 2702–2703; **20**: 2854–2856, 2860–2862, 2865–2867; **21**: 2893–2895, 2942–2944, 2945–2947, 2948–2950, 2975–2976, 3009–3011, 3016–3017, 3018–3019, 3020–3021
Reptiles
  1: 30–31; **2**: 264–265; **3**: 418–420; **4**: 480–482; **5**: 603–606; **6**: 748–749; **7**: 980–981; **8**: 1042–1044; **9**: 1190–1191; **11**: 1530–1531; **12**: 1638–1640; **13**: 1735–1736; **15**: 2058–2059, 2079–2081, 2101–2103; **17**: 2308–2309, 2442–2443; **19**: 2711–2713, 2732–2733; **20**: 2795–2799

ALASKA
Birds
  1: 22–23, 123–125; **3**: 319–320; **5**: 626–628, 704–705; **7**: 996–998; **8**: 1084–1086; **9**: 1255–1257; **12**: 1599–1601; **14**: 1931–1933, 1965–1966, 1988–1990; **15**: 2137–2138; **16**: 2233–2234; **17**: 2437–2439; **20**: 2790–2792, 2857–2859; **21**: 2890–2892
Fish
  1: 34; **8**: 1047–1048, 1121–1122; **13**: 1840–1842
Invertebrates
  3: 316–318; **4**: 558–560; **16**: 2225–2226
Mammals
  1: 134–135; **2**: 198–200, 222–223; **4**: 436–438; **5**: 659–661; **8**: 1075–1077; **12**: 1703–1705; **15**: 2129–2131; **16**: 2282–2284; **21**: 2958–2960
Alberta **2**: 198–200; **12**: 1622–1624
Albert, Lake **13**: 1751–1752
Aldabra Island **18**: 2565–2566
Aleutian Islands **8**: 1094–1096, 1132–1134
Algeria **1**: 136–137; **7**: 872–874; **8**: 1104–1105; **16**: 2191–2193
Alps **3**: 421–422; **4**: 551–552; **9**: 1272–1273; **15**: 2055–2057, 2137–2138
Amazon Basin **14**: 2006–2007; **15**: 2060–2061; **17**: 2356–2357; **21**: 2987–2988
Amazonia **19**: 2624–2626; **20**: 2756–2757
Amazon River **6**: 762–763; **11**: 1534–1535; **14**: 2006–2007
Amsterdam Island **15**: 2065–2067
Andaman Islands **14**: 1941–1942
Andes Mountains **4**: 443–445, 478–479, 490–491, 516–518, 523–525; **5**: 680–681; **7**: 898–899; **13**: 1838–1839; **14**: 1965–1966; **15**: 2087–2088
Angola **1**: 18–19; **12**: 1595–1597, 1635–1637, 1676–1677

ANTARCTIC
Birds
  1: 28–29; **6**: 779–781; **10**: 1307–1309; **14**: 1965–1966; **15**: 2045–2046; **17**: 2337–2338, 2435–2436; **18**: 2554–2555; **19**: 2661–2663
Invertebrates
  2: 158–159
Mammals
  2: 246–248; **4**: 564–565; **11**: 1449–1450; **16**: 2210–2211; **21**: 2899–2900
Antigua **19**: 2671–2673
Arabian Peninsula **3**: 385–386; **5**: 593–594; **6**: 724–725; **7**: 946–948; **8**: 1097–1098, 1123–1125; **9**: 1226–1228, 1274–1276; **10**: 1345–1347; **15**: 2058–2059, 2106–2108; **19**: 2612–2613, 2702–2703
see also Middle East
Arabian Sea **2**: 257

ARCTIC
Birds
  7: 913–915; **8**: 1110–1112; **11**: 1464–1465, 1485–1487,
1524–1526; **14**: 1931–1933, 1988–1990; **15**: 2055–2057, 2125–2126; **16**: 2233–2234; **19**: 2661–2663
Invertebrates
  2: 158–159
Mammals
  1: 83–84; **2**: 194–194, 246–248; **12**: 1717–1719; **14**: 1998–2000; **16**: 2169–2170
Arctic Ocean **2**: 246–248; **6**: 754–756; **8**: 1052–1053, 1121–1122, 1139–1141; **9**: 1219–1221; **15**: 2065–2067; **20**: 2851–2853
Argentina **2**: 252; **3**: 382–384; **4**: 443–445, 553–555, 561–563, 566–567; **5**: 680–681; **6**: 748–749; **11**: 1470–1472; **12**: 1678–1679; **15**: 2150–2151; **16**: 2233–2234; **20**: 2854–2856
Arizona **2**: 228–229; **4**: 516–518, 534–535; **7**: 972–974; **17**: 2369–2370
Arkansas **1**: 38–40
Armenia **21**: 2951–2953
Aru Islands **4**: 483–485
Ascension Island **13**: 1755–1756; **18**: 2575–2577; **20**: 2748–2750

ASIA
Amphibians
  6: 831–832; **13**: 1732–1734; **18**: 2465–2466; **19**: 2723–2725
Birds
  1: 22–23, 105–106; **2**: 151–153, 220–221, 230–232; **3**: 319–320, 349–350; **4**: 499–500, 536–538, 541–542, 549–550; **5**: 584–587, 593–594, 607–609, 610–613, 617–620, 626–628, 645–647, 680–681, 704–705, 712–713; **6**: 750–753, 842–843, 844–847; **7**: 872–874, 889–890, 911–912, 996–998; **8**: 1012–1013, 1016–1017, 1061–1063, 1064–1065, 1073–1074, 1145–1146; **9**: 1179–1181, 1198–1199, 1226–1228, 1247–1249, 1274–1276, 1289–1290; **10**: 1329–1330, 1345–1347, 1355–1357, 1367–1369; **11**: 1519–1521, 1524–1526, 1544–1545; **12**: 1590–1592, 1602–1603, 1614–1615, 1649–1651, 1714–1716; **13**: 1742–1744, 1760–1762, 1835–1837, 1849–1850; **14**: 1878–1880, 1886–1887, 1908–1910, 1921–1923, 1965–1966, 1988–1990, 1993–1995; **15**: 2041–2042, 2055–2057, 2082–2083, 2106–2108, 2122–2124, 2137–2138; **16**: 2204–2205, 2206–2207, 2229–2230; **17**: 2361–2363, 2427–2429; **18**: 2467–2469, 2484–2485, 2540–2542, 2545–2546, 2551–2553, 2565–2566; **19**: 2598–2600, 2618–2619, 2644–2646, 2647–2649, 2661–2663, 2678–2680, 2693–2695; **20**: 2793–2794, 2839–2842, 2843–2845; **21**: 2888–2889, 2896–2898, 2966–2967, 2980–2982, 2993–2995, 2996–2997
Fish
  6: 803–804; **7**: 993–995; **8**: 1018–1020; **10**: 1421–1422; **11**: 1473–1474; **12**: 1618–1619;

ASIA
Fish (*continued*)
16: 2182–2183; **17**: 2412–2413;
18: 2489–2490, 2533–2534,
2569–2570, 2585; **19**: 2704–2705;
20: 2751–2753
Invertebrates
1: 76; **2**: 211–212, 235–236,
243–244; **4**: 525–526; **7**: 898–899,
991–992; **8**: 1117–1118;
9: 1156–1157, 1166–1168;
10: 1428–1429; **11**: 1478–1480;
14: 1991–1992; **15**: 2071–2072,
2076–2077; **17**: 2376–2378;
20: 2876–2877
Mammals
1: 60–63, 83–84, 120–122, 132–133,
134–135; **2**: 198–200; **3**: 292–295,
305–307; **5**: 699–701; **6**: 767–770;
7: 884–886, 906–907; **8**: 1054–1057,
1129–1131; **9**: 1171–1173,
1239–1240, 1272–1273, 1291–1293;
10: 1323–1324, 1365–1366,
1432–1433; **11**: 1511–1513;
12: 1616–1617, 1625–1627,
1635–1637, 1701–1702;
13: 1862–1823; **14**: 1943–1944,
1962–1964; **15**: 2099–2100,
2132–2134, 2152–2155;
16: 2177–2178, 2179–2181;
17: 2374–2375; **18**: 2491–2492;
19: 2612–2613; **20**: 2740–2742,
2872–2873; **21**: 2893–2895,
2942–2944
Reptiles
1: 26–27, 30–31; **4**: 480–482;
5: 603–606; **8**: 1042–1044;
12: 1638–1640; **14**: 1972–1974;
15: 2101–2103; **17**: 2442–2443;
19: 2711–2713, 2732–2733;
20: 2837–2838; **21**: 2919–2920
see also Asia Minor, Central Asia;
Eurasia; Southeast Asia
Asia Minor **2**: 218–219; **3**: 421–422;
19: 2598–2600; **21**: 2884–2885,
2966–2967, 2975–2976
Assumption Island **2**: 261–263;
18: 2565–2566
ATLANTIC OCEAN
Birds
2: 261–263; **6**: 754–756; **7**: 889–890,
913–915, 923–925; **10**: 1307–1309,
1370–1371; **13**: 1755–1756;
15: 2065–2067, 2125–2126;
18: 2554–2555; **20**: 2748–2750
Fish
1: 45–46, 49–50, 51–52, 99–101;
2: 154–155, 258; **3**: 342, 389–390;
4: 472–473, 495–496, 521–522;
5: 634–635; **6**: 729–731, 745–747;
7: 875–876, 880–881, 896–897,
989–990; **8**: 1081–1083, 1113–1114,
1119–1120, 1121–1122;
9: 1184–1186; **10**: 1327–1328,
1403–1404, 1413–1414;
11: 1455–1456, 1499–1500,
1517–1518; **12**: 1598, 1646,
1692–1693, 1724–1725;
13: 1767–1768, 1774–1775,
1790–1791, 1840–1842;
14: 1948–1949, 2004–2005;
15: 2020–2021, 2062–2064,
2148–2149; **16**: 2238–2239,
2242–2243, 2260–2261;
17: 2379–2380, 2383–2385,
2410–2411, 2423–2424, 2430–2431,
2444–2445; **18**: 2519–2520,
2533–2534, 2575–2577;
19: 2636–2637, 2674–2675,
2698–2699; **20**: 2743–2744,

2758–2759, 2773–2775, 2782–2783;
21: 2901–2902, 2909, 2915–2916,
2934–2935, 2991–2992, 3007–3008
Invertebrates
4: 488–489; **5**: 631–633, 651–652;
9: 1176–1178; **11**: 1451–1452,
1475–1477; **12**: 1711–1713;
13: 1763–1764; **14**: 1926–1928;
16: 2269–2270; **17**: 2364–2366;
18: 2475–2476, 2493–2494,
2506–2507
Mammals
2: 170–171; **8**: 1049–1051;
15: 2024–2026
Reptiles
8: 1070–1072; **11**: 1481–1482
Atlas Mountains **8**: 1104–1105
Auckland Island **5**: 684–685;
16: 2282–2284; **21**: 3003–3004
Australasia **5**: 591–592, 645–647;
7: 877–879, 996–998; **8**: 1018–1020,
1117–1118; **11**: 1544–1545;
13: 1811–1813; **14**: 1878–1880;
15: 2082–2083; **20**: 2795–2799,
2805–2806
see also Australia; Fiji; New
Caledonia; New Guinea; New
Zealand; Pacific Ocean, islands;
Papua New Guinea; Solomon
Islands; Tasmania
AUSTRALIA
Amphibians
11: 1557–1558; **19**: 2723–2725;
20: 2868–2869
Birds
1: 113–114; **2**: 190–191, 205–207,
220–221, 224–225, 269–271;
3: 302–304, 308–309, 337–339,
340–341, 397–398; **4**: 483–485,
536–538, 549–550; **5**: 584–587,
617–620, 621–622, 626–628,
712–713; **6**: 750–753, 782–784;
7: 868–869, 900–901, 911–912,
996–998; **8**: 1030–1031, 1061–1063,
1064–1065, 1145–1146;
9: 1215–1216, 1247–1249,
1274–1276, 1289–1290;
10: 1355–1357, 1367–1369,
1385–1386; **11**: 1488–1489,
1509–1510, 1522–1523, 1527–1529,
1544–1545; **12**: 1590–1592,
1649–1651; **13**: 1737–1738,
1742–1744, 1755–1756, 1823–1824,
1835–1837, 1849–1850, 1857–1858;
14: 1908–1910, 1921–1923,
1965–1966, 1970–1971, 1978–1979,
1988–1990; **15**: 2045–2046,
2106–2108, 2122–2124;
16: 2204–2205, 2233–2234;
17: 2334–2336; **18**: 2484–2485,
2521–2523, 2537–2539, 2540–2542,
2545–2546, 2551–2553, 2586–2589;
19: 2601–2602, 2618–2619,
2669–2670, 2678–2680;
20: 2790–2792; **21**: 2923–2924,
2925–2927, 2930–2931, 2983–2984, 3012–3013
Fish
1: 77–78; **3**: 395–396; **8**: 1058–1060;
11: 1501–1502; **12**: 1684–1686;
14: 1948–1949; **15**: 2027–2028;
16: 2182–2183, 2238–2239,
2290–2291; **17**: 2379–2380;
18: 2547–2548; **20**: 2751–2753
Invertebrates
1: 20–21; **2**: 211–212; **4**: 462–463;
5: 631–633; **7**: 898–899, 930–932;
9: 1156–1157, 1166–1168,
1204–1205; **12**: 1632–1634;
13: 1832–1834; **15**: 2043–2044,

2047–2048; **16**: 2275–2276;
19: 2706–2707; **21**: 2989–2990
Mammals
1: 107–108, 128–129, 130–131;
2: 170–171; **3**: 300–301, 361–364;
5: 629–630, 677–679, 702–703;
6: 724–725; **7**: 906–907, 964–965,
984–986; **9**: 1239–1240, 1286–1287;
10: 1333–1336, 1378–1380;
11: 1449–1450, 1571–1572,
1575–1576, 1577–1578;
13: 1847–1848; **14**: 1929–1930,
1962–1964, 1985–1987;
15: 2089–2090, 2091–2093,
2115–2116, 2132–2134;
16: 2197–2198, 2282–2284;
18: 2486–2487, 2535–2536;
19: 2702–2703, 2726–2727;
20: 2846–2848, 2854–2856;
21: 2961–2963, 3018–3019
Reptiles
1: 30–31; **2**: 172–173; **5**: 603–606;
7: 894–895, 943–945;
12: 1630–1631, 1638–1640;
13: 1866–1867; **15**: 2079–2081;
17: 2308–2309, 2416–2417;
21: 2919–2920
see also Australasia; Tasmania
Austria **12**: 1667–1668
Azerbaijan **21**: 2951–2953
Azores **3**: 368–369, 389–390;
18: 2475–2476; **19**: 2706–2707;
21: 2893–2895, 2966–2967

B

Baffin Island **12**: 1599–1601;
14: 1988–1990
Bahamas **8**: 1066–1067; **12**: 1622–1624;
14: 1957–1959; **17**: 2423–2424;
20: 2831–2833
Baikal, Lake **9**: 1222–1223;
12: 1618–1619; **16**: 2169–2170
Baja California **1**: 45–46, 113–114;
4: 436–438; **7**: 916–918; **8**: 1087–1088,
1121–1122; **10**: 1430–1431;
16: 2171–2172
Bald Island **15**: 2089–2090
Bali **10**: 1315–1316; **12**: 1714–1716
Balkans **3**: 292–295; **4**: 551–552;
7: 980–981; **20**: 2837–2838
Baltic Sea **1**: 105–106; **4**: 472–473,
495–496; **8**: 1049–1051; **9**: 1176–1178,
1184–1186; **11**: 1517–1518;
13: 1774–1775; **16**: 2169–2170;
17: 2383–2385, 2410–2411;
20: 2782–2783
Bangka **2**: 201–202
Bangladesh **7**: 889–890, 936–938,
954–956, 978–979; **10**: 1410–1412;
13: 1857–1858; **19**: 2702–2703;
21: 3014–3015
Barents Sea **11**: 1499–1500
Bawean Island **1**: 107–108;
10: 1315–1316
Bear Island **11**: 1485–1487
Belarus **2**: 215–217; **5**: 666–667
Belize **3**: 387–388; **6**: 829–830;
7: 996–998
Bengal, Bay of **2**: 257
Bering Sea **7**: 916–918; **13**: 1840–1842;
15: 2065–2067
Bermuda **3**: 368–369; **11**: 1499–1500,
1557–1558; **16**: 2238–2239;
21: 2989–2990
Bhutan
7: 954–956; **17**: 2403–2405,
2432–2434; **19**: 2622–2623
Biscay, Bay of **6**: 754–756;
14: 2004–2005

BLACK SEA
Birds
1: 105–106; **6**: 754–756;
8: 1073–1074
Fish
1: 45–46, 51–52; **2**: 258; **4**: 521–522;
6: 860–861; **10**: 1327–1328;
11: 1517–1518; **12**: 1692–1693;
14: 1948–1949; **16**: 2242–2243,
2292–2293; **17**: 2444–2445;
18: 2556–2557; **20**: 2773–2775;
21: 2901–2902, 2934–2935
Invertebrates
13: 1832–1834; **14**: 1926–1928
Mammals
12: 1647–1648; **15**: 2024–2026
Reptiles
20: 2849–2850
BOLIVIA
4: 553–555; **5**: 680–681;
20: 2756–2757
Invertebrates
4: 490–491
Mammals
4: 443–445, 561–563;
11: 1470–1472; **18**: 2513–2514
Reptiles
2: 249–251; **3**: 358–360
Borneo **2**: 201–202; **4**: 460–461;
6: 732–733, 790–791, 803–804;
7: 882–883, 884–886, 900–901,
982–983; **8**: 1115–1116;
9: 1217–1218, 1286–1287;
10: 1315–1316, 1329–1330; **11**: 1459;
13: 1797–1799; **14**: 1891–1893,
1941–1942; **18**: 2562–2564;
19: 2601–2602, 2730–2731
Bosphorus **17**: 2444–2445
Botswana **12**: 1595–1597, 1635–1637,
1676–1677
Brazil **1**: 49–50, 53–54, 57–59, 90–92,
93–94; **2**: 249–251, 255–256;
3: 358–360; **4**: 490–491, 553–555,
561–563, 566–567; **9**: 1196–1197;
14: 2006–2007; **15**: 2150–2151;
17: 2356–2357; **18**: 2462–2464,
2513–2514; **19**: 2667–2668;
20: 2756–2757; **21**: 2888–2889
Britain *see* British Isles
British Columbia **1**: 118–119,
126–127; **2**: 198–200, 249–251;
3: 323–325; **12**: 1622–1624
BRITISH ISLES
Amphibians
6: 743–744; **12**: 1720–1721
Birds
3: 319–320, 351–353; **4**: 523–525,
551–552; **5**: 584–587, 607–609,
617–620, 626–628; **7**: 872–874;
8: 1027–1029, 1094–1096;
11: 1485–1487; **14**: 1878–1880,
1886–1887, 1955–1956, 1988–1990;
15: 2041–2042, 2137–2138;
17: 2361–2363; **20**: 2793–2794,
2843–2845; **21**: 2993–2995
Fish
2: 218–219; **3**: 314–315, 395–396,
402–403; **8**: 1089–1090,
1121–1122
Invertebrates
19: 2706–2707
Mammals
1: 120–122; **4**: 446–447;
11: 1579–1581; **12**: 1696–1697;
14: 1919–1920, 1962–1964;
16: 2169–2170
Reptiles
16: 2231–2232; **17**: 2406–2407
Burma *see* Myanmar
Burundi **9**: 1182–1183

C

Calamian Islands **1**: 107–108
California **1**: 105–106, 123–125;
 2: 249–251, 255–256, 261–263;
 3: 319–320, 407–408; **4**: 488–489,
 516–518; **8**: 1045–1046, 1052–1053,
 1087–1088, 1119–1120;
 10: 1430–1431; **11**: 1519–1521;
 15: 2106–2108; **17**: 2369–2370;
 21: 2958–2960
California, Gulf of **11**: 1517–1518;
 14: 1888–1890; **15**: 2024–2026
Cambodia **1**: 95–96; **2**: 197, 284–285;
 3: 305–307; **7**: 936–938
Cameroon **8**: 1023–1026;
 11: 1536–1538; **14**: 1970–1971
Campbell Island **16**: 2282–2284;
 21: 3003–3004
CANADA
Amphibians
 3: 312–313
Birds
 1: 105–106, 123–125, 126–127;
 2: 230–232; **3**: 365–367, 387–388,
 399–400; **4**: 523–525, 536–538,
 545–548; **5**: 607–609, 626–628;
 6: 754–756; **7**: 923–925, 996–998;
 8: 1084–1086, 1094–1096;
 9: 1187–1188; **11**: 1485–1487;
 12: 1599–1601; **13**: 1760–1762,
 1805–1807; **14**: 1891–1893,
 1921–1923, 1931–1933, 1988–1990,
 2012–2013; **15**: 2036–2037,
 2106–2108, 2125–2126, 2137–2138;
 16: 2233–2234, 2235–2237;
 17: 2437–2439; **18**: 2460–2461;
 19: 2627–2629, 2671–2673;
 20: 2790–2792, 2800–2802;
 21: 2890–2892, 2970–2972
Fish
 3: 402–403; **6**: 860–861;
 8: 1047–1048, 1121–1122;
 10: 1403–1404; **14**: 1917–1918
Invertebrates
 3: 316–318; **16**: 2225–2226
Mammals
 1: 120–122, 134–135; **2**: 198–200,
 222–223, 253–254; **4**: 448–450,
 561–563; **8**: 1075–1077,
 1139–1141; **9**: 1158–1160;
 10: 1432–1433; **12**: 1703–1705,
 1706–1708; **15**: 2038–2040,
 2049–2051, 2068–2070,
 2096–2098; **16**: 2179–2181,
 2294–2296; **18**: 2528–2529;
 21: 2958–2960
Reptiles
 9: 1202–1203, 1231–1232;
 10: 1361–1362; **13**: 1851–1853;
 15: 2101–2103
Canary Islands **3**: 337–339, 368–369;
 10: 1367–1369; **12**: 1632–1634;
 13: 1835–1837; **15**: 2065–2067;
 19: 2706–2707; **21**: 2966–2967
Cape Verde Islands **10**: 1367–1369
CARIBBEAN ISLANDS
Amphibians
 8: 1066–1067; **11**: 1557–1558
Birds
 3: 308–309, 323–325, 399–400;
 4: 523–524, 536–538; **5**: 610–613;
 6: 844–847; **7**: 919–920;
 9: 1213–1214, 1274–1276,
 1290–1291; **10**: 1348–1349;
 11: 1453–1454; **12**: 1622–1624;
 14: 1937–1938, 1988–1990;
 15: 2032–2033; **18**: 2537–2539,
 2540–2542; **19**: 2671–2673,
 2700–2701; **20**: 2831–2833;
 21: 2985–2986

Fish
 4: 458–459; **7**: 928–929;
 8: 1106–1107; **19**: 2704–2705
Invertebrates
 2: 255–256; **12**: 1660–1661
Mammals
 6: 840–841; **9**: 1263–1264;
 12: 1635–1637; **15**: 2096–2098,
 2156–2157
Reptiles
 1: 53–54; **5**: 603–606; **6**: 812–814;
 9: 1281–1282; **19**: 2650
 see also Bahamas; Cuba;
 Dominican Republic; Haiti;
 Hispaniola; Jamaica; Trinidad
Caribbean Sea **6**: 745–747;
 8: 1081–1083; **11**: 1534–1535;
 13: 1755–1756; **14**: 1957–1959;
 15: 2022–2023; **17**: 2423–2424,
 2440–2441; **20**: 2748–2750;
 21: 2915–2916
Caspian Sea **1**: 105–106; **8**: 1073–1074;
 14: 1886–1887; **16**: 2169–2170;
 18: 2556–2557
Cat Island **8**: 1066–1067
Caucasus **14**: 2001–2003;
 17: 2406–2407
Celebes *see* Sulawesi
Central African Republic
 8: 1023–1026; **14**: 1970–1971
CENTRAL AMERICA
Amphibians
 3: 326, 356–357; **9**: 1229–1230;
 11: 1557–1558; **19**: 2723–2725
Birds
 1: 55–56, 105–106; **2**: 220–221;
 3: 323–325, 379–381, 399–400;
 4: 523–524, 543–544, 545–548;
 5: 623–625, 645–647, 680–681;
 6: 788–789, 823–824; **7**: 919–920;
 8: 1016–1017, 1032–1034,
 1061–1063, 1142–1144, 1145–1146;
 9: 1213–1214, 1247–1249,
 1255–1257, 1289–1290;
 10: 1345–1347; **11**: 1453–1454,
 1485–1487, 1514–1516, 1532–1533;
 12: 1590–1592, 1622–1624,
 1649–1651, 1665–1666, 1699–1700;
 13: 1805–1807; **14**: 1878–1880,
 1883–1885, 1937–1938, 1965–1966,
 1988–1990; **15**: 2032–2033,
 2087–2088, 2122–2124,
 2141–2143; **16**: 2184–2185,
 2188–2189, 2235–2237;
 18: 2484–2485, 2537–2539,
 2540–2542, 2545–2546,
 2551–2553, 2567–2568,
 2586–2589; **19**: 2627–2629,
 2661–2663, 2691–2692,
 2714–2716; **20**: 2745–2747,
 2787–2789, 2831–2833;
 21: 2925–2927, 2968–2969,
 2977–2979, 2985–2986
Fish
 4: 458–459; **7**: 887–888, 928–929;
 10: 1302–1303, 1376–1377,
 1421–1422; **14**: 1948–1949,
 1967–1969, 1983–1984;
 17: 2379–2380; **19**: 2608–2609,
 2704–2705; **20**: 2751–2753
Invertebrates
 2: 243–244, 255–256; **4**: 558–560;
 10: 1417–1418; **12**: 1660–1661
Mammals
 1: 57–59, 85–87; **4**: 478–479,
 561–563; **5**: 659–661; **6**: 840–841;
 8: 1045–1046; **9**: 1158–1160,
 1252–1254, 1263–1264;
 10: 1310–1311, 1313–1314,
 1363–1364; **12**: 1689–1691;

 13: 1776–1778, 1792–1794,
 1838–1839; **14**: 1903–1905,
 1996–1997; **15**: 2068–2070,
 2096–2098, 2156–2157;
 17: 2393–2395, 2400–2402;
 19: 2630–2632; **20**: 2807–2809
Reptiles
 2: 156–157; **3**: 358–360; **5**: 603–606;
 6: 812–814; **7**: 933–935;
 9: 1281–1282; **12**: 1687–1688;
 13: 1745–1746; **14**: 1972–1974;
 15: 2101–2103, 2119–2121;
 17: 2308–2309, 2425–2426;
 19: 2650, 2732–2733;
 21: 2919–2920
CENTRAL ASIA
Amphibians
 13: 1732–1734
Birds
 1: 105–106; **2**: 180–182; **3**: 310–311,
 337–339, 351–353, 414–415;
 4: 541–542, 551–552; **5**: 584–587,
 593–594, 607–609; **6**: 844–847;
 7: 872–874, 980–981, 1001–1003;
 8: 1012–1013; **9**: 1222–1223;
 10: 1300–1301, 1355–1357,
 1401–1402, 1415–1416;
 11: 1457–1458; **12**: 1714–1716;
 13: 1737–1738, 1740–1741,
 1757–1759, 1760–1762;
 14: 1891–1893, 1908–1910,
 1934–1936, 1993–1995;
 15: 2082–2083, 2106–2108,
 2122–2124, 2139–2140;
 16: 2206–2207, 2229–2230,
 2233–2234; **17**: 2341–2343,
 2396–2397; **18**: 2496–2498,
 2521–2523; **19**: 2721–2722
Fish
 3: 395–396
Invertebrates
 7: 898–899
Mammals
 4: 561–563; **7**: 939–942, 949–951;
 8: 1078–1080; **11**: 1444–1446,
 1506–1508, 1559–1560, 1568–1570;
 12: 1701–1702; **21**: 2939–2941
Reptiles
 16: 2231–2232
Chad **1**: 24–25; **13**: 1808–1810
Chad, Lake **2**: 195–196
Chad River **6**: 803–804;
 13: 1749–1750, 1751–1752
Chatham Islands **7**: 919–920;
 15: 2045–2046
Chile **1**: 53–54; **2**: 261–263; **3**: 407–408;
 4: 443–445, 536–538, 561–563;
 8: 1119–1120; **11**: 1470–1472;
 12: 1678–1679; **13**: 1835–1837;
 15: 2150–2151; **18**: 2554–2555;
 20: 2854–2856
CHINA
Amphibians
 4: 513–514; **7**: 982–983;
 9: 1174–1175, 1229–1230;
 18: 2465–2466
Birds
 1: 113–114; **2**: 224–225; **4**: 541–542;
 5: 584–587, 593–594, 626–628,
 712–713; **7**: 868–869, 900–901,
 999–1000; **8**: 1145–1146;
 9: 1198–1199, 1222–1223,
 1274–1276; **10**: 1300–1301,
 1317–1319, 1329–1330, 1367–1369,
 1401–1402, 1415–1416;
 11: 1519–1521, 1524–1526;
 12: 1599–1601, 1614–1615,
 1714–1716; **13**: 1742–1744,
 1757–1759, 1811–1813, 1835–1837,
 1857–1858; **14**: 1878–1880,

 1886–1887, 1891–1893,
 1934–1936, 1937–1938;
 15: 2106–2108; **17**: 2361–2363;
 18: 2496–2498, 2545–2546,
 2551–2553; **19**: 2644–2646,
 2717–2718; **20**: 2745–2747;
 21: 2890–2892, 2925–2927,
 2983–2984
Fish
 3: 395–396; **4**: 468–469; **7**: 978–979;
 8: 1018–1020; **11**: 1517–1518;
 13: 1845–1846; **16**: 2290–2291;
 17: 2412–2413; **18**: 2489–2490;
 19: 2596–2597
Invertebrates
 1: 102; **7**: 966–967
Mammals
 1: 95–96, 120–122; **3**: 361–364;
 4: 446–447, 460–461; **5**: 671–672;
 7: 884–886, 960–963, 968–971;
 8: 1021–1022; **9**: 1158–1160;
 10: 1323–1324; **11**: 1459;
 12: 1635–1637, 1696–1697,
 1709–1710; **13**: 1855–1856;
 14: 1919–1920; **15**: 2024–2026,
 2052–2054, 2099–2100, 2135–2136;
 16: 2201–2203; **17**: 2432–2434;
 18: 2562–2564; **19**: 2622–2623,
 2683–2685, 2702–2703, 2730–2731;
 20: 2854–2856; **21**: 2893–2895,
 2945–2947
Reptiles
 1: 38–40; **2**: 197; **8**: 1042–1043
Christmas Island **2**: 261–263;
 7: 889–890
Cocos Island **5**: 648–650
Colombia **1**: 126–127; **2**: 156–157,
 233–234; **3**: 382–384; **4**: 478–479,
 523–525, 566–567; **5**: 603–606,
 659–661; **8**: 1081–1083;
 9: 1158–1160; **14**: 1972–1974;
 17: 2356–2357; **18**: 2513–2514;
 19: 2667–2668; **21**: 3014–3015
Colorado **2**: 198–200
Comoros Islands **4**: 497–498;
 16: 2186–2187
Congo **4**: 439–442; **8**: 1023–1026;
 11: 1536–1538
Congo, Democratic Republic of
 (Zaire)
 2: 282–283; **4**: 439–442;
 8: 1023–1026; **9**: 1182–1183;
 13: 1786–1787
Congo River **2**: 195–196; **3**: 347–348;
 6: 803–804; **13**: 1751–1752
Corsica **2**: 198–200; **4**: 551–552;
 12: 1667–1668; **13**: 1760–1762
Costa Rica **3**: 326; **18**: 2513–2514
Cozumel Island **4**: 478–479
Crete **18**: 2491–2492
Crozet Islands **14**: 1957–1959
Cuba **1**: 53–54; **3**: 312–313;
 6: 853–855; **7**: 928–929;
 8: 1066–1067; **9**: 1179–1181,
 1255–1257; **12**: 1622–1624;
 13: 1745–1746; **18**: 2452;
 19: 2700–2701; **21**: 2970–2972,
 2993–2995
Cyprus **7**: 889–890

D

Danube River **12**: 1680–1681
Denmark **1**: 140–141; **14**: 1955–1956
Djibouti **2**: 189–190
Dneiper River **18**: 2556–2557
Dniester River **12**: 1680–1681
Dominica **2**: 249–251
Dominican Republic **19**: 2700–2701
Don River **18**: 2556–2557

E

Easter Island **13**: 1755–1756
East Timor **4**: 549–550
Ecuador **2**: 156–157; **4**: 478–479;
  8: 1081–1083; **9**: 1158–1160;
  11: 1470–1472; **17**: 2356–2357,
  2425–2426
Egypt **2**: 255–256; **9**: 1222–1223;
  13: 1849–1850; **14**: 1908–1910
Elbe River **5**: 177–179
Ellesmere Island **12**: 1703–1705;
  14: 1988–1990
Equatorial Guinea **11**: 1536–1538
Ethiopia **1**: 115–117; **2**: 189–190;
  3: 331–332; **4**: 551–552; **5**: 584–587,
  673–674; **7**: 952–953; **12**: 1628–1629;
  13: 1765–1766; **14**: 1993–1995;
  21: 2939–2941
EURASIA
  Birds
    5: 607–609; **8**: 1084–1086;
    10: 1307–1309; **11**: 1519–1521;
    12: 1590–1592; **13**: 1811–1813;
    15: 2055–2057; **16**: 2212–2214,
    2233–2234; **18**: 2521–2523,
    2537–2539; **20**: 2790–2792;
    21: 2977–2979
  Fish
    3: 395–396; **12**: 1618–1619
  Invertebrates
    3: 321–322, 327–328; **4**: 504–505;
    5: 591–592; **15**: 2047–2048
  Mammals
    1: 120–122; **6**: 785–787; **7**: 884–886;
    12: 1654–1656; **15**: 2091–2093,
    2129–2131; **16**: 2201–2203
  Reptiles
    1: 30–31; **20**: 2837–2838
EUROPE
  Amphibians
    4: 510–511, 512–513; **6**: 743–744,
    825–826, 831–832; **12**: 1609–1610,
    1720–1721; **13**: 1732–1734;
    18: 2465–2466; **19**: 2723–2725
  Birds
    1: 22–23, 105–106; **2**: 151–153,
    220–221, 224–225, 230–232;
    3: 310–311, 319–320, 337–339,
    351–353, 365–367, 376–378,
    414–415; **4**: 499–500, 536–538,
    541–542, 551–552; **5**: 593–594,
    610–613, 617–620, 626–628,
    680–681, 704–705; **6**: 750–753,
    754–756, 844–847; **7**: 872–874,
    919–920, 923–925, 996–998,
    999–1000, 1001–1003;
    8: 1012–1013, 1016–1017,
    1027–1029, 1061–1063,
    1064–1065, 1073–1074, 1084–1086,
    1094–1096, 1145–1146;
    9: 1179–1181, 1187–1188,
    1198–1199, 1211–1212, 1222–1223,
    1247–1249, 1274–1276;
    10: 1300–1301, 1317–1319,
    1345–1347, 1355–1357, 1367–1369,
    1401–1402, 1415–1416;
    11: 1457–1458, 1485–1487,
    1519–1521, 1524–1526;
    12: 1590–1592, 1602–1603,
    1649–1651; **13**: 1737–1738,
    1740–1741, 1742–1744, 1757–1759,
    1760–1762, 1835–1837;
    14: 1886–1887, 1891–1893,
    1908–1910, 1921–1923, 1934–1936,
    1957–1959, 1965–1966, 1988–1990,
    1993–1995; **15**: 2041–2042,
    2055–2057, 2082–2083, 2106–2108,
    2122–2124, 2137–2138, 2139–2140;
    16: 2191–2193, 2204–2205,
    2206–2207, 2212–2214, 2229–2230;

    17: 2341–2343, 2361–2363,
    2396–2397, 2427–2429;
    18: 2467–2469, 2496–2498,
    2521–2523, 2540–2542, 2545–2546,
    2551–2553; **19**: 2598–2600,
    2644–2646, 2647–2649, 2661–2663,
    2678–2680, 2693–2695, 2721–2722;
    20: 2793–2794, 2839–2842,
    2843–2845, 2857–2859;
    21: 2890–2892, 2966–2967,
    2980–2982, 2993–2995, 2996–2997
  Fish
    1: 99–101; **2**: 218–219, 274–275;
    3: 314–315, 395–396, 402–403;
    6: 745–747, 860–861; **8**: 1018–1020,
    1047–1048, 1089–1090;
    10: 1403–1404; **11**: 1473–1474;
    12: 1618–1619, 1680–1681;
    16: 2182–2183, 2292–2293;
    17: 2379–2380; **18**: 2533–2534,
    2569–2570; **20**: 2751–2753
  Invertebrates
    1: 64–65, 76; **2**: 235–236, 243–244;
    3: 296–297; **4**: 488–489, 525–526,
    539–540; **6**: 741–742, 801–802,
    805–806; **7**: 898–899, 926–927,
    930–932, 991–992; **8**: 1117–1118;
    9: 1166–1166, 1208–1210,
    1233–1234; **10**: 1428–1429;
    11: 1475–1477; **13**: 1795–1796,
    1832–1834; **14**: 1915–1916,
    1926–1928, 1991–1992;
    15: 2043–2044, 2071–2072,
    2074–2075, 2076–2077, 2127–2128;
    16: 2275–2276, 2299–2301;
    17: 2353–2355, 2364–2366;
    18: 2530–2532; **19**: 2633–2635,
    2706–2707; **20**: 2876–2877;
    21: 2917–2918, 2964–2965
  Mammals
    1: 83–84, 120–122, 134–135;
    2: 177–179; **3**: 292–295, 305–307,
    421–422; **4**: 561–563; **5**: 699–701;
    6: 794–796, 818–820; **8**: 1049–1051,
    1054–1057, 1078–1080, 1129–1131,
    1132–1134; **9**: 1158–1160,
    1171–1173, 1239–1240, 1291–1293;
    10: 1432–1433; **11**: 1506–1508,
    1568–1570, 1579–1581;
    12: 1616–1617, 1625–1627;
    14: 1962–1964, 2001–2003;
    15: 2096–2098, 2099–2100,
    2132–2134; **16**: 2179–2181,
    2201–2203; **17**: 2374–2375;
    20: 2740–2742, 2854–2856,
    2872–2873; **21**: 2884–2885,
    2893–2895, 2942–2944,
    2945–2947, 2975–2976
  Reptiles
    1: 26–27, 30–31, 97–98;
    8: 1042–1044, 1068–1069;
    15: 2101–2103; **17**: 2406–2407;
    19: 2711–2713; **20**: 2837–2838,
    2849–2850; **21**: 2919–2920
  see also Eurasia

F

Faeroe Islands **1**: 35–37; **8**: 1061–1063;
  10: 1307–1309; **11**: 1485–1487;
  14: 1988–1990, 2004–2005; **17**:
  2444–2445; **18**: 2554–2555
Falkland Islands **5**: 684–685;
  7: 996–998; **10**: 1307–1309;
  13: 1737–1738; **16**: 2282–2284;
  17: 2337–2338; **18**: 2524–2525,
  2586–2589; **20**: 2800–2802;
  21: 2993–2995
Fernando de Noronha **20**: 2831–2833
Fiji **1**: 103–104; **3**: 308–309; **4**: 462–463;

  9: 1281–1282; **13**: 1861;
  21: 2983–2984
Finland **7**: 884–886; **10**: 1432–1433;
  14: 1965–1966; **17**: 2437–2439
Flinders Island **21**: 2961–2963
Flores Island **10**: 1383–1384
Florida **1**: 20–21, 38–40, 49–50,
  105–106, 123–125; **2**: 266–268;
  3: 308–309, 323–325; **4**: 534–535;
  5: 603–606; **6**: 788–789;
  12: 1622–1624; **21**: 3014–3015
France **1**: 97–98; **3**: 314–315;
  4: 446–447, 551–552; **5**: 666–667;
  7: 946–948; **8**: 1068–1069;
  11: 1503–1505; **13**: 1667–1668,
  1696–1697; **20**: 2849–2850
Franklin Island **18**: 2535–2536
French Guiana **4**: 490–491;
  9: 1196–1197
French Polynesia **7**: 911–912;
  16: 2186–2187; **18**: 2547–2548

G

Gabon **8**: 1023–1026; **11**: 1536–1538
Galapagos Islands **1**: 35–37;
  4: 536–538; **5**: 648–650; **6**: 844–847;
  7: 889–890; **9**: 1179–1181,
  1281–1282; **11**: 1555–1556;
  12: 1622–1624; **14**: 1888–1890,
  1957–1959; **16**: 2282–2284;
  18: 2540–2542; **19**: 2711–2713;
  20: 2748–2750, 2795–2799,
  2800–2802
Gambia River **13**: 1749–1750
Georgia (*former U.S.S.R.*) **21**: 2951–2953
Germany **12**: 1599–1601, 1667–1668;
  14: 1955–1956
Gibraltar **1**: 136–137
Gonave Island **14**: 1937–1938
Gough Island **5**: 684–685;
  12: 1649–1651; **15**: 2045–2046;
  17: 2334–2336
Greater Antilles **19**: 2700–2701
Greater Sunda Islands **12**: 1714–1716
Great Lakes **14**: 1945–1947
Greece **20**: 2849–2850; **21**: 2951–2953
Greenland **1**: 83–84, 140–141;
  3: 365–367; **4**: 525–526; **8**: 1054–1057,
  1094–1096, 1121–1122, 1132–1134;
  10: 1307–1309; **11**: 1485–1487,
  1499–1500; **12**: 1599–1601,
  1703–1705; **14**: 1931–1933,
  1955–1956, 1965–1966, 1975–1977,
  1988–1990; **15**: 2055–2057,
  2122–2124, 2125–2126, 2137–2138;
  17: 2437–2439; **20**: 2790–2792
Guam **12**: 1709–1710
Guatemala **3**: 387–388, 402–403;
  6: 829–830; **9**: 1231–1232;
  15: 2101–2103; **20**: 2784–2786
Guiana **1**: 43–44
Guinea **8**: 1023–1026; **14**: 1970–1971
Guinea, Gulf of **3**: 389–390
Gulf, the **4**: 568–569
Guyana **1**: 93–94; **4**: 490–491;
  9: 1196–1197; **18**: 2513–2514

H

Hainan **1**: 95–96
Haiti **12**: 1622–1624; **21**: 2896–2898
Hawaii **3**: 312–313; **5**: 610–613;
  9: 1164, 1213–1214, 1215–1216;
  10: 1315–1316; **11**: 1544–1545,
  1557–1558; **12**: 1631–1634,
  1652–1653; **13**: 1737–1738,
  1755–1756; **17**: 2442–2443;
  18: 2575–2577; **20**: 2748–2750;
  21: 2888–2889

Hawaiian Islands **12**: 1647–1648
  see also Hawaii; Kauai; Maui
Heard Island **11**: 1449–1450
High Atlas Mountains **1**: 136–137
Himalayan Mountains **1**: 95–96;
  14: 1934–1936, 1937–1938,
  1972–1974; **15**: 2106–2108,
  2139–2140; **18**: 2467–2469;
  19: 2612–2613, 2644–2646;
  21: 3000–3002
Hispaniola **2**: 151–153; **13**: 1854;
  14: 1937–1938; **18**: 2452;
  19: 2700–2701
Holland **1**: 140–141
Hungary **12**: 1667–1668

I

Iberia **3**: 292–295
ICELAND
  Birds
    1: 140–141; **5**: 626–628;
    8: 1061–1063; **9**: 1187–1188;
    10: 1307–1309; **11**: 1485–1487;
    12: 1599–1601; **14**: 1955–1956,
    1965–1966, 1988–1990, 1993–1995;
    15: 2106–2108, 2125–2126,
    2137–2138; **17**: 2437–2439;
    18: 2554–2555
  Fish
    8: 1119–1120, 1121–1122;
    11: 1499–1500; **14**: 1975–1977,
    2004–2005
  Invertebrates
    5: 651–652; **7**: 926–927;
    17: 2353–2355
  Mammals
    1: 83–84; **8**: 1049–1051;
    16: 2169–2170
Inaccessible Island **15**: 2106–2108
INDIA
  Amphibians
    3: 356–357; **4**: 513–514; **7**: 982–983
  Birds
    1: 103–104, 113–114; **2**: 151–153,
    224–225; **3**: 308–309; **4**: 549–550,
    551–552; **5**: 593–594, 617–620,
    645–647, 712–713; **6**: 790–791;
    7: 868–869, 889–890, 900–901,
    999–1000; **8**: 1012–1013,
    1145–1146; **9**: 1226–1228,
    1289–1290; **10**: 1315–1316,
    1329–1330, 1355–1357, 1367–1369;
    11: 1544–1545; **12**: 1614–1615,
    1714–1716; **13**: 1760–1762;
    14: 1878–1880, 1898–1900,
    1908–1910, 1937–1938, 1957–1959,
    1970–1971, 1993–1995;
    16: 2204–2205, 2212–2214;
    17: 2386–2388, 2427–2429;
    18: 2496–2498, 2545–2546,
    2551–2553; **19**: 2717–2718;
    20: 2745–2747, 2839–2842;
    21: 2925–2927
  Fish
    1: 77–78; **4**: 458–459, 468–469;
    7: 978–979; **10**: 1421–1422;
    12: 1684–1686; **21**: 3014–3015
  Invertebrates
    1: 102; **2**: 255–256; **10**: 1331–1332
  Mammals
    1: 107–108, 132–133; **2**: 226–227,
    284–285; **3**: 385–386; **4**: 460–461;
    5: 671–672; **6**: 724–725; **7**: 877–879,
    884–886, 936–938, 939–942;
    8: 1021–1022; **9**: 1206–1207,
    1267–1269, 1286–1287;
    10: 1410–1412; **11**: 1459,
    1460–1463, 1490–1492;
    12: 1696–1697, 1709–1710;

13: 1753–1754, 1855–1856;
14: 1941–1942; **15**: 2024–2026,
2132–2134, 2135–2136;
16: 2179–2181; **17**: 2403–2405,
2432–2434; **18**: 2562–2564;
19: 2612–2613, 2622–2623,
2683–2685, 2702–2703,
2730–2731; **21**: 2945–2947
Reptiles
3: 418–420; **4**: 480–482; **7**: 954–956;
15: 2079–2081; **17**: 2308–2309
INDIAN OCEAN
Birds
2: 261–263; **4**: 568–569;
13: 1755–1756; **20**: 2748–2750,
2790–2792
Fish
1: 45–46, 77–78; **2**: 257; **3**: 389–390;
4: 497–498; **5**: 706–707; **7**: 880–881,
896–897, 989–990; **8**: 1081–1083;
10: 1327–1328; **11**: 1517–1518,
1561–1562; **12**: 1652–1653,
1684–1686; **13**: 1767–1768,
1790–1791; **15**: 2020–2021,
2062–2064, 2094–2095;
16: 2219–2220, 2238–2239,
2255–2256, 2260–2261, 2290–2291;
17: 2430–2431; **18**: 2547–2548,
2575–2577; **19**: 2674–2675;
20: 2743–2744; **21**: 2909,
2915–2916, 2954–2955,
2991–2992, 3007–3008
Invertebrates
6: 777–778, 805–806, 815–817;
17: 2367–2368, 2440–2441;
18: 2493–2494; **19**: 2706–2707
Mammals
6: 724–725
Reptiles
8: 1070–1072; **11**: 1481–1482;
17: 2308–2309
Indian Ocean, islands **7**: 877–879,
906–907, 911–912, 964–965, 966–967;
16: 2186–2187; **19**: 2601–2602,
2702–2703, 2711–2713
Indian subcontinent **13**: 1859–1860;
21: 2983–2984, 3018–3019
Indochina **9**: 1229–1230, 1286–1287;
17: 2386–2388
INDONESIA
Amphibians
7: 982–983; **9**: 1229–1230
Birds
5: 610–613; **7**: 900–901, 911–912;
10: 1329–1330, 1367–1369;
11: 1488–1489, 1493–1494,
1544–1545; **13**: 1737–1738;
15: 2045–2046; **20**: 2748–2750;
21: 2930–2931
Fish
12: 1684–1686; **16**: 2290–2291
Invertebrates
1: 102; **2**: 211–212; **7**: 966–967;
8: 1117–1118; **13**: 1861;
21: 2989–2990
Mammals
3: 305–307; **5**: 671–672; **6**: 724–725;
7: 884–886, 968–971;
10: 1365–1366; **11**: 1444–1446,
1459; **13**: 1855–1856;
19: 2638–2640, 2702–2703
Reptiles
10: 1383–1384; **15**: 2079–2081;
17: 2308–2309
Iran **1**: 113–114; **2**: 198–200;
3: 402–403, 427–429; **6**: 794–796;
7: 889–890, 949–951; **9**: 1211–1212;
11: 1524–1526; **13**: 1740–1741;
15: 2139–2140; **18**: 2551–2553;
19: 2644–2646, 2678–2680,

2683–2685, 2693–2695;
21: 2884–2885, 2951–2953
Iraq **3**: 409–410; **7**: 889–890;
19: 2683–2685; **21**: 2980–2982
Ireland **1**: 120–122, 140–141;
4: 551–552; **12**: 1599–1601,
1720–1721; **14**: 1955–1956
Irian Jaya **19**: 2702–2703,
2726–2727
Israel **18**: 2565–2566; **19**: 2689–2690;
21: 2884–2885
Italy **1**: 97–98; **3**: 312–313;
4: 551–552; **5**: 595–597;
9: 1286–1287; **11**: 1503–1505;
12: 1667–1668; **13**: 1788–1789

J

Jamaica **2**: 151–153; **3**: 312–313;
4: 543–544; **8**: 1066–1067
JAPAN
Amphibians
3: 312–313; **4**: 512–513;
7: 982–983; **9**: 1174–1175
Birds
1: 22–23; **3**: 308–309, 310–311,
351–353; **4**: 523–525; **5**: 584–587;
7: 999–1000; **8**: 1094–1096;
10: 1317–1319; **12**: 1590–1592,
1614–1615; **13**: 1737–1738,
1757–1759, 1760–1762;
14: 1934–1936, 1970–1971,
1988–1990; **15**: 2055–2057,
2106–2108; **16**: 2206–2207;
17: 2396–2397; **18**: 2467–2469;
19: 2647–2649; **20**: 2839–2842;
21: 2890–2892, 2930–2931,
2966–2967
Fish
3: 395–396; **7**: 978–979;
8: 1018–1020, 1058–1060,
1121–1122; **12**: 1652–1653;
13: 1840–1842; **14**: 1948–1949;
16: 2290–2291; **19**: 2596–2597,
2704–2705
Invertebrates
4: 525–526; **5**: 591–592; **7**: 898–899,
991–992; **13**: 1832–1834;
15: 2071–2072, 2076–2077;
16: 2225–2226
Mammals
1: 95–96; **3**: 292–295;
4: 448–450, 561–563;
8: 1132–1134; **12**: 1625–1627,
1709–1710; **14**: 1962–1964;
15: 2099–2100; **16**: 2179–2181,
2282–2284, 2294–2296;
17: 2323–2324, 2374–2375;
21: 2893–2895
Reptiles
7: 943–945; **14**: 1972–1974;
17: 2442–2443
see also Ryukyu Islands
Japan, Sea of **8**: 1121–1122;
11: 1517–1518
Java **2**: 201–202; **6**: 790–791,
803–804; **8**: 1115–1116;
10: 1315–1316, 1329–1330;
11: 1459; **14**: 1898–1900
Jordan **13**: 1808–1810; **19**: 2689–2690,
2693–2695

K

Kashmir **8**: 1021–1022; **10**: 1410–1412;
14: 1962–1964
Kauai **9**: 1164, 1215–1216
Kazakhstan **2**: 274–275;
15: 2041–2042; **17**: 2432–2434
Kentucky **4**: 525–526

Kenya **1**: 60–63; **2**: 213–214; **7**: 952–953;
9: 1217–1218; **11**: 1541–1543;
12: 1628–1629; **14**: 1970–1971
Kerguelen Island **11**: 1449–1450;
14: 1957–1959
Korea **1**: 95–96; **4**: 446–447, 448–450;
5: 607–609; **7**: 978–979; **8**: 1094–1096;
13: 1757–1759, 1840–1842,
1868–1869; **14**: 1962–1964;
19: 2596–2597, 2644–2646;
20: 2839–2842, 2843–2845
Kurile Islands **16**: 2179–2181
Kyrgyzstan **17**: 2432–2434
Kyushu **1**: 95–96

L

Labrador **8**: 1139–1141
Laos **1**: 95–96; **2**: 197, 284–285;
3: 305–307; **7**: 968–971; **8**: 1021–1022;
11: 1566–1567; **14**: 1714–1716
Lapland **20**: 2837–2838
Lesser Antilles **2**: 249–251
Lesser Sunda Islands **21**: 2930–2931
Libya **8**: 1104–1105
Lithuania **2**: 215–217
Lord Howe Island **12**: 1652–1653
Louisiana **3**: 323–325; **10**: 1417–1418

M

Macquarie Island **21**: 2930–2931
MADAGASCAR
Amphibians
15: 2146–2147
Birds
1: 113–114; **2**: 180–182; **5**: 645–647,
712–713; **8**: 1097–1098, 1123–1125;
11: 1495–1496, 1544–1545;
12: 1590–1592, 1604–1605;
13: 1849–1850; **14**: 1921–1923;
15: 2082–2083, 2106–2108;
16: 2229–2230; **19**: 2598–2600;
20: 2813–2814; **21**: 2925–2927
Fish
4: 458–459, 497–498;
8: 1018–1020
Invertebrates
5: 591–592; **6**: 777–778;
20: 2805–2806
Mammals
1: 111–112; **3**: 335–336; **4**: 460–461;
7: 877–879, 906–907;
10: 1435–1437; **12**: 1635–1637,
1709–1710; **14**: 1962–1964;
17: 2371–2373; **19**: 2656–2657
Reptiles
1: 31; **3**: 418–420; **5**: 603–606;
9: 1190–1191, 1281–1282
Madeira **3**: 368–369; **15**: 2045–2046;
21: 2966–2967
Maine **2**: 266–268
Malay Archipelago **1**: 77–78;
4: 468–469; **9**: 1289–1290;
12: 1614–1615; **14**: 1962–1964,
1970–1971
Malay Peninsula **2**: 201–202;
6: 790–791; **7**: 884–886, 900–901,
911–912, 936–938, 978–979, 982–983;
8: 1115–1116; **11**: 1459, 1566–1567;
12: 1614–1615, 1635–1637,
1714–1716; **14**: 1891–1893,
1937–1938, 1941–1942, 1970–1971;
17: 2323–2324; **19**: 2601–2602,
2730–2731
Malaysia **1**: 102; **4**: 549–550;
5: 621–622, 626–628, 712–713;
6: 724–725; **7**: 868–869, 966–967;
9: 1217–1218, 1222–1223,
1229–1230; **10**: 1365–1366;

12: 1684–1686; **13**: 1855–1856;
14: 1937–1938; **21**: 2923–2924
see also Malay Archipelago;
Malay Peninsula
Mali **1**: 24–25
Manchuria **20**: 2843–2845;
21: 2958–2960
Manitoba **3**: 323–325; **12**: 1622–1624
Marion Island **5**: 684–685
Mariout, Lake **13**: 1751–1752
Marmara, Sea of **16**: 2242–2243;
17: 2444–2445
Martinique **6**: 812–814; **19**: 2671–2673
Maui **9**: 1164
Mauritania **1**: 24–25
Mauritius **1**: 103–104; **3**: 308–309;
11: 1544–1545; **12**: 1635–1637
Mayotte Island **3**: 335–336
Mediterranean region **3**: 314–315,
319–320, 418–420; **5**: 584–587,
607–609, 626–628; **6**: 794–796,
844–847; **7**: 889–890, 930–932,
999–1000; **9**: 1156–1157;
11: 1524–1526; **12**: 1599–1601,
1632–1634; **15**: 2125–2126;
17: 2341–2343; **19**: 2704–2705
MEDITERRANEAN SEA
Birds
1: 105–106; **8**: 1073–1074;
15: 2065–2067; **18**: 2554–2555
Fish
1: 45–46, 51–52; **2**: 154–155, 258;
3: 389–390; **4**: 521–522; **5**: 638–640;
6: 745–747, 860–861; **7**: 875–876,
880–881; **8**: 1081–1083, 1119–1120;
10: 1327–1328; **11**: 1455–1456,
1517–1518; **12**: 1646, 1692–1693;
13: 1774–1775, 1790–1791;
14: 1948–1949, 1960–1961,
1975–1977; **15**: 2020–2021;
16: 2219–2220, 2238–2239,
2242–2243; **17**: 2383–2385,
2444–2445; **18**: 2519–2520;
20: 2743–2744, 2773–2775,
2782–2783, 2829–2830;
21: 2901–2902, 2934–2935
Invertebrates
1: 20–21; **4**: 488–489; **5**: 631–633;
6: 801–802; **9**: 1176–1178;
11: 1475–1477, 1551–1552;
13: 1763–1764, 1832–1834;
14: 1926–1928; **15**: 2127–2128;
18: 2475–2476, 2493–2494,
2506–2507; **19**: 2706–2707;
21: 2989–2990
Mammals
12: 1647–1648; **15**: 2024–2026
Reptiles
8: 1070–1072; **11**: 1481–1482;
20: 2849–2850
MEXICO
Amphibians
1: 109–110; **3**: 312–313, 326,
356–357; **17**: 2381–2382
Birds
1: 55–56, 126–127; **2**: 252;
3: 387–388, 399–400; **4**: 553–555;
5: 607–609, 623–625, 680–681;
8: 1142–1144; **9**: 1179–1181,
1213–1214, 1255–1257,
1289–1290; **10**: 1370–1371;
11: 1514–1516, 1524–1526,
1532–1533; **12**: 1599–1601,
1622–1624, 1699–1700; **13**:
1805–1807; **14**: 2012–2013;
15: 2032–2033, 2060–2061,
2082–2083; **16**: 2235–2237;
17: 2361–2363; **18**: 2460–2461;
19: 2627–2629, 2691–2692,
2714–2716; **20**: 2784–2786,

MEXICO
  Birds (*continued*)
    20: 2787–2789, 2800–2802;
    21: 2968–2969
  Fish
    2: 233–234; **3**: 407–408, 409–410;
    6: 829–830; **7**: 887–888, 928–929;
    8: 1087–1088, 1121–1122;
    14: 1888–1890, 1983–1984;
    19: 2704–2705
  Invertebrates
    1: 90–92; **2**: 255–256; **4**: 504–505;
    9: 1204–1205; **10**: 1417–1418
  Mammals
    1: 120–122; **2**: 198–200, 222–223,
    253–254; **4**: 448–450; **6**: 840–841;
    8: 1052–1053, 1078–1080;
    10: 1304–1305, 1310–1311,
    1313–1314, 1339–1340, 1363–1364;
    12: 1689–1691; **13**: 1838–1839,
    1843–1844; **15**: 2038–2040,
    2049–2051; **16**: 2171–2172,
    2282–2284; **20**: 2807–2809
  Reptiles
    2: 156–157, 249–251; **3**: 358–360;
    4: 453–454, 534–535; **5**: 603–606;
    6: 807–808, 812–814; **7**: 972–974,
    980–981; **9**: 1202–1203, 1231–1232;
    13: 1851–1853; **15**: 2117–2118;
    17: 2369–2370; **19**: 2732–2733
  see also Baja California
Mexico, Gulf of **1**: 49–50; **2**: 266–268;
  8: 1081–1083; **16**: 2238–2239
MIDDLE EAST
  Birds
    1: 105–106; **2**: 180–182; **3**: 308–309,
    337–339, 414–415; **5**: 584–587,
    617–620, 626–628, 704–705;
    6: 844–847; **7**: 996–998, 1001–1003;
    9: 1222–1223, 1247–1249,
    1274–1276; **10**: 1355–1357,
    1415–1416; **11**: 1457–1458,
    1544–1545; **12**: 1649–1651;
    13: 1742–1744, 1760–1762,
    1811–1813, 1857–1858;
    15: 2139–2140; **17**: 2396–2397;
    18: 2545–2546, 2551–2553;
    19: 2678–2680; **20**: 2839–2842
  Fish
    1: 77–78; **4**: 458–459
  Invertebrates
    7: 898–899; **9**: 1166–1168
  Mammals
    1: 60–63; **3**: 361–364; **7**: 906–907,
    939–942; **8**: 1054–1057;
    9: 1206–1207, 1239–1240,
    1267–1269, 1270–1271, 1272–1273;
    10: 1323–1324; **11**: 1444–1446;
    12: 1709–1710
  Reptiles
    1: 30–31; **3**: 418–420; **4**: 480–482;
    12: 1638–1640
  see also Arabian Peninsula
Midway Islands **1**: 35–37; **3**: 368–369
Minnesota **3**: 323–325
Mississippi River **13**: 1845–1846;
  17: 2379–2380
Missouri River **13**: 1845–1846
Moluccan Islands **2**: 205–207;
  10: 1355–1357; **11**: 1493–1494;
  13: 1823–1824, 1861
Mongolia **3**: 361–364; **4**: 448–450;
  5: 584–587, 671–672; **7**: 884–886,
  949–951; **8**: 1012–1013, 1042–1043;
  10: 1323–1324, 1401–1402;
  11: 1519–1521; **15**: 2052–2054,
  2139–2140; **16**: 2201–2203,
  2217–2218; **17**: 2432–2434;
  18: 2528–2529; **20**: 2854–2856;
  21: 2958–2960

Montana **2**: 228–229; **8**: 1047–1048
Morocco **1**: 136–137; **3**: 389–390;
  7: 872–874; **8**: 1104–1105;
  10: 1367–1369; **13**: 1832–1834;
  14: 1962–1964, 1975–1977;
  16: 2191–2193
Mozambique **4**: 497–498
MYANMAR (BURMA)
  6: 750–753; **7**: 868–869, 889–890;
  12: 1714–1716; **14**: 1878–1880,
  1886–1887, 1891–1893, 1898–1900,
  1934–1936, 1937–1938, 1970–1971;
  19: 2717–2718
  Fish
    7: 978–979; **16**: 2238–2239
  Invertebrates
    4: 462–463
  Mammals
    1: 107–108, 132–133; **2**: 201–202,
    284–285; **7**: 882–883, 936–938;
    8: 1021–1022; **11**: 1566–1567;
    12: 1635–1637; **13**: 1859–1860;
    14: 1941–1942; **15**: 2135–2136;
    19: 2622–2623, 2702–2703
  Reptiles
    2: 197; **7**: 954–956

N

Namibia **12**: 1595–1597, 1635–1637,
  1676–1677; **19**: 2608–2609;
  20: 2795–2799
Natal **19**: 2608–2609
Nebraska **2**: 198–200
Nepal **2**: 226–227; **9**: 1286–1287;
  10: 1410–1412; **11**: 1459;
  12: 1635–1637; **13**: 1753–1754;
  14: 1886–1887; **15**: 2135–2136;
  17: 2323–2324, 2403–2405,
  2432–2434; **21**: 3014–3015
Netherlands **14**: 1955–1956;
  18: 2496–2498
Nevada **17**: 2369–2370
New Britain **7**: 964–965; **17**: 2334–2336
New Caledonia **9**: 1204–1205;
  13: 1823–1824; **19**: 2601–2602
New England **12**: 1622–1624
Newfoundland **8**: 1049–1051,
  1139–1141
NEW GUINEA
  Amphibians
    11: 1557–1558
  Birds
    1: 113–114; **2**: 180–182, 205–207;
    3: 397–398; **4**: 483–485, 549–550;
    5: 645–647; **7**: 900–901, 911–912;
    8: 1145–1146; **9**: 1215–1216;
    10: 1355–1357, 1367–1369;
    11: 1488–1489, 1493–1494,
    1544–1545; **12**: 1714–1716;
    13: 1823–1824; **14**: 1988–1990;
    15: 2106–2108; **17**: 2361–2363;
    19: 2601–2602, 2669–2670;
    21: 2923–2924
  Fish
    12: 1684–1686; **17**: 2379–2380
  Invertebrates
    2: 211–212; **7**: 898–899, 966–967;
    9: 1204–1205; **10**: 1331–1332
  Mammals
    1: 130–131; **5**: 629–630, 677–679,
    702–703; **7**: 964–965, 984–986;
    11: 1571–1572, 1577–1578;
    12: 1709–1710; **13**: 1847–1848;
    18: 2486–2487; **20**: 2846–2848;
    21: 2942–2944
  Reptiles
    7: 894–895; **15**: 2079–2081;
    17: 2308–2309, 2416–2417,
    2442–2443

New Mexico **1**: 118–119; **2**: 198–200;
  4: 534–535
New York **2**: 218–219
NEW ZEALAND
  Amphibians
    9: 1200–1201
  Birds
    1: 35–37; **2**: 220–221, 224–225;
    3: 308–309, 365–367; **4**: 536–538;
    5: 617–620, 621–622, 684–685;
    6: 750–753; **7**: 872–874, 911–912,
    919–920, 996–998; **8**: 1061–1063,
    1064–1065, 1145–1146;
    9: 1215–1216; **10**: 1343–1344,
    1355–1357, 1372–1373;
    13: 1835–1837; **14**: 1934–1936,
    1988–1990, 1993–1995;
    15: 2045–2046, 2106–2108,
    2137–2138; **16**: 2164–2165,
    2206–2207, 2233–2234;
    17: 2334–2336; **18**: 2484–2485,
    2540–2542; **19**: 2620–2621,
    2669–2670, 2678–2680;
    20: 2790–2792; **21**: 2886–2887,
    2930–2931, 3003–3004
  Fish
    3: 395–396; **8**: 1058–1060,
    1119–1120; **14**: 1948–1949;
    15: 2027–2028
  Invertebrates
    1: 20–21; **12**: 1632–1634;
    18: 2530–2532
  Mammals
    3: 300–301; **6**: 785–787;
    7: 916–918; **9**: 1171–1173;
    11: 1449–1450; **15**: 2091–2093;
    20: 2854–2856; **21**: 2893–2895
  Reptiles
    7: 943–945; **20**: 2762–2764
Nicaragua **1**: 93–94
Nicobar Islands **19**: 2601–2602
Niger **1**: 24–25
Nigeria **8**: 1023–1026; **11**: 1536–1538;
  14: 1970–1971
Niger River **3**: 347–348; **6**: 803–804;
  13: 1749–1750
Nightingale Island **17**: 2334–2336
Nile River **2**: 195–196; **6**: 803–804;
  13: 1749–1750, 1751–1752
NORTH AMERICA
  Amphibians
    11: 1447–1448, 1503–1505,
    1557–1558; **12**: 1682–1683;
    13: 1732–1734; **18**: 2465–2466;
    19: 2723–2725
  Birds
    2: 151–153, 220–221; **3**: 319–320,
    323–325; **4**: 536–538, 553–555;
    5: 584–587, 607–609, 610–613,
    626–628, 645–647, 680–681,
    704–705; **6**: 754–756, 853–855;
    7: 1001–1003; **8**: 1016–1017,
    1027–1029, 1032–1034, 1061–1063,
    1094–1096, 1145–1146;
    9: 1179–1181, 1247–1249,
    1255–1257; **10**: 1307–1309,
    1317–1319, 1345–1347, 1348–1349,
    1367–1369; **11**: 1485–1487,
    1519–1521, 1524–1526;
    12: 1590–1592, 1599–1601,
    1602–1603; **13**: 1737–1738,
    1757–1759, 1811–1813, 1819–1820,
    1835–1837; **14**: 1891–1893,
    1934–1936, 1957–1959, 1965–1966,
    1988–1990, 1993–1995;
    15: 2055–2057, 2122–2124,
    2137–2138, 2141–2143;
    16: 2188–2189, 2233–2234;
    17: 2361–2363, 2386–2388,
    2396–2397, 2427–2429;

    18: 2484–2485, 2521–2523,
    2537–2539, 2540–2542, 2586–2589;
    19: 2598–2600, 2647–2649,
    2661–2663, 2678–2680, 2693–2695,
    2714–2716, 2721–2722;
    20: 2787–2789, 2790–2792,
    2831–2833, 2839–2842;
    21: 2966–2967, 2977–2979,
    2985–2986, 2993–2995
  Fish
    1: 99–101; **2**: 163–164; **3**: 395–396;
    6: 745–747; **7**: 928–929;
    8: 1018–1020, 1058–1060,
    1119–1120; **10**: 1302–1303;
    12: 1618–1619, 1680–1681;
    13: 1840–1842, 1845–1846;
    14: 1945–1947, 1948–1949,
    1960–1961; **17**: 2327–2328,
    2410–2411; **18**: 2533–2534,
    2556–2557, 2569–2570, 2573–2574;
    20: 2751–2753, 2754–2755
  Invertebrates
    1: 20–21, 90–92, 118–119;
    2: 235–236, 255–256; **4**: 488–489,
    504–505, 525–526, 539–540,
    558–560; **5**: 591–592; **7**: 898–899;
    8: 1117–1118; **9**: 1156–1157,
    1176–1178, 1233–1234;
    10: 1353–1354, 1428–1429;
    12: 1632–1634; **14**: 1991–1992;
    16: 2275–2276; **17**: 2353–2355;
    19: 2706–2707; **21**: 2917–2918,
    3005–3006
  Mammals
    1: 83–84, 107–108, 120–122;
    2: 177–179, 198–200, 215–217;
    3: 292–295, 305–307, 391–394;
    4: 436–438; **5**: 659–661; **6**: 771–773,
    785–787; **7**: 884–886; **8**: 1045–1046,
    1052–1053, 1054–1057, 1078–1080,
    1132–1134; **9**: 1158–1160;
    10: 1304–1305, 1339–1340,
    1432–1433; **11**: 1506–1508,
    1568–1570, 1579–1581;
    12: 1616–1617, 1625–1627,
    1654–1656, 1674–1675, 1689–1691;
    13: 1792–1794, 1843–1844;
    14: 1943–1944, 1962–1964,
    1996–1997; **15**: 2091–2093,
    2129–2131, 2132–2134, 2156–2157;
    16: 2294–2296; **17**: 2393–2395;
    19: 2728–2729; **20**: 2740–2742,
    2854–2856, 2872–2873
  Reptiles
    1: 38–40; **4**: 534–535; **7**: 933–935;
    9: 1281–1282; **12**: 1687–1688;
    14: 1972–1974; **15**: 2101–2103,
    2119–2121; **17**: 2308–2309,
    2425–2426, 2442–2443;
    19: 2664–2666, 2711–2713;
    20: 2795–2799; **21**: 2919–2920
  see also Canada; United States
North Carolina **1**: 38–40, 53–54;
  2: 233–234; **4**: 534–535
North Dakota **2**: 198–200
North Sea **2**: 170–171; **17**: 2444–2445;
  20: 2743–2744
Norway **4**: 488–489; **5**: 584–587,
  651–652; **10**: 1432–1433;
  14: 1965–1966, 2004–2005;
  18: 2554–2555; **20**: 2837–2838
Nosy Mangabe Island **1**: 111–112
Nova Scotia **1**: 126–127; **4**: 553–555,
  558–560; **6**: 754–756; **10**: 1370–1371
Nubian Desert **6**: 809–811

O

Ogasawara Islands **18**: 2547–2548
Ohio River **13**: 1845–1846

Okhotsk Sea **13**: 1840–1842;
15: 2065–2067
Oklahoma **1**: 38–40
Oman **13**: 1808–1810; **21**: 2951–2953
Oman, Gulf of **4**: 568–569
Oregon **1**: 113–114
Orinoco Basin **17**: 2356–2357
Orinoco River **6**: 762–763
Otijikoto, Lake **19**: 2608–2609

P
PACIFIC OCEAN
Birds
1: 35–37; **2**: 261–263; **6**: 754–756;
7: 889–890, 913–915; **8**: 1094–1096;
10: 1307–1309, 1370–1371;
13: 1755–1756; **15**: 2065–2067;
18: 2554–2555; **20**: 2748–2750,
2790–2792
Fish
1: 45–46, 77–78; **3**: 389–390;
5: 706–707; **6**: 729–731; **7**: 880–881,
896–897, 989–990; **8**: 1081–1083,
1119–1120, 1121–1122;
10: 1327–1328; **11**: 1517–1518,
1561–1562; **12**: 1646, 1652–1653,
1684–1686, 1692–1693;
13: 1767–1768, 1774–1775,
1790–1791, 1840–1842;
14: 1888–1890, 1960–1961;
15: 2020–2021, 2062–2064,
2094–2095; **16**: 2219–2220,
2238–2239, 2255–2256, 2260–2261,
2290–2291; **17**: 2410–2411,
2430–2431; **18**: 2533–2534,
2547–2548, 2573–2574, 2575–2577;
19: 2596–2597, 2636–2637,
2674–2675; **20**: 2743–2744,
2765–2766; **21**: 2909, 2915–2916,
2954–2955, 2991–2992, 3007–3008
Invertebrates
6: 777–778, 805–806, 815–817;
11: 1475–1477; **12**: 1722–1723;
13: 1832–1834, 1861;
16: 2269–2270, 2275–2276;
17: 2364–2366, 2367–2368,
2440–2441; **18**: 2493–2494;
19: 2706–2707; **21**: 2989–2990
Mammals
6: 724–725; **15**: 2024–2026
Reptiles
8: 1070–1072; **11**: 1481–1482;
17: 2308–2309
Pacific Ocean, islands **1**: 20–21;
5: 610–613, 626–628; **7**: 877–879,
889–890, 906–907, 911–912, 964–965,
966–967; **11**: 1488–1489, 1493–1494,
1544–1545; **12**: 1649–1651,
1714–1716; **16**: 2186–2187;
17: 2308–2309, 2334–2336;
19: 2601–2602; **21**: 2923–2924,
2930–2931
Pakistan **1**: 95–96; **2**: 226–227;
7: 889–890, 954–956, 978–979;
13: 1753–1754; **14**: 1878–1880,
1898–1900; **15**: 2041–2042,
2139–2140; **17**: 2432–2434;
19: 2683–2685, 2702–2703;
21: 2896–2898, 2951–2953,
3014–3015
Palawan **6**: 790–791
Panama **1**: 57–59; **3**: 382–384;
4: 558–560; **5**: 680–681;
13: 1784–1785; **15**: 2060–2061;
18: 2513–2514
Papua New Guinea **13**: 1866–1867;
19: 2726–2727
Paraguay **1**: 43–44, 57–59; **2**: 249–251;
4: 553–555, 561–563, 566–567;

15: 2150–2151; **16**: 2233–2234;
18: 2462–2464
Persian Gulf **20**: 2790–2792
Peru **1**: 90–92; **2**: 261–263;
3: 358–360; **4**: 443–453, 536–538;
11: 1470–1472; **14**: 1888–1890,
2006–2007; **17**: 2356–2357;
18: 2513–2514; **19**: 2667–2668;
20: 2756–2757; **21**: 2987–2988
PHILIPPINES
Amphibians
9: 1229–1230; **11**: 1557–1558
Birds
5: 621–622; **6**: 790–791; **7**: 868–869,
900–901, 911–912; **9**: 1226–1228,
1289–1290; **10**: 1315–1316,
1329–1330, 1367–1369;
12: 1614–1615; **14**: 1908–1910,
1970–1971; **19**: 2601–2602;
20: 2745–2747; **21**: 2923–2924,
2925–2927
Fish
1: 77–78; **4**: 468–469;
16: 2238–2239
Invertebrates
1: 102; **7**: 966–967
Mammals
1: 120–122; **3**: 305–307;
6: 724–725; **7**: 882–883;
8: 1115–1116; **14**: 1962–1964;
19: 2638–2640, 2702–2703,
2730–2731
Reptiles
19: 2732–2733
Pine Islands **8**: 1066–1067
Poland **2**: 215–217
Polynesia **10**: 1355–1357;
11: 1488–1489; **13**: 1737–1738
Portugal **2**: 180–182; **3**: 389–390;
5: 651–652, 666–667; **7**: 946–948;
10: 1367–1369; **11**: 1506–1508,
1519–1521
Puerto Rico **2**: 249–251; **3**: 368–369;
11: 1557–1558; **21**: 2936–2938
Pyrenees Mountains **3**: 376–378,
421–422; **4**: 551–552;
5: 666–667; **12**: 1609–1610;
15: 2055–2057

Q
Quebec **4**: 436–438

R
Rapa Island **12**: 1652–1653
Red Sea **4**: 568–569; **6**: 777–778;
7: 880–881; **16**: 2219–2220,
2238–2239; **17**: 2440–2441;
18: 2506–2507, 2547–2548;
20: 2748–2750, 2790–2792
Rhône River **2**: 177–179
Rocky Mountains **2**: 198–200;
5: 607–609; **15**: 2055–2057;
16: 2199–2200
Roti **7**: 911–912
Rottnest Island **15**: 2089–2090
Rudolf, Lake **13**: 1749–1750,
1751–1752
RUSSIA
Amphibians
6: 743–744; **12**: 1720–1721
Birds
1: 140–141; **3**: 365–367, 376–378;
4: 551–552; **5**: 584–587, 607–609,
626–628; **7**: 872–874, 996–998,
999–1000, 1001–1003;
8: 1027–1029; **13**: 1757–1759,
1835–1837; **14**: 1965–1966;
16: 2206–2207; **17**: 2437–2439

Fish
2: 274–275; **3**: 314–315, 402–403;
8: 1047–1048, 1089–1090;
13: 1840–1842
Mammals
2: 177–179, 215–217; **3**: 391–394;
5: 666–667, 671–672; **6**: 818–820;
7: 884–886; **10**: 1323–1324,
1432–1433; **14**: 2001–2003;
15: 2099–2100; **16**: 2217–2218;
17: 2432–2434; **21**: 2951–2953
Reptiles
16: 2231–2232; **17**: 2406–2407
Rwanda **8**: 1023–1026; **9**: 1182–1183
Ryukyu Islands **3**: 405; **13**: 1868–1869;
18: 2547–2548

S
Sahara Desert **6**: 809–811; **7**: 949–951
St. Helena **21**: 2888–2889
St. Lucia **2**: 249–251; **6**: 812–814;
19: 2671–2673
Sakhalin Island **1**: 26–27;
19: 2721–2722; **20**: 2837–2838
Samoa **12**: 1632–1634; **13**: 1861
San Cristobel Islands **12**: 1649–1651
Sarawak **7**: 966–967
Sardinia **2**: 198–200; **4**: 551–552;
11: 1503–1505; **12**: 1667–1668
Sargasso Sea **6**: 745–747; **14**: 1960–1961
Saskatchewan **2**: 198–200;
12: 1622–1624
Saudi Arabia **1**: 115–117;
9: 1274–1276; **13**: 1808–1810
Scandinavia **1**: 76, 83–84; **2**: 177–179;
3: 292–295, 310–311, 314–315,
316–318, 376–378, 391–394;
5: 607–609, 626–628, 704–705;
7: 996–998; **8**: 1061–1063,
1094–1096, 1119–1120;
11: 1506–1508; **12**: 1599–1601;
14: 1926–1928; **15**: 2099–2100,
2127–2128, 2137–2138; **17**: 2410–2411,
2437–2439; **20**: 2790–2792;
21: 2890–2892, 2958–2960
Scotia Sea **17**: 2435–2436
Scotland **1**: 140–141; **5**: 607–609,
704–705; **8**: 1061–1063;
12: 1599–1601
Seine River **17**: 2410–2411
Semau **7**: 911–912
Senegal **3**: 331–332; **4**: 439–442
Senegal River **13**: 1749–1750,
1751–1752
Sentani, Lake **17**: 2379–2380
Seychelles **3**: 356–357; **11**: 1544–1545;
15: 2146–2147; **16**: 2186–2187;
18: 2565–2566
Shetland Islands **10**: 1307–1309
SIBERIA
Birds
3: 376–378; **5**: 584–587, 607–609,
626–628, 704–705; **6**: 754–756;
8: 1027–1029, 1073–1074,
1094–1096; **9**: 1187–1188,
1211–1212; **10**: 1300–1301,
1317–1319, 1415–1416;
11: 1485–1487; **12**: 1599–1601;
13: 1835–1837; **14**: 1886–1887,
1931–1933, 1934–1936, 1965–1966,
1988–1990, 1993–1995;
15: 2106–2108, 2137–2138;
16: 2233–2234; **17**: 2396–2397,
2437–2439; **19**: 2721–2722;
20: 2790–2792, 2857–2859;
21: 2890–2892
Fish
1: 34; **3**: 314–315; **7**: 978–979;
8: 1047–1048; **12**: 1618–1619

Invertebrates
7: 898–899
Mammals
1: 95–96; **4**: 448–450; **8**: 1021–1022,
1078–1080; **9**: 1158–1160;
10: 1432–1433; **11**: 1506–1508,
1579–1581; **16**: 2201–2203;
19: 2683–2685; **20**: 2854–2856;
21: 2884–2885, 2893–2895,
2958–2960
Sicily **4**: 551–552; **5**: 595–597
Sikkim **10**: 1410–1412
Singapore **1**: 103–104; **3**: 308–309
Snares Island **16**: 2282–2284
Solomon Islands **2**: 211–212;
4: 483–485, 549–550; **5**: 617–620,
712–713; **7**: 900–901, 964–965;
9: 1226–1228; **10**: 1367–1369;
16: 2204–2205; **17**: 2308–2309;
19: 2601–2602
Somalia **1**: 60–63, 115–117; **2**: 189–190;
3: 331–332; **5**: 673–674; **7**: 952–953;
12: 1628–1629; **21**: 2939–2941
Sonoran Desert **7**: 972–974
South Africa **1**: 18–19, 60–63, 107–108;
2: 170–171, 180–182, 259–260;
3: 331–332; **4**: 497–498, 536–538;
5: 626–628; **7**: 919–920, 957, 996–998;
8: 1058–1060; **10**: 1329–1330;
11: 1449–1450; **12**: 1595–1597,
1635–1637, 1676–1677;
13: 1835–1837; **14**: 1970–1971,
1993–1995; **16**: 2290–2291;
20: 2758–2759, 2795–2799
SOUTH AMERICA
Amphibians
1: 93–94; **3**: 356–357; **9**: 1229–1230;
11: 1557–1558, 1573–1574;
14: 1876–1877; **18**: 2578–2579;
19: 2723–2725
Birds
1: 55–56, 105–106; **2**: 151–153,
220–221, 252; **3**: 323–325, 379–381;
4: 523–525, 543–544, 545–548,
553–555; **5**: 623–625, 645–647,
684–685; **6**: 788–789, 823–824,
844–847, 853–855; **7**: 919–920,
996–998; **8**: 1016–1017, 1032–1034,
1061–1063, 1142–1144, 1145–1146;
9: 1179–1181, 1213–1214,
1247–1249, 1255–1257, 1274–1276;
10: 1307–1309, 1345–1347,
1348–1349, 1415–1416;
11: 1453–1454; **12**: 1590–1592,
1622–1624, 1649–1651, 1665–1666,
1699–1700; **13**: 1737–1738,
1784–1785, 1805–1807, 1819–1820,
1835–1837; **14**: 1878–1880,
1883–1885, 1931–1933, 1937–1938,
1957–1959, 1988–1990;
15: 2032–2033, 2060–2061,
2106–2108, 2141–2143;
16: 2262–2263; **17**: 2319–2320,
2321–2322, 2334–2336, 2337–2338,
2386–2388, 2427–2429;
18: 2484–2485, 2524–2525,
2537–2539, 2540–2542, 2545–2546,
2551–2553, 2567–2568, 2583–2584,
2586–2589; **19**: 2627–2629,
2661–2663, 2691–2692, 2708–2710,
2714–2716; **20**: 2756–2757,
2787–2789, 2790–2792, 2800–2802,
2831–2833; **21**: 2925–2927,
2968–2969, 2977–2979, 2985–2986
Fish
1: 88–89; **3**: 402–403; **4**: 458–459;
6: 762–763; **7**: 887–888, 993–995;
8: 1081–1083, 1106–1107;
10: 1302–1303, 1376–1377,
1421–1422; **11**: 1501–1502;

SOUTH AMERICA
Fish (*continued*)
14: 1888–1890, 1948–1949, 1967–1969, 2006–2007;
16: 2244–2245; 19: 2704–2705;
20: 2751–2753; 21: 2998–2999
Invertebrates
1: 20–21, 90–92, 118–119;
2: 203–204, 235–236, 255–256;
3: 316–318; 8: 1117–1118;
9: 1156–1157; 10: 1417–1418;
12: 1632–1634, 1660–1661;
16: 2225–2226; 20: 2805–2806
Mammals
1: 32–33, 57–59, 85–87, 107–108;
3: 333–334, 382–384; 4: 478–479,
561–563; 5: 659–661; 6: 840–841;
7: 916–918; 8: 1045–1046,
1099–1101; 9: 1252–1254;
10: 1310–1311, 1313–1314,
1363–1364; 11: 1449–1450,
1534–1535, 1539–1540, 1563–1565;
12: 1689–1691; 13: 1747–1748,
1776–1778, 1792–1794, 1838–1839;
14: 1903–1905; 15: 2024–2026,
2068–2070, 2091–2093, 2096–2098,
2156–2157; 16: 2177–2178,
2221–2222, 2282–2284;
17: 2393–2395, 2400–2402;
18: 2470–2471, 2477–2479;
19: 2630–2632, 2696–2697;
20: 2771–2772, 2803–2804,
2807–2809, 2834–2836
Reptiles
1: 43–44; 2: 156–157, 249–251;
3: 358–360; 4: 534–535; 5: 603–606;
6: 812–814; 9: 1190–1191,
1281–1282; 12: 1687–1688;
14: 1972–1974; 15: 2119–2121;
17: 2308–2309; 19: 2650,
2711–2713, 2732–2733;
20: 2795–2799; 21: 2919–2920
South Carolina 2: 233–234;
4: 436–438; 12: 1622–1624
South China Sea 8: 1121–1122
South Dakota 2: 228–229
SOUTHEAST ASIA
Amphibians
3: 356–357; 9: 1229–1230;
18: 2465–2466
Birds
1: 103–104, 113–114; 2: 151–153;
3: 308–309; 5: 618–620, 645–647,
712–713; 6: 790–791, 823–824;
7: 900–901, 911–912, 996–998;
10: 1315–1316, 1355–1357,
1415–1416; 11: 1519–1521,
1544–1545; 12: 1649–1651;
13: 1742–1744, 1760–1762,
1811–1813; 14: 1878–1880,
1934–1936, 1937–1938, 1970–1971;
18: 2545–2546, 2565–2566;
19: 2601–2602; 20: 2857–2859;
21: 2925–2927, 2983–2984
Fish
6: 803–804, 821–822;
11: 1473–1474; 12: 1684–1686;
13: 1868–1869; 17: 2412–2413;
18: 2489–2490; 19: 2704–2705
Invertebrates
3: 316–318; 5: 591–592;
9: 1208–1210; 10: 1331–1332;
15: 2043–2044
Mammals
5: 671–672; 7: 877–879, 884–886,
906–907, 968–971; 9: 1239–1240;
11: 1490–1492, 1566–1567;
12: 1625–1627, 1628–1629,
1635–1637, 1696–1697, 1709–1710;
16: 2179–2181; 17: 2323–2324,

2374–2375; 19: 2630–2632,
2683–2685, 2702–2703
Reptiles
4: 480–482; 7: 943–945;
12: 1638–1640; 14: 1972–1974;
15: 2079–2081; 17: 2308–2309;
19: 2732–2733
Southern Ocean 6: 771–773;
10: 1387–1388
South Georgia Island 5: 684–685;
11: 1449–1450; 14: 1957–1959;
17: 2337–2338; 21: 2899–2900
South Orkney Islands 21: 2899–2900
South Sandwich Islands 5: 684–685;
17: 2435–2436; 21: 2899–2900
South Shetland Islands
17: 2337–2338; 21: 2899–2900
Soviet Union (*former*) 2: 198–200;
8: 1047–1048; 18: 2556–2557
Spain 1: 22–23, 97–98, 103–104;
2: 180–182; 3: 337–339, 349–350;
4: 523–525, 551–552; 5: 666–667,
704–705; 7: 946–948; 8: 1068–1069;
9: 1211–1212, 1272–1273;
10: 1367–1369; 11: 1499–1500,
1506–1508, 1519–1521;
18: 2496–2498; 20: 2839–2842,
2849–2850
Spitsbergen 1: 140–141; 11: 1485–1487;
14: 1931–1933, 1955–1956, 1988–1990
Sri Lanka 1: 102, 107–108; 3: 418–420;
4: 458–459, 468–469, 549–550;
6: 790–791; 9: 1286–1287;
10: 1329–1330, 1331–1332,
1410–1412; 11: 1490–1492;
12: 1684–1686, 1714–1716;
14: 1898–1900, 1970–1971;
17: 2403–2405; 19: 2702–2703,
2732–2733; 21: 2896–2898
Stewart Island 21: 3003–3004
Sudan 1: 18–19, 115–117; 5: 584–587
Sulawesi (Celebes) 2: 180–182;
4: 483–485, 497–498; 8: 1115–1116;
10: 1315–1316, 1355–1357;
19: 2601–2602
Sumatra 2: 201–202; 6: 790–791,
803–804; 7: 882–883, 900–901;
8: 1115–1116; 11: 1459;
13: 1797–1799; 14: 1937–1938,
1941–1942; 17: 2323–2324;
18: 2560–2561, 2562–2564;
19: 2601–2602, 2730–2731;
21: 2896–2898, 2925–2927
Sunda Islands 10: 1355–1357,
1383–1384
Surinam 1: 93–94; 4: 490–491;
9: 1196–1197
Sweden 10: 1367–1369, 1432–1433;
14: 1965–1966; 20: 2837–2838
Switzerland 12: 1667–1668
Syria 19: 2689–2690, 2693–2695;
21: 2884–2885

T
Tahiti 12: 1632–1634
Taiwan 1: 95–96; 7: 884–886;
9: 1158–1160; 11: 1544–1545;
17: 2308–2309, 2323–2324;
19: 2596–2597
Tajikistan 17: 2432–2434
Tana, Lake 13: 1751–1752
Tanzania 1: 18–19; 2: 213–214;
4: 439–442; 7: 952–953;
11: 1495–1496, 1541–1543;
14: 1970–1971
Tasmania 1: 130–131; 3: 340–341;
7: 868–869, 900–901, 911–912,
943–945, 984–986; 10: 1333–1336,
1385–1386; 11: 1509–1510,

1571–1572, 1577–1578;
13: 1847–1848; 14: 1985–1987;
15: 2115–2116; 18: 2486–2487;
19: 2641–2643, 2669–2670;
21: 2961–2963
Tatra Mountains 3: 421–422
Texas 1: 38–40, 105–106; 2: 233–234;
3: 326; 4: 458–459, 534–535;
10: 1417–1418
Thailand 1: 95–96; 2: 197, 284–285;
6: 724–725; 7: 936–938, 966–967,
968–971, 978–979; 8: 1021–1022;
9: 1217–1218; 10: 1365–1366;
11: 1459, 1566–1567; 14: 1886–1887,
1934–1936, 1937–1938;
16: 2238–2239; 19: 2730–2731
Tibet 8: 1027–1029; 10: 1410–1412;
11: 1524–1526; 17: 2432–2434;
21: 2939–2941
Tibetan plateau 21: 3000–3002
Tierra del Fuego 4: 516–518
Timor 4: 549–550; 7: 911–912
Tobago 1: 55–56; 5: 623–625;
6: 812–814; 21: 2968–2969
Tonga 12: 1632–1634
Transvaal 19: 2608–2609
Trinidad 1: 43–44, 55–56; 2: 249–251,
252; 3: 379–381; 5: 623–625;
6: 812–814; 9: 1274–1276;
11: 1532–1533; 13: 1776–1778,
1784–1785; 18: 2583–2584;
19: 2691–2692; 21: 2968–2969
Tristan da Cunha 5: 684–685;
15: 2045–2046, 2106–2108;
17: 2334–2336
Tunisia 7: 872–874; 8: 1104–1105;
16: 2191–2193
Turkestan 7: 949–951
Turkey 5: 595–597; 6: 794–796;
13: 1740–1741; 16: 2206–2207;
17: 2442–2443; 18: 2491–2492;
19: 2683–2685, 2693–2695;
2721–2722; 21: 2951–2953
Turkmenistan 21: 2951–2953

U
Uganda 2: 282–283; 4: 439–442;
8: 1023–1026; 9: 1182–1183;
14: 1970–1971
Ukraine 5: 666–667; 8: 1068–1069;
18: 2528–2529
UNITED STATES
Amphibians
3: 312–313; 8: 1066–1067;
9: 1174–1175; 17: 2381–2382
Birds
1: 105–106, 123–125, 126–127;
3: 308–309, 365–367, 387–388,
399–400; 4: 499–500, 543–544;
6: 748–749; 7: 919–920, 996–998;
8: 1084–1086; 9: 1187–1188,
1274–1276; 10: 1355–1357;
11: 1453–1454, 1485–1487;
12: 1590–1592, 1622–1624,
1649–1651; 13: 1760–1762;
14: 1891–1893, 1908–1910,
1921–1923, 1931–1933, 1965–1966,
1988–1990, 2012–2013;
15: 2036–2037, 2082–2083,
2106–2108, 2125–2126;
16: 2184–2185, 2235–2237;
18: 2460–2461, 2551–2553;
19: 2627–2629, 2671–2673;
20: 2745–2747, 2784–2786,
2800–2802; 21: 2890–2892,
2925–2927, 2970–2972
Fish
2: 272–273; 3: 395–396, 402–403;
6: 860–861; 10: 1403–1404;

14: 1917–1918, 1945–1947;
17: 2379–2380; 19: 2704–2705
Invertebrates
1: 64–65; 2: 243–244, 255–256;
3: 327–328; 7: 926–927, 930–932;
9: 1204–1205; 11: 1546–1547;
14: 1926–1928, 1991–1992;
15: 2043–2044; 16: 2225–2226
Mammals
1: 85–87, 134–135; 2: 198–200,
215–217, 253–254; 4: 448–450,
478–479, 561–563; 8: 1049–1051,
1075–1077; 10: 1313–1314,
1337–1338; 11: 1534–1535;
12: 1706–1708; 13: 1753–1754,
1776–1778; 14: 1903–1905;
15: 2038–2040, 2049–2051,
2068–2070, 2096–2098;
16: 2171–2172, 2179–2181,
2199–2200, 2201–2203,
2282–2284; 18: 2528–2529;
21: 2958–2960, 3018–3019
Reptiles
1: 53–54; 2: 249–251; 4: 453–454;
6: 807–808; 7: 980–981;
9: 1202–1203, 1231–1232;
10: 1361–1362; 13: 1745–1746,
1851–1853; 14: 1972–1974;
15: 2101–2103, 2117–2118;
19: 2650, 2732–2733
Ural Mountains 3: 296–297;
4: 510–511; 15: 2139–2140;
16: 2191–2193, 2233–2234;
17: 2406–2407
Uruguay 4: 553–555, 561–563;
15: 2150–2151; 18: 2462–2464
Utah 17: 2369–2370
Uzbekistan 17: 2432–2434

V
Vancouver Island 1: 45–46
Venezuela 1: 126–127; 2: 156–157,
233–234; 3: 382–384; 4: 478–479,
490–491, 523–524, 566–567;
5: 659–661; 9: 1196–1197;
17: 2356–2357
Vietnam 1: 95–96; 2: 197, 284–285;
3: 305–307; 7: 882–883, 936–938,
966–967; 11: 1566–1567;
12: 1714–1716; 13: 1859–1860;
14: 1891–1893, 1937–1938,
1941–1942
Virginia 1: 38–40; 3: 319–320;
4: 525–526
Volga River 18: 2556–2557
Volta River 13: 1749–1750,
1751–1752

W
White Sea 4: 472–473, 495–496;
5: 634–635; 8: 1113–1114;
9: 1184–1186; 18: 2556–2557
Wyoming 2: 228–229

Y
Yangtze River 13: 1845–1846
Yellow Sea 11: 1517–1518
Yemen 1: 115–117; 9: 1274–1276
Yugoslavia (*former*) 13: 1788–1789

Z
Zaire *see* Congo, Democratic
Republic of
Zambia 1: 18–19
Zanzibar 6: 837–839; 18: 2571–2572
Zimbabwe 14: 1970–1971

Index of Animal Behaviors

Many of the species described in the *International Wildlife Encyclopedia* are listed here according to their typical behavior or behavior-related characteristics.

Use this index to search for animals by looking up certain aspects of their lifestyle, such as their diet and habits, for example, whether they live in groups or are nocturnal (active at night).

Numerals in **bold** refer to volume numbers; those in *italics* refer to picture captions. Index entries in **bold** refer to pages within guidepost articles, where several closely related species are discussed.

A

Active at night *see* nocturnal
Aggression *see* display (territorial); fighting for mates; fighting for territory; infanticide; sexual aggression; siblicide
Altruism *see* helping
Ambush
  alligator **1**: 39
  anglerfish **1**: 51–52
  ant lion **1**: 64, *64*, 65
  black widow **2**: 235–236
  boomslang **2**: 264, *264*
  caiman **3**: 359
  chameleon **3**: 418–419, *419, 420*
  crab spider **5**: 580–581, *580, 581*
  crocodile **5**: 605, *606*
  horned frog **9**: 1229
  mantis **11**: 1546–1547
  orb spider **13**: 1800–1802
  pike **14**: 1945–1946
  python **15**: 2080
  sheetweb spider **17**: 2339–2340
  snakes 17: 2422, *2422*
  snapping turtle **17**: 2425–2426
  spiders 18: *2481, 2482*
  stargazer **18**: *2520*
  trapdoor spider **19**: 2719
  viperfish **20**: 2830
  water spider **20**: 2876
  wobbegong **21**: 2954
  wolf spider **21**: 2956
Armor
  armadillo **1**: 85, 87
  armored catfish **1**: 89
  bichir **2**: 195
  caiman **3**: 358
  coelacanth **4**: 497
  crabs and lobsters 4: 572
  crocodile **5**: *603, 604*
  edible crab **6**: 741
  fish 6: 833
  gar **7**: 929
  glass lizard **7**: 980
  gurnard **8**: 1108–1109, *1108*
  horned lizard **9**: 1231
  leatherback turtle **10**: *1423*
  lobster **11**: 1477
  loggerhead turtle **11**: 1481
  mantis shrimp **11**: 1552
  millipede **12**: 1611, *1611*
  mud turtle **12**: 1688
  pangolin **13**: 1862, *1862*

sea horse **16**: *2278*
sea moth **16**: *2290*
sea urchin **17**: 2314
sharks 17: 2330
shore crab **17**: 2353
tortoise **19**: 2711–2713, *2712*
turtles 20: 2795, 2796
velvet ant **20**: 2815
Asexuality
  amoeba **1**: 41–42
  anemone **1**: 48
  aphid **1**: 74, *74*, 75
  chalcid wasp **4**: 416–417
  hydra **9**: *1265*, 1266
  paramecium **14**: 1882
  purse sponge **15**: 2075
  salp **16**: 2223, 2224
  sea fir **16**: 2273
  snakes 17: 2422
  sponge **18**: 2495
Attracting a mate with color
  anole lizard **1**: 54, *54*
  bird of paradise **2**: 206–207
  bishop bird **2**: 213
  butterflies and moths 3: 344
  cuttlefish **5**: 633
  emperor penguin **6**: 780
  frigate bird **7**: 893
  jewel fish **10**: 1326
  octopus **13**: 1781
  peafowl **14**: *1898*, 1900
  redstart **15**: 2140
  squid **18**: 2510
  wrasse **21**: 2991–2992
Attracting a mate with movement
  Adélie penguin **1**: 28
  albatross **1**: 37; **16**: 2268
  avocet **1**: 106
  bearded lizard **2**: 172–173
  booby **2**: 263, *263*
  bowerbird **2**: 270–271
  bustard **3**: 338–339, *339*
  buzzard **3**: 353
  capercaillie **3**: *376*, 378
  crowned eagle **5**: 616
  curassow **5**: 624
  curlew **5**: 627–628
  damselfly **5**: 641
  darter **5**: 647
  dragonet **5**: 707
  duiker **6**: 728
  false scorpion **6**: 799
  fiddler crab **6**: 816, 817
  godwit **7**: 998
  grebe **8**: *1061*, 1063
  grouse **8**: 1086
  kestrel **10**: 1346–1347
  kob **10**: 1382
  lyrebird **11**: 1509, *1509*
  mallard **11**: 1525–1526
  merganser **12**: 1601
  mudskipper **12**: 1684
  muscovy duck **12**: 1700
  night heron **13**: 1739
  ostrich **13**: 1816
  painted snipe **13**: 1850
  pintail **14**: 1958
  pitta **14**: 1971
  raven **15**: 2122–2123
  rhea **15**: 2150
  riflebird **2**: 207
  roadrunner **16**: 2185
  roller **16**: 2205, *2205*
  secretary bird **17**: 2317

seriema **17**: 2321
sparrow hawk **18**: 2469
stickleback **18**: 2534
sun bittern **18**: 2567
tarantula **19**: 2635
tooth-carp **19**: 2704
tragopan **19**: 2717
trumpeter **20**: 2757
wolf spider **21**: 2956
wood pigeon **21**: 2981
Attracting a mate with scent
  alligator **1**: 39–40
  axoltl **1**: 109–110
  brown butterflies **3**: 296
  owlet moth **13**: 1821–1822
  peppered moth **14**: 1915
  spiders 18: 2483
  seals 16: 2287
Attracting a mate with sound
  alligator **1**: 39
  Baltimore oriole **1**: 126, *127*
  bird of paradise **2**: 206
  butcherbird **3**: 340
  button quail **3**: 349
  canary **3**: 368
  cardinal **3**: 387
  cicada **4**: 455, *455*, 456
  cock-of-the-rock **4**: 490–491
  cricket **5**: 600
  frogs and toads 7: 903
  fruit bat **7**: 908
  gibbon **7**: *71*
  grasshopper **8**: 1037
  grouse **8**: *1084, 1085*
  moose **12**: 1656
  mouth-brooding frog **12**: 1679
  rorqual **16**: *2209*
  shearwater **17**: 2336
  snipe **17**: 2429, *2429*
  songbirds 18: 2457, 2458
  spadefoot toad **18**: 2465–2466
  trogon **20**: 2746–2747
  walrus **21**: *2853*
  see also calling (communicating with sound)

B

Basking *see* temperature control
Breeding *see* asexuality; attracting a mate; hermaphrodite; monogamy; polygamy; sexual aggression
Browsing
  agouti **1**: 32
  antelopes 1: 61
  axis deer **1**: 108
  beira antelope **2**: 190
  bighorn sheep **2**: 198
  brow-antlered deer **2**: 284
  bushbuck **3**: 332
  caribou **3**: 392
  chamois **3**: 422
  deer 5: 657
  dik-dik **5**: 676
  duiker **6**: 728
  eland **6**: 758
  fallow deer **6**: 796
  gazelle **7**: 940–941
  giraffe **7**: 976, 977
  sika **17**: 2374
  takin **19**: 2622
  tapir **19**: 2631, *2632*
  wapiti **20**: 2855

Burrowing
  aardvark **1**: 16, *16*
  ant lion **1**: 64
  armadillo **1**: 85
  banded anteater **1**: 128
  bandicoot **1**: 130–131
  bee-eater **2**: 180–181
  burrowing owl **3**: 324
  burying beetle **3**: 327–328, *327, 328*
  caecilian **3**: 356
  coypu **4**: 562–563
  crab plover **4**: 569
  crested porcupine **5**: 597
  desman **5**: 666
  earless monitor **6**: 732
  earthworm **6**: 734, 735, 736
  elephant tusk shell **6**: 777, 778, *778*
  fiddler crab **6**: 815
  gaper **7**: 926
  gerbil **7**: 950
  golden mole **7**: 1004
  groundhog **8**: 1075
  hamster **8**: 1129
  hunting wasp **9**: 1261, 1262
  jerboa **10**: 1323
  legless lizard **10**: 1430
  lugworm **11**: 1497, *1497*, 1498, *1498*
  meerkat **12**: 1595
  mole **12**: 1625–1626, *1625*
  mole rat **12**: 1628, 1629
  mudskipper **12**: 1686
  pedipalpi **14**: 1906
  piddock **14**: 1939–1940, *1940*
  platypus **14**: 1986–1987
  pocket gopher **14**: 1996, 1997, *1997*
  prairie dog **15**: 2038–2039
  puffin **15**: 2067, *2067*
  rabbit **15**: 2091–2092, 2093
  ragworm **15**: 2104
  razor shell **15**: 2127–2128, *2128*
  sand dollar **16**: 2226
  sea cucumber **16**: 2271
  sea mouse **16**: 2292
  sea urchin **17**: 2314–2315
  shearwater **17**: 2335, 2336
  shipworm **17**: 2349, *2349*
  skink **17**: 2389
  spadefoot toad **18**: 2465
  spiny anteater **18**: 2487, *2487*
  tenrec **19**: *2656*
  trapdoor spider **19**: 2719
  tuatara **20**: 2762–2763, *2762*
  tuco-tuco **20**: 2771, *2771, 2772*
  wolf spider **21**: 2956
  wombat **21**: 2961–2962

C

Calling (communicating with sound)
  apes 1: 73
  babbler **1**: 113
  bellbird **4**: 543
  beluga whale **2**: 193, *193*
  bittern **2**: 220
  bullfrog **3**: 313
  bunting **3**: 319
  burying beetle **3**: *327*, 328
  bushbaby **3**: 329
  butcherbird **3**: 340
  capercaillie **3**: *376*, 378
  cats 3: 406
  chaffinch **3**: 415
  chimpanzee: **4**: 439

Calling (*continued*)
 corncrake 4: 541
 coyote 4: *559*, 560
 crane 5: 584
 croaker 5: 601–602, *601*
 dolphin 2: 268
 fruit bat 7: 908
 howler monkey 9: 1252
 hyena 9: 1268
 katydid 10: 1342
 king penguin 10: 1358, 1360
 kookaburra 10: 1385–1386
 leopard frog 11: 1447
 limpkin 11: 1453
 loon 11: *1485*, 1486
 loris 11: 1492
 midwife toad 12: 1609
 mockingbird 12: *1622*, 1624
 monkeys 12: 1645
 mynah 12: 1716
 oilbird 13: 1784, *1784*
 orangutan 13: 1798
 owls 13: 1828, 1829
 parrot 14: 1883
 pika 14: 1943, 1944
 potoo 15: 2033
 scorpion 16: 2258
 screamer 16: 2262
 seals 16: 2287
 squirrel monkey 18: 2515
 tamarin 19: 2626
 tawny owl 19: 2644
 tree frog 19: 2723–2724
 trumpeter 20: 2756, 2757
 whales 21: 2913–2914
 see also attracting a mate with
  sound
Camouflage
 abalone 1: 21, *21*
 anteater 1: 57
 blenny 2: 238
 boa constrictor 2: 249, *249*
 boomslang 2: 264, *264*
 burrowing owl 3: *324*
 bushbuck 3: 332
 butterflies and moths 3: 344
 button quail 3: 349
 caracal 3: 386
 cats 3: 304, *304*
 cat shark 3: *408*
 courser 4: *551*
 crab spider 5: 580, *580*, *581*
 crested lark 5: *593*, 594
 dab 5: *634*
 dolphin 4: 506
 flounder 4: 860
 frogmouth 7: *900*, 901
 genet 7: *947*
 herring gull 9: 1187
 horned frog 9: 1229, *1229*
 kallima butterflies 10: 1331, *1332*
 katydid 10: *1341*
 leaf fish 10: 1421, *1421*
 leopard 3: *404*
 mantis 11: 1546–1547, 1548
 morpho 12: 1661
 nightjar 13: *1742*, 1743
 owls 13: 1827–1828
 plume moth 14: 1991
 poorwill 14: *2012*
 puff adder 15: 2058
 scalare 16: *2245*
 sea horse 16: *2278*
 sloth 17: 2401
 snow leopard 17: 2432–2433
 sole 17: 2444, *2444*
 spider crab 18: 2476
 stonefish 18: *2547*
 wobbegong 21: 2954, *2955*
 woodcock 21: 2966, *2966*

Cannibalism
 black widow 2: 236
 boomslang 2: 265
 chimpanzee 4: 442
 cockroach 4: 494
 firefly 6: 828
 flatworm 6: 849
 hake 8: 1119
 mantis 11: 1547
 midge 12: 1608
 pike 14: 1947
 sharks 17: 2333
 spiders 18: 2483
Carnivore *see* meat eating
Colony *see* living in complex
  societies; living in groups
Color, warning *see* warning coloration
Color change 6: 860
 anole lizard 1: 53, *53*
 Atlantic salmon 1: 99, 100
 chameleon 1: 54; 3: 420
 conger eel 4: 521, *522*
 cuttlefish 5: 631, 632, *632*, 633
 eel 6: 745
 ermine 6: 787, *787*
 fiddler crab 6: 815
 gliding frog 7: 982
 goatfish 7: 989, 990
 grouper 8: 1082, 1083
 halibut 8: 1121
 hare 8: *1137*, 1138
 leaf fish 10: 1422
 lumpsucker 11: *1500*
 moloch 12: 1630
 ocean sunfish 13: 1774
 octopus 13: 1780, 1781
 olm 13: 1789
 Pacific salmon 13: 1840, *1841*
 parrotfish 14: 1888
 plaice 14: *1975*
 rabbitfish 15: 2094
 spider crab 18: 2476
 squid 18: 2508, 2510
 stick insect 18: 2532
 see also attracting a mate with
  color; camouflage; threat display
Combat *see* display (territorial);
  fighting for mates; fighting for
  territory; infanticide; sexual
  aggression; siblicide
Commercially important
 abalone 1: 21
 alligator 1: 38, 40
 anchovy 1: 45, *46*
 Atlantic salmon 1: 101
 basking shark 2: 162
 bass 2: 163–164
 bees and wasps 2: 185
 blesbok 2: 240
 camel 3: 361, 364
 carp 3: 395–396
 chalcid wasp 3: 416–417
 clam 4: 462, *463*
 cod 4: 496
 coypu 4: 561
 crocodile 5: 606
 ducks, geese and swans 5: 717
 eider 6: 756
 fur seal 7: 918
 herring 9: 1185–1186
 honeybee 9: 1208–1210
 Indian buffalo 9: 1286–1288
 jungle fowl 10: 1329–1330
 ladybug 10: 1395–1396
 leech 10: 1429
 llama 11: 1470–1472
 lobster 11: 1475, *1475*, 1476
 mackerel 11: 1518
 mink 12: 1616–1617
 musk deer 12: 1702, *1702*

 mussel 12: 1713
 oyster 13: 1834
 Pacific salmon 13: 1842
 palolo worm 13: 1861
 pilchard 14: 1949
 prawn 15: 2044
 scale insect 16: 2247, *2247*
 silk moth 17: 2376–2378, *2376*, *2378*
 snakes 17: 2422
 snapper 17: 2423
 swordfish 19: 2607
 tuna 20: 2775, *2775*
 wild boar 21: 2942–2944
 wild goat 21: 2953
 yak 21: 3002
 zebu 21: 3018–3019
Communal living *see* living in
  complex societies; living in
  groups
Cooperation *see* helping; hunting in
  groups/packs
Courtship *see* attracting a mate;
  displaying in groups

D

Dangerous to humans
 barracuda 2: 154, 155
 bears 2: 176
 black widow 2: 235
 boomslang 2: 265
 brown bear 2: 294
 buffalo 3: 305
 Cape buffalo 3: 305, 372
 cobra 4: 481–482, *481*
 coral snake 4: 534
 crocodile 5: 605
 fer de lance 6: 812
 flies 6: 856, 859
 great white shark 8: 1058, *1058*,
  1059–1060
 grouper 8: 1082
 hammerhead shark 8: 1128
 jellyfish 10: 1322
 leech 10: 1428
 lion 11: 1462
 mamba 11: *1531*
 mosquito 12: 1662–1664, *1664*
 night adder 13: 1735
 porbeagle 15: 2021
 puff adder 15: 2058
 scorpion 16: 2257, 2259
 sharks 17: 2332
 stingray 18: 2543–2544
 stonefish 18: 2547–2548
 tiger 19: 2685
 tsetse fly 20: 2760–2761
 venomous animals 20: 2817–2820
 wobbegong 21: 2955
 Zambezi·shark (bull shark)
  21: 3007, *3007*
 see also poison; stinging
Defense *see* armor; camouflage;
  electrical discharge; mimicry
  (appearance); poison; threat
  display; warning coloration
Dimorphism, sexual *see* sexual
  difference
Display (sexual) *see* attracting a
  mate; displaying in groups;
  sexual aggression
Display (territorial)
 angelfish 1: 50
 anole lizard 1: 54
 bearded lizard 2: 172
 bunting 3: 319–320
 cichlid 4: 458
 fiddler crab 6: 816
 golden eagle 7: 1002
 mockingbird 12: 1623

 mudskipper 12: 1686
 prairie dog 15: 2038
 red-winged blackbird 15: 2143
 robin, European 16: 2192–2193
 swan 18: 2589
Displaying in groups
 bird of paradise 2: 206–207
 bustard 3: 338–339
 capercaillie 3: 378
 cockatoo 4: 483
 cock-of-the-rock 4: 490, 491
 flamingo 6: *844*, 846–847
 fruit bat 7: 908
 grouse 8: 1086
 guillemot 8: 1096
 magpie lark 11: 1522
 mallard 11: 1525
 manakin 11: 1532
 prairie chicken 15: 2036
 pratincole 15: 2041–2042
 ruff 16: 2213, *2213*, 2214
 see also attracting a mate
Dormancy *see* hibernation (winter
  sleeping); torpor, long-term
  (extended period of dormancy);
  torpor, short-term (brief period
  of dormancy)

E

Eating meat *see* meat eating
Eating plants *see* browsing; grazing
Echolocation
 bat 13: 1822; 19: 2688
 bats 2: 167–168
 blue whale 2: 247–248
 dolphin 4: 508
 dolphins and porpoises 5: 697
 fish-eating bat 6: 840–841
 fruit bat 7: 906, *906*, 907
 horseshoe bat 9: 1240
 oilbird 13: 1784, 1785
 penguins 14: 1913
 pipistrelle bat 14: 1963
 river dolphin 16: 2177
 sea lion 16: 2283
 seals 16: 2287
 shrew 17: 2358
 swiftlet 19: 2601, *2601*, 2602
 whales 21: 2911
Electrical discharge
 electric catfish 6: 760, *760*, 761
 electric eel 6: 762–763, *762*, *763*
 electric ray 6: 764–766, *764*;
  17: 2332
 knife fish 10: 1377
 Nile fish 13: 1749, *1749*
 skate 17: 2385, *2385*
 stargazer 18: 2519, 2520
Estivation *see* torpor, long-term
  (extended period of dormancy)

F

Fast moving
 Atlantic salmon 1: 99
 barracuda 2: 154
 basilisk 2: 156, *156*
 blackbuck 2: 227
 camel 3: 361
 Canada goose 3: 367
 caribou 3: 392
 cassowary 3: 397
 cheetah 3: 403, 427, *427*
 cottontail 4: 546
 coyote 4: 560
 dolphin 4: 507
 dragonfly 5: 708
 ducks, geese and swans 5: 717
 emu 6: *782*, 783

flying fish **7**: 876
hawkmoth **3**: 346
jack rabbit **10**: 1305
kangaroo **10**: 1333–1334, *1336*
mackerel **11**: *1517*
mamba **11**: 1530
maned wolf **11**: 1540
marlin **11**: 1561, 1562
ostrich **13**: 1816
penguins 14: 1913, *1913*
peregrine falcon **14**: 1923
pronghorn antelope **15**: 2049–2050
roadrunner **16**: 2184
sailfish **16**: 2219, *2219*
swift **19**: 2599
swift moth **19**: 2604
swordfish **19**: 2606, 2607
whip snake **21**: 2919
wild ass **21**: *2941*
Feeding *see* ambush; browsing;
    cannibalism; electrical discharge;
    filter feeding; grazing; hunting
    in groups/packs; meat eating;
    parasite; poison; robbery;
    scavenging; stalking; stinging
Feeding at night *see* nocturnal
Fighting for mates
    adder **1**: 26–27, *27*
    asp **1**: 98
    bighorn sheep **2**: *199*
    bison **2**: 217
    brow-antlered deer **2**: 284–285
    buffalo 3: 306
    chamois **3**: 422
    Chinese water deer **4**: 446
    coati **4**: 479
    fighting fish **6**: 821, *821*, 822, *822*
    ibex **9**: 1273
    jungle fowl **10**: 1329
    lechwe **10**: 1426
    muntjac **12**: 1697–1698
    musk-ox **12**: 1704
    narwhal **12**: 1719
    nilgai **13**: 1754
    Père David's deer **14**: 1919
    rhea **15**: 2150
    rhino **15**: 2155
    tahr **19**: 2612
    wapiti **20**: 2856
    whales 21: 2913
    yak **21**: 3002
Fighting for territory
    agama lizard **1**: 31
    antelopes 1: 62
    cichlid **4**: 458
    deer 5: 657
    elephant seal **6**: 772–773, *772*
    fur seal **7**: 917
    Jack Dempsey **10**: 1302, *1302*
    kob **10**: 1381
    marine iguana **11**: 1555
    mouflon **12**: 1667–1668
    musk deer **12**: 1701
    sand lizard **16**: 2231
    waterbuck **20**: 2866
Filter feeding
    anchovy **1**: 45, *45*
    barnacle **1**: 138
    basking shark **2**: 160, *160*
    blue whale **2**: 247
    brine shrimp **2**: 276
    clam **4**: 462
    clawed frog (*larva or tadpole*) **4**: 465
    daphnia **5**: 643
    devilfish **5**: 670
    fairy shrimp **6**: 792
    fanworm **6**: 801
    featherstar **6**: 806
    flamingo **6**: 845
    lamp shell **10**: 1405–1406

lancelet **10**: 1407
menhaden **12**: 1598
mussel **12**: 1711–1712
opossum shrimp **13**: 1795–1796
oyster **13**: 1833
paddlefish **13**: 1845
piddock **14**: 1940
right whale **16**: 2167–2168
rorqual **16**: 2209
salp **16**: 2223
sea lily **16**: 2281
sea squirt **17**: 2312–2313
sponge **18**: 2494
whales 21: 2910, 2913
whale shark **21**: 2915
worm shell **21**: 2989
Flocking *see* living in groups;
    migration

G

Grazing
    addax **1**: 25
    antelopes 1: *60*, 61
    axis deer **1**: 108
    bison **2**: 217
    blesbok **2**: 239
    bontebok **2**: 260
    brow-antlered deer **2**: 284
    buffalo 3: 305, *307*
    Cape buffalo **3**: 371, 372
    chamois **3**: 422
    deer 5: 657
    dugong **6**: 724
    eland **6**: 758
    fallow deer **6**: 796
    gazelle **7**: 940–941
    kangaroo **10**: 1334–1335, *1335*
    lechwe **10**: 1427, *1427*
    sika **17**: 2374
    wildebeest **21**: 2948
    zebra **21**: 3010
Groups, hunting in *see* hunting in
    groups/packs
Groups, living in *see* living in groups

H

Harem *see* living in groups
Heat control *see* temperature control
Helping
    burying beetle **3**: 327
    butcherbird **3**: 341
    Cape hunting dog **3**: 375
    cheetah **3**: 406
    dogs 5: 692
    dolphin **2**: 266
    elephant **6**: 770
    hippopotamus **9**: 1195
    hoatzin **9**: 1196
    honeyguide **9**: 1217–1218
    hornbill **9**: 1228
    jay **10**: 1319
    killer whale **10**: 1351
    kookaburra **10**: 1385
    lion **3**: 406
    magpie **11**: *1521*
    magpie lark **11**: 1523, *1523*
    moorhen **12**: 1651
    night monkey **13**: 1748
    pelican **14**: *1910*
    tamarin **19**: 2626
    tarsier 19: 2640
    tiger **3**: 406
    wren **21**: 2995
    see also hunting in groups/packs
Herbivore *see* browsing; grazing
Herd *see* living in groups; migration
Hermaphrodite
    anemone **1**: 48

anglerfish **1**: 52
barnacle **1**: 139
brittle star **2**: 281
clam **4**: 463
earthworm **6**: 737
garden snail **7**: 932
hydra **9**: 1266
nutshell **13**: 1764
pond snail **14**: 2011
purse sponge **15**: 2074–2075
sea squirt **17**: 2313
slug **17**: 2409, *2409*
soft coral **17**: 2441
tubifex worm **20**: 2769
Hibernation (winter sleeping)
    bats 2: 169
    bear **2**: 175, 176
    black bear **2**: 222, 223
    brown bear **3**: 294
    dormouse **5**: *700*, 701
    groundhog **8**: 1075–1076, *1076*,
        1077
    ground squirrel **8**: 1078, 1080
    hamster **8**: 1130
    hedgehog **9**: 1173, *1173*
    horseshoe bat **9**: 1239
    marmot **11**: 1570
    pack rat **13**: *1843*
    pipistrelle bat **14**: 1963, 1964
    polar bear **14**: 1999
    poorwill **14**: 2013
    raccoon dog **15**: 2099, *2099*
    tomb bat **19**: 2703
    see also torpor, long-term
        (extended period of dormancy);
        torpor, short-term (brief period
        of dormancy)
Hiding *see* ambush; burrowing;
    camouflage; shy behavior;
    stalking
Hunting in groups/packs
    Cape hunting dog **3**: 373–374
    coyote **4**: 558, 560
    dhole **5**: 671
    dingo **5**: 678, *679*
    gray wolf **8**: 1055–1056, *1057*
    hyena **9**: 1268
    killer whale **10**: 1351–1352
    lion **11**: 1462
    piranha **14**: 1968
    see also helping
Hybrid
    Atlantic salmon **1**: 101
    black molly **2**: 234
    canary **3**: 368
    crocodile **5**: *605*
    dingo **5**: 677, *677*, 679
    hartebeest **8**: 1149
    muscovy duck **12**: 1700
    peafowl **14**: 1898
    pig-tailed macaque **14**: 1942
    Przewalski's horse **15**: 2052
    roach **16**: 2182
    sika **17**: 2375
    stilt **18**: 2542
    swordtail **19**: 2609
    tilapia **19**: 2689–2690
    wapiti **20**: 2854–2855
    wild ass **21**: 2941
    yak **21**: 3000
    zebu **21**: 3018, 3019

I

Infanticide
    bullhead **3**: 315
    chameleon **3**: *418*
    gerbil **7**: 951
    hake **8**: 1120
    lammergeier **10**: 1401

K

Killing of siblings *see* siblicide
Killing of young *see* infanticide

L

Life span *see* long living; short living
Living in complex societies
    ants and termites 1: 66–67
    army ant **1**: 90–92
    bees and wasps 2: 185
    bumblebee **3**: 317
    burying beetle **3**: 327
    coral **4**: 527
    dead man's fingers **5**: 651
    harvesting ant **9**: 1156–1157, *1156*
    honeybee **9**: 1208–1210
    mole rat **12**: 1628
    pyrosoma **15**: 2078, *2078*
    sea squirt **17**: 2312
    soft coral **17**: 2440
    termite **19**: 2658–2660
    vespula wasp **20**: 2824–2826
    wood ant **21**: 2964–2965
Living in groups
    Adélie penguin **1**: 28–29, *29*
    anchovy **1**: 45–46
    antelopes 1: 61
    apes 1: 73
    babbler **1**: 113–114
    baboon **1**: 115–116
    bats 2: 168–169
    beaver **2**: 177
    bee-eater **2**: 181
    beira antelope **2**: 189–190
    beluga whale **2**: 193; **21**: *2914*
    bighorn sheep **2**: 199, *200*
    bishop bird **2**: 213
    bison **2**: 216
    blackbuck **2**: 226–227
    black-headed gull **2**: 231
    blesbok **2**: 239
    bluefish **2**: 245
    budgerigar **3**: 302–303, *303*
    buffalo 3: 305–306
    bushbaby **3**: 329
    bushpig **3**: 335
    Canada goose **3**: 365, *367*
    Cape buffalo **3**: 370–371
    Cape hunting dog **3**: 373
    capuchin monkey **3**: 379–380
    capybara **3**: 382–383
    caribou **3**: 391, 393; **20**: *2778*
    cats 3: 406
    cheetah **3**: 406, *428*, 429
    chimpanzee **4**: 440–441
    chinchilla **4**: *443*
    coati **4**: 478
    cockatoo **4**: 483–484
    colobus monkey **4**: 501, 502
    dogs 5: 691–692
    dolphin **2**: 266, *266*; **4**: 507–508, *508*
    dolphins and porpoises 5: 696
    dugong **6**: 724
    eider **6**: 756
    eland **6**: 757
    elephant seal **6**: 771
    emperor penguin **6**: 780
    emu **6**: 782
    false killer whale **6**: 797–798
    flamingo **6**: 846–847
    flying fox **7**: 879, *879*
    fruit bat **7**: 908, *908*
    gannet **7**: 923
    gaur **7**: 936
    gibbon **7**: 970
    giraffe **7**: 975
    godwit **7**: 996–997, *996*
    gorilla **8**: 1025, 1026
    grackle **8**: 1032, *1033*, 1034

Index entries in **bold** refer to pages within guidepost articles, where several closely related species are discussed.

3109

Living in groups (*continued*)
greylag goose **8**: 1073
guenon **8**: 1091–1092
hammerhead shark **8**: 1127, *1128*
harrier **8**: 1145
hartebeest **8**: 1147, 1148
herring **9**: 1184, *1185*
herring gull **9**: 1188
hippopotamus **9**: 1193
house sparrow **9**: 1247–1248
howler monkey **9**: 1253
hyrax **9**: 1271, *1271*
ibis **9**: 1276, *1276*
Indian buffalo **9**: 1286, *1287*
killer whale **10**: 1350–1351
king penguin **10**: *1360*
krill **10**: 1387
kudu **10**: 1390, *1391*
langur **10**: 1410–1411, 1412
lemur **10**: 1435
lion **3**: 406; **11**: 1460–1461, *1460*
little auk **11**: 1464
locust **11**: 1478–1480
macaque **11**: 1511–1513
macaw **11**: 1515
mackerel **11**: 1517
mangabey **11**: 1543
marine iguana **11**: 1556
marmoset **11**: 1563–1564
martin **12**: 1591
meerkat **12**: *1595*, 1596–1597
monkeys 12: 1644
mouflon **12**: 1667
narwhal **12**: 1718
night heron **13**: 1739
night monkey **13**: 1748
nilgai **13**: 1753, *1754*
oropendola **13**: 1807
oryx **13**: 1808, *1810*
ostrich **13**: 1814
oystercatcher **13**: *1837*
parakeet **14**: 1879, *1880*
parrotfish **14**: 1888
partridge **14**: 1891, *1891*
peccary **14**: 1904–1905, *1905*
pilchard **14**: 1948
pilot whale **14**: 1952
pipistrelle **14**: 1964
piranha **14**: 1968
prairie dog **15**: 2038
pratincole **15**: 2041
quail **15**: 2082, 2083
quelea **15**: 2084–2085, *2084, 2085*, 2086, *2086*
red-winged blackbird **15**: 2143, *2143*
reedbuck **15**: 2145
rhea **15**: 2150
river dolphin **16**: 2177
Rocky Mountain goat **16**: 2199
rook **16**: 2206, *2207*
rorqual **16**: 2208
sable antelope **16**: 2215, *2216*
saiga **16**: 2217
sand eel **16**: *2227*
sand grouse **16**: 2229
sea butterfly **16**: 2269
sifaka **17**: 2371
silverside **17**: 2379, 2380
smelt **17**: 2411
sperm whale **18**: 2472
spider monkey **18**: 2477
spoonbill **18**: 2497, *2497*, 2498
springbok **18**: 2500
squid **18**: 2508
squirrel monkey **18**: 2513–2514, 2515
stork **18**: 2552, 2553, *2553*
tuna **20**: 2773, *2774*
vampire bat **20**: 2808

viscacha **20**: 2834
walrus **20**: *2851, 2852*
wapiti **20**: 2855–2856
waxwing **21**: 2890–2891
weaver **21**: 2897, *2898*
whales 21: 2913–2914, *2914*
whale shark **21**: *2915*
wild ass **21**: 2939
wild boar **21**: 2942
wood swallow **21**: 2983, *2983*
zebra **21**: 3011
Long living
albatross **1**: 37
anemone **1**: 47
apes **1**: 73
centipede **3**: 413
cicada **4**: 457
flea **6**: 852
giant snail **7**: 966
lobster **4**: 572
moose **12**: 1656
mud turtle **12**: 1688
pelican **14**: 1910
perch **14**: 1918
rorqual **16**: 2209
tortoise **19**: 2713
tuatara **20**: 2764

M

Marking with scent *see* scent marking
Mate guarding
goby **7**: 994
greylag goose **8**: 1073–1074
multimammate rat **12**: 1694
parakeet **14**: 1880
pink-footed goose **14**: *1955*, 1956
secretary bird **17**: 2318
Mating *see* attracting a mate; hermaphrodite; monogamy; polygamy; sexual aggression
Meat eating
adder **1**: 26
alligator **1**: 39
anaconda **1**: 43, *43*
anglerfish **1**: 51–52
army ant **1**: 91–92
asp **1**: 97
bald eagle **1**: 124–125
barn owl **1**: 149–150
bears 2: 176
binturong **2**: 201, *201*
black bear **2**: 222
bullfrog **3**: 312
burrowing owl **3**: 324
buzzard **3**: 352
caiman **3**: 359
Cape hunting dog **3**: 373, 374
caracal **3**: 385–386
cheetah **3**: 427–428
chimpanzee **4**: 442
condor **4**: *516*, 517
cone shell **4**: 519–520
cowrie **4**: 556
crocodile **5**: 605, *606*
crowned eagle **5**: 614–615, *614*
dragonfly **5**: 710
ermine **5**: 786
fennec fox **6**: 809–810
fer de lance **6**: 812–813, *813*
goshawk **8**: *1027*, 1028–1029, *1029*
grass snake **8**: 1043
gray wolf **8**: 1055–1056, *1057*
great white shark **8**: 1058, 1059–1060, *1059*
gyrfalcon **8**: 1111–1112, *1112*
harpy eagle **8**: 1143–1144
hyena **9**: 1267–1269
jaguar **10**: 1311

jellyfish **10**: 1321
kestrel **10**: 1345–1346
Komodo dragon **10**: 1383, *1384*
leopard **11**: 1445–1446, *1446*
leopard seal **11**: 1449–1450
lion **11**: 1462, *1462, 1463*
lungfish **11**: 1501
lynx **11**: 1507
mantis **11**: 1546–1547
mantis shrimp **11**: 1551, *1552*
marabou **11**: 1553, 1554, *1554*
marsupial cat **11**: 1572
martial eagle **12**: 1588, *1589*
merlin **12**: 1602
mink **12**: 1616
monitor lizard **12**: 1640
peregrine falcon **14**: 1921–1922
piranha **14**: 1968–1969
pit viper **14**: 1973
polar bear **14**: 1998–1999
polecat **14**: 2002–2003
puff adder **15**: 2058–2059
puma **15**: 2068, *2069*, 2070
python **15**: 2080, *2080*
raccoon **15**: 2097
raccoon dog **15**: 2099
rattlesnake **15**: 2120–2121
scad **16**: 2242
sea otter **16**: 2294, *2294*, 2295
sea slug **16**: 2300
secretary bird **17**: 2316–2317
serval **17**: 2325
shrike **17**: 2361–2363, *2361*
sidewinder **17**: 2369–2370
skunk **17**: 2395
snow leopard **17**: 2433
solifugid **18**: 2453–2454
spiders 18: 2482
starfish **18**: 2517
tarsier **19**: 2639
Tasmanian devil **19**: 2642–2643
tawny owl **19**: 2644–2645
tiger **19**: 2683–2684
water spider **20**: 2876
weasel **21**: 2893–2894
whelk **21**: 2917–2918
wildcat **21**: 2946–2947
wolverine **21**: 2960
zorilla **21**: 3020
Metamorphosis
backswimmer **1**: 119
beetles and bugs 2: 186–187
butterflies and moths 3: 346
edible frog **6**: 744
frogs and toads 7: 904
greenhouse frog **8**: *1067*
mayfly **12**: 1593–1594
salamander **11**: 1505
see also neoteny
Migration
Atlantic salmon **1**: 99–100
Baltimore oriole **1**: 126
barnacle goose **1**: 140
barn swallow **2**: 152–153
beaked whale **2**: 170–171
birds of prey 2: 210
blue whale **2**: 246, 248
Bombay duck **2**: 257
butterflies and moths 3: 346, *346*
Canada goose **3**: 365, 367
caribou **3**: 391, 392, *392, 394*; **20**: 2778, *2778*, 2779
catbird **3**: 399
cod **4**: 495
common toad **4**: 512
conch **4**: *514*
crane **5**: *584*, 586, 587
cuckoo **5**: 617–618
dolphin **4**: 506–507
dragonfly **5**: 709–710

ducks, geese and swans 5: 717
eel **6**: 745, 746–747
flycatcher **7**: 872–873
godwit **7**: 996–997, *997*
gray whale **8**: 1052–1053, *1052*
green turtle **8**: 1070
hammerhead shark **8**: 1127
harp seal **8**: 1139, 1140–1141
hawkmoth **9**: 1166
hummingbird **9**: 1255–1256
humpback whale **9**: 1258, 1259–1260, *1260*
lamprey **10**: 1403
lantern fish **10**: 1413
lemming **10**: 1432–1433, *1434*
lobster **4**: 573
locust **11**: 1479
martin **12**: 1590–1591
millipede **12**: 1613
monarch **3**: *346*; **12**: 1632, *1634*
mule deer **12**: 1689–1690
narwhal **12**: 1717
osprey **13**: 1811
owlet moth **13**: 1821
Pacific salmon **13**: 1840, *1841*, 1842
pilchard **14**: 1948, *1948*, 1949
pilot whale **14**: 1953
porpoise **15**: 2024–2025
red-winged blackbird **15**: 2143, *2143*
right whale **16**: 2167
Rocky Mountain goat **16**: 2199
rorqual **16**: 2208, *2208*
seabirds 16: 2264–2265
sea hare **16**: *2275*
shearwater **17**: 2335, *2336*
songbirds 18: *2456*, 2458–2459
squid **18**: *2510*
stork **18**: 2553
sturgeon **18**: 2556
swift **19**: 2598
swordfish **19**: 2606
tanager **18**: *2456*; **19**: 2628
tern **19**: 2661–2663
vireo **20**: 2832, 2833
walrus **20**: 2851
warbler **20**: 2857–2858
waxwing **21**: 2890, 2892
whales 21: *2912*, 2913
wildebeest **21**: 2948, *2949, 2950*
wood warbler **21**: 2985
Mimicry (appearance)
blue butterflies **2**: 244
butterflies and moths 2: 244; 3: 345
copper butterflies **4**: 525
coral snake **4**: 535
egg-eating snake **6**: 749
flies 6: *857*, 858
hawkmoth **9**: 1167, *1167*, 1168
hoverfly **6**: *857*; **9**: 1250, *1251*
long-horned beetle **11**: 1484, *1484*
sawfly **16**: *2240*
swallowtail **18**: 2582
tiger moth **19**: 2688
Monogamy
accentor **1**: 23
beaver **2**: 179
fulmar **7**: *915*
greylag goose **8**: 1073
partridge **14**: 1893
songbirds 18: 2457
Mutualism (symbiosis)
aphid **1**: 74–75
bitterling **2**: 218–219
clam **2**: 463
copper butterfly **4**: 525
coral **4**: 529
damselfish **5**: 638, 640, *640*
flowerpecker **7**: 868
hermit crab **9**: 1178, *1178*

honey badger **9**: 1206
honeyguide **9**: 1217, *1217*, 1218, *1218*
leafcutter ant **10**: 1417
nutcracker **13**: 1758
oxpecker **13**: 1830, *1830*
yucca moth **21**: 3005, *3005*, 3006

N

Neoteny
axolotl **1**: 109, 110, *110*
emu **6**: 782
hellbender **9**: 1175
midge **12**: 1608
mudpuppy **12**: 1683
olm **13**: 1789
siren **17**: 2382
see also metamorphosis
Nocturnal
agouti **1**: 32
anteater **1**: 58, *58*
armadillo **1**: 85
aye-aye **1**: 111
badger **1**: 120–121
bandicoot **1**: 130
barn owl **2**: 149
bat-eared fox **2**: 165
bats 2: 169
binturong **2**: 201, *201*
boa constrictor **2**: 250
bobcat **2**: 253, *253*
bristletail **2**: 279
bushbaby **3**: 330
bushpig **3**: 335
butterfly fish **3**: 348
caracal **3**: 386
cardinal fish **3**: 390
civet **4**: *460*
coral **4**: 529
crested porcupine **5**: 596, *596*
diving-petrel **5**: 684
fish owl **6**: 842
flying squirrel **7**: 884, *885*
frogmouth **7**: 900
gecko **7**: 945
genet **7**: 948, *948*
hamster **8**: 1129
horseshoe bat **9**: 1239
mudpuppy **12**: 1683
night heron **13**: 1737
night monkey **13**: 1747
oilbird **13**: 1784
owlet-nightjar **13**: 1823
owls 13: 1825–1829
paca **13**: 1838, *1838*
phalanger **14**: 1929, 1930
poorwill **14**: 2012, *2013*
potoo **15**: 2032
red fox **15**: 2134, *2134*
slater **17**: 2398, *2398*
springhare **18**: 2502, *2503*
stone curlew **18**: 2546
tarsier **19**: 2639–2640
Tasmanian devil **19**: 2641
tawny owl **19**: 2644, *2645*
vesper bat **20**: 2823
wild boar **21**: 2942
Nomadism
budgerigar **3**: 302–303, *304*
desert-living birds **5**: 665
flamingo **6**: 844
fruit bat **7**: 908
Gouldian finch **8**: 1030–1031
pelican **14**: 1910
plains-wanderer **14**: 1978
polar bear **14**: 1998
pronghorn antelope **15**: 2050
spoonbill **18**: *2498*
swan **18**: *2588*
waxwing **21**: 2891

O

Odor *see* attracting a mate with
scent; scent marking

P

Packs, hunting in *see* hunting in
packs
Parasite
basket star **2**: 159
blue butterfly **2**: 243–244
catfish **3**: 402
chalcid wasp **3**: 416, 417; **21**: 2928,
2933
cowbird **1**: 127; **3**: 400; **4**: 553, *553*,
554–555; **15**: 2143; **18**: 2459, 2461
cuckoo **5**: 617, 619–620, *619*, *620*
deep-sea anglerfish **5**: 653–654
flea **6**: 850, *850*, 851–852
flies 6: 859
fluke **7**: 870, *870*
honeyguide **9**: 1218
horsehair worm **9**: 1237, 1238,
1238
ichneumon wasp **9**: 1279–1280,
1279; **19**: 2611
lamprey **10**: 1403, *1404*; **17**: *2332*
leech **10**: 1428, *1428*
mantis fly **11**: 1549, *1550*
mite **7**: 988; **12**: 1620, *1620*
oil beetle **2**: 188
pearlfish **14**: 1902
stifftail **18**: 2539
tachina fly **17**: 2345; **19**: 2610–2611
tick **19**: 2681
velvet ant **20**: 2815, 2816, *2816*
whydah **21**: 2889, 2937, 2938
Pest
aphid **1**: 74, *75*
bandicoot rat **1**: 133
barnacle **1**: 138
beetles and bugs **2**: 187
blowfly **2**: 241
bollworm **2**: 255–256
brown rat **3**: 298, *299*
brush tail possum **3**: 301
bushpig **3**: 336
capybara **3**: 383
click beetle (*larva*) **4**: 467
clothes moth **4**: 470, *470*, 471
cockatoo **4**: 485
cockroach **4**: 494
Colorado beetle **4**: 504, 505
corn borer **4**: 539–540
cottontail **4**: *545*
coypu **4**: 563
earwig **6**: 739
flies 6: 859
giant snail **7**: 966, *966*, 967
hopper **9**: 1224
jack rabbit **10**: 1305
locust **11**: 1478
millipede **12**: 1611–1612
muskrat: **12**: 1708
oil beetle **13**: 1783
owlet moth **13**: 1822
plume moth **14**: 1991
quelea **15**: 2084, 2085, *2085*, *2086*
raccoon **15**: 2097
rice rat **15**: 2156
sawfly **16**: 2241
scale insect **16**: 2246, *2246*
shieldbug **17**: 2344
ship rat (black rat) **17**: 2348, *2348*
shipworm **17**: 2349, *2349*, 2350
springtail **18**: 2504
termite **1**: 68; **19**: 2660
weevil **21**: 2904
white butterflies **21**: 2928, *2929*

whitefly **21**: 2932, 2933
Pheromone *see* attracting a mate
with scent; scent marking
Plant eating *see* browsing; grazing
Poison
adder **1**: 26, 27
anemone **1**: 47
arrow-poison frog **1**: 93, 94; **7**: 905
asp **1**: 98
bees and wasps 2: 183
black widow **2**: 235
boomslang **2**: 264, 265
bumblebee **3**: 318
burnet moth **3**: 322
butterflies and moths 3: 345
centipede **3**: 412–413
cobra **1**: 98; **4**: 480–481, *481*
common toad **4**: 512, 513
cone shell **4**: 519, 520
coral snake **4**: 534
dogfish **5**: 688
eagle ray **6**: 731
false scorpion **6**: 799, *799*, *800*
fer de lance **6**: 812, 814
fire salamander **6**: 831–832
gila monster **7**: 972, *973*, 974
jellyfish **10**: 1322
marine toad **11**: 1557–1558
midwife toad **12**: 1609
millipede **12**: 1612–1613
monarch **12**: 1633
moray eel **12**: 1659
newt **13**: 1734
night adder **13**: 1735, *1735*, 1736
octopus **13**: 1781; **20**: *2819*
platypus **14**: 1985
Portuguese man-of-war
15: 2030–2031
puff adder **15**: 2058
pufferfish **15**: 2064
rabbitfish **15**: *2095*
rattlesnake **15**: 2120, *2121*
scorpion **16**: 2257–2258, *2257*;
20: 2818
scorpionfish **16**: 2260, 2261;
20: *2820*
sea snake **17**: 2308
snakes 17: 2420, 2421
solifugid **18**: 2454
spider crab **5**: 580
spiders 18: 2482
squid **18**: 2510
stargazer **18**: 2519
stingray **18**: 2543–2544
stonefish **18**: 2547–2548, *2548*
swallowtail **18**: 2581, 2582
tarantula **19**: 2633, 2634, 2635
trunkfish **20**: 2758–2759
venomous animals 20: 2817–2820
water shrew **20**: 2872
weever **21**: 2901–2902
see also dangerous to humans;
stinging; warning coloration
Polygamy (*including* polyandry;
polygyny)
accentor **1**: 23
agama lizard **1**: 31
Baltimore oriole **1**: 127
bishop bird **2**: 214
black widow **2**: 236
bunting **3**: 320
button quail **3**: 350
deer 5: 656–657
flycatcher **7**: 874
harrier **8**: 1146
lion **11**: 1463
ostrich **13**: 1814–1815
painted snipe **13**: 1849
rabbit **15**: 2092
red-winged blackbird **15**: 2142

rhea **15**: 2150
songbirds 18: 2457–2458
starling **18**: 2523
turkey **20**: 2785

R

Robbery
bald eagle **1**: 124
black-headed gull **2**: 231
crow **5**: 613
frigate bird **2**: 263; **7**: 891–892
grackle **8**: 1034
hawkmoth **9**: 1168
hyena **3**: 374–375; **9**: 1268
jaeger **10**: 1309
magpie **11**: *1519*, 1520
sheathbill **17**: 2338

S

Scavenging
basket star **2**: 158
edible crab **6**: 741
hyena **9**: 1269, *1269*
marabou **11**: 1553
mussel **12**: 1712
opossum **13**: 1793
red fox **15**: 2134
sheathbill **17**: 2337
slater **17**: 2398, *2398*
spider crab **18**: 2475, *2476*
toadfish **19**: 2699
vespula wasp **20**: 2826
vultures 20: 2839, 2839, 2840
wood louse **21**: 2973
Scent marking
aardwolf **1**: 19
Asiatic black bear **1**: 95
badger **1**: 120
brown rat **3**: 298
bushbaby **3**: 329
bushdog **3**: 334
capybara **3**: 382
cats 3: 406
civet **4**: 461
hartebeest **8**: 1147
house mouse **9**: 1245
jaguar **10**: 1310
lemur **10**: 1435
marten **11**: 1580
muntjac **12**: 1697
musk deer **12**: 1701
musk shrew **12**: 1710
nyala **13**: 1765
oribi **13**: 1803
peccary **14**: 1905
polecat **14**: 2002
potto **15**: 2034
rabbit **15**: *2093*
red panda **15**: 2135
rhino **15**: 2154, 2155
roe deer **16**: 2203
sloth **17**: 2402
titi **19**: 2696
water vole **21**: 2885
weasel **21**: 2894–2895
wildcat **21**: 2945–2946
School *see* living in groups
Secretive behavior *see* shy behavior
Sedentary behavior *see* stationary
behavior
Sex change
fanworm **6**: 801
grouper **8**: 1083
limpet **11**: 1452
oyster **13**: 1834
parrotfish **14**: 1890
purple sea snail **15**: 2073
shipworm **17**: *2350*

Index entries in **bold** refer to pages within guidepost articles, where several closely related species are discussed.

Sexual activity or display *see*
    attracting a mate with color;
    attracting a mate with scent;
    attracting a mate with sound;
    sexual aggression
Sexual aggression
    bighorn sheep **2**: 199
    chimpanzee **4**: 441
    chinchilla **4**: 444
    eland **6**: 758–759
    fallow deer **6**: 796
    fighting fish **6**: 821, *821*, 822
    gaur **7**: 938
    giraffe **7**: 977
    haddock **8**: 1114
    hamster **8**: 1131
    oryx **13**: 1809
    swordtail **19**: 2608
Sexual difference
    bird of paradise **2**: 206
    bullfrog **3**: 312, *313*
    copper butterfly **4**: 526
    deep-sea anglerfish **5**: 653
    dolphin **4**: 506
    dragonet **5**: 706
    earwig **6**: 738, *738*
    eider **6**: 754, *755*
    elephant seal **6**: 771
    fur seal **7**: 916
    gharial **7**: *955*, 956
    golden oriole **8**: 1012, *1013*
    green lizard **8**: 1068
    guppy **8**: 1106
    lion **11**: 1460
    mallard **11**: 1524
    mandrill **11**: 1536
    parakeet **14**: *1879*
    parrot **14**: *1883*
    peafowl **14**: 1898
    pheasant **14**: 1934, 1935
    pintail **14**: *1959*
    quelea **15**: 2084
    ruff **16**: 2212, *2212*, 2213
    songbirds 18: 2457
    Surinam toad **18**: 2578
    terrapin **19**: 2664
    tooth-carp **19**: 2704, *2704*
    wattlebird **21**: *2886*, 2887
    wrasse **21**: 2991
Shoal *see* living in groups
Short living
    glowworm **7**: 987
    mayfly **12**: 1593, *1593*
    sawfly **16**: 2240
    water shrew **20**: 2872
Shy behavior
    agouti **1**: 32
    boat-billed heron **2**: 252
    broadbill **2**: 283
    bulbul **3**: 310
    bullhead **3**: 314–315
    bushdog **3**: 333, 334
    bushpig **3**: 335
    button quail **3**: 349
    colobus monkey **4**: 501
    corncrake **4**: 541, *541*
    cottontail **4**: 545–546
    coucal **4**: 549
    crake **5**: 583
    dibatag **5**: 674
    dik-dik **5**: 676
    finfoot **6**: 823
    francolin **7**: 890
    golden oriole **8**: 1013
    goshawk **8**: 1027–1028
    hero shrew **9**: 1182
    nightingale **13**: 1740
    pitta **14**: 1970
    wildcat **21**: *2946*
    zebra mouse **21**: 3016

Siblicide
    crowned eagle **5**: 616
    electric eel **6**: 763
    golden eagle **7**: 1003
Singing *see* attracting a mate with
    sound; calling (communicating
    with sound)
Sleeping *see* hibernation (winter
    sleeping)
Slow moving
    beluga whale **2**: 193
    boa constrictor **2**: 250
    chameleon **3**: 418
    cone shell **4**: 519
    dragonet **5**: 706
    dugong **6**: 724
    fire salamander **6**: 831
    gila monster **7**: 972
    giraffe **7**: 975–976
    loris **11**: 1491
    manatee **11**: 1534
    mantis **11**: 1547
    porcupine fish **15**: 2022
    potto **15**: 2034
    pufferfish **15**: 2063
    right whale **16**: 2167, 2168
    sloth **17**: 2400
    sunfish **13**: 1775
    tortoise **19**: 2711–2712
    trunkfish **20**: 2758
Sociable behavior *see* living in
    groups
Speed *see* fast moving; slow
    moving
Stalking
    bittern **2**: 221
    cheetah **3**: 428
    cuttlefish **5**: 633
    darter **5**: 647
    gar **7**: 928
    John Dory **10**: 1327–1328, *1327*
    lion **11**: 1462
    marsupial cat **11**: 1572
    night heron **13**: 1738
    ocelot **13**: 1778
    puma **15**: 2069–2070
Stationary behavior
    anemone **1**: 47
    barnacle **1**: 138–139
    basket star **2**: 158
    dead man's fingers **5**: 651
    fanworm **6**: 801
    mussel **12**: 1711
    oyster **13**: 1832, 1833, *1833*, *1834*
    sea lily **16**: 2280
    sea squirt **17**: 2312
    sponge **18**: 2493, 2494
    worm shell **21**: 2989
Stinging
    anemone **1**: 47–48
    bees and wasps 2: 183
    bumblebee **3**: 318
    dead man's fingers **5**: 652
    hydra **9**: 1265, *1266*
    jellyfish **10**: 1322
    Portuguese man-of-war **15**: 2029,
        2030–2031, *2031*
    scorpion **16**: *2258*, 2259
    scorpionfish **16**: 2261
    sea pen **16**: *2297*
    spiders 18: 2480–2483
    stingray **17**: *2331*; **18**: 2543–2544
    stonefish **18**: 2547–2548
    velvet ant **20**: 2815
    venomous animals 20: 2817–2820
    vespula wasp **20**: 2826
    weever **21**: 2901–2902
    see also dangerous to humans;
        poison
Symbiosis *see* mutualism

T
Temperature control
    Arctic fox **1**: 83
    bat-eared fox **2**: 165, *165*
    bearded lizard **2**: 173
    beaver **2**: 177
    buffalo 3: 305
    bumblebee **3**: 318
    camel **3**: 363
    caribou **3**: 391
    chuckwalla **4**: 453
    crocodile **5**: 603–604, *603*, *604*
    dolphin **2**: 267
    elephant **3**: *767*, 768
    emperor penguin **6**: 781, *781*
    fence lizard **6**: 808
    fennec fox **2**: 165; **6**: 809, *811*
    ground squirrel **8**: 1080
    hedgehog **9**: 1173
    jack rabbit **10**: *1305*, 1306
    jerboa **10**: 1323
    leatherback turtle **10**: 1423
    legless lizard **10**: 1430
    lizards 11: 1468–1469
    marine iguana **11**: 1556
    meerkat **12**: 1596, *1597*
    morpho **12**: *1661*
    oil beetle **13**: 1782
    oryx **13**: 1810
    painted turtle **13**: *1851*, 1853
    poorwill **14**: 2012
    ptarmigan **15**: 2057
    python **15**: 2081
    quokka **15**: *2089*, 2090
    sheathbill **17**: *2337*
    sidewinder **17**: 2369, *2370*
    slowworm **17**: 2406
    termite **19**: 2660
    turtles 20: 2798–2799
    velvet ant **20**: 2816
    see also torpor, long-term
        (extended period of dormancy);
        torpor, short-term (brief period
        of dormancy)
Territoriality *see* calling
    (communicating with sound);
    display (territorial); fighting for
    territory; scent marking
Thermoregulation *see* temperature
    control
Threat display
    bearded lizard **2**: 172, *172*
    big-headed turtle **2**: 197
    black molly **2**: 233
    bullhead **3**: 315
    chameleon **3**: 419–420
    hairstreak butterflies **8**: 1117–1118
    horned lizard **9**: 1232, *1232*
    mandrill **11**: 1538
    mantis shrimp **11**: 1552
    pit viper **14**: *1973*
    puss moth **15**: 2077, *2077*
    skunk **17**: 2395, *2395*
    snapping turtle **17**: 2425
    snowy owl **17**: 2439
    swallowtail **18**: 2582
    see also display (territorial)
Tool using
    apes 1: 72, 72
    capuchin monkey **3**: 380
    chimpanzee **4**: 442, *442*
    Darwin's finches **5**: 649–650, *650*
    sea otter **16**: 2295, 2296
    vultures 20: 2842
Torpor, long-term (extended period
    of dormancy)
    adder **1**: 26
    asp **1**: 97
    bullfrog **3**: 312
    bumblebee **3**: 317

butterflies and moths 3: 346
    chuckwalla **4**: 453
    common frog **4**: 510
    dead man's fingers **5**: 652
    edible frog **6**: 743
    fire salamander **6**: 831, 832
    garden snail **7**: 931–932
    garter snake **7**: 934
    giant snail **7**: 966
    lacewing **10**: 1392
    ladybug **10**: 1395
    lungfish **11**: 1501
    natterjack toad **12**: 1720
    newt **13**: 1733
    painted turtle **13**: 1853
    pond skater **14**: 2009
    racer **15**: 2101–2102
    vanessa butterflies **20**: *2811*
    viviparous lizard **20**: 2837
    wood ant **21**: 2964
    see also hibernation (winter
        sleeping)
Torpor, short-term (brief period of
    dormancy)
    chipmunk **4**: 449
    common frog **4**: 510
    dormouse possum **5**: 702
    hummingbird **9**: 1257
    lemur **10**: 1436
    poorwill **14**: 2013
    swift **19**: 2598, 2600
    water-holding frog **20**: 2868
    see also hibernation (winter
        sleeping)
Toxin *see* poison

V
Valuable financially *see*
    commercially important
Vegetarian *see* browsing; grazing
Venom *see* poison
Vocalization *see* calling
    (communicating with sound)

W
Warning coloration
    arrow-poison frog **1**: 93, *93*, 94
    bees and wasps 2: 184
    birdwing butterfly **2**: 212
    burnet moth **3**: *321*, 322
    butterflies and moths 3: 345
    caecilian **3**: 356
    centipede **3**: 413
    coral snake **4**: 534–535, *535*
    fire-bellied toad **6**: 825, *825*
    gila monster **7**: 972, *974*
    ladybug **10**: *1394*
    monarch **12**: 1633, *1633*
    sawfly **16**: 2241
    shieldbug **17**: 2344, *2344*, 2345
    trunkfish **20**: *2758*
    venomous animals 20: 2819–2820
    see also mimicry (appearance);
        poison; threat display

Y
Young, killing of *see* infanticide

Comprehensive Index

Numerals in **bold** refer to volume numbers; those in *italics* refer to picture captions. Index entries in **bold** refer to guidepost or biome and habitat articles.

A

Aardvark (ant bear) **1**: 16–17
Aardwolf (werewolf) **1**: 18–19
Aba *see* Nile fish
Abalone (ormer; sea ear; earshell) **1**: 20–21
    Black **1**: 20, 21
    Green **1**: 21
    Red **1**: 20, 21
Abarenicola **11**: 1498
Abastor erythrogrammus **17**: 2382
Abominable Snowman **17**: 2324
Aborigines **3**: 300; **9**: 1224; **13**: 1847–1848; **18**: 2536, 2548; **19**: 2604; **20**: 2847–2848, 2869; **21**: 2963
Abramis brama **2**: 275; **16**: 2182
Abudefduf saxatilis **5**: 638
Abu markub *see* Stork, Shoebill
abyssopelagic zone **13**: 1772
Acanthaster planci **18**: 2518
Acanthisitta chloris **16**: 2165
Acanthiza **19**: 2669, 2670
Acanthochromis polyacanthus **5**: 639
Acanthocinus **11**: 1483, 1484
    aedilis **11**: 1483, *1483*
Acanthomyops nigra **14**: *1894*
Acanthorhynchus **9**: 1215
Acanthostracion quadricornis **20**: 2758, 2759
Acanthoxyla prasina **18**: 2530
Acanthurus
    bahianus **18**: 2577
    coeruleus **18**: 2576
    triostegus **18**: 2575–2576
Acara, Platinum **10**: 1366
Accentor **1**: 22–23
    Alpine **1**: 22, 23, *23*
    Hedge (dunnock; hedge sparrow; shuffle-wing) **1**: 22, *22*, 23
    Himalayan **1**: 22
    Siberian **1**: 22
Accipiter **2**: 208; **5**: 620; **8**: 1027, 1028; **18**: 2467, 2468; **19**: 2617
Acehlia australiensis **17**: 2311
Aceros nipalensis **9**: 1226
Achatina
    fulica **7**: 966, 967
    marginata **7**: 966, 967
Acherontia atropos **3**: 346; **9**: 1166, 1167, 1210
Acheta domesticus **5**: 598–599
Achorutes nivicola **18**: 2504
acid rain **21**: 2908
Acilius **5**: 683
Acinonyx jubatus **3**: 403, 428; **8**: 1040
Acipenser
    ruthenus **18**: 2556
    sturio **18**: 2556
    transmontanus **18**: 2556, 2557
Acmaea tessulata **11**: 1451, 1452
Acomys **18**: 2492
    cahirinus **18**: 2491
    capensis **18**: 2491
    dimidiatus **18**: 2491, 2492
Acouchi **1**: 32
Acridotheres
    cristatellus **12**: 1714
    ginginianus **12**: 1714
    tristis **12**: 1714, 1715

Acris **19**: 2723
    crepitans **19**: 2724
Acrocephalus
    arundinaceus **5**: 620
    palustris **18**: 2458
    schoenobaenus **18**: 2459; **20**: *2858*
Acrochordus javanicus **20**: 2874, 2875
Acropora **4**: 528
Acryllium vulturinum **8**: 1097
Actinia **1**: 48
    equina **1**: 48
Actitis hypoleucos **16**: 2233
Actophilornis africanus **10**: *1398*
Acyrthosiphon **1**: 75
Adalia
    bipunctata **10**: 1394
    conglomerata **10**: *1395*
    decempunctata **10**: 1394
adaptive radiation **4**: 459; **5**: 650; **9**: 1214; **10**: 1398
Addax (white antelope; screwhorn antelope) **1**: 24–25; **5**: 665
Addax nasomaculatus **1**: 25; **5**: 665
Adder **1**: 26–27
    Cape Mountain **15**: 2059
    Desert mountain **15**: 2058
    European **7**: 935
    Horned **15**: 2059; **17**: 2370
    Many-horned **15**: 2059
    Namaqua dwarf **15**: 2058
    Night *see* Night adder
    Péringuey's **15**: 2058
    Puff (*African*) **15**: 2058–2059
    Puff (*North American*) *see* Hognose snake
    Spread *see* Hognose snake
Admiral
    Blue **20**: 2810
    European white **12**: 1634
    Italian red **20**: 2810
    Red **6**: 811; **20**: 2810, *2810*, 2811
    White **20**: 2810, 2811, 2812
Aechmophorus occidentalis **8**: 1063
Aedes aegypti **12**: 1663, 1664
Aegolius funereus **13**: 1829
Aegotheles
    albertisii **13**: *1823*
    cristatus **13**: 1823, 1824
Aegypius monachus **20**: 2840
Aenictus **1**: 91
Aeolidia papillosa **16**: 2301
Aeoliscus **17**: 2367
    punctulatus **17**: 2368
    strigatus **17**: 2367, *2367*, 2368
Aepyceros melampus **1**: 61; **9**: 1285
Aepyprymnus rufescens **15**: 2115, 2116
Aequidens latifrons **10**: 1366
Aequipecten **16**: 2248
Aeronautes saxatilis **19**: 2598
Aeshna **5**: 710–711
    cyanea **5**: 711
    mixta **5**: 709
Aethiomastacembelus **18**: 2490
Aetobatus **6**: 731
    narinari **6**: 730, *730*
African devil's flower **11**: 1547, 1548
Afrixalus **15**: 2146
    fornasinii **15**: 2146, 2147
    spinifrons **15**: 2146, 2147
Afronandus **10**: 1422

Afropavo **14**: 1935
    congensis **14**: 1898
Agalychnis **7**: 982
    callidryas **7**: *903*; **19**: *2723*
    spurrellii **7**: *983*
Agama **1**: 30, *30*, 31
Agama lizard **1**: 30–31
    Common (rainbow lizard) **1**: 30, 31, *31*
    Desert **1**: 30, 31
    Flying **11**: *1466*
    Ground **1**: *30*
    Starred (hardun; dragon lizard) **1**: 30, 31
Aganippe subtristis **19**: 2720
Agapornis **11**: 1495, 1496
Agdistis **14**: 1991
    bennetii **14**: 1992
Agelaius phoeniceus **2**: 224; **15**: 2142; **18**: 2458
Agelastes niger **8**: 1097
Agelena labyrinthica **17**: 2339
aggression *see* display (territorial); fighting for mates; fighting for territory; infanticide; sexual aggression; siblicide
Agkistrodon
    contortrix **14**: 1972, 1973
    himalayanus **14**: 1973
    piscivorus **14**: 1972, 1973
Aglais urticae **20**: 2812
Agouti **1**: 32–33
    Black **1**: 33
    Central American **1**: 32, 33
Agouti (genus *Agouti* *see* Paca
Agouti
    paca **13**: 1838, 1839
    takzanowski **13**: 1838, 1839
agriculture, affecting
    burrowing owl **3**: 323
    chough **4**: 451
    corncrake **4**: 542
    racer **15**: 2103
Agriocharis ocellata **20**: 2784
Agrion **5**: 642
    splendens **5**: 641
Agriotes **4**: 467
    lineatus **4**: 467
    obscurus **4**: 467
Agrippa, Giant **13**: 1821, 1822
Agrius convolvuli **9**: 1166
Agrostichthys parkeri **13**: 1767
Ahaetulla **17**: 2420
    mycterizans **21**: 2920
Ailuroedus dentirostris **3**: 399
Ailuropoda **2**: 174
    melanoleuca **7**: 961
Ailurus fulgens **10**: 1363
Aipysurus laevis **17**: 2309
air sacs **16**: 2263
Aix
    galericulata **5**: 716–717; **21**: 2925, 2971
    sponsa **5**: 716–717; **21**: 2925, 2970, 2971
Ajaya ajaja **21**: 2907
Akule *see* Scad, Bigeye
Alabama argillacea **13**: 1822
Alaria mustelae **7**: 871
Alarmbird *see* Kookaburra
alarm sounds *see* warning sounds
Alaska blackfish (Alaska fish; devilfish) **1**: 34
Alaska fish *see* Alaska blackfish

alates **1**: 69
Alauda
    arvensis **17**: 2396, 2397; **18**: 2459
    gulgula **17**: 2396
Albacore **20**: 2773, 2774, 2775
    Yellow-finned *see* Tuna, Yellowfin
Albatross **1**: 35–37; **16**: 2264, 2267, 2268
    Black-browed **1**: 35, 36, *36*
    Buller's **1**: 36
    Laysan **1**: 37
    Short-tailed **16**: 2268
    Sooty **1**: 35, 36
    Wandering **1**: *35*, 36, 37; **16**: *2267*
    Waved **1**: 35
Alburnus alburnus **12**: 1619
Alca torda **15**: 2126
Alcedo
    atthis **10**: 1355, 1356
    cristata **10**: *1355*
Alcelaphus **1**: 62
    buselaphus **8**: 1148
Alces
    alces **12**: 1655
    alces gigas **5**: 655
Alcides **20**: 2805
    zodiaca **20**: 2805, 2806
Alcyonium **5**: 651
    digitatum **5**: 652; **17**: 2440
Aldabrachelys elephantina **20**: 2799
Alectoris
    graeca **14**: *1891*
    rufa **14**: 1891
Alectura lathami **11**: 1527
Alepidomus evermanni **17**: 2379
Alepisaurus
    brevirostris **10**: 1409
    ferox **10**: 1409
Aleuterus scriptus **20**: 2743
alevins, salmon **1**: *100*, 101; **13**: 1842
Alewife **17**: 2327
Aleyrodes proletella **21**: 2932, 2933
Alfonsino **18**: 2512
Allactaga **10**: 1324
Alle alle **11**: 1465; **16**: 2265
Allenopithecus nigroviridis **12**: 1643
Alligator
    mississippiensis **1**: 38, 39; **21**: *2906*
    sinensis **1**: 38
Alligator **1**: 38–40
    American **1**: 38, *38*, 40; **21**: *2906*
    Chinese **1**: 38, *39*, 40
Allocebus **10**: 1436
Allogalathea elegans **18**: *2506*
Allolobophora longa **6**: 737
allopreening **1**: 104, 114
Allurus fulgens **15**: 2136
Alopex **5**: 690
    lagopus **1**: 84; **5**: 691
Alopias
    pelagicus **19**: 2674, 2675
    superciliosus **17**: 2333; **19**: 2674, 2675
    vulpinus **19**: 2674, 2675
Alosa
    alosa **17**: 2327
    chrysochloris **17**: 2328
    fallax **17**: 2327
    pseudoharenpus **17**: 2327
    sapidissima **9**: 1184; **17**: 2327, 2328
Alouatta **9**: 1252, 1253
Alpaca **11**: 1470, 1471, 1472; **12**: 1672
Alpheus **17**: 2364
    pachychirus **17**: 2366

Alps 12: 1672
Alsophis antiguae 15: *2103*
alternation of generations 16: 2223, 2274
altruism *see* helping
Alucita hexadactyla 14: 1991
Alytes 7: 904
    cisternasii 12: *1609*
    obstetricans 12: 1610
Amandava
    amandava 1: 103, 104
    formosa 1: 103
    subflava 21: 2888
Amandina
    erythrocephala 11: 1544
    fasciata 11: 1544
Amazon 14: 1883
    Orange-winged 14: 1884, *1884*, 1885, *1885*
    White-fronted 14: 1883
    Yellow-headed 14: 1883
Amazona
    albifrons 14: 1883
    amazonica 14: 1884, 1885
    ochrocephala 14: 1883
Amazon River 16: 2176, *2176*
ambergris 18: 2474
Amberjack 16: 2242
    Greater 16: 2242
    Yellowtail 16: 2242
Amblonyx cinerea 16: 2179
Amblyeleotris sungami 7: *994*
Amblyomma americanum 19: 2682
Amblyrhynchus cristatus 11: 1556
Amblysomus 7: 1005
ambush
    anglerfish 1: 51–52
    cone shell 4: 519, 520
    horned frog 9: 1229
    mantis 11: 1546–1547
    pike 14: 1945–1946
    python 15: 2080
    snapping turtle 17: 2425–2426
    stargazer 18: *2520*
    trapdoor spider 19: 2719
    viperfish 20: 2830
    water spider 20: 2876
    wobbegong 21: 2954
    wolf spider 21: 2956
Ambystoma 11: 1504
    dumerilii 1: 109
    gracile 1: 109
    jeffersonianum 11: *1505*
    mexicanum 1: 109, 110
    tigrinum 1: 109; 11: *1504*
Ameles spallanzania 11: 1547
Amia calva 2: 273; 5: 686
ambergris
Amitermes
    atlanticus 19: 2658–2659
    meridionalis 19: 2660
ammocoete larvae 10: 1403–1404
Ammodorcas clarkei 5: 674; 7: 952
Ammodramus savannarum 18: 2460
Ammodytes
    hexapterus 16: 2228
    tobianus 16: 2227
Ammomanes deserti 5: 594
Ammophila 9: 1262
    pubescens 9: 1261
Ammospermophilus 8: 1079, 1080
    leucurus 8: 1079, *1079*
Amoeba 1: 41–42
Amoeba
    discoides 1: *42*
    proteus 1: 41–42
Ampedus cardinalis 4: 467
Ampeliceps coronatus 12: 1714
Amphioxus *see* Lancelet

Amphiprion 5: 640
    clarkii 5: 640
    ocellaris 5: 640
    percula 5: 639
    perideraion 5: 640
Amphiura 2: 281
amplexus 1: 93; 4: 510, *513*; 7: 904; 11: *1558*; 12: 1609; 18: 2578
Anabas testudineus 4: 468, 469, *469*
Anableps
    anableps 7: 888
    dowi 7: 888
    microlepis 7: 888
Anaconda (water boa) 1: 43–44; 17: 2419
    Paraguayan (southern anaconda) 1: 43
Analcidometra caribbea 16: *2281*
Anaplectes 21: 2896
Anaplophora glabripennis 11: 1484
Anarhynchus frontalis 14: 1989
Anarta myrtilli 13: 1821, 1822
Anas 5: 715, 717; 11: 1525; 14: 1957, 1958, 1959; 18: 2551; 19: 2647, 2648, 2649, *2649*;
Anastomus
Anatis ocellata 10: 1394–1395
Anax junius 5: 710
Anchovy 1: 45–46
    California (northern anchovy) 1: 45, 46
    European 1: 45, 46
Andes 12: 1669, 1672
Andrena 2: 183
Andrias
    davidianus 9: 1175
    japonicus 9: 1175
Andricus 7: 922
    kollari 7: 921
androconia 3: 296
androgen 14: 1933
Aneides 11: 1504
    lugubris 11: 1503–1504
Anemone (sea anemone) 1: 47–48; 4: 533
    Beadlet 1: 48, *48*
    Gem 1: *47*
    Opelet (snake locks anemone) 16: 2301
Anemonefish *see* Clownfish
Anemonia 1: 48
    mammilifera 9: 1178
    sulcata 16: 2301
Aner *see* Serval
Anergates atratulus 1: 69
Angel, Black 16: 2245
Angelfish (*freshwater bony fish*) *see* Scalare
Angelfish (*marine bony fish*) 1: 49–50; 4: 532
    Emperor 1: *50*
    French 1: 49, 50
    French old woman 1: 49
    King 6: *833*
    Queen 1: 50
Angelfish (*shark species*) *see* Monkfish (*shark species*)
Angelshark *see* Monkfish (*shark species*)
Angler
    Sargassum 1: 51
    Warty (handfish) 1: 51
Anglerfish 1: 51–52; 6: 833
    Deep-sea 1: 51; 5: 653–654
Anguilla
    anguilla 6: 745, 746; 14: 1980–1981
    rostrata 6: 745, 746
Anguis fragilis 10: 1430; 17: 2407
Angwantibo *see* Potto (*African*), Golden
Anhima cornuta 16: 2262

Anhinga, American (snakebird) 5: 645, 646–647, *646*; 21: 2907
Anhinga
    anhinga 5: 645, 646; 21: 2907
    melanogaster 5: 645
    novaehollandiae 5: 645
    rufa 5: 645
animal experimentation, macaques 11: 1513
Anisognathus flavinucha 19: 2628
Anniella
    geronimensis 10: 1431
    pulchra 10: 1431
    pulchra nigra 10: *1430*
Anoa 9: 1288
    Lowland 3: 305, 306, 307; 9: 1286
    Mountain 3: 305, 306, 307; 9: 1286
Anodonta 12: 1711
Anodorhynchus
    glaucus 11: 1516
    hyacinthinus 11: 1514
    leari 11: 1516
Anole lizard 1: 53–54
    Brown 1: *54*
    Green (American chameleon) 1: 53, 54
    Knight 1: 53
Anolis
    barkeri 1: 53
    carolinensis 1: 53
    equestris 1: 53
    lionotus 1: 53
    poecilopus 1: 53
    sagrei 1: 54
Anomalure *see* Scaly-tailed squirrel
Anomalurus
    beecrofti 16: 2251
    peli 16: 2250
Anomia 13: 1832
Anomma 1: 91
Anopheles 12: 1662, 1663
    darlingi 12: 1663
    gambiae 12: 1663
Anoplichthys jordani 3: 409, 410
Anous
    albivittus 13: 1755
    minutus 13: 1755, 1756
    stolidus 13: 1755
    tenuirostris 13: 1755
Anoxypristis cuspidata 16: 2238
Anser 5: 717; 14: 1955
    anser 5: 717; 8: 1074; 14: 1955
    brachyrhynchus 14: 1956
    cygnoides 5: 717
    fabalis 14: 1955
    indicus 5: 717
    serrirostris 14: 1955
Anseranas semipalmata 5: 714
Ant 19: 2660
    Army (driver ant; foraging ant; legionary ant; safari ant; soldier ant) 1: 90–92; 21: 2969
    Black 14: *1894, 1895*; 21: 2964–2965
    Harvesting (harvester ant) 9: 1156–1157, *1225*
    Honey (honey-pot ant) 1: 67; 9: 1204–1205
    Leafcutter (parasol ant) 1: 68; 10: 1417–1418
    Meadow *see* Wood ant, Hairy
    Pastoral (*ecological group*) 1: 68; 14: 1894–1895
    Slavemaker 21: 2965
    Velvet *see* Velvet ant
    White *see* Termite
    Wood *see* Wood ant
    see also Ants and termites
Ant bear *see* Aardvark; Anteater, Giant

Antbird 1: 55–56; 21: 2969
    Black-faced 1: 56
    Spotted 1: 55
Ant cow *see* Aphid
Anteater 1: 57–59
    Banded (numbat; marsupial anteater) 1: 128–129
    Giant (ant bear) 1: 57, *57*, 58, 59
    Scaly *see* Pangolin
    Silky (two-toed anteater) 1: 57, 58, *58*, 59
    Spiny *see* Spiny anteater
Antechinus 11: 1577
    stuartii 11: 1578
Antedon bifida 6: 806
Antelope
    Beira 1: 61; 2: 189–190
    Blue 1: 63
    Dwarf *see* Dwarf antelope
    Four-horned *see* Chousingha
    Indian *see* Blackbuck
    Pronghorn 1: 60; 15: 2049–2051
    Roan 16: 2215, 2216
    Royal 1: 60; 18: 2571, 2572
    Sable 1: 61; 13: *1830*; 16: 2215–2216
    Saiga *see* Saiga
    Screwhorn *see* Addax
    Tibetan *see* Chiru
    White *see* Addax
    Zebra *see* Duiker, Zebra
    see also Addax; Antelopes; Blackbuck; Blesbok; Bontebok; Bushbuck; Dibatag; Dik-dik; Duiker; Eland; Gazelle; Gerenuk; Hartebeest; Impala; Klipspringer; Kob; Kudu; Lechwe; Nilgai; Nyala; Oribi; Oryx; Reedbuck; Springbok; Steenbok; Suni; Waterbuck; Wildebeest
Antelopes 1: 60–63
Antheraea 17: 2377
    mylitta 17: 2377
    paphia 17: 2377
    pernyi 17: 2377
    yamamai 17: 2377
Anthias 8: 1081
Anthocharis cethura 21: 2928
Anthonomus grandis 21: 2904
Anthornis melanura 21: 191
Anthracoceros coronatus 9: 1227
anthrax 9: 1235; 20: 2789
Anthreptes
    collaris 18: *2566*
    platurus 18: 2455, 2565, 2566
Anthus
    antarcticus 14: 1965
    bogotensis 14: 1965
    chloris 14: 1965
    pratensis 5: 620; 14: 1965
    rubescens 14: 1966
    spinoletta 14: 1965
    spraguei 14: 1965
    trivialis 14: 1965
    vervinus 14: 1965
Antidorcas marsupialis 18: 2500
antifreeze, in blood 9: 1278
Antilocapra americana 1: 60; 15: 2050
Antilope cervicapra 1: 60; 2: 227
antivenin, snake 20: 2820
antlers 5: 655–656
    axis deer 1: 107
    brow-antlered deer 2: 284
    caribou 3: 391, *391*, 392–393, *393*
    fallow deer 6: 794, *794*, 796
    moose 12: 1654, *1654*, 1655
    mule deer 12: 1689, *1689*
    muntjac 12: 1698
    Père David's deer 14: 1919
    roe deer 16: 2201, *2202*, 2203

Schomburgk's deer (*extinct*) **2**: 285
sika **17**: 2374, *2375*
Ant lion (doodlebug) **1**: 64–65
Antpitta **1**: 55
Antrozous pallidus **20**: 2822, 2823
ants
   antbirds and **1**: 56, *56*
   anteaters and **1**: 59
   aphids and **1**: 68, 74; **14**: 1894, *1894, 1895*
   armadillos and **1**: 86
   flickers and **6**: 853–854
   molochs and **12**: 1630
   pangolins and **13**: 1862, *1862*
   spiny anteaters and **18**: 2487
Antshrike **1**: 56
   Amazonian **1**: *55*
   Barred **1**: 55
   Black-crested **1**: 55
   Great **1**: *56*
Ants and termites 1: 66–69
Ant-tanager **19**: 2629
Ant-thrush **1**: 55, 56
   Black-faced **1**: 55, *56*
Antvireo **1**: 56
Antwren **1**: 56
   Pygmy **1**: 55
   White-flanked **1**: 55
Anuraphis maidi-ridicis **14**: 1895
Anurida maritima **18**: 2504
Anvilhead *see* Hamerkop
Aonyx capensis **16**: 2179, 2180
Aotus **12**: 1642; **13**: 1748
Apaloderma narina **20**: 2745
Apalone
   ferox **17**: 2443
   mutica **17**: 2443
   spinifera **17**: 2443
   spiniferus spiniferus **17**: *2442*
Apalopteron familiaris **9**: 1215
Apanteles glomeratus **9**: 1280; **21**: 2928
Apantesis **19**: 2687
Apara *see* Armadillo, Three-banded
Apatura
   ilia **15**: *2071*
   iris **15**: 2072
Ape **1**: 70–73
   Barbary (Barbary macaque)
      1: 136–137; **12**: 1643
   see also Apes; Chimpanzee;
      Gibbon; Gorilla; Orangutan
Apeltes quadracus **18**: 2533
Apes 1: 70–73
Aphelocoma coerulescens **3**: 426;
      10: 1317
Aphid (plant louse; greenfly;
      blackfly; ant cow) **1**: 74–75;
      21: 2965
   Black bean **1**: 74, 75
   Peach-potato **1**: 74
   Woolly (American blight) **1**: 74
Aphidius **9**: 1279
Aphis **1**: 75; **21**: 2965
   fabae **1**: 74
Aphomia sociella **20**: 2826
Aphredoderus sayanus **20**: 2755
Aphriza virgata **20**: 2790
Aphrodite aculeata **16**: 2293
Apion ulicis **21**: 2903, 2904
Apis
   dorsata **9**: 1208, 1209
   florea **9**: 1208, 1209
   indica **9**: 1208, 1209
   mellifera **9**: 1208, 1209
Aplochiton zebra **21**: 3015
Aplodinotus **5**: 602
Aplodontia rufa **12**: 1675
Aplysia
   californica **16**: 2276
   punctata **16**: 2275, 2276

Apocephalus **10**: 1418
Apodemus
   flavicollis **21**: 2975, 2976
   sylvaticus **21**: 2975, 2976
Apogon **3**: 390
   ellioti **3**: 390
   imberbis **3**: 390
Apostlebird **11**: 1522, *1523*
appeasement ceremonies **12**: 1668
Appendicularian **17**: 2313, *2313*
Apristurus **3**: 407
Aptenodytes
   forsteri **1**: 82; **6**: *773*, 780; **14**: 1911
   patagonicus **6**: 779; **10**: 1359;
      14: 1912
Apterodromia evansi **11**: 1550
Apteronotus albifrons **10**: *1376*
Apteryx
   australis **10**: 1372, 1373
   haasti **10**: 1372
   oweni **10**: 1372
Aptychotrema **8**: 1103
Apulus muralis **13**: 1783
Apus
   apus **19**: 2598, 2599
   caffer **19**: 2598
   melba **19**: 2598
   pacificus **19**: 2598
   pallidus **19**: 2598
aquaculture
   abalone **1**: 21
   farmed salmon **1**: 101
   mullet **12**: 1692
aquarium fish
   angelfish **1**: 49
   fighting fish **6**: 821, *821*, 822, *822*
   guppy **8**: *1107*
   Jack Dempsey **10**: 1302
   kissing gourami **10**: 1365
   paradise fish **13**: 1868, 1869
   piranha **14**: 1968, *1968*
   platy **14**: 1983, *1983*, 1984
   pompadour fish **14**: *2006*
   scalare **16**: 2244
   scat **16**: 2256
   spiny eel **18**: 2490
   swordtail **19**: 2608
   tetra **19**: 2667, 2668; **21**: 2998, *2998*, 2999
   tooth carp **19**: 2704, *2704*
   zebra fish **21**: 3014, *3014*
aquifers **21**: 2908
Aquila
   audax **5**: 679; **6**: 784
   chrysaetos **7**: 1002
   heliaca **7**: 1001
   pomarina **21**: 2906
   verreauxii **7**: 1001; **12**: 1589
Ara
   ararauna **11**: 1514
   chloroptera **11**: 1514
   glaucogularis **11**: 1516
   macao **11**: 1514, 1515
   militaris **11**: 1514
   rubrogenys **11**: 1516
Aracari **19**: 2714
   Collared **19**: 2714, 2715
Aramus guarauna **6**: 824; **11**: 1454
Araneus **13**: 1801; **18**: 2482
Araschnia levana **20**: 2811
Arbanitis **19**: *2720*
Arborophila **14**: 1891
Arcella **1**: 41
Archaeolemur **17**: 2373
Archamia **3**: 390
   fucata **3**: *389*
Archboldia papuensis **2**: 271
Archerfish **1**: 77–78
   Banded **1**: 78
   Spotted **1**: *77*

Archidoris
   flammea **16**: 2300
   pseudargus **16**: 2300
   pseudoargus **16**: 2299
Archilochus colubris **9**: 1255, 1256
Archispirostreptus **12**: *1613*
Archoplites interruptus **18**: 2569
Archteuthis **18**: 2509
   dux **18**: 2509
   longimanus **18**: 2509
   princeps **18**: 2509
Arctia **19**: 2686
   caja **19**: 2686, 2687
   hebe **19**: 2687
   quenseli **19**: 2687
   villica **19**: 2686–2687
Arctic and Antarctic 1: 79–82
Arctictis binturong **2**: 202
Arctocebus calabarensis **8**: 1015
Arctocephalus **7**: 916, 917
   australis **7**: 918
   gazella **7**: 916
   pusillus **7**: 917
Arctogalidia trivergata **13**: 1855
Arctonetta fischeri **6**: 754
Arctonyx collaris **1**: 120
Arctosa perita **21**: 2957
Ardea
   cinerea **9**: 1179
   herodias **9**: 1179, 1180; **21**: *2906*
   purpurea **9**: 1179
Ardeola
   ibis **6**: 750
   ralloides **9**: 1179
Ardeotis
   australis **3**: 338
   kori **3**: *337*
   nigriceps **3**: 338
Arenaria
   interpres **20**: 2790, 2791
   melanocephala **20**: 2790
Arenicola **11**: 1498
   cristata **11**: 1497
   ecaudata **11**: 1497
   marina **11**: 1497, *1497*, 1498
Arenicolides **11**: 1498
Argali *see* Sheep, Marco Polo
Argiope **13**: 1801, *1802*
   bruennicchi **18**: 2481
Argonaut *see* Nautilus, Paper
Argonauta argo **12**: 1722
Argopteron **17**: 2391
Argus, Brown **2**: 243
Argusfish *see* Scat
Argynnis
   aglaia **7**: 898
   cydippe **7**: 898
   paphia **7**: 898
Argyreus hyperbius **7**: 898
Argyropelecus **9**: 1162
   aculeatus **9**: 1162
   hemigymnus **9**: 1162
Argyroneta aquatica **18**: 2481–2482;
      20: 2877
Ariolimax **17**: 2409
   columbianus **17**: 2408
Arion **17**: 2409
Aricia agestis **2**: 243
Arizona elegans **17**: 2370
Armada, Spanish **17**: 2350
Armadillidium **21**: 2973, 2974
   vulgare **21**: 2973
Armadillo **1**: 85–87
   Fairy **1**: 85, 86
   Giant **1**: 85, 86, 87
   Greater fairy **1**: 87
   Hairy (peludo) **1**: 86, *86*

Lesser fairy (pichiciago) **1**: 85, 86, 87
   Naked-tailed **1**: 85, 86
   Nine-banded **1**: 85, *85*, 86, 87
   Pygmy (pichi) **1**: 85, 86, 87
   Six-banded **1**: 85, 86, 87
   Three-banded (apara) **1**: 85, 86, 87
Armandia lidderdalei **18**: 2582
armor
   armadillo **1**: 85, 87
   armored catfish **1**: 89
   bichir **2**: 195
   coelacanth **4**: 497
   crabs and lobsters **4**: 572
   edible crab **6**: 741
   gar **7**: 929
   glass lizard **7**: 980
   leatherback turtle **10**: *1423*
   lobster **11**: 1477
   loggerhead turtle **11**: 1481
   mantis shrimp **11**: 1552
   millipede **12**: 1611, *1611*
   mud turtle **12**: 1688
   pangolin **13**: 1862, *1862*
   sea horse **16**: *2278*
   sea moth **16**: *2290*
   sea urchin **17**: 2314
   sharks **17**: *2330*
   shore crab **17**: 2353
   turtles **20**: 2795, 2796
   velvet ant **20**: 2815
   see also shells
Army worm **13**: 1822
Arniocera erythropyga **3**: 322
Aromia **11**: 1484
Aromobates **1**: 94
Arothron
   meleagris **15**: *2062*
   stellatus **15**: 2063
Arrau *see* Turtle, Amazon river
Arrow-poison frog (dart frog;
      poison-dart frog) **1**: 93–94;
      7: 905; **20**: 2819
   Blue **1**: *93*
   Golden **1**: 93, 94
   Orange and black **1**: *93*
   Strawberry **1**: 93
Artactisteus
   tristoechus **7**: 929
   tropicus **7**: 929
Artamus
   cinereus **21**: *2983*
   cycanopterus **21**: 2983
   fuscus **21**: 2983, 2984
   leucorhynchus **21**: 2983
   minor **21**: 2983, 2984
   monachus **21**: 2984
   personatus **21**: 2983
   superciliosus **21**: 2984
Artemia salina **2**: 276, 277
Arthropleona **18**: *2504*
Arvicola
   sapidus **21**: 2884, 2885
   terrestris **21**: 2884, 2885
Ascaphus truei **9**: 1200
asexuality
   anemone **1**: 48
   hydra **9**: *1265*, 1266
   purse sponge **15**: 2075
   salp **16**: 2223, 2224
   sea fir **16**: 2273
   sponge **18**: 2495
   see also budding
Asilus crabroniformis **6**: *859*
Asio
   flammeus **5**: 650; **7**: 893; **13**: 1828
   otus **13**: 1826
Asity, Sunbird **12**: 1605
Asp (aspic viper; June viper) **1**: 97–98

Aspidelaps **4**: 534
Aspidertes nigricans **17**: 2443; **20**: 2798
Aspidites **15**: 2080
   melanocephalus **15**: 2079
   ramsayi **15**: 2079
Aspredinichthys tibicen **3**: 401
Ass
   Domestic (donkey) **21**: 2939, *2940*
   Wild *see* Wild ass
assemblies, ceremonial **11**: 1521
Astacopis gouldi **5**: 591, 592
Astacus
   astacus **5**: 592
   pallipes **5**: 592
Astatatilapia **4**: 459
Asterias **18**: 2517, 2518
Asterina gibbosa **18**: 2518
Asthenes **13**: 1819
Astraea inaequalis **19**: 2706
Astrapia, Princess Stephanie's **2**: 205
Astrapia stephaniae **2**: 205
Astropecten **18**: 2518
   irregularis **18**: 2517
Astrophytum **2**: 159
Astroscopus **18**: 2520
   guttatus **18**: 2519
Atacama Desert **5**: 662–663
Ateles
   geoffroyi **12**: *1641*
   paniscus **18**: *2477*, 2477, 2478
Atelocynus microtis **17**: 2357
Athene
   blewitti **13**: 1826
   cunicularia **3**: 324; **13**: 1828
   noctua **13**: *1827*
Atherinopsis californiensis **17**: 2379
Athous **4**: 467
Atilax paludinosus **12**: 1636
Atlantisia rogersi **15**: 2108
atolls **4**: 528, 532, *532*
Atrax **20**: 2820
   robustus **18**: 2482
Atretochoana **3**: 357
Atta **10**: 1417, 1418
   sexdens **10**: 1417
Attacus
   atlas **1**: 102
   ricini **17**: 2377
Attagis
   gayi **17**: 2319
   malouinus **17**: 2319
attenuated attitude **19**: 2646
attracting a mate with color
   agouti **1**: 32
   anole lizard **1**: 54, *54*
   bird of paradise **2**: 206–207
   bulbul **3**: 308
   butterflies and moths 3: 344
   cuttlefish **5**: 633
   emperor penguin **6**: 780
   frigate bird **7**: 893
   jewel fish **10**: 1326
   king penguin **10**: 1359
   octopus **13**: 1781
   redstart **15**: 2140
   squid **18**: 2510
   wrasse **21**: 2991–2992
attracting a mate with movement
   Adélie penguin **1**: 28
   albatross **1**: 37; **16**: 2268
   avadavat **1**: 103
   avocet **1**: 106
   bearded lizard **2**: 172–173
   bishop bird **2**: 213
   booby **2**: 263, *263*
   bowerbird **2**: 270–271
   buzzard **3**: 353
   capercaillie **3**: *376*, 378
   caribou **3**: 393
   crowned eagle **5**: 616

curassow **5**: 624
curlew **5**: 627–628
damselfly **5**: 641
darter **5**: 647
dragonet **5**: 707
duiker **6**: 728
false scorpion **6**: 799
fiddler crab **6**: 816, 817
godwit **7**: 998
Gouldian finch **8**: 1031
great bustard **3**: 338–339, *339*
grebe **8**: *1061*, 1063
grouse **8**: 1086
kestrel **10**: 1346–1347
kob **10**: 1382
lyrebird **11**: 1509, *1509*
mallard **11**: 1525–1526
merganser **12**: 1601
mudskipper **12**: 1684
muscovy duck **12**: 1700
night heron **11**: 1739
ostrich **13**: 1816
painted snipe **13**: 1850
palm dove **13**: 1857
peafowl **14**: 1900
pintail **14**: 1958
pitta **14**: 1971
quail **15**: 2082
raven **15**: 2122–2123
redpoll **15**: 2137
rhea **15**: 2150
riflebird **2**: 207
roadrunner **16**: 2185
roller **16**: 2205, *2205*
sandpiper **16**: 2234
secretary bird **17**: 2317
seriema **17**: 2321
sparrow hawk **18**: 2469
stickleback **18**: 2534
sun bittern **18**: 2567
swallow-tanager **18**: 2583
tarantula **19**: 2635
tooth-carp **19**: 2704
tragopan **19**: 2717
trumpeter **20**: 2757
wolf spider **21**: 2956
wood pigeon **21**: 2981
attracting a mate with scent
   alligator **1**: 39–40
   axolotl **1**: 109–110
   brown butterflies **3**: 296
   owlet moth **13**: 1821–1822
   peppered moth **14**: 1915
   spiders **18**: 2483
attracting a mate with sound **3**: *376*, 378
   Adélie penguin **1**: 28
   alligator **1**: 39
   avadavat **1**: 103
   Baltimore oriole **1**: 126, *127*
   bird of paradise **2**: 206
   button quail **3**: 349
   canary **3**: 368
   cardinal **3**: 387
   cicada **4**: 455–456, *455*
   cock-of-the-rock **4**: 490–491
   cricket **5**: 600
   frogs and toads 7: 903
   gibbon **1**: *71*
   grasshopper **8**: 1037
   grouse **8**: *1084*, *1085*
   moose **12**: 1656
   mouth-brooding frog **12**: 1679
   rorqual **16**: *2209*
   shearwater **17**: 2336
   snipe **17**: 2429, *2429*
   spadefoot toad **18**: 2465–2466
   stifftail **18**: *2538*, 2539
   trogon **20**: 2746–2747
Augur *see* Buzzard, Jackal

Auk **8**: 1094, 1095, 1096; **16**: 2265, 2267
   Great (*extinct*) **14**: 1911; **16**: 2268
   Little (dovekie) **11**: 1464–1465; **16**: 2265
   Razorbilled *see* Razorbill
Aulichthys japonicus **20**: 2766
Aulorhynchus flavidus **20**: 2766
Aurelia aurita **10**: 1320, 1321–1322
auricularian larvae **14**: 1981; **16**: 2272
Australophocaena dioptrica **15**: 2024, 2025
Austropotamobius pallipes **5**: 592; **16**: 2175
Autographa gamma **13**: 1821
Automolus **13**: 1819
autotomy (severance of body part) by worms **6**: 736, 805
   see also limbs, loss; tail, loss
autozooids **17**: 2440
Avadavat (red munia; strawberry finch; strawberry waxbill; tiger finch) **1**: 103–104
   Green **1**: 103
   Red **1**: 103, *103*, 104, *104*
Avicularia **19**: *2634*
Avocet **1**: 105–106
   American **1**: 105, *105*
   Australian (red-necked avocet) **1**: 105
   Chilean **1**: 105
   Eurasian (pied avocet; black-capped avocet) **1**: 105, 106, *106*
Awoura **16**: 2251
Axis
   axis **1**: 107–108
   calamianensis **1**: 107
   kuhlii **1**: 107
   porcinus **1**: 107
Axolotl (axototl) **1**: 109–110
   Mexican **1**: 109, *109*, 110
   Tiger **1**: 109
Axototl *see* Axolotl
Aye-aye **1**: 111–112
Ayo *see* Sweetfish
Aythya **5**: 717; **14**: 1993, 1994, 1995

B

Babbler **1**: 113–114; **18**: 2456
   Common **1**: 114
   Jungle **1**: 113
   Lowland mouse **1**: *113*
   Rufous **1**: *114*
   Scimitar **1**: 113
   Song **1**: 113
   Tit **1**: 113
   Wren **1**: 113
Baboon **1**: 115–117; **12**: 1644
   Gelada **1**: 115, 117; **12**: *1643*, *1644*, 1645
   Guinea **1**: 115, 116
   Hamadryas (sacred baboon) **1**: 115, *115*, 116; **12**: 1644
   Savanna (chacma baboon; olive baboon; yellow baboon) **1**: 115, *116*, *117*
Bacillus
   rossii **18**: 2530
   thuringiensis **4**: 540
backbone *see* vertebrae
Backswimmer (boat-fly; water boatman; wherry-man) **1**: 118–119; **2**: 188
bacteria, chemosynthetic **13**: 1773
Badger **1**: 120–122
   American **1**: 120, *120*, 122
   Eurasian (brock) **1**: 120, 121, *121*, 122, *122*; **19**: 2653
   Ferret (tree badger; pahmi) **1**: 120

   Hog (sand badger; hog-nosed badger) **1**: 120
   Honey (ratel) **1**: 120; **9**: 1206–1207
   Malaysian stink **1**: 120
   Palawan stink (Calamanian stink badger) **1**: 120
   Stink **1**: 120
Badis **10**: 1422
   badis **10**: 1421
Baikal, Lake **10**: 1397, 1398
Baji *see* River dolphin, Chinese
Balaena mysticetus **16**: 2166, 2167; **21**: 2910
Balaeniceps rex **8**: 1123; **17**: 2352; **18**: 2551
Balaenoptera
   acutorostrata **16**: 2208, 2209; **21**: 2914
   borealis **16**: 2208; **21**: 2914
   edeni **16**: 2208; **21**: 2911
   musculus **2**: 247; **5**: 695; **14**: 1954; **16**: 2208; **21**: 2911
   physalus **16**: 2208; **21**: 2914
Balanus nubilis **1**: 138, 139
Balearica paronina **5**: 584
baleen **2**: 246, 247; **8**: 1052; **9**: 1258–1259; **16**: 2166, 2167–2168, 2209; **21**: 2913
Balistapus undulatus **20**: 2744
Balistes carolinensis **20**: 2744
Balistoides conspicillum **20**: *2743*, 2744
Balloonfish *see* Pufferfish
ballooning **18**: 2481
Bambusicola **14**: 1891
Bandicoot **1**: 130–131
   Desert (*extinct*) **1**: 130
   Eastern barred **1**: *130*
   Giant **1**: 130, 131
   Long-nosed **1**: 131
   Northern brown **1**: *131*
   Pig-footed (*extinct*) **1**: 130
   Rabbit-eared (greater bilby) **1**: 130–131
   Short-nosed **1**: 131
   Western barred **1**: 130
Bandicoot rat **1**: 132–133
   Greater (large bandicoot rat; Indian bandicoot rat) **1**: 132, *132*, 133
   Lesser (Indian mole rat) **1**: 132, 133
   Savile's **1**: 132, 133
   Short-tailed (pest rat) **1**: 132
Bandicota
   bengalensis **1**: 132, 133
   indica **1**: 132, 133
   savilei **1**: 132, 133
Band-tailed cat *see* Ringtail
Banteng (tsaine) **3**: 306, *306*, 307; **7**: 936, 937; **21**: 3018
Barb, Zebra (striped barb) **21**: 3015
Barba amarillo *see* Fer de lance
Barbatula barbatula **11**: 1473
Barbel **8**: 1089, 1090
barbels
   carp **3**: 395
   catfish **1**: 88; **3**: 401, *401*
   flying fish **7**: 875
   goatfish **7**: 989, *989*, *990*
   gudgeon **8**: *1089*, 1090
   sturgeon **18**: 2556, *2556*
Barber *see* Hornero, Rufous
Barbfish **16**: 2260
Barbourisia rufa **21**: 2909
Barbus barbus **8**: 1089
Barfani chita *see* Snow leopard
Barley bird *see* Mannikin, Chestnut-breasted
Barleyfinch *see* Mannikin, Chestnut-breasted

Barnacle **1**: 138–139; **14**: 1981
  American **1**: 138, 139
  Common acorn **1**: 138, 139, *139*
  Goose **1**: 138, 139
Barnea
  candida **14**: 1939–1940
  parva **14**: 1939, 1940
Barracouta *see* Snoek
Barracuda **2**: 154–155; **4**: 532
  California (Pacific barracuda)
    2: 155
  European **2**: 154, 155
  Great (picuda; becuna) **2**: *154*, 155
  Northern **2**: 154, 155
Bartramia longicauda **16**: 2234
Baryphthengus martii **12**: 1665
Basilicus
  basilicus **2**: 156, 157
  plumifrons **2**: 157
Basilisk
  Common (Jesus lizard) **2**: 156
  Plumed (leafy plumb-tree basilisk)
    2: 157
Basket star (Gorgon's head)
  2: 158–159
basking *see* temperature control
Bass **2**: 163–164
  Common (European bass) **2**: 163,
    164
  Largemouth **2**: 164, *164*; **8**: 1063;
    18: 2569
  Potato *see* Grouper, Potato
  Sea *see* Sea bass
  Smallmouth **2**: 164
  Stone (wreckfish) **8**: 1082
  Striped **8**: 1082
  White **8**: 1082
  Yellow **8**: 1082
Bassaricyon **10**: 1363
  gabbii **10**: 1363
Bassariscus
  astutus **16**: 2171, 2172
  sumichrasti **16**: 2171, 2172
Bat
  Bumblebee **2**: 167
  Desert **20**: 2822, 2823
  False vampire **6**: 840
  Fish-eating (greater bulldog bat)
    2: 167; **6**: 840–841
  Fishing **2**: 167; **20**: 2822, 2823
  Frog-eating **2**: 167
  Fruit *see* Fruit bat
  Fruit-eating **2**: 169
  Ghost **2**: *169*; **19**: 2702
  Hairless (giant naked bat)
    8: 1115–1116
  Hoary **20**: 2821, 2822
  Horseshoe *see* Horseshoe bat
  Long-eared *see* Long-eared bat
  Mexican fishing **6**: 840
  Mouse-tailed (rat-tailed bat; long-
    tailed bat) **19**: 2702, *2702*, 2703
  Pallid **20**: *2821*
  Pipistrelle *see* Pipistrelle
  Red **20**: 2821, 2822, *2822*
  Sheath-tailed (sac-winged bat)
    19: 2702, 2703
  Spotted (pinto bat) **2**: 168; **20**: 2821
  Tomb **19**: 2702–2703
  Townsend's big-eared **2**: 168;
    20: 2821
  Vampire *see* Vampire bat
  Vesper **20**: 2821–2823
  Whiskered **20**: 2822, 2823
  see also Bats
Batfish (*anglerfish*) **1**: 51, 52
Batfish (*sea moth*) *see* Sea moth,
  Longtail
Bathothauma **18**: 2509
Bathycrinus carpenterii **16**: 2281

Bathylagus callorhinus **7**: 917
Bathylychnops exilis **7**: 888
bathypelagic zone **13**: 1772
Bathyraja abyssicola **17**: 2383
Bathysolea profundicola **17**: 2444
Batrachoseps **11**: 1504
  attenuatus **11**: 1503
Batrachostomus
  auritus **7**: 901
  hodgsoni **7**: 900
batrachotoxin **1**: 94; **20**: 2819
Bats 2: 167–169
Bdelloura **6**: 849
beaches **4**: 476
Beaked whale **2**: 170–171; **21**: 2911
  Blainville's **2**: *170*
  Cuvier's **2**: *171*
  Shepherd's (Tasman beaked
    whale) **2**: 170
Beak-head (*extinct*) **20**: 2762
Bean goose **14**: 1955
  Taiga **14**: 1955
  Tundra **14**: 1955
Bear
  Asiatic black (Himalayan black
    bear; moon bear) **1**: 95–96; **2**: 176
  Black (American black bear;
    Kermode bear; glacier bear;
    Everglades bear) **2**: 176, *176*,
    222–223; **12**: 1672; **19**: 2653
  Brown (grizzly bear; Kodiak bear;
    European brown bear; Tibetan
    brown bear) **2**: 174–175;
    3: 292–295; **12**: 1672; **19**: 2616,
    2617, 2653
  Cave (*extinct*) **3**: 295
  Honey (*kinkajou*) *see* Kinkajou
  Koala *see* Koala
  Polar (ice bear) **1**: 82; **2**: 174, *175*,
    176; **14**: 1998–2000
  Sloth (Indian bear; bhalu) **2**: 176;
    17: 2403–2405
  Spectacled (Andean bear) **2**: 174,
    176; **12**: 1672; **18**: 2470–2471
  Sun (honey bear; Malaysian bear;
    bruang) **18**: 2562–2564
  see also Bears; Panda, Giant
Bears 2: 174–176
Beatragus hunteri **1**: 63
Beaver **2**: 177–179; **19**: 2616
  Canadian **2**: 177, 178, 179, *179*
  European **2**: 177, *177*, 178, 179;
    21: 2906
  Mountain *see* Mountain beaver
Becard **4**: 543
  Jamaican **4**: 543
  White-winged **4**: 543, 544
Becuna *see* Barracuda, Great
Bee
  Cuckoo **2**: 185; **3**: 318
  Hairy-legged mining **2**: *184*
  Hive *see* Honeybee, European
  Leafcutter (alfalfa bee)
    10: 1419–1420
  Mason **2**: 185
  see also Bees and wasps;
    Bumblebee
Beebe, William **10**: 1413
Bee-eater **2**: 180–182
  Carmine **2**: 181
  European **2**: 180, 181, *181*
  Little **2**: *182*
  Rainbow (rainbow bird) **2**: 180
Bees and wasps 2: 183–185
Beetle
  Black *see* Cockroach, Common
  Burying *see* Burying beetle
  Click *see* Click beetle
  Colorado (Colorado potato beetle;
    potato bug) **4**: 504–505; **17**: 2344

Cucujo **6**: 827
Diving *see* Diving beetle
Dung **2**: 187; **16**: 2252–2254
Fruit **19**: *2611*
Glowworm *see* Glowworm
Goliath **2**: 186; **16**: 2252
Hercules **2**: 186; **16**: 2252
June *see* Cockchafer
Lady *see* Ladybug
Long-horned *see* Long-horned
  beetle
Longicorn *see* Long-horned beetle
May *see* Cockchafer
Oil **2**: 188; **13**: 1782–1783
Rhinoceros **16**: 2252
Scarab *see* Scarab beetle
Stag **19**: *2652*
Striped elaterid **4**: 467
Tiger **2**: *186*
Wasp **11**: 1484
Water *see* Water beetle
Whirligig *see* Whirligig beetle
Wood-boring *see* Long-horned
  beetle
  see also Firefly; Glowworm;
    Ladybug; Weevil
Beetles and bugs 2: 186–188
Beira *see* Antelope, Beira
Beishung *see* Panda, Giant
Bellbird **2**: 191–192; **4**: 543
  Bare-throated **2**: 191
  Bearded (mossy-throated bellbird;
    black-winged bellbird) **2**: 191;
    4: 543, 544
  Crested (pan-pan-panella; Dick-
    Dick, the devil) **2**: 191–192, *191*
  New Zealand **2**: 191
  Three-wattled **2**: 191; **4**: 543
  White (snowy cotinga) **2**: 191;
    4: 543
Belone belone **12**: 1725
Beluga *see* Sturgeon, Russian
Beluga whale *see* Whale, Beluga
Belukha *see* Whale, Beluga
benthic zone **4**: 475; **13**: 1772–1773
Berardius **2**: 170
  bairdii **2**: 170
Bergmann's rule **21**: 2930
Besra **18**: 2467
Betcherrygah *see* Budgerigar
Betta, Penang **6**: 822
Betta
  picta pugnax **6**: 821
  pugnax **6**: 821, 822
  splendens **6**: 821, 822
Bettong **15**: 2116
  Brush-tailed (woylie) **15**: 2116,
    2116
Bettongia **15**: 2116
  lesueuri **15**: 2115, 2116
  pencillata **15**: 2116, *2116*
Bezoar *see* Goat, Wild
Bhalu *see* Bear, Sloth
Bhutan glory **18**: 2582
Bib-finch *see* Mannikin, Madagascar
Bighorn sheep **2**: 198–200
  Mexican (desert bighorn sheep)
    2: 198, 199
  Nelson's **2**: 199
  Rimrock **2**: 199
  Rocky Mountain **2**: 199
  Weems' **2**: 199
Bilby
  Greater *see* Bandicoot,
    Rabbit-eared
  Lesser (*extinct*) **1**: 130
bilharzia (schistosomiasis) **7**: 870,
  871; **12**: 1695
bill
  avocet **1**: 105, *105*

chaffinch **3**: 414
crossbill **5**: *607*, 608–609
Darwin's finches **5**: 648–649, *648*,
  649, 650
herring gull **9**: 1189
hornbill **9**: 1226
kea **10**: 1343
kiwi **10**: 1372, *1372*
nutcracker **13**: 1757, *1758*
passerine **18**: 2456
raven **15**: *2123*
shoebill **17**: 2351
skimmer **17**: 2386, 2387, 2388, *2388*
snipe **17**: 2428, *2428*
spoonbill **18**: 2496, 2497
stilt **18**: 2540, 2542
sunbird **18**: *2458*, 2565
toucan **19**: 2714, *2714*, 2715, 2716
vanga **20**: 2813
wattlebird **21**: *2886*, 2887
Billet *see* Coalfish
Billfish *see* Gar, Longnose
binary fission **1**: 42; **14**: 1882
Binneya **17**: 2408, 2409
Binturong **2**: 201–202
biological clock **2**: 237; **17**: 2353
biological pest control
  of bollworms **2**: 256
  of clothes moths **4**: 471
  of Colorado beetles **4**: 504
  of corn borers **4**: 540
  of fer de lance **6**: 814
  of giant snails **7**: 967
  of horntails **9**: 1280
  of sawflies **16**: 2241
  of spider mites **12**: 1621
  using ladybugs **10**: 1396
  using mongooses **6**: 814
  using tachina flies **19**: 2611
  using wasps **2**: 185; **3**: 416, *417*;
    4: 504–505; **21**: 2928, 2933
bioluminescence *see* light production
Biorhiza **7**: 922
Bipalium kewense **6**: 849
bipinnaria larvae **18**: 2518
Bichir **2**: 195–196
  Ornate **2**: 195
Bird of paradise **2**: 205–207
  Crested **2**: *206*
  Emperor **2**: 205, 206, 207, *207*
  Greater **2**: 206
  King of Saxony **2**: 205
  Lesser **2**: *205*
  Magnificent **2**: 206
  Prince Rudolf's blue **2**: 205, 207
  Raggiana **2**: 206–207
birds
  mating for life **2**: 214, 263; **3**: 366;
    10: 1385; **11**: 1487; **18**: 2589;
    19: 2645, 2710
  perching (passerines)
    18: 2455–2456
birds, cage **10**: 1316; **11**: 1516;
  14: 1880, 1883; **21**: *2930*, 3012,
  3013
Birds of prey 2: 208–210
  see also Buzzard; Eagle; Falcon;
    Goshawk; Gyrfalcon; Harrier;
    Hawk; Hobby; Kestrel; Kite;
    Merlin; Osprey; Secretary bird
Birdwing **2**: 211–212; **15**: *2110*;
  18: 2582; **20**: 2806
  Cairns **2**: 211
  Common **2**: 211, 212
  Queen Alexandra's **2**: 211; **3**: 344
  Queen Victoria's **2**: 211, 212
  Rajah Brooke's **2**: 211, *211*, 212
  Tailed **2**: 211
Birgus latro **4**: 571; **9**: 1177; **16**: 2186,
  2187

Bishop bird **2**: 213–214
    Burton's black-and-yellow
      (yellow-shouldered widow
      bird) **2**: 213
    Fire-crowned (black-winged
      bishop; black-winged red
      bishop; red-crowned bishop)
      2: 213, 214
    Golden-backed **2**: 213
    Red **2**: *214*
    Yellow-crowned **2**: 213
Bison
    bison **2**: 215, 216; **8**: 1040
    bison bison **2**: 215, 216
    bison bison athabascae **2**: 215, 216
    bonasus **2**: 215, 216
Bison **2**: 215–217
    American (buffalo) **2**: *215*, 216,
      217; **3**: 305, 306, 307; **8**: *1038*,
      1040
    European (wisent) **2**: 215, 216;
      3: 305, 306, 307
    Indian *see* Gaur
    Plains **2**: 215, 216
    Wood **2**: 215, 216
Biston
    betularia **14**: 1915–1916
    betularia carbonaria **14**: 1915–1916
    betularia insularia **14**: 1915
Bitis **15**: 2058, 2059; **17**: 2370, *2422*
Bitterling **2**: 218–219
Bittern **2**: 220–221; **9**: 1179
    American (bog-bull) **2**: *220*, 221
    Australian **2**: 220
    Eurasian **2**: 220, *221*
    Least **2**: 220
    Little **2**: 220
    Pinnated **2**: 220
    Sun **18**: 2567–2568
bivouacs (balls of ants) **1**: 92
Biziura lobata **18**: 2538
Blackbird (Eurasian blackbird;
    European blackbird) **2**: 224–225;
    19: 2678, 2679
    Austral **13**: 1805
    Red-winged **2**: 224; **15**: 2141–2142;
      18: 2458
    Yellow-headed **2**: 224
Blackbuck (Indian antelope) **1**: 60;
    2: 226–227; **10**: 1382
Blackcap **20**: 2857
Black Death **6**: 850; **17**: 2346–2347
Black-eye, Olive **21**: 2930
Blackfish (*freshwater fish*), Alaska
    1: 34; **12**: 1681
Blackfish (*whale*) *see* Killer whale
Blackfly *see* Aphid
Black snapper *see* Massasauga
Black widow (hourglass spider; shoe
    button spider; red mark; red
    back) **2**: 235–236
Bladderfoot *see* Thrips
Bladdernose *see* Seal, Hooded
Blaniulus guttulatus **12**: 1611–1612
Blarina brevicauda **17**: 2359
Blastocerus dichotomus **18**: 2462, 2463
Blatta **4**: 493
    germanica **4**: 493
    orientalis **4**: 492–493, 493
Blattella **4**: 493
Bleak **12**: 1619
Bleekeria mitonkurii **16**: 2227
Blennius
    nigriceps **2**: 238
    ocellaris **2**: 238
    pholis **2**: 237, 238
Blenny **2**: 237–238
    Black-headed **2**: 238
    Butterfly **2**: 238
    Fanged **2**: 237

    Four-eyed **7**: 888
    Viviparous (eelpout) **2**: 238
Blepharipoda **18**: 2507
Blesbok **2**: 239–240, 259, 260; **6**: 759;
    16: 2216
Blesmol **12**: 1628, 1629
blind, cave fish **3**: 409, *409*, 410
Blindfish, Kentucky **3**: 409
Blind worm *see* Slowworm
blood, antifreeze in **9**: 1278
blood squirting (reflex bleeding)
    horned lizard **9**: 1232
    oil beetle **13**: 1783
Bloodworm *see* Midge, Non-biting;
    Tubifex worm
Bloody Henry *see* Starfish, Blood
    sea star
Blowfish *see* Pufferfish
Blowfly (bluebottle; greenbottle;
    cluster fly; attic fly; screwworm
    fly) **2**: 241–242; **6**: 859
blubber **6**: 773; **16**: 2289
Blue arrow *see* Damselfly
Bluebelly *see* Fence lizard
Bluebird
    Eastern **18**: *2455*, 2457
    Fairy *see* Fairy bluebird
Bluebottle (*fly*) *see* Blowfly
Bluebottle (*Portuguese man-of-war*)
    15: *2029*
Bluebuck (*extinct*) **1**: 63; **16**: 2216
Blue buck *see* Nilgai
Blue bull *see* Nilgai
Blue butterfly **2**: 243–244; **3**: 344
    Alcon large **2**: *244*
    Chalk-hill **2**: 243
    El Segundo **2**: 244
    Holly **2**: 243
    Large **2**: 243–244
    Long-tailed **2**: 243
    Mission **2**: 244
    Pygmy **2**: 243
    Short-tailed **2**: 243
    Silver-studded **2**: 243, *243*
    Small **2**: 243
    Western pygmy **2**: 244
Bluefin *see* Tuna, Northern
    bluefin
Bluefish **2**: 245–246
Bluehead **21**: 2991, 2992
Bluemouth *see* Rosefish,
    Blackbelly
Bluet *see* Damselfly
Bluethroat *see* Rosefish,
    Blackbelly
Boa **17**: 2418, 2419
    Common *see* Boa constrictor
    Dwarf **17**: 2419
    Emerald tree **17**: *2418*
    Puerto Rican **2**: 251
    Rosy **2**: 249
    Rubber **2**: 249
    Virgin Islands tree **2**: 251
    Water *see* Anaconda
Boa constrictor **2**: 250; **17**: 2422
Boa constrictor (common boa)
    2: 249–251; **17**: 2422
Boar, Wild (wild pig) **19**: 2653, *2655*;
    21: 2907, 2942–2944
Boat-fly *see* Backswimmer
Boatswain bird *see* Tropic bird
Bobak (Himalayan marmot; steppe
    marmot) **11**: 1568, 1569
Bobbing gnat *see* Crane fly
Bobcat **2**: 253–254
Bobolink **13**: 1805, 1807; **18**: *2457*;
    20: 2833
Bobwhite, Northern **15**: 2082,
    2083
Boers **16**: 2216

Bog-bull *see* Bittern, American
bogs **21**: 2905, 2908
Bolbonota **9**: *1225*
Bolborhynchus lineola **14**: 1880
Boleophthalmus **12**: 1684, 1685, 1686
Bollworm **2**: 255–256; **13**: 1822
    American (New World bollworm;
      corn earworm; tomato
      fruitworm; tomato grub)
      2: 255, 256
    Cotton **2**: 255, 256
    Pink **2**: 255, 256
Bombay duck (bummalo;
    bummalaw) **2**: 257
Bombina **20**: 2819
    bombina **6**: 826
    maxima **6**: 825
    orientalis **6**: 825
    variegata **6**: 825
Bombus **3**: 317
    lucorum **3**: 317
    terrestris **3**: 318
Bombycilla
    cedrorum **21**: 2890
    garrulus **21**: 2890, 2891
    japonica **21**: 2890
Bombyliopsis abrupta **19**: 2610
Bombyx mori **3**: 345; **17**: 2376, 2377
Bonasa umbellus **8**: 1084, 1085
bones, dropping **10**: 1401, *1401*, 1402
Bongo **1**: 63; **10**: 1389–1390
Bonito **2**: 258
Bonobo (pygmy chimpanzee) **1**: *70*,
    71, *72*; **4**: 439–440, 441
Bontebok **1**: 63; **2**: 239, 259–260;
    6: 759; **16**: 2216
Bonxie *see* Skua, Great
Bonyfish *see* Menhaden, Atlantic
Booby **2**: 261–263
    Abbott's **2**: 261, 262
    Blue-footed **2**: 261, *263*
    Brown **2**: 261, 262, 263
    Masked (blue-faced booby) **2**: *261*,
      262, *262*, 263
    Peruvian (piquero) **2**: 261, 262, 263
    Red-footed **2**: 261, 262
Boodie **15**: 2115, 2116
book lungs **18**: 2481
Book scorpion *see* False scorpion
booming grounds **15**: 2036, 2037
Boomslang **2**: 264–265; **6**: 749;
    13: 1736
Boophilus **19**: 2682
Boreogadus saida **4**: 495
Bos **7**: 936–938; **21**: 3000, 3001, 3018,
    3019
Boselaphus tragocamelus **1**: 60;
    13: 1754
Bo'sun bird *see* Tropic bird
Botaurus
    lentiginosus **2**: 220
    pinnatus **2**: 220
    poiciloptilus **2**: 220
    stellaris **2**: 220
Botfly, Human **6**: 858
Bothriocyrtum californicum **19**: 2720
Bothrops
    ammodytoides **14**: 1974
    asper **20**: 2817
    atrox **6**: 813
    bilineata **17**: 2421
Botia, Green tiger *see* Loach, Tiger
Botia
    hymenophysa **11**: 1473
    macracanthus **11**: *1473*
Boto *see* River dolphin, Amazon
Botryllus **17**: 2312
Bottlehead *see* Whale, Northern
    bottlenose
Boulengerochromis microlepis **4**: 459

Bowerbird (satinbird) **2**: 269–271
    Archbold's **2**: 271
    Fire-maned **2**: 271
    Golden **2**: 271
    Great gray **2**: *270*
    Regent **2**: 271
    Satin **2**: *270*, 271, *271*
    Spotted **2**: 271
Bowfin (American mudfish; spotfin;
    freshwater dogfish; freshwater
    wolf) **2**: 272–273; **5**: 686
Boxfish *see* Trunkfish
brachiation **1**: 71
brachiolaria larvae **18**: 2518
Brachiones przewalskii **7**: 949
Brachycope **21**: 2896
Brachypelma boehmei **19**: *2633*
Brachyteles arachnoides **18**: 2477, 2478
Braconid **9**: 1279, 1280
Bradypus **17**: 2400, 2402
    infuscatus **15**: *2113*
    torquatus **17**: 2401
    tridactylus **17**: 2401
    variegatus **17**: 2401
branchial basket **10**: 1407
Branchiocerianthus **16**: 2273
Branchiomaldane **11**: 1498
Branchiostoma **10**: 1407
    californiense **10**: *1408*
    lanceolatum **10**: 1407, 1408
Brandling **6**: 734
Branta **5**: 717
    canadensis **3**: 366
    leucopsis **1**: 141
    ruficollis **5**: *716*
    sandvicensis **5**: 717; **9**: 1164
breaching **2**: 161; **5**: 698, *698*; **15**: *2024*;
    18: 2472; **21**: *2913*, 2914
Bream (common bream; bream flat)
    2: 274–275; **16**: 2182
    Freshwater *see* Tilapia
    Sea **2**: 274
    White (silver bream) **2**: 274
breeding *see* asexuality; attracting a
    mate; hermaphrodite;
    monogamy; polygamy; sexual
    aggression
Brephidium **2**: 243
    exilis **2**: 244
Brevoortia tyrannus **2**: 245; **9**: 1184;
    12: 1598
Brill **20**: 2782
Brimstone **21**: 2929
brine pools **13**: 1773
Bristlemouth (lightfish) **2**: 278
Bristletail **2**: 279
Brittle star **2**: 280–281
Broadbill (*duck*) *see* Scaup, Greater
Broadbill (*fish*) *see* Swordfish
Broadbill (*Old World tropical bird*)
    2: 282–283
    Black-and-red **2**: *282*
    Blue-crested **2**: 283
    Dusky **2**: 283
    Grauer's (green broadbill; African
      green broadbill) **2**: 283
    Green **2**: 282
    Long-tailed **2**: 282
    Red-sided **2**: 282
Brock *see* Badger, Eurasian
Brocket **18**: 2463, 2464
Brolga **5**: 584, 587
Bronze-cuckoo, Shining **5**: 617–618,
    619; **19**: 2670
brood patches **2**: 232; **15**: 2125
brood pouches **13**: 1796
Brookesia **3**: 418
    setaroi **3**: 419
    taeniabronchum **3**: 419
Brotogeris tirica **14**: 1878

Brownbul **3**: 308
Brown butterfly (brown) **3**: 296–297
  Meadow **3**: 296–297
browsing
  agouti **1**: 32
  antelope **1**: 61
  axis deer **1**: 108
  beira antelope **2**: 190
  bighorn sheep **2**: 198
  brow-antlered deer **2**: 284
  bushbuck **3**: 332
  caribou **3**: 392
  chamois **3**: 422
  colobus monkey **4**: 502–503
  deer **5**: 657
  dik-dik **5**: 676
  duiker **6**: 728
  eland **6**: 758
  fallow deer **6**: 796
  gazelle **7**: 940–941
  giraffe **7**: 976, 977
  howler monkey **9**: 1254
  sika **17**: 2374
  takin **19**: 2622
  tapir **19**: 2631, *2632*
  wapiti **20**: 2855
Bruang *see* Bear, Sun
brucellosis **3**: 307
Brush turkey **11**: 1528
  Australian **11**: 1527
Bubalus
  bubalis **9**: 1286, 1287
  depressicornis **9**: 1286
  mindorensis **9**: 1286
  quarlesi **9**: 1286
bubble nests *see* nests, "bubble"
bubble rafts **15**: 2073
Bubo bubo **3**: 1825
bubonic plague **3**: 299; **6**: 850; **7**: 951;
  8: 1079; **12**: 1695
Buccinum see Whelk, Common
Buccinum undatum **21**: 2917, 2918
Bucephala clangula **5**: 717
Buceros rhinoceros **9**: 1227
Buckie *see* Whelk, Common
buckskin **12**: 1691
Bucorvus
  cafer **9**: 1226
  leadbeateri **9**: 1226
budding
  coral **4**: 529
  hydra **9**: *1265*, 1266
  pyrosoma **15**: 2078
  salp **16**: 2223, 2224
  sea fir **16**: 2273
Budgerigar (budgerygah;
  betcherrygah; canary parrot;
  lovebird; flightbird) **3**: 302–304
Budgerygah *see* Budgerigar
Budorcas taxicolor **12**: 1703–1705;
  19: *2622*, 2623
Buffalo 3: 305–307
Buffalo
  Cape (African buffalo; forest
    buffalo; bushcow) **3**: 305, 306, 307,
    307, 370–372
  Indian (Asian water buffalo)
    3: 305, 306, 307; **9**: 1286–1288
  see also Bison, American; Gaur;
    Yak; Zebu
Buffalo bird *see* Cowbird, Brown-
    headed
Bufo
  bufo **4**: 510, 513
  calamita **12**: 1721
  marinus **7**: *905*; **11**: 1558
  viridis **12**: 1720
Bug
  Ambush **2**: 188
  Assassin **2**: 188, *188*

Bed **2**: 188
Bee assassin **2**: 188
Butterfly *see* Butterfly bug
  Forest **17**: 2345
  Green vegetable **17**: 2344
  June *see* Cockchafer
  May *see* Cockchafer
  Parent **17**: 2345
  Potato *see* Beetle, Colorado
  Tortoise **17**: 2344
  see also Aphid; Backswimmer;
    Beetles and bugs; Cicada;
    Hopper; Pond skater;
    Shieldbug; Water scorpion
Bugfish *see* Menhaden, Atlantic
Bulbul **3**: 308–309
  Black **3**: 308
  Brown-eared **3**: 308
  Common (African bulbul) **3**: *308*,
    309
  Red-vented **3**: 309
  Red-whiskered **3**: 309, *309*
  White-cheeked **3**: 308
Bulinus **7**: 870
Bullfinch (common bullfinch;
  Eurasian bullfinch) **3**: 310–311
Bullfrog (American bullfrog; jug
  o'rum) **3**: 312–313; **6**: 744; **7**: 903
Bullhead **3**: 314–315
  European (sculpin; miller's-
    thumb) **3**: 315
Bullhead shark
  Crested **15**: 2027
  Japanese **15**: 2027
Bumblebee (humblebee; carderbee)
  2: 185; **3**: 316–318
  White-tailed **3**: *317*
Bummalaw *see* Bombay duck
Bummalo *see* Bombay duck
Bunaka **7**: 993
Bungarus **2**: 265; **17**: 2308
Bunodactis verrucosa **1**: *47*
Bunting **3**: 319–320
  Corn **3**: 320
  House **3**: 319
  Lapland **3**: 319
  Lark **18**: 2458
  McKay's **3**: 319
  Ortolan **3**: 319, *320*
  Painted **3**: *319*
  Reed **3**: 320; **18**: 2457
  Snow **3**: 319–320
Buphagus
  africanus **13**: 1830
  erythrorhynchus **13**: 1830, 1831
Burbot (eel pout) **4**: 495; **11**: 1455
Burhinus
  bistriatus **18**: 2545
  grallarius **18**: 2545
  oedicnemus **18**: 2545, 2546
  superciliaris **18**: 2545
  vermiculatus **18**: 2545
Burnet moth **3**: 321–322, 344
  Doherty's longtail **3**: 321
  Eastern grapeleaf skeletonizer
    3: 321
  Fire grid **3**: 322
  Mountain **3**: 321–322
  Provence **3**: 322
  Six spot **3**: *321*
  Variable **3**: 322
Burnett salmon *see* Lungfish,
    Australian
Burramys (mountain pygmy
  possum) **5**: 703
Burramys **5**: 702, 703
Burro *see* Wild ass, African
burrowing
  aardvark **1**: 16, *16*
  ant lion **1**: 64

armadillo **1**: 85
banded anteater **1**: 128
bandicoot **1**: 130–131
bandicoot rat **1**: 132–133
bee-eater **2**: 180–181
burrowing owl **3**: 324
caecilian **3**: 356
chickaree **4**: 436
chipmunk **4**: 448–449
coypu **4**: 562–563
crab plover **4**: 569
crested porcupine **5**: 597
desman **5**: 666
earless monitor **6**: 732
earthworm **6**: 734, 735, 736
elephant tusk shell **6**: 777, 778, *778*
fennec fox **6**: 809, 811
fiddler crab **6**: 815
gaper **7**: 926
gerbil **7**: 950
golden mole **7**: 1004
in grasslands **8**: 1040
groundhog **8**: 1075
hamster **8**: 1129
hunting wasp **9**: 1261, 1262
jerboa **10**: 1323
legless lizard **10**: 1430
lugworm **11**: 1497, *1497*, 1498,
  1498
meerkat **12**: 1595
mole **12**: 1625–1626, *1625*
mole rat **12**: 1628, 1629
motmot **12**: 1665–1666
mountain beaver **12**: 1674
mudskipper **12**: 1686
nutshell **13**: 1763
otter shrew **13**: 1818
paca **13**: 1838
pedipalpi **14**: 1906
piddock **14**: 1939–1940, *1940*
platypus **14**: 1986–1987
pocket gopher **14**: 1996, 1997, *1997*
prairie dog **15**: 2038–2039
prion **15**: 2045–2046
puffbird **15**: 2061
puffin **15**: 2067, *2067*
rabbit **15**: 2091–2092, 2093
ragworm **15**: 2104
razor shell **15**: 2127–2128, *2128*
sand dollar **16**: 2226
sea cucumber **16**: 2271
sea mouse **16**: 2292
sea urchin **17**: 2314–2315
shearwater **17**: *2335*, 2336
shipworm **17**: 2349, *2349*
shrimp **17**: 2366
skink **17**: 2389
spadefoot toad **18**: 2465
spiny anteater **18**: 2487, *2487*
springhare **18**: 2502
steppe lemming **18**: 2528
swallow-tanager **18**: 2583
tenrec **19**: *2656*
tortoise **19**: 2711
trapdoor spider **19**: 2719
tuatara **20**: 2762–2763, *2762*
tuco-tuco **20**: 2771, *2771*, 2772
viscacha **20**: 2834–2835
water vole **21**: 2884
wolf spider **21**: 2956
wombat **21**: 2961–2962
Burying beetle (sexton beetle;
  gravedigger) **3**: 327–328
  American **3**: 328
Bushbaby **3**: 329–330
  Greater **3**: 330
  Needle-clawed **3**: 330
  Senegal **3**: 329, 330
  South African **3**: 329, 330
  Thick-tailed greater **3**: 329, 330

Bushbuck **1**: 63; **3**: 331–332
Bushcow *see* Buffalo, Cape
Bushdog **3**: 333–334; **17**: 2357
Bushmaster **14**: 1972, 1973, 1974
Bushpig (African bush pig; Red
  River hog) **3**: 335–336
Bush-shrike **17**: 2361
Bustard **3**: 337–339
  Australian **3**: 338
  Great **3**: 337, 338
  Great Indian **3**: 338
  Houbara **3**: 339
  Kori **3**: 337
Bustard quail
  Barred **3**: 350
  Little *see* Button quail,
    common
Busycon canaliculatum **21**: 2917
Butcherbird **3**: 340–341; **17**: 2363
  Black **3**: 340
  Black-backed **3**: 340
  Black-headed **3**: 340
  Gray (silver-backed butcherbird;
    Derwent jackass; Tasmanian
    jackass; whistling Jack) **3**: 340,
    340, 341
  Pied **3**: 340, *341*
  White-rumped **3**: 340
Buteo **9**: 1211
  buteo **3**: 352
  galapagoensis **5**: 650
  jamaicensis **3**: *353*, 426
  lagopus **2**: 209; **3**: 351
  rufofuscus **3**: 351
Buthus occitans **16**: *2257*
Butterfish (gunnell; rock gunnell;
  nine-eyes) **3**: 342
Butterflies and moths 3: 343–346
Butterfly
  African leaf **10**: 1332
  Apollo (European Apollo;
    parnassian) **1**: 76
  Birdwing *see* Birdwing
  Blue *see* Blue butterfly
  Brimstone **21**: 2929
  Brown *see* Brown butterfly
  Cabbage white *see* White butterfly,
    Small white
  Clouded yellow **21**: 2929
  Comma **3**: 345
  Common tiger *see* Monarch
  Copper *see* Copper butterfly
  European cabbage *see* White
    butterfly, Small White
  Grayling **3**: 297
  Hairstreak *see* Hairstreak butterfly
  Indian leaf (orange oakleaf
    butterfly) **10**: 1332
  Kallima **10**: 1331–1332
  Map **20**: 2811
  Milkweed *see* Monarch
  Question-mark **20**: 2811
  Small white **3**: 416
  Swallowtail **2**: 212
  Thistle (cosmopolitan; painted
    lady) **20**: 2812
  Vanessa **20**: 2810–2812
  White *see* White butterfly
  Yellow *see* Sulphur butterfly
  see also Birdwing; Fritillary;
    Monarch; Morpho; Purple
    emperor; Skipper; Swallowtail
Butterfly bug **9**: 1225
Butterfly cod *see* Lionfish, Red
Butterfly fish (*freshwater fish*)
  3: 347–348
Butterflyfish (*marine fish*) **1**: 49, *49*
  Copperband **1**: 49
butterfly flight **17**: 2435–2436
Butter hamlet **8**: 1081, 1083

Button quail 3: 349–350; 15: 2082
    Australian black-breasted 3: 350
    Common (little button quail; little
        bustard quail; kurrichane
        button quail; Andalusian
        hemipode) 3: 349, 350
    Yellow-legged 3: 349
Button shell 19: 2707
Buzzard 2: 209; 3: 351–353
    Black-breasted (black-breasted
        kite) 3: 352–353
    Common (Eurasian buzzard;
        steppe buzzard) 3: 351–352, 351,
        352
    Honey see Honey buzzard
    Jackal (augur) 3: 351
    Rough-legged 2: 209; 3: 351
Bycanistes brevis 9: 1226–1227
byssus threads 12: 1711, 1713;
    13: 1832; 16: 2248

C
Cabassous 1: 85
Cacajao
    calvus 20: 2803, 2804
    melanocephalus 20: 2803, 2804
Cacatua 4: 483, 483, 484, 485
Cachalote, Brown 13: 1820
Cacomistle see Ringtail
Cactospiza 5: 649
Caddis fly (sedge fly) 3: 354–355;
    16: 2175
Caecilia 3: 357
Caecilian (naked snake) 3: 356–357
    Sticky 3: 356
Caecobarbus geertsi 3: 410
Caecomastacembelus 18: 2490
Cahow 16: 2268
Caiman
    crocodilus 3: 358, 359, 359
    latirostris 3: 358
    yacare 3: 358
Caiman 3: 358–360
    Black 3: 358, 359–360, 360
    Broad-snouted 3: 358
    Common (spectacled caiman)
        3: 358, 359, 359, 360
    Cuvier's dwarf 3: 358
    Paraguayan (yacare caiman)
        3: 358, 359
    Smooth-fronted 3: 358
Cairina moschata 5: 717; 12: 1700
Cake urchin see Sand dollar
Calamospiza melanocorys 18: 2458
Calcarius lapponicus 3: 319
calcivirus 15: 2093
Calcochloris 7: 1005
Calfbird (capuchinbird) 4: 543
Calicalicus
    madagascariensis 20: 2813
    rufocarpalis 20: 2813, 2814
Calidris
    alba 16: 2233
    alpina 16: 2233
    canutus 16: 2233
    ferruginea 16: 2234
    maritima 16: 2233
    melanotos 20: 2779, 2833
Callaeas
    cinerea 21: 2886, 2887
    cinerea cinerea 21: 2886
    cinerea wilsoni 21: 2886
Calliactis parasitica 9: 1178
Callianassa subterranea 18: 2507
Callicebus
    gigot 19: 2696, 2697
    moloch 19: 2696, 2697
    personatus 19: 2696, 2697
    torquatus 19: 2696, 2697

Callichthys callichthys 1: 89; 20: 2755
Callinectes sapidus 4: 573
calling
    babbler 1: 113
    bellbird 4: 543
    bittern 2: 220
    bullfrog 3: 313
    bunting 3: 319
    bushbaby 3: 329
    capercaillie 3: 376, 378
    chaffinch 3: 415
    corncrake 4: 541
    coucal 4: 549
    coyote 4: 559, 560
    crane 5: 584
    dolphin 2: 268
    francolin 7: 889
    fruit bat 7: 908
    gibbon 1: 73
    house sparrow 9: 1247–1248
    howler monkey 9: 1252
    jackdaw 10: 1300
    katydid 10: 1342
    king penguin 10: 1358, 1360
    kookaburra 10: 1385–1386
    leopard frog 11: 1447
    limpkin 11: 1453
    loon 11: 1485, 1486
    loris 11: 1492
    mannikin 11: 1544
    midwife toad 12: 1609
    monkey 12: 1645
    oilbird 13: 1784, 1784
    orangutan 13: 1798
    owls 13: 1828, 1829
    pika 14: 1943, 1944
    potoo 15: 2033
    screamer 17: 2262
    seriema 17: 2321, 2321
    sika 17: 2374
    squirrel monkey 18: 2515
    tamarin 19: 2626
    tawny owl 19: 2644
    tree frog 19: 2723–2724
    trumpeter 20: 2756, 2757
    see also attracting a mate with
        sound
Calling hare see Pika
Callionymus
    leucopoecilus 5: 706
    lyra 5: 706, 706, 707
Calliostoma zizyphinum 19: 2706, 2707
Callipepla gambelii 15: 2082
Callistege ni 13: 1821
Callithis 18: 2509
Callithrix 11: 1563, 1563, 1564, 1565,
    1565
Calliurichthys 5: 707
Callocephalon fimbriatum 4: 483
Callophrys 8: 1118
    rubi 8: 1118
Callorhinus 7: 916, 917
Callorhynchus milii 17: 2330
Calma glaucoides 16: 2300
Calonectris diomedea 17: 2334
Caloprymnus campestris 15: 2115, 2116
Calyptorhynchus
    baudinii 4: 485
    funereus 4: 485
Calyptura, Kinglet 4: 544
Calyptura cristata 4: 544
Camarhynchus 5: 649
    pallidus 5: 648
Camberwell beauty see Mourning
    cloak
Camel 3: 361–364; 5: 665
    Arabian (one-humped camel;
        dromedary) 3: 361, 362, 364
    Bactrian (two-humped camel)
        3: 361, 362, 363, 364

Camel spider see Solifugid
Camelus 5: 665
camouflage
    abalone 1: 21, 21
    anteater 1: 57
    big-headed turtle 2: 197
    blenny 2: 238
    boa constrictor 2: 249, 249
    boomslang 2: 264, 264
    burrowing owl 3: 324
    bushbuck 3: 332
    butterflies and moths 3: 344
    button quail 3: 349
    caracal 3: 386
    cat shark 3: 408
    courser 4: 551
    crab spider 5: 580, 580, 581
    crested lark 5: 593, 594
    dab 5: 634
    dolphin 4: 506
    edible crab 6: 742
    flounder 6: 860
    frogmouth 7: 900, 901
    genet 7: 947
    hairstreak butterflies 8: 1117–1118
    herring gull 8: 1187
    mantis 11: 1546–1547, 1548
    morpho 12: 1661
    nightjar 13: 1742, 1743
    owl 13: 1827–1828
    plume moth 14: 1991
    poorwill 14: 2012
    puff adder 15: 2058
    scalare 16: 2245
    sea horse 16: 2278
    sloth 17: 2401
    snow leopard 17: 2432–2433
    sole 17: 2444, 2444
    spider crab 18: 2476
    stonefish 18: 2547
    wobbegong 21: 2954, 2955
    woodcock 21: 2966, 2966
Campephaga phoenicea 5: 621, 622
Campephilus
    imperialis 21: 2977
    principalis 21: 2977
Campodea 2: 279
Camponotus 9: 1205
    inflatus 9: 1205
    ligniperda 1: 66
Camposcia retusa 18: 2476
Campostoma anomalum 12: 1618
Camptorhynchus labradorius 5: 717
Campylorhamphus 21: 2968
Campylorhynchus
    brunneicapillus 21: 2993, 2995
    chiapensis 21: 2993
Canary 3: 368–369
Cancer pagurus 6: 741, 742
cancer research 14: 1984
Candirú 3: 402
Canecutter see Rabbit, Swamp
Canis 3: 426; 4: 559; 5: 678, 689, 690,
    692; 8: 1054–1057; 9: 1291, 1292;
    10: 1336; 19: 2641, 2653; 21: 2906
cannibalism
    black widow 2: 236
    boomslang 2: 265
    chimpanzee 4: 442
    cockroach 4: 494
    eel 6: 745
    firefly 6: 828
    flatworm 6: 849
    hake 8: 1119
    horned frog 9: 1229, 1230
    Jack Dempsey 10: 1303
    mantis 11: 1547
    midge 12: 1608
    morpho 12: 1660
    pike 14: 1947

sharks 17: 2333
spiders 18: 2483
    X-ray tetra 21: 2999
    zebra fish 21: 3014
cantharidin 13: 1783
Canvasback 5: 717; 14: 1993–1994,
    1995
Capercaillie (capercailzie; horse of
    the woods) 3: 376–378
    Black-billed 3: 376, 377
    Western 3: 377
Capercailzie see Capercaillie
Caperea marginata 16: 2166, 2167;
    21: 2910
Capra 9: 1272, 1273; 11: 1560;
    12: 1672; 21: 2951, 2951, 2952
Capreolus
    capreolus 5: 657; 16: 2201, 2202;
        17: 2375
Caprimulgus 13: 1742, 1743, 1744;
    14: 2012, 2013
captive breeding
    addax 1: 24
    antelope 1: 63
    binturong 2: 202
    black-footed ferret 2: 228–229
    ducks, geese and swans 5: 717
    fighting fish 6: 821, 821, 822
    gharial 7: 954–955
    night monkey 13: 1748
    okapi 13: 1786–1787
    oryx 13: 1810
    panda 7: 962
    Przewalski's horse 15: 2053, 2053
    spectacled bear 18: 2470–2471
    turtles 20: 2799
    wood duck 21: 2972
Capuchinbird see Calfbird
Capuchin monkey (sapajou; ring-tail
    monkey) 3: 379–381; 12: 1645
    Tufted (weeping capuchin) 3: 379,
        380, 381, 381
    Wedge-capped 3: 379, 380
    White-faced 3: 379, 379, 380
    White-fronted 3: 379, 380, 380
Capybara (water pig; water cavy)
    3: 382–384; 8: 1040
Caracal (desert lynx; rock cat)
    3: 385–386, 404
carapace 1: 85, 87; 17: 2353; 20: 2795
Carapus 14: 1901
    acus 14: 1901–1902
    bermudensis 14: 1901–1902
    homei 14: 1902
Carassius
    auratus 8: 1018, 1019
    auratus gibelio 8: 1018
    carassius 3: 396; 8: 1018
Carausius morosus 18: 2530, 2531
Carcharhinus
    leucas 17: 2329; 21: 3007, 3008
    limbatus 17: 2332
    longimanus 14: 1950
    perezi 15: 2148; 17: 2329
Carcharias taurus 17: 2332
Carcharodon carcharias 8: 1059;
    17: 2331
Carcinoscorpio 10: 1354
Carcinus maenas 4: 572; 17: 2353, 2354
Carderbee see Bumblebee
Cardinal (northern cardinal; red
    bird) 3: 387–388
Cardinal fish 3: 389–390; 4: 532
    Brownspot 3: 390
    Coral 3: 390
    Orange-lined 3: 389
Cardinalis cardinalis 3: 388
Cardium edule see Cerastoderma edule
Carduelis 3: 368; 8: 1016, 1017, 1065;
    11: 1457, 1458; 15: 2137, 2138

Caretta caretta **11**: 1482; **20**: 2797
Carettochelys insculpta **13**: 1867; **20**: 2796
Cariama *see* Seriema
Cariama cristata **17**: 2321, 2322
Caribou (reindeer) **3**: 391–394; **5**: 655, 657; **20**: *2776, 2778, 2778*; **21**: 2958
carnassials **2**: 166, 174
Carnegiella **9**: 1161
carnivores *see* meat eating
Carollia perspicillata **7**: *907*
carotenoids **1**: 103
Carp (minnow) **3**: 395–396
    Common **3**: 396
    Crucian (Prussian carp) **3**: *396*; **8**: 1018
    Gibel *see* Goldfish
    Golden *see* Goldfish
    Prussian *see* Goldfish
    see also Minnow
Carpodacus erythrinus **4**: 499
Carpophilus hemipterus **19**: *2611*
carrion
    and blowflies **2**: 241
    and burying beetles **3**: 327–328
    and condors **4**: *516, 517, 518*
    and golden eagles **7**: *1002*
    and hyenas **9**: 1269
    and marabou storks **11**: 1553, *1554*
    and ravens **15**: 2123, *2124*
    and sea mice **16**: 2292–2293
    and Tasmanian devils **19**: 2643
    and turkey vultures **20**: 2787–2788, *2787, 2789*
    and turnstones **20**: 2792
    and vultures **20**: *2839, 2839, 2840, 2842*
Caspiomyzon **10**: 1404
casques **5**: 623; **9**: 1226
Cassiopeia **10**: 1321
Cassowary **3**: 397–398
    Dwarf **3**: 397
    Northern **3**: 397
    Southern (Australian cassowary; double-wattled cassowary; two-wattled cassowary) **3**: 397, 398, *398*
castes
    ant **1**: 67, 90
    termite **1**: 67
Castlebuilder *see* Spinetail
Castor
    canadensis **2**: 178; **19**: 2616
    fiber **2**: 178; **21**: 2906
Castor-oil fish *see* Escolar
Casuarius
    bennetti **3**: 397
    casuarius **3**: 398; **6**: 782
    unappendiculatus **3**: 397
Cat
    Band-tailed *see* Ringtail
    Coon *see* Ringtail
    Desert **3**: 404
    Iriomote **3**: 405
    Margay *see* Margay cat
    Marsupial *see* Marsupial cat
    Miner's *see* Ringtail
    Mountain *see* Ringtail
    Native *see* Marsupial cat
    Otter *see* Jaguarundi
    Pallas' **3**: 404
    Ring-tailed *see* Ringtail
    Rock *see* Caracal
    Sand **3**: 403, 404, *406*
    Servaline *see* Serval
    Weasel *see* Jaguarundi
    see also Bobcat; Caracal; Cats; Cheetah; Civet; Genet; Jaguar; Jaguarundi; Leopard; Lion;

Lynx; Ocelot; Palm civet; Puma; Serval; Tiger; Wildcat
catadromous species **7**: 993
Catagonus wagneri **14**: 1903, 1904
Cat-bear *see* Panda, Red
Catbird **3**: 399–400
    Abyssinian **3**: 399
    Black **3**: 399
    Gray (common catbird; northern catbird) **3**: 399, *399*, 400, *400*; **4**: 555
    Tooth-billed (stagemaker) **3**: 399
caterpillars (*lepidopteran larvae*) **3**: 345–346
    Atlas moth **1**: 102
    blue butterfly **2**: 243
    bollworm **2**: 255–256
    burnet moth **3**: 322
    cannibalistic **12**: 1660
    copper butterfly **4**: 525, 526
    corn borer **7**: 539, 540
    fritillary **7**: 898
    goat moth **7**: 991, 992, *992*
    hairy **15**: 2048; **19**: 2687–2688
    hawkmoth **9**: 1166–1167
    honeydew secretion **2**: 243, *243*; **8**: 1118
    plume moth **14**: 1991–1992
    processionary moth **15**: 2047–2048, *2048*
    purple emperor **15**: 2072
    puss moth **15**: 2076, 2077
    silk moth **17**: 2377–2378, *2377*
    tiger moth **19**: 2687–2688
    urania moth **20**: 2806
    vanessa **20**: 2810
Catfish **3**: 401–402
    Armored **1**: 88–89; **20**: 2755
    Banjo **3**: 401
    Blue **3**: 402
    Channel **3**: 401–402
    Electric *see* Electric catfish
    European (sheatfish; wels; silurus) **3**: 401–402
    Flathead **3**: 401
    Loricariid **1**: 88
    Mailed **1**: 88–89, *88, 89*
    Sharptooth **3**: *401*
    Talking **1**: 88
    Thorny **1**: 88
Catharacta
    chilensis **10**: 1307
    maccormicki **10**: 1307–1308
    skua **10**: 1307
    skua antarctica **10**: *1308*
Catharsius **16**: 2254
Cathartes **4**: 518; **20**: 2841
    aura **4**: 518; **20**: 2788, 2840
    burrovianus **4**: 518; **20**: 2787
    melambrotus **4**: 518; **20**: 2787
Catharus
    fruscescens **19**: 2678
    guttatus **19**: 2678
    minimus **19**: 2678
Cats 3: 403–406
Cat shark **3**: 407–408; **17**: 2333
    Variegated **3**: *408*
Cat-squirrel *see* Ringtail
Cattle
    Brahman *see* Zebu
    Western **21**: 3018
Causus
    defilipii **13**: 1735
    lichtensteinii **13**: 1735
    resimus **13**: 1735
    rhombeatus **13**: 1735, 1736
cave-dwellers
    anole lizard **1**: 53
    cave fish **3**: 409–410
    crayfish **5**: 591

lungless salamander **11**: 1504
olm **13**: 1788
pedipalpi **14**: 1907
swiftlet **19**: 2601, 2602
Cave eel, Blind **18**: 2585
Cave fish **3**: 409–410
    African **3**: 410
    Iraqi **3**: 410
    Mexican (Mexican cave characin) **3**: 409, 410
Cavia **8**: 1100
    porcellus **8**: 1099, 1100
caviar **18**: 2556
Cavy **8**: 1099, 1100
    Common **8**: 1100, *1101*
    Mountain **8**: 1099, 1100
    Rock (moco) **8**: 1099, 1100, 1101
    Spotted *see* Paca
    Yellow toothed (cui) **8**: 1099, 1100, *1100*
Cebuella pygmaea **11**: 1563, 1564
Cebus
    albifrons **3**: 379, 380
    apella **3**: 379, 380
    capucinus **3**: 379, 380, 381
    olivaceous **3**: 379, 380
Celastrina argiolis **2**: 243
Celeus brachyurus **21**: 2979
cellulose, digestion **1**: 69; **4**: 547; **9**: 1254; **11**: 1483; **12**: 1644; **17**: 2402; **19**: 2727; **21**: 2952
cement glands, bowfin **2**: 273
cementum **1**: 17
Centipede **3**: 411–413
Centriscus **17**: 2367
    cristatus **17**: 2367, 2368
    scutatus **17**: 2368
Centrocercus urophasianus **8**: 1084
Centropomus **13**: 1752
    parallelus **17**: 2430
Centropus **4**: 549, 550
Centruroides
    gertschi **16**: 2257
    sculpturatus **16**: 2257
Cepaea hortensis **7**: 930
Cephalopholis **17**: 1083
Cephalophus **1**: 61; **6**: 726, *727*
    adersi **6**: 726, 727
    jentinki **6**: 726, 727
    monticola **6**: 726
    natalensis **6**: 727, 728
    siviicultor **6**: 726
    zebra **6**: 726, 727
Cephalopterus **4**: 543
Cephalorhynchus **5**: 698
Cephaloscyllium ventriosum **3**: 407, 408
cephalothorax **18**: 2481
Cephea cephea **10**: *1321*
Cephus pygmaeus **16**: 2241
Cepphus grylle **8**: 1094
Cerastes cerastes **17**: *2419*
Cerastoderma edule **4**: 488, 489
Ceratias holboelli **5**: 653, 654
Ceratophrys **9**: 1229
    calcarata **9**: *1230*
    ornata **9**: 1229, 1230
    varia **9**: 1230
Ceratophyllus
    columbae **6**: 850
    gallinae **6**: 850
    styx **6**: 850
Ceratotherium simum **15**: 2152, 2153
cercariae **7**: 870, 871
Cercartetus
    caudatus **5**: 703
    concinnus **5**: 703
    lepidus **5**: 702, 703
    nanus **5**: *703*
cerci **8**: 1035; **18**: 2549
Cercocebus **11**: 1541

agilis **11**: 1541, 1542
galeritus **11**: 1541, 1542
torquatus **11**: 1541, 1542
Cercopithecus **8**: 1091, 1092; **12**: 1643; **14**: 1897
Cerdocyon thous **4**: 567; **17**: *2357*
cere **14**: 1883
Ceresa bubalis **9**: 1225
Certhia
    americana **19**: 2721
    brachydactyla **19**: 2721
    familiaris **19**: 2721, 2722
Certhidea **5**: 649
    olivacea **5**: 648
Cerura vinula **2**: 284, 285; **3**: 345; **15**: 2076, 2077
Cervus **2**: 284, 285; **5**: 656, *657*, 658; **10**: 1355; **17**: 2375; **19**: 2653; **20**: 2854, 2855; **21**: 2907, *2909*
Cetonia **4**: 487
    aurata **4**: 487
Cetorhinus maximus **2**: 161; **17**: 2331
Chachalaca, Plains **20**: 2757
Chad, Lake **10**: 1397; **16**: 2176
Chaenocephalus **9**: 1278
    aceratus **9**: 1277, 1278
Chaenodraco **9**: 1278
    wilsoni **9**: 1277
Chaeradodus rhombicollis **11**: *1546*
Chaetophractus villosus **1**: 86
Chaetopterus **20**: 2767
    variopedatus **20**: 2767, *2767*, 2768
Chaetura **18**: 2484
    boehmi **18**: 2484
    gigantea **18**: 2484
    pelagica **18**: 2484, 2485; **19**: 2598
    vauxi **18**: 2484; **19**: 2598
Chafer **4**: 486, 487, *487*
    Garden **4**: 487
    Rose **4**: 487
Chaffinch **3**: 414–415
Chaimarrornis leucocephalus **15**: 2139
Chalcides ocellatus **17**: 2390
Chalcolepidius zonatus **4**: 466
Chalcopsitta **11**: 1493
Chama **16**: 2248
Chamaea fasciata **1**: 113; **3**: 426
Chamaelo
    chamaeleon **3**: 418, 419
    dilepis **3**: 418, 419
    jacksoni **3**: 418, 419
    oweni **3**: 419
Chameleon **3**: 418–420; **11**: 1467
    American *see* Anole lizard, Green
    Common **3**: 418, 419
    Flap-necked **3**: 418, 419
    Jackson's **3**: 418, 419, *419*
    Setaro's dwarf **3**: 419
    Smith's dwarf **3**: 419
    Stump-tailed **3**: 418
    Three-horned **3**: 419
Chamois **3**: 421–422; **19**: 2623
Champsocephalus **9**: 1278
    esox **9**: 1277, 1278
Chanda
    buruensis **7**: 979
    nama **7**: 979
Channa **17**: 2412
    gachua **17**: 2412
    marulius **17**: 2412, 2413
Channichthys **9**: 1278
Chaoborus **12**: 1606
Chaos carolinensis **1**: 41, 42
Chaparral 3: 423–426
Chaparral cock *see* Roadrunner, Greater
Characin, Mexican cave *see* Cave fish, Mexican
Charadrius
    alexandrinus **14**: 1989

Charadrius (continued)
  dubius **14**: 1988
  hiaticula **14**: 1988
  montanus **14**: 1989
  morinellus **5**: 705
  semipalmatus **14**: 1988
  vociferus **10**: 1348–1349
Charina bottae **2**: 249
charming, by ermine **6**: 785
Charonia
  rubicunda **21**: 2917
  tritonis **18**: 2518
Chat **18**: 2458; **19**: 2678
  Yellow-breasted **21**: 2985
Chauliodus **10**: 1413; **20**: 2829
  barbatus **20**: 2829, *2829*
  danae **20**: 2829
  macouni **20**: 2829
  minimus **20**: 2829
  pammelas **20**: 2829
  schmidti **20**: 2829
  sloani **20**: 2829, 2830
  vasnetzovi **20**: 2829
Chauna
  chavaria **16**: 2262, 2263
  torquata **16**: 2262
Cheetah **3**: 403, 406, 427–429; **8**: 1040
Cheilinus undulatus **21**: *2991*
Cheilopogon heterurus **7**: 876
Cheirogaleus **10**: 1436
Cheiromeles
  parvidens **8**: 1115, 1116
  torquata **8**: 1115, 1116
chelicerae **14**: 1906, 1907; **17**: 2310;
  18: 2453
Chelidonichthys lucerna **8**: 1108, 1109
Chelidoptera tenebrosa **15**: 2060
Chelifer
  cancroides **6**: 799
  lattleillei **6**: *800*
Chelmon rostratus **1**: 49
Chelodina
  expansa **17**: 2417
  longicollis **17**: *2416*, 2417
Chelonia mydas **8**: 1071; **20**: 2798
Chelon labrosus **12**: 1692, 1693
Chelostoma **2**: 183
Chelydra serpentina **12**: 1688; **17**: 2425,
  2426
chemical defense
  fire-bellied toad **6**: 825, *826*
  pedipalpi **14**: *1907*
  puss moth larvae **15**: 2077
  shieldbug **17**: 2344, 2345
  spadefoot toad **18**: 2465
  wood ant **21**: 2965
  see also smell (for defense)
chemosynthesis **13**: 1773
Chennai *see* Dhole
Chenonetta jubata **21**: 2970, 2971
Cherogil (daman) **9**: 1270
Chevrotain **18**: 2572
Chiasmodon niger **13**: *1773*
Chickadee *see* Tit
Chickaree **4**: 436–438
Chicken, Prairie *see* Prairie chicken
Chiffchaff **20**: 2857
  Iberian **20**: 2859
Chiggers **12**: 1621
Chimaera, Plow-nosed **17**: *2330*
Chimpanzee **1**: 70, 71, 72, 73;
  4: 439–442
  Common **4**: 439, 440, 442
  Pygmy *see* Bonobo
Chinactis **4**: 535
Chinchilla **12**: 1672
  brevicaudata **4**: 443–444
  laniger **4**: 443, 444
Chinchilla **4**: 443–445; **12**: 1672
  Long-tailed **4**: 444

  Mountain *see* Viscacha, Mountain
  Short-tailed **4**: 444
Chinook *see* Salmon, Chinook
Chionaspis pinifoliae **16**: 2246, 2247
Chionis
  alba **17**: 2337, 2338
  minor **17**: 2337
Chionobathyscus **9**: 1278
Chionodraco **9**: 1278
  rastrospinosus **9**: 1277
Chipmunk **4**: 448–450
  Eastern **4**: 448, *448*, 449, 450;
    19: 2653
  Least **4**: 449
  Siberian **4**: 448, 449
  Townsend's **4**: 449, *449*
  Western **4**: 448, 450
Chirocephalus diaphanus **6**: 792, 793
Chironectes minimus **13**: 1792, 1793
Chironex fleckeri **10**: 1322;
  20: 2818–2819
Chironomus **12**: 1607; **20**: 2769
Chiroptes **16**: 2221
  albinasus **16**: 2221, 2222
  satanas **16**: 2221, 2222
Chiroteuthis veranyi **18**: 2509
Chiroxiphia
  linearis **11**: 1533
  pareola **11**: 1532
Chiru (Tibetan antelope) **16**: 2217,
  2218, *2218*
Chital *see* Deer, Axis
Chitala **6**: 803, 804
  chitala **6**: 803, *803*, 804
  lopis **6**: 803, 804
Chitra indica **17**: 2442
Chlamydera
  maculata **2**: 271
  nuchalis **2**: 269
Chlamydochaera jefferyi **5**: 621
Chlamydosaurus kingii **7**: 895
Chlamydoselachus anguineus **7**: 897;
  17: 2330
Chlamydotis undulata **3**: 339
Chlamyphorus
  retusus **1**: 87
  truncatus **1**: 85
Chlamys **16**: 2248
  asperrimus **16**: *2249*
  spercularis **16**: *2248*
Chlidonias
  hybridus **19**: 2661
  leucopterus **19**: 2661
  niger **19**: 2661
Chloebia gouldiae **8**: 1031
Chloephaga **5**: 714, 716
Chlorocebus **8**: 1091, 1092
  pygerythrus **8**: 1091; **12**: 1643
  sabaeus **8**: 1091; **12**: 1643
Chloroceryle
  amazona **10**: 1355
  americana **10**: 1355
Chlorocharis emiliae **21**: 2930
Chlorohydra viridissima **9**: 1265, 1266
Chlorophanes spiza **9**: 1213, 1214
Chlorophonia **19**: 2627
Chlorophonia **19**: 2627
chloroplastids **16**: 2301
Chlorotalpa **7**: 1005
Chlorurus microrhinos **14**: 1889
choanocytes **15**: 2074
Choeropsis liberiensis **9**: 1192
Choloepus **17**: 2400, 2402
  didactylus **17**: 2401
  hoffmani **17**: 2401
Chondestes grammacus **18**: 2460
Chondropython **15**: 2080
  viridis **15**: 2079, 2080
Chorthippus **8**: 1036
Chordeiles minor **14**: 2013

  brunneus **8**: 1037
  parallelus **8**: 1037
Chortoicetes terminifera **11**: 1479
Chough **4**: 451–452
  Alpine (yellow-billed chough)
    4: 451; **12**: 1673
  Red-billed (common chough;
    Cornish chough) **4**: 451, *451*, 452,
    452; **12**: 1673
  White-winged **4**: 451; **11**: 1522
Chousingha (four-horned antelope)
  1: 60; **13**: 1753, 1754
chromatophores **5**: *631*, 632; **13**: 1780,
  1795; **17**: 2399; **18**: 2508
Chromis **17**: 2379
  cyanea **5**: 639
Chromis **17**: *2379*
  Blue **5**: 639
Chromodoris
  bullocki **16**: *2300*
  quadricolor **16**: *2301*
chromosomes **7**: 910
Chrotogale owstoni **13**: 1855
Chrysaora **10**: 1320
  hyoscella **10**: *1320*
Chrysemys
  picta **13**: 1851, 1852
  picta belli **13**: 1851, 1852
  picta dorsalis **13**: 1851, 1852
  picta marginata **13**: 1851, 1852
  picta picta **13**: 1851, 1852
Chrysiridia **20**: 2805
  ripharea **20**: 2805, 2806
Chrysobatrachus **15**: 2146
Chrysochloris **7**: 1005
Chrysococcyx
  americanus **5**: *617*
  caprius **5**: 617; **21**: 2898
  lucidus **5**: 617–618
Chrysocyon **5**: 690
  brachyurus **11**: 1540
Chrysolophus
  amherstiae **14**: 1935
  pictus **14**: 1935
Chrysopa carnea **10**: 1392
Chrysopalax **7**: 1005
  trevelyani **7**: 1004
Chrysopelea
  ornata **19**: 2732, 2733
  paradisi **19**: 2732
Chrysops **9**: 1235, 1236
  caecutiens **9**: *1235*
Chrysotoxum **9**: 1251
  cautum **9**: *1251*
Chub **12**: 1618
  Common (European chub)
    12: 1618; **16**: 2182
  Flathead **12**: 1618
Chuck (*lizard*) *see* Chuckwalla
Chuck (*mammal*) *see* Groundhog
Chuckwalla (chuck) **4**: 453–454
  San Esteban Island **4**: 454, *454*
Chuck-will's-widow **13**: 1744;
  14: 2012
Chum *see* Salmon, Chum
Chuna burmeisteri **17**: 2321
Chunkhead *see* Copperhead
Churchill, Winston **1**: 137
Churn owl *see* Nightjar, Eurasian
Cicada **4**: 455–457
  Malaysian empress **4**: 455, 456
  Periodic (17-year cicada) **4**: 456,
    457
Cicada bird, Black-bibbed **5**: 622
Cicadetta **4**: 456
Cichlasoma
  citrinellum **4**: *458*
  meeki **6**: 830; **10**: 1303
  nigrofasciatum **21**: 3015
  octofasciatum **10**: 1303

Cichlid **4**: 458–459; **10**: 1398, *1400*
  Jewel *see* Jewel fish
  Pike **4**: 459
  Red devil **4**: *458*
  Rio Grande **4**: *459*
  Zebra (convict cichlid) **21**: 3015
Cicindela hybrida **2**: *186*
Cicinnarus magnificus **2**: 206
Ciconia **18**: 2551
  ciconia **18**: 2551, 2552
  maguari **18**: 2551
  nigra **18**: 2551
ciguatera **2**: 155; **12**: 1659;
  17: 2424
Cinclocerthia
  gutturalis **19**: 2671
  ruficauda **19**: 2671
Cinclodes **13**: 1819
Cinclodes **13**: 1819, 1820
Cinclus
  cinclus **5**: 680
  leucocephalus **5**: 680
  mexicanus **5**: 680, 681
  pallasii **5**: 680
  shulzi **5**: 680
Ciona intestinalis **17**: 2313
Cionus scrophulariae **21**: 2903, 2904
Circaetus cinereus **12**: 1589
Circus
  aeruginosus **8**: 1145
  cyaneus **8**: 1146
  macrourus **8**: 1145
  melanoleucus **8**: 1145
  pygargus **8**: 1145
  ranivorus **8**: 1145
cirri **1**: 138, *138*; **16**: 2280
Cirrothauma **13**: 1779
Cirrothauma **13**: 1779
Cissa
  chinensis **11**: 1519
  ornata **11**: 1519–1520
Cissopis leveriana **19**: 2627
Cisticola **19**: 2618
  exilis **19**: 2618
Cisticola, Golden-headed **19**: 2618,
  2618
Cistothorus
  palustris **21**: 2995
  platensis **21**: 2994
Civet **4**: 460–461
  African **4**: 460, *460*, 461
  Indian (oriental civet) **4**: 460, 461
  Malabar large-spotted **4**: 461
  Malagasy (fanaloka) **4**: 460, 461
  Otter **4**: 460, 461
  Palm *see* Palm civet
  Two-spotted *see* Palm civet,
    African
Civettictis civetta **4**: 460, 461
Cladorhynchus leucocephalus **18**: 2540
Clam **4**: 462–463
  Giant **4**: 462–463, *462*, *463*
  Hard (quahog) **4**: 462
  High-neck *see* Gaper, North
    American
  Horse *see* Gaper, North American
  Jackknife *see* Razor shell, Pod
  Otter-shell *see* Gaper, North
    American
  Rubberneck *see* Gaper, North
    American
  Soft-shelled *see* Gaper, Sand
  Summer *see* Gaper, North
    American
  Washington *see* Gaper, North
    American
Clamator
  coromandus **5**: 617
  glandarius **5**: 617
Clam-cracker *see* Stingray, Southern

Clam worm *see* Ragworm
Clangula hyemalis **5**: 717
Clarias gariepinus **3**: 401
Clavelina lepadiformis **17**: 2313
Cleg *see* Horsefly
Cleidopus gloriamaris **18**: 2512
Clelia clelia **6**: 814
Cleopatra (*queen of Egypt*) **1**: 98
Clethrionomys **15**: 2129
   californicus **1**: 134; **15**: 2129
   centralis **15**: 2129
   gapperi **1**: 134; **15**: 2129, 2130
   glareolus **1**: 134–135; **15**: 2129
   rufocanus **15**: 2129
   rutilus **1**: 134; **15**: 2129
Clevelandia ios **7**: 995
Click beetle (skipjack; *larva:* wireworm) **4**: 466–467
   Elegant **4**: 466
   Harlequin **4**: 466
Cliff-springer *see* Klipspringer
Climacteris **19**: 2721
Climbing fish *see* Perch, Climbing
Clinocardium fucanum **4**: 488
Clione **16**: 2269, 2270
   limacina **16**: 2269, 2270, *2270*
Clitarchus hookeri **18**: 2530, 2531
clitellum **6**: 735, *736*; **20**: 2769
cloaca **3**: 420
Clock bird *see* Motmot, Blue-crowned
Cloeon **12**: 1593
Clonorchis **7**: 871
   sinensis **7**: 871
Clossiana
   chariclea **7**: 898
   euphrosyne **7**: 898
   selene **7**: 898, *899*
Clothes moth **4**: 470–471
   Casemaking **4**: 470, 471
   Common (webbing clothes moth; miller) **4**: 470, 471
Clouded yellow **21**: 2929
Clownfish (anemonefish) **4**: 533; **5**: 638, 639, *640*
clumps
   avadavat **1**: 103
   babbler **1**: 113–114
Clupea
   harangus **9**: 1185
   pallasii **9**: 1184
Clypeaster rosaceus **16**: 2226
Clytoceyx rex **10**: 1357
Clytus **11**: 1484
   arietis **11**: 1484
Cnemidophorus
   neomexicanus **19**: 2650
   sexlineatus sexlineatus **19**: 2650
   sexlineatus viridis **19**: 2650
   tigris **19**: 2650
Cnemophilus macgregorii **2**: *206*
Coachwhip (prairie runner; red racer) **21**: 2919–2920
Coalfish (saithe; coley; billet; glassan; black pollack) **4**: 472–473, 495
Coast 4: 474–477
Coati **4**: 478–479
   Island **4**: 479
   Mountain **4**: 479
   Ring-tailed (coatimundi) **4**: 479
   White-nosed (coatimundi) **4**: 480
Coatimundi *see* Coati, Ring-tailed; Coati, White-nosed
Cobitis taenia **11**: 1473
Cobra **2**: 265; **4**: 480–482; **17**: 2418; **20**: 2817
   Cape **4**: 480, 481, *481*, 482
   Egyptian **4**: 480, 481
   Forest **4**: 480, 481

Indian **4**: 480, *480*, 481–482
   King **17**: 2420
   Mozambique spitting **4**: 480, 481, *482*
   Spitting **4**: 480, 481; **20**: 2818
Coccidula rufa **10**: 1394
Coccinella septempunctata **10**: 1394
cochineal **16**: 2247, *2247*
Cochineal insect **16**: 2247, *2247*
Cochlearius cochlearius **2**: 252; **8**: 1123; **9**: 1179
Cockatiel **4**: 483, 485
   Moluccan **4**: *484*
Cockatoo **4**: 483–485
   Gang-gang **4**: 483
   Palm (great palm cockatoo; great black cockatoo; goliath cockatoo; Cape York cockatoo) **4**: 483, *485*
   Philippine **4**: 485
   Pink **4**: *483*
   Rose-breasted *see* Galah
   Sulfur-crested (white cockatoo) **4**: 483, 485
   White-tailed black **4**: 485
   Yellow-crested **4**: 485
   Yellow-tailed black **4**: 485
Cockchafer (May beetle; May bug; June beetle; June bug; common cockchafer; green fig-eater) **4**: 486–487; **16**: 2252
Cockle **4**: 488–489
   Common (edible cockle) **4**: 488, 489
   Common Pacific egg **4**: 488
   Fucan **4**: 488
   Giant Pacific **4**: 488
   Jumping **4**: 488
   Prickly **4**: 488
   Smooth giant **4**: 488
   Yellow **4**: 488
Cock-of-the-rock **4**: 490–491
   Andean **4**: 490, 491, 543
   Guianan **4**: 490, 491, 543
Cockroach **4**: 492–494
   American **4**: *492*, 493, 494
   Common (black beetle) **4**: 492–493, *494*
   German (steamfly; shiner) **4**: 493, 494
   Giant Madagascar hissing **4**: 493
Cod (Atlantic cod; codling; codfish) **4**: 472, 495–496; **6**: 836
   Arctic (polar cod) **4**: 495
Codfish *see* Cod
Codling *see* Cod
Coelacanth (kombessa) **4**: 497–498
Coendou **19**: 2729
   prehensilis **19**: 2728, *2729*
coenenchyme **17**: *2441*
Coenobita **9**: 1176
Coenocorypha **17**: 2427
   pusilla **17**: 2429
Cofferfish *see* Trunkfish
Coho *see* Salmon, Coho
Colaptes **6**: 853
   auratus **6**: 853, 854
   campestris **6**: 853
   chrysoides **6**: 853
   fernandinae **6**: 853
   pitius **6**: 853
Coleonyx
   brevis **7**: 943
   variegatus **7**: 943
Coley *see* Coalfish
Colias
   croceus **21**: 2929
   eurydice **21**: 2929
   eurytheme **21**: 2929

Colibri **9**: 1257, *1257*
Colinus virginianus **15**: 2082
Colius
   colius **12**: 1676, 1677
   indicus **12**: 1676
   leucocephalus **12**: 1676
   macrourus **12**: 1676
   striatus **12**: 1676
collecting
   butterfly **2**: 212; **4**: 526
   topshells **19**: 2707
   see also feathers, collected by man
Collessendeis colossea **17**: 2311, *2311*
Collocalia
   esculenta **19**: 2601, 2602
   fuciphaga **19**: 2601
   maxima **19**: 2601
   vestita **19**: 2601
Colluricincla
   rectirostris **21**: 2923
   tenebrosa **21**: 2923
Colobus **4**: 501, 502
   angolensis **4**: 501, 502
   guereza **4**: 501, 502
   satanus **4**: 501, 502
Colobus monkey **4**: 501–503; **12**: 1644
   Angola pied **4**: 501, 502
   Black **4**: 501, 502, 503
   Black and white (pied colobus) **4**: 502, 503
   Central African red **4**: 502
   Guereza **4**: 501, 502, *502*, *503*
   Olive **4**: 501, 502, 503
   Red **4**: 501, *501*, 502, 503
   Tana River red **4**: 503
   Western red **4**: 502
   Zanzibar red **4**: 503
colonies *see* living in complex societies; living in groups
color
   countershading **13**: 1771; **19**: *2605*
   cryptic **20**: 2812
   dichromatism **7**: 900
   flash **1**: 93; **7**: 880; **10**: 1332; **13**: 1821; **18**: 2531; **19**: *2687*, 2688, 2725
   structural coloration **20**: 2806; **21**: 2929
   warning *see* warning coloration
   see also attracting a mate with color; camouflage; color change; warning coloration
color change **6**: 860
   anole lizard **1**: 53, *53*
   Atlantic salmon **1**: 99, 100
   bream **2**: *274*
   chameleon **1**: 54; **3**: 420
   clawed frog **4**: 464
   common toad **4**: 512
   conger eel **4**: 521, *522*
   cuttlefish **5**: *631*, 632, *632*, 633
   eel **6**: 745
   ermine **6**: 787, *787*
   fiddler crab **6**: 815
   gliding frog **7**: 982
   goatfish **7**: 989, 990
   grouper **8**: *1082*, 1083
   halibut **8**: 1121
   hare **8**: *1137*, 1138
   heron **9**: 1181
   leaf fish **10**: 1422
   lumpsucker **11**: *1500*
   moloch **12**: 1630
   ocean sunfish **13**: 1774
   octopus **13**: 1780, 1781
   olm **13**: 1789
   Pacific salmon **13**: 1840, *1841*
   parrotfish **14**: 1888
   plaice **14**: *1975*
   rabbitfish **15**: 2094

slater **17**: 2399
   spider crab **18**: 2476
   squid **18**: 2508, 2510
   stick insect **18**: 2532
   see also attracting a mate with color; camouflage; threat displays
Colostethus **1**: 94
colostrum **21**: 2948
Coluber **15**: 2421
   constrictor **15**: 2101, 2102
   constrictor flaviventris **15**: 2102
   viridiflavus **21**: 2919, 2920
Columba
   livia **6**: 850; **16**: 2195
   palumbus **21**: 2981
   palumbus palumbus **21**: 2980
Columbus, Christopher **11**: 1535
combat *see* display (territorial); fighting for mates; fighting for territory; infanticide; sexual aggression; siblicide
Comma **20**: 2811–2812
   Eastern **20**: 2812
   Gray **20**: 2812
   Green **20**: 2812
   Hoary **20**: 2812
   Oreas **20**: 2812
   Satyr **20**: 2811
commensalism
   elephant shrew **6**: 776
   goby **7**: 995
   pea crab **4**: 573
   pearlfish **14**: 1901
commercially important
   abalone **1**: 21
   alligator **1**: 38, 40
   basking shark **2**: 162
   bird of paradise **2**: 207
   bream **2**: 275
   carp **3**: 395–396
   cod **4**: 496
   elephant seal **6**: 773
   gray seal **8**: 1050
   hake **8**: 1120
   halibut **8**: 1122
   herring **9**: 1185–1186
   king penguin **10**: 1360
   ling **11**: 1455
   mackerel **11**: 1518
   menhaden **12**: 1598
   opossum **13**: 1794
   oyster **13**: 1834
   Pacific salmon **13**: 1842
   pilchard **14**: 1949
   prawn **15**: 2044
   scad **16**: *2242*, 2243
   silk moth **17**: 2376–2378, *2376*, *2378*
   skate **17**: 2385
   snoek **17**: 2431
   sweetfish **19**: 2596
   swordfish **19**: 2607
   tuna **20**: 2775, *2775*
   turbot **20**: 2783
   wild goat **21**: 2953
   yak **21**: 3002
communal living *see* living in complex societies; living in groups
Compsilura concinnata **19**: 2611
Conch (fighting conch; trumpet shell; stromb shell) **4**: 514–515; **18**: 2547
   Horse **4**: 514
   Queen **4**: 514, 515, *515*
Condor **4**: 516–518
   Andean **4**: 516, *516*, *517*; **12**: 1673; **20**: 2841
   California **4**: 516, 517, 518, *518*; **20**: 2841

Condylura cristata **12**: 1625
Conepatus
　chinga **17**: 2393–2394
　humboldti **17**: 2393–2394
　leuconotus **17**: 2393–2394
　mesoleucus **17**: 2393–2394
　mesoleucus telmalestes **17**: 2394
　semistriatus **17**: 2393–2394
Cone shell 4: 519–520; **20**: 2818
　Elephant trunk **4**: *519*
　Glory of the sea **4**: 520
　Legate **4**: *520*
　Mediterranean **4**: 520
Coney *see* Hyrax
Conger
　conger **4**: 521, 522
　verrouxi **4**: *521*
conifers **19**: 2615, 2617
Conilurus albipes **18**: 2536
conjugation **14**: 1882, *1882*
Connochaetes
　gnou **16**: 2216; **21**: 2948, 2949
　taurinus **8**: 1040; **21**: 2948, 2949
Conolophus **9**: 1281, 1282
Conophilia
　alisogularis **9**: 1216
　rufogularis **9**: 1216
conquistadors, Spanish **3**: 364
conservation, rain forests **15**: 2114
continental shelves and slopes
　13: 1769
contractile vacuoles **1**: 41; **14**: 1881
Conure **14**: 1878
Conuropsis carolinensis **14**: 1880
Conus **20**: 2818
　gloriamaris **4**: 520
　legatus **4**: *520*
　mediterraneus **4**: 520
　striatus **4**: *519*
Convictfish **18**: 2576
Convoluta **6**: 849
　paradoxa **6**: 849
　roscoffensis **6**: 849
Cook, James **7**: 918
Coon cat *see* Ringtail
cooperation *see* helping; hunting in
　packs/groups
Coot **4**: 523–524
　American **4**: 523, *523*, 524
　Crested (knob-billed coot) **4**: 523
　European (common coot) **4**: 523,
　　524, *524*
　Giant **4**: 523, 524
　Hawaiian **4**: 523
　Horned **4**: 523
Cooter **13**: 1851; **20**: 2796
Copepod **14**: *1981*, 1982
Copperbelly *see* Water snake, Red-
　bellied
Copper butterfly **4**: 525–526
　Large **4**: *525*, 526
　Purple-edged **4**: 526
　Scarce **4**: 526
　Small **4**: 525–526, *526*
Copperhead (*cobra*), Australian
　7: 935
Copperhead (*pit viper*), (chunkhead;
　highland moccasin) **14**: 1972,
　1973
Copris hispanus **16**: 2253–2254
coprophagy (fecal pellet digestion)
　4: 547; **6**: 783; **14**: 1943; **15**: 2092
Coquille St. Jacques *see* Scallop
Coracias **16**: 2204
　abyssinicus **16**: 2204
　benghalensis **16**: *2205*
　caudata **16**: 2205
　garrulus **16**: 2204
Coracina
　lineata **5**: 621

　mindanensis **5**: 622
　newtoni **5**: 622
　novaehollandiae **5**: 621
　ostenta **5**: 622
　pectoralis **5**: 622
Coragyps atratus **20**: 2787, 2840
Coral **4**: 527–529
　Black **4**: 527
　Blue **4**: 527
　Elephant's ear **17**: 2441
　Fire **4**: 527
　Horny **4**: 527
　Soft **4**: 527; **17**: 2440–2441
　Stony (true coral) **4**: 527, *528*
　Thorny **4**: 527
Corallus caninus **17**: *2418*
Coral reef 4: 530–533, 531, 533;
　18: 2518; **21**: 2905, 2907
　fish **4**: 532; **14**: 1889, *1889*; **17**: 2423,
　　2423, *2424*
Coral snake **4**: 534–535
　Arizona (western coral snake;
　　Sonoran coral snake) **4**: 534, *534*,
　　535
　Common (eastern coral snake;
　　Texas coral snake) **4**: 534, 535
Corcorax melanorhamphos **4**: 451;
　11: 1522
Cordon-bleu (blue waxbill) **21**: 2888,
　2889
Cordulegaster boltoni **5**: 711
Cordylophora **16**: 2273
Corella, Little **4**: 483, 485
Coris julis **21**: 2991
Corkwing (sea pheasant) **21**: 2991
Cormorant **4**: 536–538; **16**: 2265, 2268
　Brandt's **4**: 536
　Cape **4**: 538
　Double-crested (white-crested
　　cormorant) **4**: 536, 537
　Flightless **4**: 536, 537, *537*, 538
　Great (common cormorant; black
　　cormorant) **4**: 536, *536*, 537
　Guanay **4**: 538
　King **4**: *538*
　Neotropic (olivaceous cormorant)
　　4: 536
　Pelagic **4**: 536
　Red-faced **4**: 536
Corn borer (European corn borer;
　maize moth) **4**: 539–540
Corncrake (land rail) **4**: 541–542
Cornetfish **20**: 2766
Coronula **1**: 138
Corucea zebrata **17**: 2390
Corvus **5**: 610
　albicollis **15**: 2122
　albus **5**: *611*
　brachyrhynchos **5**: 610, 611
　caurinus **5**: 610
　corax **15**: 2122, 2123; **18**: 2455
　corone **4**: 452; **5**: 610
　coronoides **15**: 2122
　crassirostris **15**: 2122
　cryptoleucus **15**: 2122
　dauuricus **10**: 1300
　frugilegus **4**: 452; **5**: 610; **10**: 1300;
　　16: 2207
　mellori **15**: 2122
　monedula **10**: 1300, 1301; **18**: 2457
　ossifragus **5**: 610
　rhipidurus **15**: 2122
　splendens **5**: *610*; **18**: 2457
　tasmanicus **15**: 2122
Corydon sumatranus **2**: 283
Corydoras armatus **1**: *88*
Corynorhinus townsendii **20**: 2821
Coryphaena
　equiselis **5**: 693, 694
　hippurus **5**: 693, 694

Coryphaeshna ingens **5**: 710
Corythaeola cristata **20**: 2780
Corythoichthys flavofasciatus **14**: 1960
Coscinoscera hercules **3**: 344
Coscoroba coscoroba **18**: 2586; **21**: 2925
Cosmopolitan *see* Butterfly, Thistle
Cossus cossus **7**: 992
Cossypha natalensis **6**: 776
coteries **15**: 2038, *2040*
Cotinga
　amabilis **4**: *543*
　maculata **4**: 544
Cotinga **4**: 543–544
　Banded **4**: 544
　Blue **4**: *543*
　Snowy (white bellbird) **4**: 543
　Swallow-tailed **4**: *544*
Cotinis **4**: 487
　mutabilis **4**: 487
Cottonmouth *see* Moccasin, Water
Cotton spinner **16**: 2272
Cottontail **4**: 545–548
　Desert **4**: 545, 546, *547*
　Dice's **4**: 546
　Eastern **4**: 545, 546
　New England **4**: 545, 546
　Nuttall's **4**: 545, 546, *546*
　Omilteme **4**: 546
　Tres Marias **4**: 546
Cottus gobio **3**: 314–315
Coturnix **15**: 2082
　coturnix **15**: 2082, 2083
Coua **4**: 549; **20**: 2780
Coucal **4**: 549–550
　Black-hooded **4**: 550
　Blue-headed **4**: 549
　Common (crow-pheasant) **4**: 549
　Coppery-tailed **4**: *550*
　Greater black **4**: 549
　Green-billed **4**: 550
　Lesser black **4**: 549
　Pheasant (common coucal; swamp
　　cuckoo) **4**: 549–550, 551
　Senegal **4**: 549, *549*
　White-browed **4**: 549, 550
Cougar *see* Puma
countershading **13**: 1771;
　19: *2605*
Courser **4**: 551–552
　Burchell's **4**: 551
　Cream-colored **4**: 551, *551*
　Double-banded **4**: *552*
　Indian **4**: 551
　Jerdon's **4**: 551
　Temminck's **4**: 551–552
courtship *see* attracting a mate;
　displaying in groups
Couta *see* Snoek
coveys **14**: *1891*, 1892
Cowbird **4**: 553–555; **18**: 2461
　Bay-winged **4**: 553, 555
　Brown-headed (buffalo bird)
　　3: 400; **4**: 553, 554, *554*, 555;
　　15: 2141, 2143; **18**: 2459
　Screaming **4**: 553
　Shiny **4**: 553, 555, *555*
Cowfish **20**: 2758
　Scrawled **20**: 2759
Cow-killer **20**: 2815
Cowrie **4**: 556–557
　Chestnut **4**: 556–557
　European **4**: 556, 557
　Mole **4**: *556*
　Money **4**: 557
Coyote **3**: 425, 426; **4**: 558–560
Coypu **4**: 561–563
Crab
　Arrow **18**: *2475*
　Blue edible **4**: 573
　Decorator *see* Spider crab

　Edible (common crab; great crab)
　　6: 741–742
　Fiddler *see* Fiddler crab
　Ghost **4**: 573
　Giant *see* Spider crab, Giant
　Harlequin **18**: 2476
　Hermit *see* Hermit crab
　Kelp **18**: 2475
　Land (*crayfish*) **5**: 591
　Mole **18**: 2507
　Pea **4**: 573
　Pigmy **4**: 571
　Porcelain **4**: *571*, 572; **18**: 2507
　Red mangrove **4**: *573*
　River **4**: 571
　Robber (coconut crab) **4**: 571, 572,
　　573; **9**: 1178; **16**: 2186–2187
　Sally Lightfoot **4**: *570*
　Sand **18**: 2507
　Shore (green crab) **4**: 572; **17**: 2353–2355
　Spider *see* Spider crab
　Stone **18**: 2507
　Tasmanian **4**: 571
　Thornback *see* Spider crab, Spiny
Crabs and lobsters 4: 570–573
Cracticus
　cassicus **3**: 340
　louisiadensis **3**: 340
　mentalis **3**: 340
　nigrogularis **3**: 340
　quoyi **3**: 340
　torquatus **3**: 340, *340*, 341
Crago franciscorum **17**: 2364
Crake **5**: 582–583
　Baillon's **5**: 582, 583
　Carolina *see* Sora
　Little **5**: 582, 583, *583*
　Ruddy **5**: 582
　Spotted **5**: 582
Crampfish *see* Electric ray
Crane **5**: 584–587
　Black-crowned **5**: 584
　Common (Eurasian crane) **5**: 584,
　　585, *586*
　Red-crowned (Japanese crane)
　　5: 584, *584*, 585–586, 587
　Sandhill **5**: 584, 586, *587*
　Siberian **5**: 585
　Whooping **5**: 584, 585; **20**: 2757
Crane fly (mosquito hawk; bobbing
　gnat; daddy longlegs; *larva*:
　leatherjacket) **5**: 588–590;
　16: 2175
Crangon crangon **17**: 2364, 2365
Crango septemspinosus **17**: 2364
Crania **13**: 1406
Cranioleuca **13**: 1819
Craseonycteris thonglongyai **2**: 167
Crassotrea **13**: 1832, 1834
　angulata **13**: 1832
　virginica **13**: 1834
Crateroscelis murina **1**: *113*
Cratogeomys **14**: 1997
Crawdad *see* Crayfish
Crawfish *see* Crayfish; Spiny lobster
Crax
　alberti **5**: 625
　alector **5**: 623
　blumenbachii **5**: 625
　daubentoni **5**: *623*
　rubra **5**: 623, 624
Crayfish (crawfish; crawdad; yabbie)
　5: 591–592
　Giant **5**: 592
　Noble **5**: 592
　Signal **5**: 592
　Tasmanian giant **5**: 591
　White-clawed **5**: 592; **16**: 2175
Creatophora cinerea **18**: 2521

Creeper (*larva*) **18**: 2550
Creeper, Spotted *see* Treecreeper
    cremaster **21**: 2928
Crematogaster **14**: 1894
Crenicichla **4**: 459
Crepidula fornicata **11**: 1452
Crest, Canary Islands **7**: 999
Crex crex **4**: 542, *542*
Cricetomys gambianus **7**: 964, 965
Cricetus cricetus **8**: 1130
Cricket **5**: 598–600
    Bush *see* Katydid
    Field **5**: *598*, 599
    House **5**: 598–599, *599*
    Mole **5**: 598, 599–600
    Stone **5**: *600*
    Wood **5**: 599, 600
Cricosaura typica **13**: 1745
Croaker (sea trout) **5**: 601–602
    Atlantic **5**: 601, 602
    Meager (redfish) **5**: 601
    Spotfin **5**: 602
    Yellowfin **5**: *601*, 602
    see also Drum
Crocidura **12**: 1710
Crocodile **5**: 603–606; **18**: 2546
    American **5**: 603, *603*, 604; **21**: 2907
    Cuban **5**: *605*
    Dwarf **5**: 603
    Indo-Pacific (saltwater crocodile;
        estuarine crocodile) **5**: 603, *604*;
        21: 2907
    Marsh (mugger) **5**: 603, 606
    Nile **5**: 603, 604, 606, *606*
    Orinoco **5**: 603
Crocodile bird *see* Plover, Egyptian
Crocodylus
    acutus **5**: 603, 604; **21**: 2907
    intermedius **5**: 603
    niloticus **5**: 603, 604
    palustris **5**: 603
    porosus **5**: 603; **21**: 2907
    rhombifer **5**: *605*
Crocuta **19**: 2610
    crocuta **3**: 374–375; **9**: 1267, 1268
crops, cereal **8**: 1041
Crossarchus obscurus **12**: 1636
Crossbill **5**: 607–609; **18**: 2457
    Common (red crossbill) **5**: 607,
        608–609, *608*
    Parrot **5**: 607, 608, *609*
    Scottish **5**: 607, 608
    Two-barred **5**: 607, 608
Crossoptilon **14**: 1934–1935
Crotalus **3**: 426, *426*; **15**: 2119, *2119*,
    2120, *2120*; **17**: 2370, 2422
Crow **5**: 610–613; **18**: 2457
    American (Western crow; Florida
        crow) **5**: 610, 611, 612, 613, *613*
    Carrion **4**: 452; **5**: 610, 611, *612*, 613
    Fish **5**: 610, 611–612
    Hooded **5**: 610, 612
    House **5**: *610*; **18**: 2457, 2459
    King *see* Drongo, Black
    Northwestern **5**: 610
    Pied **5**: *611*
    Tamaulipas (Mexican crow) **5**: 610
    Wattled *see* Kokako
Crow-pheasant *see* Coucal, Common
Crying bird *see* Limpkin
Cryodraco **9**: 1278
Cryptobranchus alleganiensis **9**: 1175
Cryptochloris **7**: 1005
Cryptolaemus montrouzieri **10**: 1396
Cryptomys hottentotus **12**: *1629*
Cryptoprocta ferox **17**: 2373
Crypturellus soui **19**: 2691
Ctenocephalides
    canis **6**: 852
    felis **6**: 850, 851

Ctenodactylus
    gundi **8**: 1104, 1105
    vali **8**: 1104, 1105
Ctenodiscus crispatus **18**: 2518
Ctenolepisma
    longicaudata **2**: 279
    quadriseriata **2**: 279
Ctenomys
    lewisi **20**: 2771, 2772
    opimus **20**: 2772
    peruanus **20**: 2772
Ctenopoma **4**: 468, *468*
Cubanichthys cubensis **19**: 2705
Cuchia **18**: 2585
Cuckoo **5**: 617–620; **21**: 2938
    Chestnut-winged **5**: 617
    Common (European cuckoo;
        Eurasian cuckoo) **5**: 617, 618, *618*,
        619, 620; **9**: 1224; **21**: 2898
    Didric **5**: 617; **21**: 2898
    Great spotted **5**: 617, 619
    Indian **5**: 713
    Striped **5**: 619
    Swamp *see* Coucal, Pheasant
    Yellow-billed **5**: *617*
Cuckoo-shrike **5**: 621–622; **21**: 2984
    Barred **5**: 621
    Black-faced (blue jay; shufflewing)
        5: 621, *621*
    Ground **5**: 621
    Red-shouldered (black cuckoo-
        shrike) **5**: 621, 622
    Réunion **5**: 622
    White-breasted **5**: 622
    White-winged **5**: 622
Cuculus
    canorus **5**: 617, 618; **9**: 1224;
        21: 2898
    glandarius **5**: 619
    micropterus **5**: 713
    varius **5**: 620
Cucumeria
    japonica **16**: 2272
    miniata **16**: 2272
    saxicola **16**: 2271
cud **3**: 305
Cui *see* Cavy, Yellow toothed
Culaea inconstans **18**: 2533
Culex **12**: 1662, *1664*
    pipiens **12**: *1662*, 1663
Culicoides **12**: 1607
Cuon alpinus **5**: 672, 677, 690
Cupido minimus **2**: 243
Curaeus curaeus **13**: 1805
curare **20**: 2804
Curassow **5**: 623–625
    Black **5**: 623
    Blue-billed **5**: 625
    Great (Mexican curassow; great
        crested curassow; globose
        curassow) **5**: 623, 624
    Helmeted **5**: 623
    Razor-billed (mutum) **5**: 623, *624*,
        625, *625*
    Red-billed **5**: 625
    Trinidad white-headed **5**: 623
    Yellow-knobbed **5**: *623*
Curculio
    glandium **21**: *2903*
    nucum **21**: 2903, 2904
Curlew **5**: 626–628
    Bristle-thighed **5**: 626, 628
    Eskimo (doughbird) **5**: 628
    Eurasian (common curlew;
        western curlew; European
        curlew) **5**: 626, 627
    Hudsonian (whimbrel) **5**: 626, 627
    Long-billed **5**: 626
    Madagascar **5**: 626
    Slender-billed **5**: 628

    Stone *see* Stone curlew
    White *see* Ibis, White
Cursorius
    coromandelicus **4**: 551
    cursor **4**: 551
    rufus **4**: 551
    temminckii **4**: 551
Cuscus **5**: 629–630
    Black-spotted **5**: 630
    Common **14**: *1930*
    Gray **5**: 630
    Obi **5**: 630
    Silky **5**: 630
    Southern common **5**: 630
    Spotted **5**: 629–630, *629*
    Woodlark Island **5**: 630
Cusimanse **12**: 1636
Cutlassfish **17**: 2415, *2415*
Cutthroat *see* Whistler, Golden
Cuttlefish **5**: 631–633
    Common (common cuttle)
        5: 631–632, *631*
    Harvest **5**: *632*
Cutworm **13**: 1822
Cyanea **10**: 1320; **14**: 1980; **21**: 2935
Cyanerpes
    careuleus **9**: *1214*
    cyaneus **9**: 1213
Cyanocitta
    cristata **10**: 1317
    stelleri **10**: 1317
Cyanocorax yncas **10**: 1318
Cyanopica cyanus **11**: 1520
Cyanopsitta spixii **11**: 1516
Cyclanorbis senegalensis **17**: 2443
Cyclocephala **4**: 487
cyclomorphosis **5**: 644
Cyclopes didactylus **1**: 57
Cyclops **12**: 1595, 1636 → (check)
Cyclopterus lumpus **11**: 1499, 1500
Cyclorana **19**: 2723
    alboguttatus **5**: 665
    platycephala **10**: 1399; **20**: 2869
Cyclothone
    microdon **2**: 278
    pallida **2**: 278
Cyclura **9**: 1281, 1282
Cygnus
    atratus **18**: 2587
    buccinator **5**: 714; **18**: 2587; **20**: 2757
    columbianus **5**: 717; **18**: 2587;
        20: 2779
    columbianus bewickii **18**: 2586
    columbianus columbianus
        18: 2586–2587
    cygnus **5**: 714; **18**: 2586
    melanocoryphus **18**: 2587
    olor **5**: 715; **18**: 2586
Cymatogaster aggregata **18**: 2574
Cymbirhynchus macrorhynchos **2**: *282*
Cynictis pencillata **12**: 1595, 1636
Cynocephalus
    variegatus **7**: 882, 883
    volans **7**: 882, 883
Cynogale bennetti **4**: 460
Cynomys **8**: 1040
    gunnisoni **15**: *2038*
    leucurus **15**: 2039
    ludovicianus **15**: 2038, 2039
    mexicanus **15**: 2040
    parvidens **15**: 2040
Cynopterus
    brachyotis **7**: 907
    sphinx **7**: 907
Cynoscion **5**: 601, 602
    regalis **5**: 602
Cyphonia clavata **9**: 1225
Cypraea
    chinensis **4**: *557*
    europaea **4**: 556, 557
    moneta **4**: 557

    spadicea **4**: 556, 557
    talpa **4**: *556*
Cyprinodon
    macularius **19**: *2705*
    variegatus variegatus **19**: 2705
Cyprinus carpio **3**: 396
cypris larvae **1**: 139
Cypseloides
    niger **19**: 2598
    senex **19**: 2599
Cypselurus **7**: 876
    californicus **7**: 875
    heterurus **7**: 875
Cypsiurus parvus **19**: 2598
Cystophora cristata **9**: 1220
cysts, of amoeba **1**: 42

D

Dab **5**: 634–635
Dace **12**: 1618–1619; **17**: 2411
    Common **12**: 1618
    Southern redbelly **12**: 1618–1619
Dacelo
    gigas **10**: 1355, 1386
    leachii **10**: 1385
Dacnis
    cayana **9**: 1213; **18**: 2558
    venusta **9**: 1213
Dacnis
    Blue **9**: 1213
    Scarlet-thighed **9**: 1213
Dacnonypha **3**: 343
Dactylopius coccus **16**: 2247
Dactyloptena **7**: 880, 881
    orientalis **7**: 880, 881
Dactylopterus **7**: 880, 881
    volitans **7**: 880, 881
Daddy longlegs (*crane fly*) *see* Crane
    fly
Daddy longlegs (harvest spider;
    harvester; harvestman)
    5: 636–637
Daddy longlegs (*spider*) *see* Spider,
    Cellar
Dallia pectoralis **1**: 34
Dama dama **5**: 656; **6**: 795
Damaliscus
    dorcas **2**: 260; **16**: 2216
    lunatus **21**: 2950
    phillipsii **2**: 260; **16**: 2216
Damaliscus dorcas **1**: 63
Daman (cherogil) **9**: 1270
Damon variegatus **14**: *1906*
dams, beaver **2**: 177–178, *178*
Damsel *see* Damselfly
Damselfish (demoiselle) **4**: 532;
    5: 638–640; **6**: 836
    Domino **5**: 638
Damselfly (demoiselle; dancer;
    damsel; bluet; firetail; sprite;
    blue arrow) **5**: 641–642
Danaus plexippus **3**: 345; **12**: 1633;
    20: 2810
Dancer *see* Damselfly
dangerous to humans
    Asiatic black bear **1**: 96
    barracuda **2**: 154, 155
    bear **2**: 176
    black widow **2**: 235
    boomslang **2**: 265
    brown bear **3**: 294
    Cape buffalo **3**: 305, 372
    cobra **4**: 481–482, *481*
    cone shell **4**: 520
    coral snake **4**: 534
    crocodile **5**: 605
    fer de lance **6**: 812
    great white shark **8**: 1058, *1058*,
        1059–1060

dangerous to humans (*continued*)
  grouper **8**: 1082
  hammerhead shark **8**: 1128
  jellyfish **10**: 1322
  leech **10**: 1428
  leopard **11**: 1446
  lion **11**: 1462
  mamba **11**: *1531*
  night adder **13**: 1735
  platypus **14**: 1985
  porbeagle **15**: 2021
  puff adder **15**: 2058
  puma **15**: 2069
  scorpion **16**: 2257, 2259
  sharks **17**: 2332
  sloth bear **17**: 2405
  stingray **18**: 2543–2544
  stonefish **18**: 2547–2548
  tiger **19**: 2685
  wobbegong **21**: 2955
  Zambezi shark **21**: 3007, *3007*
  see also poisons; stinging
Danio, Zebra *see* Zebra fish
Danio rerio **21**: 3015
Daphnia (water flea) **5**: 643–644;
  10: 1399
Daphnia magna **5**: 644
Daphnis neri **9**: 1167
Darter (*bird*) **5**: 645–647
  African **5**: 645
  Australian **5**: 645
  Oriental (Indian darter) **5**: 645
Darter (*fish*) **14**: 1917
Darwin, Charles **4**: 518, 531; **5**: 648;
  9: 1283
  and earthworms **6**: 736, 737
  and robber crabs **16**: 2186
Dascyllus albisella **5**: 638
Dassie *see* Hyrax, Cape
Dasyatis americana **18**: 2544
Dasycercus cristicauda **11**: 1577, *1578*
Dasymutilla occidentalis **20**: 2815, *2815*
Dasypeltis **6**: 748
  medici **6**: 748
  scabra **6**: 748, *748*, 749; **9**: 1203
Dasypoda hirtipes **2**: *184*
Dasyprocta
  fuliginosa **1**: *33*
  punctata **1**: 33
Dasypus **1**: 85
  novemcinctus **1**: 85, *85*
Dasyuroides byrnei **11**: *1577*
Dasyurus
  albopunctatus **11**: 1571
  geoffroii **11**: 1571
  hallucatus **11**: 1571, 1572
  maculatus **11**: 1572
  spartacus **11**: 1571
  viverrinus **11**: 1571, 1572
Daubentonia madagascariensis **1**: 112
DDT **1**: 123; **7**: 1003; **14**: 1922;
  16: 2289; **18**: 2468
Dead man's fingers (sea fingers;
  dead man's toes) **5**: 651–652;
  17: 2440
Dead man's toes *see* Dead man's
  fingers
Dealfish **13**: 1768
death, feigning
  cardinal fish **3**: 390
  click beetle **4**: 466–467
  coyote **4**: 560
  false scorpion **6**: 799
  grass snake **8**: 1044
  gundi **8**: 1105
  hognose snake **9**: 1203
  hyena **9**: 1269
  opossum **13**: 1794
deep-sea habitat **13**: 1771
  anglerfish **1**: 51

bristlemouth **2**: 278
deep-sea anglerfish **5**: 653–654
hatchet fish **9**: 1161–1162
ice fish **9**: 1278
lancet fish **10**: 1409
lantern fish **10**: 1413
oarfish **13**: 1767
sea spider **17**: 2310, 2311, *2311*
sea squirt **17**: 2312
snake mackerel **17**: 2414
viperfish **20**: 2829–2830, *2829*
Deer 5: 655–658
Deer
  Andean *see* Guemal
  Axis (chital; Ceylon deer; spotted
    deer; Ganges stag) **1**: 107–108;
    21: 2907
  Barking *see* Muntjac
  Brow-antlered (thamin; Eld's deer;
    Manipur brow-antlered deer)
    2: 284–285
  Chinese water (Korean water
    deer; water deer; Chinese river
    deer; Yangtze River deer; river
    deer) **4**: 446–447; **5**: 655
  Fallow (Persian fallow deer)
    5: 656, 657; **6**: 794–796
  Hog **1**: 107
  Marsh (swamp deer) **18**: *2462,*
    2463, 2464
  Mouse **5**: 655
  Mule (black-tailed deer) **3**: 426;
    12: 1689–1691
  Musk *see* Musk deer
  Pampas **18**: *2462–2463, 2463,* 2464
  Père David's **14**: 1919–1920
  Prince Alfred's **5**: 658
  Red *see* Wapiti
  Rib-faced *see* Muntjac
  Roe **5**: 657; **16**: 2201–2203; **17**: 2375
  Schomburgk's (*extinct*) **2**: 285; **5**: 658
  South American **18**: 2462–2464
  Swamp **5**: *657*
  Tufted **12**: 1696
  White-tailed **5**: 656, *656*; **12**: 1689,
    1690; **19**: 2653, 2682
  see also Brocket; Caribou; Pudu;
    Sika; Wapiti
Deerfly **9**: 1235, *1235*, 1236, 1237
Deer mouse (vesper mouse)
  5: 659–661
  Range **5**: 659, 660
Defassa *see* Waterbuck
defense
  armadillo **1**: 87
  bird-eating spider **2**: 204
  chuckwalla **4**: *453*, 454
  cicada **4**: 457
  crested porcupine **5**: 597
  earwig **6**: 738, *739*
  frilled lizard **7**: 894–895, *894*
  frogmouth **7**: 901
  hedgehog **9**: 1171
  impala **9**: 1285, *1285*
  kallima butterfly **10**: 1332
  kestrel **10**: 1347
  limpet **11**: 1451
  lungless salamander **11**: 1505
  meerkat **12**: 1597
  octopus **13**: 1779, 1780
  pangolin **13**: 1863, *1863*
  sea hare **16**: *2276*
  sea mouse **16**: 2293
  spider **18**: 2483
  tawny owl **19**: 2646
  tiger moth **19**: 2688
  tree porcupine **19**: 2729
  see also armor; blood squirting;
    camouflage; chemical defense;
    color, flash; death, feigning;

electrical discharge; eyespots;
  inflation; mimicry; poison; ring
  formation; smell (for defense);
  tail, loss; tail loss; threat
  displays; warning coloration;
  warning sounds
defensive display, sun bittern
  18: 2567, *2568*
Deilephila elpenor **9**: 1167
Delichon urbica **12**: 1590
Delphinapterus leucas **2**: 194; **21**: 2911
Delphinus **4**: 506
  delphis **4**: 507
deltas **21**: 2905, 2907
Demansia psammophis **21**: 2919, 2920
Demoiselle (*fish*) *see* Damselfish
Demoiselle (*insect*) *see* Damselfly
Dendragapus
  canadensis **8**: 1084
  obscurus **8**: 1084
Dendraster exentricus **16**: 2226
Dendrelaphis punctulata **19**: 2733
Dendroanthus indicus **20**: 2843
Dendroaspis **2**: 265
  angusticeps **11**: 1530, *1530,* 1531
  jamesonii **11**: 1530, 1531
  polylepis **11**: 1530, 1531; **12**: 1597;
    15: 2103
  viridis **11**: 1530, 1531
Dendrobaena **6**: 735
  rubida **6**: 734
Dendrobates **1**: 93, 94, *94*; **20**: 2819
Dendrochirus **6**: 836
Dendrocincla anabatina **21**: 2969
Dendrocoelum **6**: 849
  lacteum **6**: 848
Dendrocolaptes certhia **21**: 2968
Dendrocopus
  major **21**: 2654
  pubescens **19**: 2654
Dendrocygna **5**: 714; **21**: 2925
  arborea **21**: 2926
  arcuata **21**: 2925
  autumnalis **21**: 2926
  bicolr **21**: 2925
  eytoni **21**: 2925
  javanica **21**: 2925
  viduata **21**: 2925
Dendrogale murina **19**: 2730
Dendrohyrax **9**: 1270, 1271
  arboreus **9**: 1271
  dorsalis **9**: 1271
  validus **9**: 1271
Dendroica
  caerulescens **21**: 2985
  coronata **21**: 2986
  kirtlandii **21**: 2985
  pensylvanica **21**: 2985
  petechia **4**: *553*; **21**: 2985
  striata **21**: 2985
Dendrolagus **19**: 2726, 2727
Dendronotus **16**: 2300
Dendropoma petraea **21**: 2990
Dendropos minor **19**: 2617
Dentalium
  elephantinum **6**: 777, 778
  entalis **6**: *778*
denticles **6**: 731; **17**: 2330
Dermacentor **19**: 2682
  andersoni **19**: 2682
  variabilis **19**: 2682
Dermatobia hominis **6**: 858
Dermochelys coriacea **10**: 1424;
  20: 2795
Dermophis mexicanus **3**: 356
Deroceras **17**: 2409
  reticulatum **17**: 2408
Desert 5: 662–665
desert adaptations **5**: 665
  addax **1**: 24–25

bat-eared fox **2**: 165, *165*
camel **3**: *361, 362–363, 364*
caracal **3**: 386
eland **6**: 757–758
fennec fox **2**: 165; **6**: 809, 811, *811*
gerbil **7**: 949–950
ground squirrel **8**: 1080
jack rabbit **10**: *1305,* 1306
jerboa **10**: 1323, 1324
moloch **12**: 1630, 1631
oryx **13**: 1810
pack rat **13**: 1843
roadrunner **16**: 2184
sand grouse **16**: 2229, *2229*, 2230
sidewinder **17**: 2369, *2370*
skink **17**: 2390
tomb bat **19**: 2702
water-holding frog **20**: 2868, *2869*
Desman **5**: 666–667
  Pyrenean **5**: 666, 667
  Russian **5**: 666, *667*
Desmana moschata **5**: 666, 667
Desmocerus **11**: 1484
Desmodema lorum **13**: 1768
Desmodus rotundus **20**: 2807, 2808
Desmognathus **11**: 1504
  fuscus **11**: 1503
  ocoee **11**: 1503
  wrighti **11**: 1503
detrivores **4**: 474–475; **15**: 2112
  basket stars **2**: 158
Devil *see* Tasmanian devil
Devilfish (*freshwater*) *see* Alaska
  blackfish
Devilfish (manta ray; devil ray)
  5: 668–670; **6**: 729
  Greater (Atlantic manta ray)
    5: 668, *670*
  Japanese manta ray **5**: 669
  Pacific manta ray **5**: 668
  Prince Alfred's manta ray **5**: 668,
    669
  Pygmy Australian devil ray **5**: 668
  Southern blue **5**: 668
Devilfish (*whale*) *see* Whale, Gray
Devonia semperi **16**: 2272
Dexiosoma caninum **19**: 2610
Dhole (wild dog; Asiatic wild dog;
  red wolf; red dog; chennai)
  5: 671–672, 677, 690
Diablotin *see* Oilbird
Diadema **17**: 2368
Diaemus youngi **20**: 2807–2808
Dialeurodes **21**: 2933
  citri **21**: 2932
Diamondback *see* Rattlesnake
diapause, embryonic **10**: 1336;
  15: 2116
Diapheromera femorata **18**: 2531
Diaphus ostenfeldi **10**: *1413*
diastema **4**: 438
diatoms **1**: 82; **2**: 246
  toxic diatoms affecting fish flesh
    2: 155
Diatryma **17**: 2322
Dibatag (Clarke's gazelle)
  5: 673–674; **7**: 952; **10**: 1382
Dicaeum
  hirundinaceum **7**: 869
  trigonostigma **7**: 869
Dicerorhinus sumatrensis **15**: 2152,
  2153
Diceros bicornis **15**: 2152, 2153
dichromatism **7**: 900
Dickcissel **18**: 2458
Dick-Dick, the devil *see* Bellbird,
  Crested
Dicrostonyx **10**: 1433
  torquatus **10**: *1432*

Dicrurus
  hottentottus **5**: 712, 713
  leuciphaeus **5**: 713
  macrocercus **5**: 712
  paradiseus **5**: 712
  remifer **5**: 712
Didelphis virginiana **13**: 1792, 1793
dieldrin **7**: 1003
digitigrade walking **3**: 404
Dik-dik **1**: 61; **5**: 675–676
  Guenther's **5**: 675, 676, *676*
  Kirk's **5**: 676
  Salt's **5**: 675, 676
  Silver **5**: 676
dimorphism, sexual *see* sexual
    difference
Dineutes americanus **21**: 2922
Dingleberry *see* Trunkfish
Dingo **5**: 677–679; **10**: 1336; **19**: 2641
Dinomys branicki **13**: 1838
Dioch, Sudan *see* Quelea, Red-billed
Diodon holacanthus **15**: 2023
Diomedea
  albatrus **16**: 2268
  bulleri **1**: 36
  exulans **1**: 35; **16**: *2267*
  innutabilis **1**: 37
  irrorata **1**: 35
  melanophris **1**: 36
Diopsittaca **11**: 1514
Diphylla ecaudata **20**: 2808
Diplogale hosei **13**: 1855
Diplolepis **7**: 922
  rosae **7**: 921
Diploria **4**: 528
Dipodomys **10**: *1339, 1340, 1340*
Dipper **5**: 680–681; **18**: 2457
  American (American water ouzel)
    5: 680, 681
  Brown **5**: 680
  Rufous-throated **5**: 680
  White-capped **5**: 680
  White-throated **5**: 680, *680*
Dipsosaurus **9**: 1282
  dorsalis **9**: 1281, 1282
Dipturus batis **17**: 2383, 2384
Dipus **10**: 1324
Dipylidium caninum **6**: 852
Discus **4**: 459
  Blue **14**: 2006–2007, *2007*
  Red **14**: 2006, *2006, 2007*
diseases
  of crayfish **5**: 592
  of harbor seals **8**: 1134
  of rabbits **15**: 2093
diseases, carried by
  buffalo **3**: 307
  deerfly **9**: 1235
  flies **6**: 859
  ground squirrel **8**: 1079
  housefly **9**: 1243
  midge **11**: 1607
  mosquito **12**: 1662, 1663–1664
  multimammate rat **12**: 1695
  rat **3**: 298–299
  ship rat (black rat) **17**: 2346–2347,
    2348
  thrips **19**: 2676–2677
  tick **19**: 2681–2682, *2681*, 2682
  tsetse fly **20**: 2760–2761
  turkey vulture **20**: 2789
  vampire bat **20**: 2809
dispersal, "hitching" a ride *see*
    phoresy
Dispholidus typhus **2**: 264–265; **6**: 749;
    13: 1736
displacement activities **16**: 2203
display (sexual) *see* attracting a mate;
  displaying in groups; sexual
    aggression

display (territorial)
  angelfish **1**: 50
  anole lizard **1**: 54
  avadavat **1**: 103
  bearded lizard **2**: 172
  bunting **3**: 319–320
  chameleon **3**: 419–420
  cichlid **4**: 458
  coot **4**: 523
  curlew **5**: 627–628
  deer 5: 657
  fiddler crab **6**: 816
  firemouth **6**: 829, *829*
  golden eagle **7**: 1002
  grouse **8**: 1085
  Jack Dempsey **10**: 1302
  mockingbird **12**: 1623
  mudskipper **12**: 1686
  mynah **12**: 1715
  oystercatcher **13**: 1837
  prairie dog **15**: 2038
  red-winged blackbird **15**: 2143
  robin, European **16**: 2192–2193
  ruff **16**: 2214
  sand lizard **16**: 2231
  shrew **17**: 2360
  sugarbird **18**: 2558
  swan **18**: 2589
  titi **19**: 2696
  wagtail **20**: 2845
  waterbuck **20**: 2866
displaying in groups
  bird of paradise **2**: 206–207
  capercaillie **3**: 378
  cockatoo **4**: 483
  cock-of-the-rock **4**: 490, 491
  flamingo **6**: *844*, 846–847
  fruit bat **7**: 908
  grouse **8**: 1086
  guillemot **8**: 1096
  magpie lark **11**: 1522
  mallard **11**: 1525
  manakin **11**: 1532
  prairie chicken **15**: 2036
  pratincole **15**: 2041–2042
  see also attracting a mate
distraction displays
  dotterel **5**: 705
  killdeer **10**: 1349, *1349*
  nightjar **13**: 1744
  osprey **13**: 1813
  ostrich **13**: 1816
  painted snipe **13**: 1850, *1850*
  patas monkey **14**: 1897
  pheasant **14**: 1936
  plover **14**: 1990
  pratincole **15**: 2042
  ptarmigan **15**: 2057
  woodcock **21**: 2967
Diver
  Black-throated *see* Loon, Arctic
  Great northern *see* Loon, Common
  Red-throated *see* Loon, Red-
    throated
  White-billed *see* Loon, Yellow-
    billed
Diving beetle (*larva*: water tiger)
    5: 682–683
  Great **5**: 682, *682*
Diving-petrel **5**: 684–685; **16**: 2267
  Common (jack diver; kuaka;
    flying penguin) **5**: 684, *684*, 685
  Magellanic **5**: 684
  Peruvian (potoyunco)
    5: 684–685
  South Georgia **5**: 684
Doctor fish *see* Tench
Dodger *see* Leafhopper
Dog
  Asiatic wild *see* Dhole

Cape hunting (African wild dog;
  African hunting dog) **3**: 373–375;
  5: 689; **8**: 1041
Crab-eating *see* Fox, Crab-eating
Raccoon *see* Raccoon dog
Red *see* Dhole
Short-eared (small eared dog;
  short-eared zorro) **17**: 2356–2357
Singing **5**: 677
Wild *see* Dhole
see also Bushdog; Coyote; Dingo;
  Dogs; Fox; Jackal; Wolf
Dogface, California **21**: 2929
Dogfish (grayfish; rock salmon; rock
  eel; fluke) **5**: 686–688
  Alligator (bramble shark) **5**: 686
  Freshwater *see* Bowfin
  Greater spotted (nurse hound)
    5: 686
  Lesser spotted **5**: 686, *686*, 687
  Pacific **5**: 686
  Spiny (spur dogfish) **5**: 686,
    687–688; **17**: 2333
Dogs 5: 689–692
Dog-whelk (mud snail) **21**: 2917
  Striped **21**: *2917*
doldrums **1**: 35
Dolichonyx oryzivorus **13**: 1805;
  18: *2457*; **20**: 2833
Dolichotis **8**: 1100
  patagonum **8**: 1099
Dolichovespula **11**: 1233
doliolaria larvae **6**: 806
Doliolum **16**: 2224, *2224*
  nationalis **16**: 2224
Dollarbird **16**: 2204
Dolomedes **18**: 2482
Dolphin **5**: 695, 696; **13**: 1771–1772
  Atlantic spotted **5**: 698
  Bottlenose (bottle-nosed dolphin;
    common porpoise; gray
    porpoise; black porpoise)
    2: 266–268; **4**: 507
  Common (white-bellied porpoise;
    saddleback dolphin; crisscross
    dolphin; hourglass dolphin)
    4: 506–509, *509*
  Dusky **5**: 698, *698*
  Ganges **5**: 697; **16**: 2177, 2178
  Hector's **5**: 698
  Indus **5**: 697; **16**: 2177, 2178
  Irrawaddy **21**: 2911
  La Plata (Franciscana) **5**: 697;
    16: 2177, 2178
  Pantropical spotted **5**: 698
  River *see* River dolphin
  Spinner **5**: 698
  Spotted **5**: *697*
  Striped **5**: 698
  see also Dolphins and porpoises;
    False killer whale; Killer whale;
    Pilot whale; Porpoise; Whale,
    Beluga
Dolphinfish **5**: 693–694
  Common (Ma-hi Ma-hi; dorado)
    5: 693, 694, *694*
  Pompano **5**: 693
Dolphins and porpoises 5: 695–698
domestication
  blesbok **2**: 240
  canary **3**: 369
  dog **5**: 692
  eland **5**: 759
  gaur **7**: 937, *938*
  greylag goose **8**: 1074
  guinea fowl **8**: 1098, *1098*
  jungle fowl **10**: *1330*
  llama **11**: 1470–1471
  muscovy duck **12**: 1699–1700
  rabbit **15**: 2091

silk moth **17**: 2376
wild ass **21**: *2940*
wild boar **21**: 2944
yak **21**: 3002
zebu **21**: 3018
Donkey (domestic ass) **21**: 2939, *2940*
Doodlebug *see* Ant lion
Dor-beetle **16**: 2253
Dorcatragus megalotis **1**: 61; **2**: 190
Dorcopsis veterum **20**: 2846
Doree *see* John Dory
Dorhawk *see* Nightjar, Eurasian
dormancy *see* estivation; hibernation
Dormouse **5**: 699–701
  African **5**: 699
  Desert **5**: 699
  Edible (fat dormouse) **5**: 699,
    700–701, *701*
  Forest **5**: 699
  Garden **5**: 699, *699*, 701
  Hazel (common dormouse;
    sleeper) **5**: 699–700, *700*, 701;
    19: 2653
  Japanese **5**: 699
  Mouse-tailed **5**: 699
Dorosoma **17**: 2328
Dory *see* John Dory
Dorylus **1**: 90, 91, 92
  wilverthi **1**: 92
Doryrhompus dactyliophorus **14**: 1960
Dotterel **5**: 704–705
  Eurasian **5**: 704, 705, *705*
Douc **10**: 1410, 1411
Doughbird *see* Curlew, Eskimo
Douroucouli **19**: 2697
Dove
  Collared (Eurasian collared dove;
    collared turtle dove) **4**: 499–500
  Fruit *see* Fruit dove
  Palm (town dove; village dove;
    garden dove; laughing dove)
    13: 1857–1858
  Ring *see* Pigeon, Wood
  Rock (feral pigeon; domestic
    pigeon; racing pigeon; fancy
    pigeon) **16**: 2194–2196
Dovekie *see* Auk, Little
dove milk **4**: 500
down
  eider **6**: 754, 756
  goose **8**: 1074
  powder **2**: 220–221; **9**: 1179;
    14: 1883; **21**: 2983
Draco **11**: *1466*
Dragon
  Bearded *see* Lizard, Bearded
  Frilled *see* Lizard, Frilled
  Komodo *see* Komodo dragon
Dragonet **5**: 706–707
Dragonfish (lionfish) *see* Lionfish, Red
Dragonfish (*sea moth*) *see* Sea moth,
  Longtail
Dragonfish (*sea moth*), Short *see* Sea
  moth, Dragon
Dragonfly (horse-stinger) **5**: 708–711
  Borneo **5**: 711
  Golden-ringed **5**: 711
  Hawker **5**: 710–711
Drill (*monkey*) **1**: 115; **11**: 1536–1537,
    1538
Drill (*shellfish*) *see* Whelk
Dromas ardeola **4**: 569
Dromedary *see* Camel, Arabian
Drongo **5**: 712–713
  Ashy **5**: 713
  Black (king crow) **5**: 712, 713, *713*
  Greater racket-tailed **5**: 712
  Lesser racket-tailed **5**: 712, *712*
  Spangled (hair-crested drongo)
    5: 712, 713

Drosophila
  melanogaster **7**: 910
  sigmoides **7**: 909, 910
Drum
  Black **5**: 601, 602
  Freshwater **5**: 602
Drurya antimachus **18**: 2581, 2582
Drycopus pileatus **21**: 2979
Dryomys **5**: 699
Dryophius **19**: *2732*
Duck
  Black-headed **18**: 2538, 2539
  Blue **19**: 2710
  Blue-billed (spiny-tailed duck) **18**: 2538, 2539
  Cornfield *see* Whistling duck, Black-bellied
  Eider *see* Eider, Common
  Ferruginous **14**: 1993
  Freckled **5**: 714
  Labrador (*extinct*) **5**: 717
  Logger *see* Steamer duck, Flightless
  Long-tailed (oldsquaw) **5**: 717
  Maccoa **18**: 2538
  Mandarin **5**: 716–717; **21**: 2925, 2970, 2971
  Masked **18**: 2538
  Muscovy (musky duck; musk duck) **5**: 717; **12**: 1699–1700
  Musk (mold goose) **18**: 2538, 2539
  Painted *see* Shelduck, Paradise
  Paradise *see* Shelduck, Paradise
  Perching **21**: 2925
  Pink-headed **5**: 717
  Ruddy **18**: 2537, *2537*, 2538, *2538*, 2539
  Steamer *see* Steamer duck
  Torrent **19**: 2708–2710
  Tree *see* Duck, Wood; Whistling duck
  Tufted **14**: 1993, 1995
  Whistling *see* Whistling duck
  White-backed **18**: 2538
  White-headed **18**: 2537, 2538–2539, 2539, *2539*
  Wood (Carolina duck; summer duck; tree duck) **5**: 716–717; **21**: 2925, 2970–2972
  see also Ducks, geese and swans; Eider; Mallard; Merganser; Pintail; Pochard; Shelduck; Stifftail; Teal
Duckbill *see* Platypus
Duckmole *see* Platypus
Ducks, geese and swans 5: 714–717; 21: 2927
Dugesia **6**: 849
  tigrina **6**: 849
Dugong (sea cow; siren) **6**: 724–725
Dugong dugon **6**: 725
Duiker **1**: 61, 63; **6**: 726–728
  Ader's **6**: 726, 727
  Blue **6**: 726, 728; **18**: 2572
  Common (gray duiker) **6**: 726, *726*, 727, 728
  Jentink's **6**: 726, 727
  Red **6**: 727, 728, *728*
  Yellow-backed **6**: 726
  Zebra (zebra antelope) **6**: 727
Dulus dominicus **13**: 1854
Dumatella carolinensis **3**: 399, 400; **4**: 555
Dunckerocampus multiannulatus **14**: *1960*
Dunlin **16**: 2233
Dunnart, Fat-tailed **11**: 1577, *1578*
Dunnock *see* Accentor, Hedge
dust baths **10**: 1339; **14**: *1892*
dust bowls **8**: 1041

Dwarf antelope **18**: 2571
  Bates' **18**: 2571, 2572
dyes, from whelks **21**: 2917
dymantic displays **13**: 1780
Dynamena **16**: 2274
Dynastes neptunus **2**: 186
Dysdera crocata **21**: 2974
Dytiscus **5**: 683
  marginalis **5**: 682
Dzo *see* Yak, Domestic

E

Eagle **2**: 209
  Bald (American eagle; white-headed sea eagle) **1**: 82, 123–125; **2**: 208; **21**: 2907
  Brown harrier (brown snake-eagle) **12**: 1589
  Crowned (African crowned eagle; crowned hawk-eagle) **5**: 614–616; **8**: 1093
  Fish *see* Fish eagle
  Golden **7**: 1001–1003
  Harpy *see* Harpy eagle
  Imperial **7**: 1001
  Lesser spotted **21**: 2906
  Martial **12**: 1588–1589
  Monkey-eating **15**: 2111
  River *see* Fish eagle
  Verreaux's (black eagle) **7**: 1001, 1003; **12**: 1589
  White-tailed **6**: 837, 839
Eagle ray **6**: 729–731
  Common **6**: 729, 730
  Spotted **6**: 730, *730*
Earth-creeper **13**: 1819
Earthworm (nightcrawler) **6**: 734–737
  Common **6**: 734
  Giant Australian **6**: 734
earthworms, eaten by blowfly maggots **2**: 242
Earwig (pincerbug) **6**: 738–740
  Common **6**: 738–739
  Small **6**: 738, 739
Earworm, Corn *see* Bollworm, American
Eastern grapeleaf skeletonizer **3**: 321
eating meat *see* meat eating
eating plants *see* browsing; grazing
ecdysis *see* molting
Echeneis
  naucrates **15**: 2148, 2149
  neucratoides **15**: 2148
Echidna *see* Spiny anteater
Echidnophaga gallinacea **6**: 851
Echiichthys vipera **21**: 2901, 2902
Echinarachnius parma **16**: 2225, 2226
Echinocardium cordatum **17**: 2314
Echinocyamus pusillus **16**: 2226
Echinoprocta rufescens **19**: 2729
Echinops telfairi **19**: 2657, *2657*
Echinorhinus brucus **5**: 686
Echinostrephus molaris **17**: 2314–2315
Echinus esculentus **17**: 2315
Echiodon **14**: 1901
  drummondi **14**: 1901, 1902
echolocation
  bat **2**: 167–168; **13**: 1822; **19**: 2688
  dolphins and porpoises 4: 508; 5: 697
  fish-eating bat **6**: 840–841
  fruit bat **7**: 906, *906*, 907
  horseshoe bat **9**: 1240
  oilbird **13**: 1784, 1785
  pipistrelle **14**: 1963
  river dolphin **16**: 2177
  sea lion **16**: 2283
  shrew **17**: 2358

swiftlet **19**: 2601, *2601*, 2602
whales 2: 247–248; 21: 2911
Eciton **1**: 90, 91, *92*
  burchelli **1**: 90, 91, 92
Eclectus roratus **14**: *1883*
ecstatic display, Adélie penguin **1**: 28
Ectobius **4**: 493
Ectopistes migratorius **3**: 328
ectoplasm **2**: 41
Edovum puttleri **4**: 504–505
Eel (elver; glass eel; leptocephalus) **6**: 745–747, 836
  American **6**: 745, 746, 747
  Cave *see* Cave eel
  Conger **4**: 521–522
  Electric (electric knifefish) **6**: 762–763
  European **6**: 745, *745*, 746, *746*, 747; **14**: 1981
  Moray *see* Moray eel
  Ribbon *see* Moray eel, Leaf-nosed
  Rock *see* Dogfish
  Sand *see* Sand eel
  Spiny *see* Spiny eel
  Swamp *see* Swamp eel
Eelpout *see* Blenny, Viviparous
Eel pout *see* Burbot
Eel-skipper **12**: 1684, 1686
Egg-eater
  Common *see* Egg-eating snake, Common
  Rhombic *see* Egg-eating snake, Common
Egg-eating snake **6**: 748–749
  African **6**: 748; **9**: 1203
  Common (common egg-eater; rhombic egg-eater) **6**: 748, 749
  East African **6**: 748
  Indian **6**: 748
egg-laying, budgerigars responsive to male's song **3**: 304
eggs, synchronized hatching **15**: 2083
egg tooth **2**: 252; **3**: 360; **8**: 1044, 1067; **17**: *2421*
Egret **6**: 750–753
  Cattle **6**: 750, 752, 753
  Eastern reef (blue reef heron) **6**: 750
  Great (great white egret; great white heron; great American heron; white heron; large heron) **6**: 750, 751, 753
  Great white **9**: 1179
  Little **6**: 750–751, 753
  Snowy **6**: 750, *751*
Egretta
  alba **6**: 750, 751; **9**: 1179
  ardesiaca **9**: 1181
  garzetta **6**: 750
  sacra **6**: 750
  thula **6**: 750
Egyptians, ancient
  and the addax **1**: 24
  and the electric catfish **6**: 761
  and the Nile perch **13**: 1752
  and the scarab beetle **16**: 2254
  and wild boar **21**: 2944
Eider **6**: 754–756
  Common (eider duck) **6**: 754, 755, *756*
  King **6**: 754, 755, *755*
  Spectacled **6**: 754
  Steller's **6**: 754, *754*, 756
Eidolon
  helvum **7**: 908
  sabaeum **7**: 908
Eisenia **6**: 735
  foetida **6**: 734, *734*
Elacatinus oceanops **7**: 995
Elachistodon westermanni **6**: 748

elaiosomes **9**: 1157
Eland **1**: 62; **6**: 757–759
  Common (Cape eland) **6**: 757, 758, *758*
  Giant (Derby eland) **1**: 60; **6**: 758
Elanus
  caeruleus **10**: 1367–1368
  leucurus **2**: 209; **10**: 1368
Elaphe **15**: 2117, 2118
Elaphodus cephalophus **12**: 1696
Elaphurus davidianus **14**: 1920
Elasmucha grisea **17**: 2345
electrical discharge
  electric catfish **6**: 760, *760*, 761
  electric eel **6**: 762–763, *762*, *763*
  electric ray **6**: 764–766, *764*; **17**: 2332
  knife fish **10**: 1377
  Nile fish **13**: 1749, *1749*
  skate **17**: 2385, *2385*
  stargazer **18**: 2519, 2520
Electric catfish **6**: 760–761
  Small **6**: 760, 761
Electric ray (torpedo fish; torpedo; torpedo ray; numbfish; crampfish) **6**: 764–766; **17**: 2332
  Atlantic **6**: 764, 765–766; **17**: 2332
  Australian crampfish **6**: 764
  Pacific **6**: 766
Electrophorus electricus **6**: 763
electroreceptors **14**: *1987*; **17**: 2384, 2385
electroshock therapy **6**: 766
Eledone cirrhosa **13**: 1779
Elephant **6**: 767–770
  African **6**: 767–769, *767*, *770*; **15**: 2113
  Indian (Asian elephant) **6**: 767, 768, *768*, 769
elephantiasis **12**: 1664
Elephant-nose fish **13**: 1749, *1750*
Elephant seal **6**: 771–773
  Northern **6**: 771–772, *772*, 773
  Southern **6**: 771, 772, *773*; **16**: 2285, 2286, 2288
Elephant shrew **6**: 774–776
  Checkered **6**: 774, 775, 776
  Forest **6**: 775, *776*
  Long-eared (lesser elephant shrew) **6**: 774, 775
  Yellow-rumped (golden-rumped elephant shrew) **6**: 774, 775, 776
Elephant trunk fish *see* Spiny eel, Lesser
Elephant tusk shell **6**: 777–778
Elephantulus **6**: 775
  myurus **6**: 774
Elephas maximus **6**: 768
Eleutherodactylus planirostris **8**: 1067; **12**: 1610
Eliomys quercinus **5**: 699
Elk
  Irish (*extinct*) **5**: 656
  see also Moose; Wapiti
El tigre *see* Jaguar
elytra **2**: 186; **6**: 738
Emarginula elongata **11**: 1451
Emberiza
  hortulana **3**: 319
  schoeniclus **3**: 320; **18**: 2457
  striolata **3**: 319
Emerita **18**: 2507
Empidonax virescens **20**: 2802
Emu **6**: 782–784
Emu-wren **21**: 2993
Emydura maquarii **17**: *2417*
Encarsia formosa **2**: 185; **3**: 416, 417; **21**: 2933
Encheliophis **14**: 1901
  homei **14**: 1901

Enchelycore pardalis **12**: 1657, 1658
Endomychura hypoleuca **16**: 2264–2265
endoplasm **1**: 41
endostyle **10**: 1403, 1407
Engraulis
    encrasicolus **1**: 45
    mordax **1**: 46
Enhydra lutris **16**: 2179, 2295
Enhydrina **20**: 2820
Enicognathus
    cornutus **14**: 1879
    leptorynchus **14**: 1878
Enneacanthus
    gloriosus **18**: 2569
    obesus **18**: 2570
Ensatina **11**: 1504
    croceator **11**: 1505
Ensifera ensifera **9**: 1255
Ensis
    ensis **15**: 2128
    siliqua **15**: 2127, *2128*
Entamoeba **1**: 41
Enterochromis **4**: 459
Entomobrya mawsoni **18**: 2504
Eochoreutes **10**: 1324
Eopsetta grigorjewi **8**: 1122
Eos histrio **11**: 1494
Eotragus **1**: 60
Eozapus **10**: 1324
Ephippiorhynchus **18**: 2551
    asiaticus **18**: 2551
    senegalensis **18**: 2551
Ephydra hians **6**: 858
ephyra larvae **10**: 1322
Epicauta velata **13**: 1783
Epicopeia polydora **20**: 2806
Epicrates
    inornatus **2**: 251
    monensis granti **2**: 251
epigyne **18**: 2483
Epinephelus **8**: 1081, *1081*
    itajara **8**: 1082–1083
    lanceolatus **8**: 1082
    striatus **8**: 1083
    tukula **8**: *1081*
Epipedobates **1**: 94
epipelagic zone **13**: 1769, 1770
epiphytic plants **15**: 2111
epitokes **13**: 1861
epitotles **15**: 2104–2105
Epixerus **13**: 1860
    ebi **13**: 1859
    wilsoni **13**: 1859
Epomophorus **7**: 907–908
Epomops franqueti **7**: 908
epontic community **1**: 82
Equus **8**: 1039, 1040;**15**: 2052–2054;
    16: 2216; **21**: 2939, 2940, 2941,
    3009–3010, 3011
Eremitalpa **7**: 1005
    granti **7**: 1004
Erethizon dorsatum **19**: 2728, 2729
Eretmochelys imbricata **8**: 1170;
    20: 2798
Eriborus terebrans **4**: 540
Ericerus pe-la **16**: 2247
Erignathus barbatus **1**: *82*; **16**: 2286
Erinaceus europaeus **9**: 1170, 1172
Eriocampa limacina **16**: 2240
Eriococcus
    devoniensis **16**: 2246, 2247
    spurius **16**: 2246, 2247
Erionota **17**: 2391
    thrax **17**: 2392
Erisoma lanigerum **1**: 74
Eristalis **9**: 1251
    tenax **9**: 1251
Erithacus rubecula **16**: 2192; **19**: 2678
Ermine (short-tailed weasel; stoat)
    6: 785–787; **14**: 2003

Erynnis toges **17**: 2391
erythrocruorin **6**: 734
erythrophores **3**: 420
Esacus
    magnirostris **18**: 2545
    recurvirostris **18**: 2545
Escolar (oilfish; scourfish; castor-oil
    fish) **17**: 2414, 2415
Esox
    americanus **14**: 1946
    americanus americanus **14**: 1945
    americanus vermiculatus **14**: 1945,
        1946
    lucius **8**: 1063; **14**: 1946
    masquinongy **14**: 1945, 1946
    niger **14**: 1945, 1946
    reichertii **14**: 1945, 1946
Espinal *see* Chaparral
estivation
    chuckwalla **4**: 453
    giant snail **7**: 966
    lungfish **11**: 1501
    vanessa butterflies **20**: *2811*
Estrilda **21**: 2888
    astrild **21**: 2888, 2889
    locustella **21**: 2888
    melpoda **21**: *2888*
    troglodytes **21**: 2888
estrus, aseasonal **4**: 567
estuaries **16**: 2176
Ethmidium maculatum **12**: 1598
Eubalaena
    australis **16**: 2166, 2167; **18**: 2510;
        21: 2910
    glacialis **16**: 2166, 2167; **21**: 2910
Eublepharus macularius **7**: 943
Eubranchipus vernalis **6**: 792
Eucidaris thouarsii **17**: 2315
Eucometis penicillata **19**: 2628
Eucrossorhinus dasypogon **17**: 2331;
    21: 2955
Euderma maculatum **2**: 168; **20**: 2821
Eudia pavonia **3**: 346
Eudocimus
    albus **9**: 1274
    ruber **9**: 1274, 1275
Eudromia elegans **19**: 2691
Eudynamys scolopacea **5**: 620, 713
Eudyptes **14**: 1912, 1913
    chrysocome **14**: 1913
    pachyrhynchus **14**: 1914
    robustus **14**: 1914
    sclateri **14**: 1914
Eudyptula
    albosignata **14**: 1911
    minor **14**: 1911
Eulacestoma nigropectus **21**: 2923
Eulachon **17**: 2410, 2411
Eulemur **10**: 1436
    coronatus **10**: *1437*
Euliphera
    leucyana **2**: 244
    mirifica **2**: 244
Eumeces
    fasciatus **17**: 2390
    longirostris **17**: 2390
    multivirgatus **17**: 2390
    obsoletus **17**: 2390
Eumenes coarctatus **9**: 1261
Eumetopias jubatus **16**: 2282, 2283,
    2289
Eumomota superciliosa **12**: 1665
Eunectes
    murinus **1**: 43–44; **17**: 2419
    notaeus **1**: 43
Euoticus **3**: 330
Euphausia superba **10**: 1387, *1387*,
    1388

Euphilotes bernardino allyni **2**: 244
Eupholus quinitaenia **21**: *2904*
Euphonia **19**: 2627
Euphonia **19**: 2628
Euphractus **1**: 85
Euphydryas aurinia **7**: 898
Euplagia quadripunctaria **19**: 2687
Euplectes **2**: 213
    afer **2**: 213
    aurea **2**: 213
    capensis **2**: 213
    hordeaceus **2**: 213, 214
    orix **2**: 213
Euplectrus comstockii **3**: 417
Eupodotis afra **3**: 338
Eupoecilia **4**: 487
Euro *see* Wallaroo, Common
Euroleon **1**: 65
    nostras **1**: 64
European Apollo *see* Butterfly,
    Apollo
Eurostopodus
    argus **14**: 2013
    guttatus **13**: 1742
    mystacalis **13**: 1742
Eurycea **11**: 1504
    longicauda **11**: 1503
Euryceros prevostii **20**: 2813
Eurygaster integriceps **17**: 2344
Eurypegasus
    draconis **16**: 2291, *2291*
    papilio **16**: 2290
euryphagous **17**: 2423
Eurypyga helias **18**: 2568
Eurystomus **16**: 2204
    glaucurus **16**: 2204
    orientalis **16**: 2204
Euschemon rafflesia **17**: 2392
Eusphyra blochii **8**: 1126, 1127
Eutamias **4**: 448, 449; **19**: 2653
    minimus **4**: 449
    sibiricus **4**: 448, 449
    townsendii **4**: 449
Eutermes triodiae **19**: 2660
Eutoxeres **9**: 1255
Eutrigla gurnardus **8**: 1108, 1109
eutrophication **10**: 1398
Eve-jarr *see* Nightjar, Eurasian
Everglades **21**: *2906, 2907*
Eviota **7**: 994
    queenlandica **7**: *993*
evolution
    by natural selection **14**: 1915
    fish to amphibian **2**: 196
    parallel **14**: 1966
    see also missing links
evolution, convergent
    basilisk **2**: 157
    horned frog **9**: 1230
    kangaroo mouse **10**: 1338
    knife fish **10**: 1376
    mantis fly **11**: 1550
    marsupial mole **11**: 1576
    woodcreeper **21**: 2968
execution, by driver ants **1**: 92
Exelastis atomosa **14**: 1991
Exocoetus **7**: 876
    volitans **7**: 875
Extatosoma tiaratum **18**: 2531, *2531*
extinct animals
    antelope **1**: 63
    bandicoot **1**: 130
    brown bear **3**: 295
    great auk **16**: 2268
    honeyeater **9**: 1214
    New Zealand wren **16**: 2165
    passenger pigeon **3**: 328
    Schomburgk's deer **2**: 285
eyelid, third *see* nictitating
    membrane

eye(s)
    compound **6**: 857
    parietal **11**: 1468
    of scallops **16**: *2248*, 2249
    third (pineal eye) **20**: 2762
eyespots, butterfly and moth
    8: 1117–1118; **18**: 2582; **19**: 2604
Eyra *see* Jaguarundi

F

Fairy bluebird **6**: 790–791
    Blue-backed (Asian fairy bluebird;
        blue-mantled fairy bluebird)
        6: 790
    Philippine **6**: 790
Fairy-wren **21**: 2993
Falco **2**: *208,* 209, 210; **8**: 1029, 1110,
    1111; **9**: 1198, 1199; **10**: 1345,
    1346; **12**: 1603; **14**: 1922
Falcon
    African pygmy **2**: *209*
    Aplomado **2**: *208*
    Eleanora's **2**: 210
    Lanner **8**: 1110
    Peregrine **2**: 209, 210; **8**: 1029, 1110,
        1112; **14**: 1921–1923; **16**: 2196
    Saker **8**: 1110
Falconet, Black-thighed **2**: 208
Falculea palliata **20**: 2813
Fallfish **12**: 1619
False killer whale (false pilot whale;
    pseudorca) **6**: 797–798
False scorpion (pseudoscorpion;
    book scorpion) **6**: 799–800
False spider *see* Solifugid
Fanaloka (Malagasy civet) **4**: 460, 461
Fannia canicularis **9**: 1241, 1242
Fantail **20**: 2843
Fanworm **6**: 801–802
    Red Sea **6**: *802*
farming
    crocodile **5**: 606; **9**: 1283
    iguana **9**: 1283
    mink **12**: 1617
    mussels **12**: 1713
farming (agricultural) *see* agriculture
Fasciola **7**: 871
    hepatica **7**: 870, *870*
fast moving
    agouti **1**: 33
    Atlantic salmon **1**: 99
    barracuda **2**: 154
    basilisk **2**: 156, *156*
    blackbuck **2**: 227
    camel **3**: 361
    Canada goose **3**: 367
    capybara **3**: 382
    caribou **3**: 392
    cassowary **3**: 397
    cheetah **3**: 403, 427, *427*
    cottontail **4**: 546
    courser **4**: 551
    coyote **4**: 560
    dolphin **4**: 507
    dragonfly **5**: 708
    emu **6**: *782*, 783
    flying fish **7**: 876
    hawkmoth **3**: 346
    iguana **9**: 1282
    jack rabbit **10**: 1305
    kangaroo **10**: 1333–1334, *1336*
    lion **11**: 1461
    mackerel **11**: *1517*
    mamba **11**: 1530
    maned wolf **11**: 1540
    marlin **11**: 1561, 1562
    ostrich **13**: 1816
    penguin **14**: 1913
    peregrine falcon **14**: 1923

fast moving (continued)
pheasant **14**: 1936
pronghorn antelope **15**: 2049–2050
roadrunner **16**: 2184
sailfish **16**: 2219, *2219*
swift **19**: 2599
swift moth **19**: 2604
swordfish **19**: 2606, 2607
whip snake **21**: 2919
wild ass **21**: *2941*
wombat **21**: *2962*
fat, brown **9**: 1173
Fatback *see* Menhaden, Atlantic
Fathead **12**: 1618
Featherback (featherfin; Old World
knife fish) **6**: 803–804
Feather duster worm **6**: *801*
feathers
barn owl **2**: 150
bee-eater chicks **2**: 182
bird of paradise **2**: 205, 207
collected by man **1**: 37; **2**: 205, 207;
6: 752–753, *845*, 847; **9**: 1255;
10: 1355, 1373; **13**: 1814; **15**: 2088;
21: 2887
drongo **5**: 712, 713
ducks, geese and swans 5: 716, 717
eaten by grebes **8**: 1063
emu **6**: 782
flamingo **6**: *845*, 847
ostrich **13**: 1814
parakeet **14**: 1880
peafowl **14**: 1899–1900
quetzal **15**: 2087, *2087*, *2088*
see also down
Featherstar **6**: 805–806; **16**: 2280
Rosy **6**: 805–806
feeding *see* ambush; browsing;
cannibalism; electrical
discharge; filter feeding;
grazing; hunting in
packs/groups; meat eating;
parasite; poison; robbery;
scavenging; stalking;
stinging
semizygodactile **20**: 2780
zygodactile **16**: 2184
Felis **3**: 386, 403, *403*, 404, 405;
10: 1314; **11**: 1506, 1507, 1559;
19: 2653; **21**: 1672, 2907,
2945–2947, *2945*
Felis rufus **2**: 254
Felovia vae **8**: 1104
Fence lizard (swift; bluebelly)
6: 807–808
Eastern (plateau lizard; pine
lizard; prairie lizard; northern
fence lizard; southern fence
lizard) **6**: 807–808, *807*, *808*
Western **6**: 807, 808
Fennecus zerda **2**: 165; **5**: 665; **6**: 810
fens **21**: 2905
Fer de lance (barba amarillo)
6: 812–814
Fern owl *see* Nightjar, Eurasian
Ferret **14**: 2003
Black-footed **2**: 228–229
Ficedula
albicollis **7**: 873
hypoleuca **7**: 872–873; **18**: 2459
parva **7**: 873
semitorquata **7**: 873
Ficopomatus enigmaticus **6**: 801
Fiddle fish *see* Monkfish (*shark species*)
Fiddler crab **4**: 573; **6**: 815–817
European **6**: 817
Southwestern Atlantic **6**: 816
Fieldfare **18**: *2459*
Figbird **8**: 1012
Yellow **8**: 1012

Fig-eater, Green *see* Cockchafer
Fighting fish **6**: 821–822
Siamese **6**: 821, *821*
fighting for mates
adder **1**: 26–27, *27*
asp **1**: 98
bighorn sheep **2**: *199*
bison **2**: 217
brow-antlered deer **2**: 284–285
buffalo 3: 306
Canada goose **3**: 366
chamois **3**: 422
Chinese water deer **4**: 446
coati **4**: 479
ibex **9**: 1273
jungle fowl **10**: 1329
lechwe **10**: 1426
muntjac **12**: 1697–1698
musk-ox **12**: 1704
narwhal **12**: 1719
nilgai **13**: 1754
Père David's deer **14**: 1919
rhea **15**: 2150
rhino **15**: 2155
tahr **19**: 2612
wapiti **20**: 2856
yak **21**: 3002
fighting for territory **12**: 1667–1668
aardwolf **1**: 19
agama lizard **1**: 31
antelope 1: 62
bee-eater **2**: 182
cichlid **4**: 458
deer **5**: 657
elephant seal **6**: 772–773, *772*
fur seal **7**: 917
Jack Dempsey **10**: 1302, *1302*
kob **10**: 1381
marine iguana **11**: 1555
moorhen **12**: *1650*, 1651
musk deer **12**: 1701
red-winged blackbird
15: 2141–2142
sand lizard **16**: 2231
steamer duck **18**: 2525
waterbuck **20**: 2866
Fig insect **3**: 417
filariasis **12**: 1664
Filefish **20**: 2743, 2744
filter feeding
barnacle **1**: 138
basking shark **2**: 160, *160*
blue whale **2**: 247
brine shrimp **2**: 276
clam **4**: 462
clawed frog (*tadpole*) **4**: 465
daphnia **5**: 643
devilfish **5**: 670
fairy shrimp **6**: 792
fanworm **6**: 801
featherstar **6**: 806
flamingo **6**: 845
lamp shell **10**: 1405–1406
lancelet **10**: 1407
menhaden **12**: 1598
mussel **12**: 1711–1712
opossum shrimp **13**: 1795–1796
oyster **13**: 1833
paddlefish **13**: 1845
piddock **14**: 1940
right whale **16**: 2167–2168
rorqual **16**: 2209
salp **16**: 2223
sea lily **16**: 2281
sea squirt **17**: 2312–2313
shad **17**: 2328
sponge **18**: 2494
whales 21: 2910, 2913
whale shark **21**: 2915
worm shell **21**: 2989

Finch
Bengalese **11**: 1544
Chestnut *see* Mannikin, Chestnut-
breasted
Chestnut-breasted *see* Mannikin,
Chestnut-breasted
Cocos Island **5**: 649
Cut-throat **11**: 1544
Gouldian (rainbow finch; painted
finch) **8**: 1030–1031
Ground **5**: 649
Java *see* Sparrow, Java
Locust **21**: 2888
Mangrove **5**: 649
Melba **21**: 2938
Nutmeg *see* Mannikin, Spotted
Red-headed **11**: 1544, *1544*
Spice *see* Mannikin, Spotted
Strawberry *see* Avadavat
Tiger *see* Avadavat
Tree **5**: 649
Vegetarian **5**: 649
Zebra **21**: 3012–3013
Finches, Darwin's (Galapagos
finches) **5**: 648–650
Finfoot **6**: 823–824
African **6**: 823, *823*
American (sungrebe) **6**: 823, 824;
21: 2907
Asian (masked finfoot) **6**: 823
fins **6**: *833*
adipose **20**: 2755
limblike **4**: 497, *497*
Fireback **14**: 1934
Siamese **14**: *1935*
Firebelly *see* Toad, Fire-bellied
Firebrat **2**: 279
Firecrest **7**: 999
Taiwan **7**: 999
Fire eel **18**: 2490
Fire-finch **21**: 2938
Firefly (lightning bug; glowworm)
6: 827–828
see also Glowworm
Fire-fox *see* Panda, Red
Firemouth **6**: 829–830
fires, chaparral **3**: 425
Firetail *see* Damselfly
Fire-toad, Variegated *see* Toad,
Yellow-bellied
Fireworm **16**: 2293
Fish 6: 833–836
air-breathing **2**: 195, 272; **12**: 1685,
1685; **17**: 2412, 2413
cartilaginous and bony **6**: 833, 834,
835; **17**: 2329
cleaner **4**: 533; **18**: 2577; **21**: 2991,
2992
deep-sea *see* deep-sea habitat
fresh water **6**: 835
jawless **10**: 1403–1404; **17**: 2329
marine **6**: 835–836
mouth-brooding *see* mouth-
brooders
reef *see* coral reefs, fish
stomachless **17**: 2368
striped **21**: 3014, 3015
travelling on land **4**: 468; **12**: 1685;
17: 2412
Fish eagle (African fish eagle; river
eagle) **6**: 837–839
African **6**: 837
Madagascar **6**: 837
Pallas' **6**: 837
Fisher (pekan) **11**: 1579, 1580–1581,
1580; **19**: 2653, 2729
fisheries/fishing
and the albatross **1**: 37
and Atlantic salmon **1**: 101
bass **2**: 163

cod **4**: 496
dogfish **5**: 686, 688
and dolphins **4**: 509, *509*; **5**: 698;
20: 2775
edible crab **6**: 741, *742*
the fisher's fly **12**: 1594
goby **7**: 995
grouper **8**: 1083
gudgeon **8**: 1090
haddock **8**: *1113*
hake **8**: 1120
halibut **8**: 1122
hammerhead shark **8**: 1128
herring **9**: 1185–1186
ling **11**: 1455
mackerel **11**: 1518
menhaden **12**: 1598
pilchard **14**: 1949
pollack **14**: 2004
porbeagle **15**: 2021
prawn **15**: 2044
roach **16**: 2183, *2183*
sailfish **16**: 2219–2220
salmon **13**: 1842
sawfish **16**: 2239
scad **16**: 2242, 2243
and sea lions **16**: *2284*
and seals **16**: 2289
sharks **17**: 2333
shore crab **17**: 2355
silverside **17**: 2380
skate **17**: 2385
smelt **17**: 2410, *2411*
snoek **17**: *2430*, 2431
sturgeon **18**: 2556
sweetfish **19**: 2596, *2597*
swordfish **19**: 2607, *2607*
tarpon **19**: 2636, 2637
and thresher sharks **19**: 2675
for thresher sharks **19**: 2674, 2675
tuna **20**: 2775, *2775*
turbot **20**: 2783
using cormorants **19**: 2596, *2597*
using purse-seine nets **4**: 509
whiting **21**: 2934
see also whaling
fishing
by bald eagles **1**: 124
by fish-eating bats **6**: 840–841, *841*
by fish owls **6**: 842
by kingfishers **10**: 1356, *1357*
by night herons **13**: 1738, *1739*
by ospreys **13**: 1811–1812, *1812*
by pelicans **14**: 1909, *1910*
by skimmers **17**: 2386–2387, *2386*
fish ladders **1**: 100
Fish owl (fishing owl) **6**: 842–843;
13: 1827
Blakiston's **6**: 843
Brown **6**: *842*, 843
Buffy **6**: 842, *843*
Pel's (African fishing owl; African
fish owl) **6**: 842, 843
Rufous **6**: 842
Tawny **6**: 842–843
Vermiculated **6**: 842
Fissurella costaria **11**: 1451, 1452
Fitchew *see* Polecat
Flagfish **19**: 2705
Flamingo **6**: 839, 844–847
Andean **6**: 844, 846, *846*
Chilean **6**: 844, 846
Greater (Caribbean flamingo)
6: 844, 845–846, *845*
James' **6**: 844, 846
Lesser **6**: *844*, 846
Flapshelled turtle (flapshell) **17**: 2442
Indian **17**: 2442
Senegal **17**: 2443
Flatworm **6**: 848–849

Flea 6: 850–852
  Bird 6: 852
  Cat 6: 850, 851, 852
  Gray squirrel 6: *850*
  Hen 6: 850
  Human 6: 850, 851, 852
  Jigger 6: 851
  Mole 6: 852
  Oriental rat 6: 850
  Sticktight 6: 851
Flectonotus pygmaeus 11: 1574, *1574*
Flehmen pose 20: 2867
Flicker 6: 853–855
  Andean 6: 853
  Campo 6: 853
  Chilean 6: 853
  Fernandina's 6: 853
  Gilded 6: 853, *855*
  Northern (common flicker;
    yellow-shafted flicker; red-
    shafted flicker) 6: 853, *853*, 854,
    854, 855
Flies 6: 856–859
flight
  albatross 1: 35–36, *35*; **16**: 2267
  buzzard 3: 351, 353
  capercaillie 3: 376–377
  crane fly 5: *588*, *589*, 590
  flies 6: 856–857
  frigate bird 7: 891, *891*
  frogmouth 7: 900
  golden eagle 7: 1001
  greylag goose 8: 1073, *1074*
  hoatzin 9: 1196
  hoverfly 9: 1250
  hummingbird 9: 1256–1257, *1256*
  ibis 9: 1275
  kestrel 10: 1345
  kite 10: 1368
  marabou 11: 1553
  martin 12: 1592
  merlin 12: 1602
  pelican 14: 1910; **16**: 2267
  seabirds 16: 2267
  shearwater 17: 2334
  swan 18: 2588
  swift 19: 2598–2599
  tomb bat 19: 2702
  turkey vulture 20: 2787
  vulture 20: 2839–2840, 2842
  wagtail 20: 2844
Flightbird *see* Budgerigar
Flither *see* Limpet, Common
Flitter *see* Limpet, Common
Floating shell of the high seas *see*
  Sea snail, Purple
flocking *see* living in groups;
  migration
floodplains 21: 2905
floods 16: 2176
Florican
  Bengal 3: 338
  Lesser 3: 338
Flounder (right-eye flounder) 6: 833,
  860–861
  Kamchatka 8: 1121
Flowerpecker 7: 868–869; **18**: 2456
  Mistletoe *see* Mistletoe bird
  Orange-bellied 7: *869*
  Scarlet-backed 7: 868
  Thick-billed 7: 869
  Tickell's 7: 869
Fluke (*fish*) *see* Dogfish
Fluke (*flatworm*) 7: 870–871;
  14: 2011
  Blood 7: 870, 871
  Chinese liver 7: 871
  Intestinal 7: 870, 871
  Liver 7: 870, *870*, *871*; **14**: 2011
  Lung 7: 871

Fly
  Attic *see* Blowfly
  Brine *see* *858*
  Caddis *see* Caddis fly
  Cluster *see* Blowfly
  Drone 9: 1251
  Fruit 7: 909–910
  Harlequin *see* Midge, Nonbiting
  Hessian 6: 859; **12**: 1608
  Ichneumon *see* Wasp, Ichneumon
  Lantern *see* Lantern fly
  Mantid *see* Mantis fly
  Mantis *see* Mantis fly
  Narcissus 9: 1251
  Robber 3: 318; 6: 859, *859*
  Sand *see* Midge, Biting
  Screwworm *see* Blowfly
  Sponge *see* Sponge fly
  Spongilla *see* Sponge fly
  Stable 9: 1242–1243
  Tachina (tachinid fly) 17: 2345;
    19: 2610–2611
  Tsetse 6: 859; **20**: 2760–2761
  Window 4: 471
  see also Flies; Horsefly; Housefly;
    Hoverfly
Flycatcher (*New World*)
  Acadian 20: 2802
  Boat-billed 20: 2801
  Many-colored tyrant 20: 2800
  Northern royal 20: 2800
  Scissor-tailed 20: 2800, *2802*
  Tyrant 4: 555; 7: 872; **18**: 2455;
    19: 2616; **20**: 2800–2802
  Vermilion 20: 2800, 2801, *2801*
Flycatcher (*Old World*) 7: 872–874;
  18: 2456, 2458
  Collared 7: 873, 874
  Madagascar paradise 7: *873*
  Pied 7: 872–873, 874; **18**: 2459
  Semicollared 7: 873
  Spotted 7: 872–873, *874*
Flying fish 6: 836; 7: 875–876;
  17: 2414
  Atlantic 7: 876
  California 7: 875
  Caribbean 7: 875
  Four-winged 7: 875, *875*, 876
  Two-flying 7: 875
Flying fox 2: 167; 7: 877–879; **15**: 2110
  Gray-headed 7: 879
Flying gurnard 7: 880–881
  Atlantic 7: 880, 881, *881*
  Oriental 7: 880–881
Flying lemur 7: 882–883
Flying mouse *see* Scaly-tailed
  squirrel
Flying squirrel (*gliding possum*) *see*
  Gliding possum, Lesser
Flying squirrel (*squirrel*) 7: 884–886
  Eurasian (European flying
    squirrel) 7: 884, 885
  Kashmir 7: 884
  North American 7: 884, 885, 886;
    19: 2616
  Pigmy 7: 884
  Scaly-tailed *see* Scaly-tailed
    Squirrel
  Southern 7: *884*
Foetorepus goodenbeani 5: 706
Foliage-gleaner 13: 1819
Fondiator 7: 876
food-begging 1: 104
food hoarding
  bellbird 2: 191–192
  butcherbird 3: 341
  chipmunk 4: 450
  dung beetle 16: 2252
  flying squirrel 7: 885–886
  gerbil 7: 950–951

ground squirrel 8: 1079
  jay 10: 1319
  kangaroo rat 10: 1339
  merlin 12: 1603
  mole 12: 1626
  nutcracker 13: 1758–1759
  nuthatch 13: 1761–1762
  pika 14: 1943
  rook 16: 2206
  shrike 17: 2362
  steppe lemming 18: 2528–2529
  tree squirrel 20: 2740–2741
  water vole 21: 2884
  wood mouse 21: 2975–2976
  woodpecker 21: *2978*
  zorilla 21: 3020
food stealing *see* robbery
food vacuoles 1: 41
food washing 15: 2098
food webs, coastal 4: 475
Forcipomyia 12: 1607
forensic science, and flies 6: 859
Forficula auricularia 6: 738, 739, *740*
Formica 21: 2964
  aquilonia 21: 2965
  fusca 21: 2964–2965
  lugubris 21: 2965
  pratensis 21: 2964, 2965
  rufa 1: 68; 21: 2964
  sanguinea 21: 2965
Formicarius analis 1: 55, 56
Forpus passerinus 14: 1885
Fortune Bay Riot 9: 1186
Fossa 17: 2373
Fossa fossa 4: 460
fossil behavior 14: 1987
fossils
  earless monitor 6: 733
  living *see* living fossils
  penguin 14: 1911
  peripatus 14: 1925
  sifaka 17: 2373
  stonefly 18: 2549
  tapir 19: 2632
  termites 19: 2658
  turtle 20: 2795, 2797
Foul-marten *see* Polecat
Foumart *see* Polecat
Four-eyed fish 7: 887–888
  Finescale 7: 888
  Pacific 7: 888
Fox
  Arctic 1: 83–84; 5: 691
  Bat-eared 2: 165–166; 5: 665, 689
  Blue 1: 83
  Crab-eating (crab-eating dog;
    crab-eating zorro; wood fox)
    4: 566–567
  Culpeo 5: 689, 691
  Fennec (fennec) 2: 165; 5: 665;
    6: 809–811
  Gray (tree fox) 3: 426; 5: 689;
    8: 1045–1046
  Indian (Bengal fox) 5: *690*
  Island (beach fox) 8: 1045, 1046
  Raccoon *see* Ringtail
  Red 5: 689; **15**: 2132–2134; **19**: 2653
  Sand (pale fox) 5: 689
  San Joaquin kit fox 4: 560
  South American gray (zorro)
    5: *689*
  Swift (kit fox) 4: 560; 5: *691*, *692*
Fox-cat *see* Panda, Red
fragmentation 17: 2441
Franciscana *see* Dolphin, La Plata
Francolin 7: 889–890
  Black (black partridge) 7: 889, 890
  Chinese 7: 890
  Coqui 7: 889
  Crested 7: 890

Gray 7: 890
  Gray-winged 7: 889
  Hartlaub's 7: 889–890
  Natal 7: *889*
  Painted 7: 890
  Swamp 7: 890
Francolinus francolinus 7: 890
Franklin, Benjamin 1: 125
Frankliniella occidentalis 19: 2677
Fratercula
  arctica 15: 2065, 2066; **16**: *2228*
  cirrhata 15: 2065
  corniculata 15: 2065
Fregata
  andrewsi 7: 893
  aquila 7: 893
  magnificens 7: 891, 892
  minor 7: *892*
frenulum 17: 2392; **19**: 2603
Freshwater bream *see* Tilapia
Freshwater shrimp
  European 17: 2364
  Giant 17: 2364
Freshwater wolf *see* Bowfin
Friarbird 9: 1215
  Helmeted 9: 1215
  Noisy 9: 1215
Fricke, Hans 4: 498
Frigate bird 7: 891–893;
    16: 2266–2267
  Ascension 7: 893
  Christmas Island 7: 893
  Great 7: *892*
  Magnificent (man-o'-war bird;
    hurricane bird) 7: *891*, 892
Fringewing *see* Thrips
Fringilla coelebs 3: 415
Fritillary 7: 898–899
  Arctic 7: 898
  Dark green 7: 898–899
  Glanville 7: 898, 899
  Gulf 7: *898*
  High brown 7: 898–899
  Marsh 7: 898, 899
  Pearl-bordered 7: 898
  Queen of Spain 7: 898
  Silver-washed 7: 898–899
  Small pearl-bordered 7: 899, *899*
Frog 7: 903; **20**: 2819
  American tailed (bell toad)
    9: 1200–1201
  Archey's 9: 1200
  Arrow-poison *see* Arrow-poison
    frog
  Arum 15: 2147
  Barking 19: 2724
  Bird-voiced 19: 2724
  Bonyheaded 19: 2725
  Chorus 19: 2723, 2724
  Clawed (African clawed toad)
    4: 464–465; 7: 903
  Common (grass frog) 4: 510–511;
    6: 744
  Cricket 19: 2723–2724
  Cuban 8: 1066
  Dart *see* Arrow-poison frog
  Edible 6: 743–744
  Flying *see* Gliding frog
  Gaudy leaf 19: *2723*
  Ghost 7: 957
  Gliding *see* Gliding frog
  Goeldi's 19: 2725
  Grass *see* Leopard frog
  Green 19: 2724
  Greenhouse 8: 1066–1067; **12**: 1610
  Hamilton's 9: 1200, *1201*
  Hochstetter's 9: 1200–1201
  Horned *see* Horned frog
  Leopard *see* Leopard frog
  Marsh 6: 743, 744

Frog (*continued*)
  Marsupial *see* Marsupial frog
  Meadow *see* Leopard frog
  Mouth-brooding (Darwin's toad;
    Darwin's frog) **12**: 1678–1679
  Paradoxical **14**: 1876–1877
  Poison-dart *see* Arrow-poison frog
  Pool **6**: 743
  Reed *see* Reed frog
  Sedge *see* Reed frog
  Shad *see* Leopard frog
  Smith (blacksmith frog) **19**: 2723
  Spring peeper **7**: 903, *904*
  Stephens Island **20**: 2869
  Tongueless *see* Toad, Surinam
  Tree *see* Tree frog
  Water-holding **10**: 1399;
    20: 2868–2869
  Wood **7**: 903
  see also Bullfrog; Frogs and toads
Frogfish **1**: 51, 52
Froghopper **9**: 1224, 1225
Frogmouth **7**: 900–901
  Large **7**: 901
  Tawny (mopoke; freckled
    frogmouth) **7**: 900, *900*, 901
Frogs and toads 7: 902–905
  eaten by blowfly maggots **2**: 242
Fruit bat (*New World*) **7**: 906, *907*
  Short-tailed **7**: *907*
Fruit bat (*Old World*) **7**: 906–907
  Arabian straw-colored **7**: 908
  Epauletted **7**: 907–908
  Franquet's **7**: 908
  Hammerhead **7**: 908
  Pollen-eating (long-tongued bat)
    7: 908
  Short-eared **7**: 907
  Straw-colored **7**: 908
  Tube-nosed **7**: 908
  see also Flying fox
Fruit dove
  Jambu **7**: *911*
  Makatea **7**: *912*
Fruit pigeon **7**: 911–912
  Blue **7**: 911
  New Zealand (New Zealand
    pigeon; native pigeon) **7**: 912
  Wedge-tailed **7**: 911
  Yellow-bellied **7**: 911
Fruitworm, Tomato *see* Bollworm,
    American
fugu **15**: 2064
Fulica **4**: 523, 524
Fulmar (Atlantic fulmar; Pacific
    fulmar; northern fulmar; fulmar
    petrel) **7**: 913–915; **16**: 2265, 2268
  Antarctic (silver-gray fulmar;
    southern fulmar) **7**: 913
Fulmarus
  glacialis **7**: 914; **16**: 2268
  glacialoides **7**: 913; **17**: 2435
Funambulus **13**: 1859, 1860
  pannanti **13**: *1860*
  tristriatus **13**: *1859*
Funarius **13**: 1819, 1820
Fundulus zebrinus **21**: 3015
fungi
  farmed by ants **10**: 1417, *1417*
  temperate forest **19**: 2655
Fungia **4**: 528
Funiculini quadrangularis **16**: 2298
Funnel-web spider **18**: 2480; **20**: 2820
  Sydney **18**: 2482
furcula **18**: 2504
Fur seal **7**: 916–918; **16**: 2285
  Antarctic (Kerguelen fur seal)
    7: *916*, 917
  Northern (Pribilof fur seal; Alaska
    fur seal) **7**: 916, 917, 918; **16**: *2288*

  South African **7**: *917*
  South American **7**: 918
  Southern **7**: 916, 917, 918, *918*
furs, hunted for
  beaver **2**: 179
  brush tail possum **3**: 301
  chickaree **4**: 438
  chinchilla **4**: 445
  colobus monkey **4**: 503
  coypu **4**: 561
  desman **5**: 667
  ermine **6**: 785
  giant panda **7**: 963
  gray fox **8**: 1046
  hare **8**: 1138
  jaguar **10**: 1312
  leopard **11**: 1446
  lynx **11**: 1508
  marten **11**: 1581
  muskrat **12**: 1706, 1708
  ocelot **13**: 1777, *1777*
  polar bear **14**: 2000
  raccoon **15**: 2098
  raccoon dog **15**: 2100
  red panda **15**: 2136
  sea otter **16**: 2294, 2295
  snow leopard **17**: 2434
  wolverine **21**: 2960
Fynbos *see* Chaparral

G

Gadus morhua **4**: 496; **6**: 836
Galago
  moholi **3**: 329, 330
  senegalensis **3**: 329, 330
Galago
  Demidoff's dwarf **3**: 329, 330
  Dwarf **3**: 330
  Silver **3**: 329
Galagoides **3**: 330
  demidorff **3**: 329, 330
Galah (rose-breasted cockatoo)
    4: 483, 485
Galathea **18**: 2506, 2507, *2507*
Galatheathuria aspera **16**: 2271
Galea **8**: 1099, 1100
  musteloides **8**: *1100*
Galerida
  cristata **5**: 594
  theklae **5**: 593
Galidia elegans **12**: 1637
Galidictis striata **12**: 1635
Gallinago **17**: 2427, 2428, 2429
Gallinula **12**: 1649, 1650; **15**: 2108
Gallinule **7**: 919–920
  American (American purple
    gallinule) **7**: 919, *919*, 920
  Common *see* Moorhen, Common
  Florida *see* Moorhen, Common
  Purple (purple swamphen) **7**: 919,
    920
Gallirallus
  australis **15**: 2108
  wakensis **15**: 2108
galls **1**: 74; **2**: 183; **3**: 416; **7**: 921, 922,
    922; **12**: 1608; **16**: 2240; **19**: 2677;
    20: 2827
Gallus **10**: 1329, *1330*
Gamasellus racovitzai **1**: 82
Gammarus **17**: 2364
Gangara **17**: 2391
Ganges River **21**: 2905
Ganges stag *see* Deer, Axis
Gannet (northern gannet) **7**: 923–925;
    16: 2264, 2265, 2268, *2268*

  Australian (Pacific gannet) **7**: 923
  Cape **7**: 923, *924*; **16**: 2264
ganoin **2**: 195
Gaper **7**: 926–927
  Blunt **7**: 927
  North American (summer clam;
    rubberneck clam; high-neck
    clam; horse clam; otter-shell
    clam; Washington clam)
    7: 926, 927
  Sand (soft-shelled clam; old maid)
    4: 462; **7**: 926, 927
  Small **7**: 927
Gapper *see* Moccasin, Water
Gar **7**: 928–929
  Alligator **7**: 928, 929
  Cuban **7**: 929
  Longnose (billfish) **7**: *928*, 929
  Shortnose **7**: 928, 929
  Spotted **7**: *928*
  Tropical **7**: 929
Garden spider **18**: 2482
  European (diadem spider; cross
    spider) **13**: 1800
  Lesser **13**: *1800*
Garibaldi **5**: 638, 639
Garpike (*needlefish*) **12**: *1724*, 1725
Garrulus glandarius **10**: 1317, 1318;
    14: 1884; **18**: 2457
Garter snake (ribbon snake)
    7: 933–935
  Checkered **7**: *935*
  Common **7**: 933, 934, *934*
  Mexican **7**: 933, 934
  Mountain **7**: 933–934
  Narrow-headed **7**: 933
  Plains **7**: 933, 934
  San Francisco **17**: 2422
  Short-headed **7**: 933
  Wandering (western terrestrial
    garter snake) **7**: 933, *933*, 934
Gasteropelecus **9**: 1161
Gasterosteus aculeatus aculeatus
    18: 2533, 2534
Gastrosaccus spinifer **13**: 1795
Gastrotheca **11**: 1573; **19**: 2723
  marsupiata **11**: *1573*, 1574
  mertensi **11**: 1574
Gato *see* Jaguarundi
Gaur (Indian bison; seladang) **3**: 305,
    306; **7**: 936–938; **21**: 3018
  Domestic (gayal; mithan) **7**: 937,
    938
Gavia
  adamsii **11**: 1485
  arctica **11**: 1485; **21**: 2906
  immer **11**: 1485, 1486
  stellata **11**: 1485
Gavial *see* Gharial
Gavialis gangeticus **7**: 955; **21**: 2907
Gayal *see* Gaur, Domestic
Gazella **1**: 61; **7**: 939, 940
  dama **7**: 940
  dorcas **7**: 940
  granti **7**: 940
  soemmerringi **7**: 940
  spekei **7**: 940
  thomsoni **1**: 61; **7**: 940; **8**: 1040
  walleri **7**: 952
Gazelle **1**: 61; **7**: 939–942; **10**: 1382
  Clarke's *see* Dibatag
  Dama **7**: 939, 940
  Dorcas **7**: 939, 940, 942
  Giraffe-necked *see* Gerenuk
  Goitered **7**: 940, 941
  Grant's **1**: 62; **7**: 939, *939*, 940, 941
  Mongolian **7**: 939, 942
  Przewalski's **7**: 939
  Sand **7**: 942
  Soemmerring's **7**: 939, 940

  Speke's **7**: 939, 940
  Thomson's (red gazelle) **1**: 61, *61*;
    7: 939, 940, *940*, 941, *942*; **8**: 1040
  Tibetan **7**: 939
  Waller's *see* Gerenuk
Gecko **7**: 943–945; **11**: 1466–1467,
    1469
  Banded **7**: 943
  Fat-tailed **7**: 943
  Flying **7**: 945
  Knob-tailed **11**: *1468*
  Leopard **7**: 943
  Madagascar day **7**: *943*
  Northern leaf-tailed **7**: *945*
  Texas banded **7**: 943
  Tokay **7**: 943
  Tree **7**: 944
Gekko
  gecko **7**: 943
  vittatus **7**: *944*
Gelochelidon nilotica **19**: 2661
Gemfish **17**: 2414
  Black **17**: 2414
gemmules **18**: *2493*, 2495
Gempylus serpens **17**: 2414, 2415
Gemsbok (beisa oryx; tufted oryx;
    true gemsbok) **13**: 1808, 1809,
    1810
Genet **7**: 946–948
  Aquatic **7**: 946, 947, 948
  Blotched (tigrine genet) **7**: 946, 947
  Common (feline genet; small-
    spotted genet) **7**: 946, 947, 948
  Ethiopian **7**: 946
  Giant **7**: 946
  Rusty-spotted **7**: 946
genetic control, of bollworms **2**: 256
Genetta **7**: 946, 947
Geocapromys **15**: 1263, 1264
Geochelone **19**: 2711, 2712; **20**: 2712,
    2798, 2799
Geococcyx
  californianus **3**: 426; **16**: 2184, 2185
  velox **16**: 2184
Geocolaptes olivaceus **21**: 2979
Geoduck **4**: 462
  Pacific **4**: 462; **7**: 926
Geogale aurita **19**: 2657
Geomys **14**: 1997
Geopsittacus occidentalis **14**: 1880
Geosciurus inauris **12**: 1595
Geositta **13**: 1819
Geospiza **5**: 649
  difficilis **5**: 649
Geotria **10**: 1403, 1404
Geotrupes **16**: 2253
  stercorarius **16**: 2254
Gerbil (sand rat) **2**: 166; **5**: 665;
    7: 949–951
  Fat sand rat **7**: 950, 951
  Great **7**: 949, 951
  Greater Egyptian **7**: *949*
  Indian **7**: 950, 951
  Mongolian (clawed jird) **7**: *950*
  Przewalski's **7**: 949
Gerbillus pyramidum **7**: *949*
Gerenuk (giraffe-necked gazelle;
    Waller's gazelle) **7**: 952–953
Geronticus eremita **9**: 1274
Gerrhonotus **3**: 426
Gerris **14**: 2009, *2009*
Gesner, Conrad **3**: 369
Gharial (gavial) **7**: 954–956; **21**: 2907
  False **5**: 603; **7**: 954, 955–956
Ghostbird *see* Shearwater, Sooty
Giant rat **7**: 964–965
  African *see* Pouched rat, African
    giant
Giant salamander
  Chinese **9**: 1174, *1174*, 1175

Japanese **9**: 1174, 1175
see also Hellbender
Gibbon **1**: 71, 73; **7**: 968–971
Agile **7**: 968
Black-capped (pileated gibbon)
7: 968, 970
Bornean (Mueller's gibbon)
7: 968, *970*
Concolor (crested gibbon) **7**: 968,
970, *971*
Hoolock **7**: 968, 970
Javan (silver gibbon; moloch
gibbon) **7**: 968, 970
Kloss's (dwarf gibbon) **7**: 968
White-handed (lar gibbon) **1**: *71*;
7: 968, *968*, 969, *969*
Gibbula cineraria **19**: 2706
Gila monster **7**: 972–974; **20**: 2818
gill rakers **2**: 160, *162*; **17**: 2328;
21: 2915
Gilpinia hercyniae **16**: 2241
Gilt-tail **6**: 734
ginko trees **2**: 273
Giraffa camelopardalis **7**: 976
Giraffe **7**: 975–977
Girty *see* John Dory
Glareola
nordmanni **15**: 2041
pratincola **15**: 2041, 2042
Glassan *see* Coalfish
Glassfish **6**: *834*; **7**: 978–979
Asiatic **7**: 978, 979
Cuban **17**: 2379
Indian (Indian glassy fish)
7: 978–979, *978*
Glass lizard **6**: 733; **7**: 980–981;
10: 1430
Eastern **7**: 980, 981
European **7**: *980*, 981
Island **7**: 980, 981
Slender **7**: 980, 981
Glaucidium
minutissimum **13**: 1825
passerinum **13**: 1829
Glaucomys
sabrinus **7**: 884; **19**: 2616
volans **7**: 884, 885
glaucothoes **16**: 2186
gliding
albatross **1**: 35; **16**: 2267
flying fish **7**: 876
flying frogs **7**: 904
flying squirrel **7**: 884, *884*, 886, *886*
gliding frog **7**: 983, *983*
gliding possum **7**: 984, *984*,
985–986, *985*
lemur **7**: 882–883
scaly-tailed squirrel **16**: 2250
seabirds 16: 2267
tree snake **19**: 2732, *2733*
Gliding frog (flying frog)
7: 904, 982–983
Malayan **7**: 982, *982*
Wallace's **7**: 982
Gliding possum **7**: 984–986
Greater (greater glider) **7**: 984–985,
986
Lesser (lesser glider; sugar glider;
flying squirrel; sugar squirrel;
lesser flying phalanger)
7: 984–985, *984*, 986, *986*
Pygmy glider (feathertail glider)
7: 984, 986
Glirulus japonicus **5**: 699
Glis glis **5**: 699
global warming
and collared doves **4**: 499
and coral reefs **4**: 533
and green lizards **8**: 1068–1069,
1069

and penguins **1**: 29
in polar regions **1**: 80, 82
and wall lizards **20**: 2849–2850
Globefish *see* Pufferfish
Globicephala **14**: 1952
macrorhynchus **14**: 1952, 1953
melas **14**: 1952, 1953
glochidia larvae **2**: 219
Glory of the sea **4**: 520
Glossina **6**: 859; **20**: 2760
morsitans **20**: 2760
palpalis **20**: 2760
Glossopsitta
porphyrocephala **11**: 1488
pusilla **11**: 1488
Glowworm (glowworm beetle)
7: 987–988
European **7**: 987
see also Firefly
glutathione **15**: 2030
Glutton *see* Wolverine
Glyph **13**: 1736
Glyphorhynchus spirurus **21**: *2969*
Gnat *see* Midge
Gnathoncus **6**: 852
Gnathonemus petersii **13**: 1749
Gnow *see* Mallee fowl
Gnu *see* Wildebeest
Goanna *see* Monitor lizard
Goat **2**: 198
Domestic **21**: 2951, *2951*, 2952
Rocky Mountain **16**: 2199–2200;
19: 2623
Wild (bezoar; Sind ibex; Chiltan
goat; Cretan goat) **9**: 1272;
21: 2951–2953
goat-antelopes **3**: 421; **8**: 1021;
17: 2323; **19**: 2623
Goatfish **7**: 989–990; **12**: 1692
Bigscale **7**: 989, 990
Manybar **7**: 989, 990
Red **7**: 989, 990
Spotted **7**: 989
Yellow **7**: 989, 990, *990*
Goatsucker *see* Nightjar, Eurasian
Gobi Desert **5**: 663–664
Gobio gobio **8**: 1090
Gobiomorus dormitor **7**: 993
Gobiosoma **4**: 533; **7**: 995
Gobius cobitis **7**: 993
Goby **6**: *834*; **7**: 993–995
Arrow **7**: 995
Blind **7**: 995
Catalina (blue-banded) **7**: 993–994
Giant **7**: 993
Luzon **7**: 993
Magnus **7**: *994*
Neon **4**: 533; **7**: 995
Red striped **7**: *995*
Godwit **7**: 996–998
Bar-tailed **7**: 996, 997, *998*
Black-tailed **7**: 996, 997, *997*, 998;
21: 2906
Hudsonian **7**: 996, 997
Marbled **7**: 996, 998
Goldcrest **7**: 999–1000; **18**: 2455
Goldeneye **5**: 717
Golden mole **7**: 1004–1005
Cape **7**: 1005
Grant's **7**: 1004, *1004*
Large **7**: 1004
Golden plover **14**: 1988, 1989
American **14**: 1988, 1989, 1990
Eurasian **14**: 1988, *1990*
Pacific **14**: 1988
Goldfinch **3**: 368; **8**: 1016–1017
American **8**: 1016, 1017, *1017*
Dark-backed (lesser goldfinch)
8: 1016
European **8**: 1016, *1016*

Goldfish (golden carp; gibel carp;
Prussian carp) **8**: 1018–1020
Goliathus giganteus **2**: 186
Gomys **8**: 1040
Gonepteryx rhamni **21**: 2929
Gonodactylus guerinii **11**: 1551
Gonolek **17**: 2361
gonopodium **2**: 233; **7**: 887
gonozooids **16**: 2224
gonys **4**: 568
Goodall, Jane **4**: 442
Goosander *see* Merganser,
Common
Goose
Bar-headed **5**: 717
Barnacle (bar goose) **1**: 140–141
Bean *see* Bean goose
Black **5**: 717
Canada **3**: 365–367
Gray **5**: 717; **14**: 1955
Greylag **5**: 717; **8**: 1073–1074
Hawaiian (nene; lava goose)
5: 717; **9**: 1163–1165
Magpie **5**: 714
Maned (wood duck) **21**: 2970, 2971
Mold *see* Duck, Musk
Pink-footed **14**: 1955–1956
Pygmy **5**: 714; **21**: 2927
Ragitata *see* Sheldrake, Paradise
Red-breasted **5**: *716*
Ruddy-headed **5**: 716
Snow **5**: *715*
Spur-winged **5**: 717
Swan **5**: 717
White pygmy **5**: 714
see also Ducks, geese and swans
Goosefish **1**: 51, 52
Gopher, Pocket *see* Pocket gopher
Gopherus
agassizii **19**: 2711, 2712; **20**: 2799
berlandieri **19**: 2711, 2712
polyphemus **19**: 2711; **20**: *2795*
Goral **8**: 1021–1022; **19**: 2623
Brown (Red goral) **8**: 1021, 1022
Himalayan **8**: 1021, *1021*, 1022
Long-tailed **8**: 1021, 1022
Goraschius **13**: 1737
Gordius **9**: *1237*
villoti **9**: *1238*
Gorgonocephalus **2**: 158, 159
Gorgon's head *see* Basket star
Gorilla
gorilla beringei **1**: 73; **8**: 1024–1025
gorilla gorilla **8**: 1023–1024
gorilla graueri **1**: 71; **8**: 1024
Gorilla **1**: 72, 73; **8**: 1023–1026
Lowland (eastern lowland gorilla;
western lowland gorilla) **1**: 71;
8: 1023–1024, *1023*
Mountain **1**: 73; **8**: 1024–1025, *1026*
Gorytes **9**: 1262
Goshawk (northern goshawk) **2**: 208;
8: 1027–1029
Black **8**: 1027
Chanting **8**: 1027
Crested **8**: 1027
White **8**: 1027
Gourami **6**: 836
Climbing *see* Perch, Climbing
Kissing **10**: 1365–1366
Goureen *see* Shad, Twaite
Grackle **8**: 1032–1034
Boat-tailed **8**: 1032, 1034
Common (bronzed grackle)
8: 1032, 1033, *1033*, *1034*
Great-tailed **8**: 1032, 1033, 1034
Rothschild's *see* Mynah,
Rothschild's
Slender-billed **8**: 1032
Gracula religiosa **12**: 1714

Grallina
bruijni **11**: 1522
cyanoleuca **11**: 1522, 1523
Grammistes sexlineatus **8**: 1083
Grampus *see* Killer whale
Grape phylloxera *see* Vine pest
Graperoot louse *see* Vine pest
Graphium **18**: 2582
antiphates itamputi **18**: *2582*
Graphiurus **5**: 699
Graphocephala coccinea **9**: *1224*
Graphosoma lineatum **17**: *2344*
grappling behavior **18**: 2478
Grapsus grapsus **4**: *570*
grasses, grassland **8**: 1041
Grass grub **19**: 2603
Grasshopper **8**: 1035–1037
Common field **8**: 1037
Common green **8**: 1037
Large marsh **8**: 1036
Long-horned *see* Katydid
Lubber **8**: *1035*
Meadow **8**: 1036–1037
Mottled **8**: 1037
Stripe-winged **8**: 1037
Grassland 8: 1038–1041
Grasswren **21**: 2993
Gravedigger *see* Burying beetle
Grayfish *see* Dogfish
Grayling **8**: 1047–1048
Arctic (American grayling;
Montana grayling; Amur
grayling; Baikal grayling;
Kamchatka grayling; East
Siberian grayling) **8**: 1047
European **8**: 1048
grazing
addax **1**: 25
antelope **1**: *60, 61*
axis deer **1**: 108
bison **2**: 217
blesbok **2**: 239
bontebok **2**: 260
brow-antlered deer **2**: 284
buffalo **3**: 305, *307*
Canada goose **3**: 365
Cape buffalo **3**: 371, 372
chamois **3**: 422
conch **4**: 515
deer 5: 657
dugong **6**: 724
eland **6**: 758
fallow deer **6**: 796
gazelle **7**: 940–941
on grasslands **8**: 1041
kangaroo **10**: 1334–1335, *1335*
lechwe **10**: 1427, *1427*
sika **17**: 2374
viscacha **20**: 2836
warthog **20**: 2861
wildebeest **20**: 2948
zebra **21**: 3010
Great Barrier Reef **4**: 531, 532; **18**: 2518
Great Plague of London **17**: 2347
Grebe **8**: 1061–1063
Atitlan **19**: 2649
Eared **8**: 1061, 1063
Flightless Titicaca (short-winged
grebe) **8**: 1063
Great crested **8**: 1061, *1061*, 1062
Great pied-billed (Atitlan grebe)
8: 1063
Horned **8**: 1061, *1062*, 1063
Little **8**: 1062
Pied-billed **8**: 1061, 1063, *1063*
Puna (Junin) **8**: 1063
Western **8**: 1063
Greeks, ancient
and electric eels **6**: 766
and snakes **15**: 2118

Greenbottle *see* Blowfly
Greenbul **3**: 308
Greenfinch (green linnet) **3**: 368;
    8: 1064–1065; **11**: 1457
Greenfly *see* Aphid
Greenlet **20**: 2831
    Gray-headed **20**: 2831
Green pigeon **7**: 911
    Timor **7**: 912
Greenshank **16**: 2233
Gromphadorhina **4**: 493
    portentosa **4**: 493
grooming
    baboon **1**: 116
    chimpanzees **4**: *441*
    coati **4**: 478
    kudu **10**: 1390
    peccary **14**: 1905
    sifaka **17**: 2371
    spider monkey **18**: 2477
ground effect **16**: 2267
Groundhog (woodchuck; chuck;
        whistle pig) **8**: 1075–1077;
    11: 1568; **12**: 1672
Ground squirrel (rock squirrel;
        suslik) **8**: 1040, 1078–1080;
    12: 1595
    Antelope **8**: 1080, *1080*
    Arctic **8**: 1080
    California **8**: 1078, 1079
    Franklin's **8**: 1078
    Golden-mantled **8**: 1079
    Harris's antelope **8**: 1079, *1079*
    Thirteen-striped **8**: 1078–1079
Ground thrush *see* Pitta
Grouper **4**: 532; **8**: 1081–1083
    Giant **8**: 1082
    Golden-striped **8**: 1083
    Nassau **8**: 1083
    Potato (potato cod; potato bass)
        8: *1081*
    Spotted **8**: *1081*, 1082
groups, living in *see* living in groups
Grouse **8**: 1084–1086
    Black **8**: 1085, 1086
    Blue **8**: 1084, *1084*
    Ruffed **8**: 1084, 1085, *1085*, 1086
    Sage **8**: 1084, 1086
    Sand *see* Sand grouse
    Sharp-tailed **8**: 1084, 1086
    Spruce **8**: 1084, 1086, *1086*
    Willow (willow ptarmigan)
        8: 1084–1086; **15**: 2055, *2057*
Grunion **8**: 1087–1088
Grus **5**: 584, 585; **20**: 2757
Gryllus **5**: 599
    campestris **5**: *598*
Grysbok (Cape grysbok) **18**: 2526,
        2527
    Sharpe's **18**: 2526, 2527
Guácharo *see* Oilbird
Guaguanche **2**: 155
Guanaco **11**: 1470, 1471, 1472, *1472*;
    12: 1672
guano **2**: 263; **4**: 538; **5**: 685;
    13: 1755–1756; **16**: 2268
Guavina **7**: 993
Guayonoche *see* Ringtail, Central
        American
Gudgeon **8**: 1089–1090
Guemal (Andean deer)
        18: 2463–2464
Guenon **8**: 1091–1093; **12**: 1643
    Red *see* Monkey, Patas
Guereza *see* Colobus monkey,
        Guereza
Guillemot (murre) **8**: 1094–1096
    Black **8**: 1094, 1095, 1096, *1096*
    Brünnich's (thick-billed murre)
        8: 1094, 1095–1096

Common (common murre)
    8: 1094, 1095, *1095*, 1096
Guinea fowl **8**: 1097–1098
    Black **8**: 1097
    Crested **8**: 1097, 1098
    Helmeted (tufted guinea fowl)
        8: 1097, 1098, *1098*
    Vulturine **8**: 1097, *1097*
Guinea pig (common cavy;
        Patagonian hare; moco; cui)
        8: 1099–1101
    Domestic **8**: 1099, *1099*, 1100
Guitarfish (shovelnose ray;
        shovelnose shark; fiddle ray)
        8: 1102–1103
    Atlantic (spotted guitarfish)
        8: 1103
    Giant **8**: 1102, 1103
gular plate **2**: 272
gular pouch (gular sac) **7**: *892*, 893;
    14: 1908
Gull **16**: 2265, *2266*
    Black-headed (European gull;
        common gull) **2**: 230–232
    Herring **9**: 1187–1189
    Laughing **2**: 230
    Mediterranean **2**: 230
    Ross's **16**: *2266*
Gulo
    gulo **21**: *2959*
    gulo gulo **21**: *2959*
    gulo luscus **21**: *2959*
Gundi **8**: 1104–1105
Gunnell *see* Butterfish
Gunnell, Rock *see* Butterfish
Guppy (rainbow fish; millions fish)
        8: 1106–1107; **12**: 1664
    Veiltail **8**: *1106*
Gurnard (sea robin; armored sea
        robin; gurnet) **8**: 1108–1109
    Flying *see* Flying gurnard
    Gray **8**: 1108, 1109
    Leopard sea robin **8**: 1109
    Northern sea robin **8**: 1109
    Piper **8**: 1109
    Yellow **8**: 1108, *1109*
Gurnet *see* Gurnard
Guttera pucherani **8**: 1097, 1098
Gymnachirus nudus **17**: 2444
Gymnammodytes semisquamatus
        16: 2227
Gymnarchus niloticus **10**: 1377;
    13: 1749, 1750
Gymnetrus russelii **13**: 1767
Gymnocephalus cernua **14**: 1917, 1918
Gymnocorymbus ternetzi **19**: 2667
Gymnogyps californianus **4**: 516;
    20: 2841
Gymnomuraena zebra **12**: 1658
Gymnothorax **12**: 1659
    dovii **12**: *1659*
    funebris **12**: 1657–1658
Gymnotus carapo **10**: 1377
Gypaetus barbatus **10**: 1402; **20**: 2840
Gypohierax angolensis **20**: 2841
Gyps **20**: 2841
    africanus **20**: *2839*
    bengalensis **10**: 1401
    fulvus **10**: 1401
Gyrfalcon **2**: 210; **8**: 1110–1112
Gyrinophilus **11**: 1504
Gyrinus **21**: 2921
    natator **21**: 2922

H

habitat loss
    albatross **1**: 37
    antelope **1**: 63
    ape **1**: 73

Asiatic black bear **1**: 96
    bat **2**: 169
    bear **2**: 176
    birdwing butterfly **2**: 212
    boa constrictor **2**: 251
    bumblebee **3**: 317
    bushdog **3**: 334
    coucal **4**: 550
    jaguar **10**: 1312
    leopard **11**: 1446
    marmoset **11**: 1565
    marmoset rat **11**: 1567
    panda **7**: 963
    songbird **18**: 2459
    spoonbill **18**: 2498
    stilt **18**: 2542
    tamarin **19**: 2626
    tapir **19**: 2632
    tortoise **19**: 2713
    turkey **20**: 2784–2785
Habu (Japanese pit viper) **14**: 1973
hadal zones **13**: 1772
Haddock **4**: 495; **6**: 836;
    8: 1113–1114
Hadropithecus **17**: 2373
Haemaphysalis **19**: 2682
Haematopota **9**: 1235, 1236
Haematopus **13**: 1835, 1836
Haemopis sanguisuga **10**: 1428
Hagfish **17**: 2329
Hairstreak butterfly **8**: 1117–1118
    Brown **8**: *1117*
    Green **8**: *1118*
    Hewitson's **8**: 1117
Hake **4**: 495; **8**: 1119–1120
    European **8**: 1119, 1120
    Silver **8**: 1119, 1120
Halcyon
    chelicuti **10**: 1357
    sancta **10**: 1355
Halfbeak **12**: 1724, 1726
Haliaeetus **1**: 124; **2**: 208; **6**: 837, 838;
    21: 2907
Haliastur indus **10**: 1367
Halibut **7**: 927; **8**: 1121–1122
    Atlantic **8**: 1121, *1121*, 1122
    Greenland (black halibut) **8**: 1121
    Pacific **8**: 1121
    Shotted **8**: 1122
Halichoerus grypus **8**: 1050; **16**: 2289
Halichondria **16**: 2299
Haliotis **1**: 20, *20*, 21
Halobaena caerulea **15**: 2045
Halobates **14**: 2008, 2009
halteres **5**: *588*, 590; **6**: 856; **7**: 909, 910
Hamerkop (hammerhead;
        anvilhead) **8**: 1123–1125; **18**: 2551
Hamirostra melanosternon **6**: 784
Hammerhead *see* Hamerkop
Hammerhead shark **8**: 1126–1128
    Common (smooth hammerhead)
        8: 1127
    Great **8**: 1127
    Scalloped **8**: 1127, *1128*
    Smalleye **8**: 1127
    Whitefin **8**: 1127
    Winged (winged shark) **8**: 1126
Hamster **8**: 1129–1131
    Common (European hamster)
        8: 1129, *1129*, 1130
    Dwarf **8**: 1129, 1130, *1130*
    Golden **8**: 1129, 1130, 1131, *1131*
    Gray **8**: 1129, 1130–1131
    Romanian **8**: 1131
hamula **18**: 2504
Handfish *see* Angler, Warty
Hanseatic League **9**: 1186
Hapalemur **10**: 1436
Hapalochlaena maculosa **13**: 1781;
    20: 2819–2820

Hapalomys
    delacouri **11**: 1566, 1567
    longicaudatus **11**: 1566, 1567
Hardhead (*bird*) **14**: 1993, 1995
Hardhead (*fish*) **12**: 1618
Hardun *see* Agama lizard, Starred
Hare **8**: 1135–1138
    Belgian **8**: 1135
    Blue **8**: 1137
    Brown (common hare)
        8: 1135–1136, *1135*
    Calling *see* Pika
    Jumping *see* Springhare
    Northern (mountain hare; blue
        hare; varying hare) **8**: 1136, 1137
    Piping *see* Pika
    Plains *see* Jack rabbit, White-tailed
    Prairie *see* Jack rabbit, White-tailed
    Snowshoe **8**: 1136, *1137*, 1138,
        1138; **15**: 1508
    Sumatran *see* Rabbit, Sumatran
    Whistling *see* Pika
harems *see* living in groups
Harpedon nehereus **2**: 257
Harpia harpyja **2**: 208; **8**: 1143
Harpy eagle **2**: 208, 209; **8**: 1142–1144
    New Guinea **8**: 1142
Harpyia **15**: 2076
Harpyopsis novaeguineae **8**: 1142
Harrier **8**: 1145–1146
    African marsh **8**: 1145
    Black **8**: 1145
    European marsh **8**: 1145, 1146
    Marsh (swamp harrier) **8**: 1145
    Montagu's **8**: 1145, *1146*
    Northern (marsh hawk; hen
        harrier) **8**: 1145, *1145*, 1146
    Pallid **8**: 1145
    Pied **8**: 1145
Harrisina americana **3**: 321
Hartebeest **1**: 62; **8**: 1147–1149
    Common (bubal hartebeest; red
        hartebeest; North African
        hartebeest) **8**: 1147, *1147*, 1148,
        1148, 1149, *1149*
    Hunter's **1**: 63
    Lichenstein's **8**: 1147, 1148
Harvest bug *see* Mite, Harvest
Harvestfly, Dogday **4**: 456
Harvest mouse **9**: 1158–1160
    American **9**: 1158, 1159–1160
    Eastern **9**: 1158
    Old World **9**: 1158–1159, *1158*,
        1160, *1160*
    Salt marsh **9**: 1158
    Western **9**: 1158
Hassar **1**: 88
Hatchet fish **9**: 1161–1162
Hawk
    Bat **2**: 210
    Cooper's **18**: 2467
    Fish *see* Osprey
    Galapagos **6**: 650
    Harrier **17**: 2373
    Harris **3**: 426
    Marsh *see* Harrier, Northern
    Pigeon *see* Merlin
    Red-tailed **3**: *353*, 426
    Savanna **2**: *210*
    Sharp-shinned **18**: 2467
    Snail *see* Kite, Everglade snail
    Sparrow *see* Sparrow hawk
Hawk-cuckoo, Indian **5**: 620
Hawk-eagle
    African **12**: 1589
    Ayres' **12**: 1589
    Crowned *see* Eagle, Crowned
Hawkmoth **9**: 1166–1168
    Bee **9**: 1168
    Convolvulus **9**: 1166, 1167, 1168

Death's head **3**: 346; **9**: 1166–1167, 1168, *1168*, 1210
Elephant **9**: *1166*, 1167
Eyed **9**: 1168
Lime **9**: 1166
Oleander **9**: *1167*
Popular **9**: 1166
Privet **9**: 1166
Small hummingbird hawk **9**: 1166
heat control *see* temperature control
Heath hen (*extinct*) **15**: 2036
heat sensors, snake **17**: 2420
hectocotylus **18**: *2509*
Hedgehog (hedgepig; urchin) **1**: 27; **2**: 232; **9**: 1171–1173
Eurasian **9**: 1171–1173, *1172*
Long-eared **9**: *1171*
Hedgepig *see* Hedgehog
Hedge sparrow *see* Accentor, Hedge
Hediste diversicolor **15**: 2104, *2104*, 2105, *2105*
Helarctos malayanus **18**: 2563
Heleophryne **7**: 957, *957*
Helicoverpa **2**: 256
Heliocolenus dactylopterus **16**: 2260, 2261
Heliocopris **16**: 2254
Heliopais personata **6**: 823
Heliopsyche **3**: 354
Heliornis fulica **6**: 823, 824; **21**: 2907
Heliothrips haemorrhoidalis **19**: 2676
Helix aspersa **7**: 930–932
Hellbender **9**: 1174–1175
Helmetbird *see* Vanga, Helmet
Helmet-shrike **17**: 2361
Heloderma
horridum **7**: 972, 974
suspectum **7**: 972, 973, 974; **20**: 2818
Helostoma temminckii **10**: 1366
helping
burying beetle **3**: 327
butcherbird **3**: 341
Cape hunting dog **3**: 375
cheetah **3**: 406
dogs **3**: 692
dolphin **2**: 266
elephant **6**: 770
hamerkop **8**: 1125
hippopotamus **9**: 1195
hoatzin **9**: 1196
honeyguide **9**: 1217–1218
hornbill **9**: 1228
jay **10**: 1319
kelp gull **2**: 263
killer whale **10**: 1351
kookaburra **10**: 1385
lion **3**: 406
moorhen **12**: 1651
night monkey **13**: 1748
pelican **14**: *1910*
spiny mouse **18**: *2492*
tamarin **19**: 2626
tarsier **19**: 2640
tiger **3**: 406
wren **21**: 2995
see also hunting in packs/groups
Hemachatus **2**: 265
Hemaris
fuciformis **9**: 1168
tityus **9**: 1168
Hemicentetes semispinosus **19**: 2656, 2657
Hemichromis bimaculatus **10**: 1326
Hemidactylium **11**: 1504
scutatum **11**: 1503
Hemiechinus auritus **9**: *1171*
Hemigalus derbyanus **13**: 1855, 1856
Hemigrammus
caudovittatus **19**: 2667
erythrozonus **19**: 2667

Hemigrapsus oregonensis **17**: 2353
Hemimerus **6**: 740
hemipenis **1**: 44
Hemiphaga novaeseelandiae **7**: 912
Hemipode, Andalusian *see* Button quail, Common
Hemipus picatus **5**: 621
Hemitaurichthys **1**: 49
Hemitheconyx caudicinctus **7**: 943
Hemitragus **19**: 2612, 2613
hemocyanin **18**: 2481
hemoglobin **2**: 276; **5**: 644; **12**: 1608
hemolymph **18**: 2481
Hen fish *see* Lumpsucker
Henicopernis longicauda **9**: 1211
Henricia sanguinolenta **18**: 2518
Hepialus **19**: 2603, 2604
herbivores *see* browsing; grazing
herds *see* living in groups; migration
Herichthys cyanoguttatum **4**: *459*
hermaphrodites
anemone **1**: 48
anglerfish **1**: 52
barnacle **1**: 139
brittle star **2**: 281
clam **4**: 463
earthworm **6**: 737
fish **6**: 836
garden snail **7**: 932
hydra **9**: 1266
nutshell **13**: 1764
pond snail **14**: 2011
protogynous **8**: 1083
purse sponge **15**: 2074–2075
rhythmical **13**: 1834
sea squirt **17**: 2313
slug **17**: 2409, *2409*
soft coral **17**: 2441
tubifex worm **20**: 2769
Hermit crab **4**: 572; **9**: 1176–1178; **18**: 2507
Common **9**: 1176, 1177, *1177*, 1178
Prideaux's **9**: 1178
Hermodice carunculata **16**: 2293
Heron **9**: 1179–1181
Black **9**: 1181
Blue reef *see* Egret, Eastern reef
Boat-billed **2**: 252; **8**: 1123
Gray **9**: 1179, *1180*, 1181
Great American *see* Egret, Great
Great blue **9**: 1179, 1180, 1181; **21**: *2906*
Great white *see* Egret, Great
Green-backed (green heron) **9**: 1179, *1180*, 1180–1181
Large *see* Egret, Great
Night *see* Night heron
Purple **9**: 1179
Squacco **9**: 1179, 1181
White *see* Egret, Great
Herpestes
auropunctatus **12**: 1637
ichneumon **12**: 1635
urva **12**: 1635
Herring **9**: 1184–1186; **16**: 2270
Atlantic **9**: 1184–1186, *1184*, *1186*
Pacific **9**: 1184, 1185
Herring hopper *see* Leopard frog
Hesperoleon **1**: 65
Heteractis crispa **5**: 640
Heteralocha acutirostris **21**: 2886
Heterocephalus glaber **12**: 1629
heterocotylus **5**: 633
Heterodon
nasicus **9**: 1203
platyrhinos **9**: 1203
simus **9**: 1203
Heterodontus
francisci **15**: 2027

galeatus **15**: 2027
japonicus **15**: 2027
portusjacksoni **15**: 2027, 2028
Heterogale parvula **12**: 1635
Heterohyrax **9**: 1270, 1271
Heterometrus **16**: *2258*
Heteronetta atricapilla **18**: 2538
Heteropsammia **4**: 528
Heterospizias meridionalis **2**: *210*
Heteroxenia **17**: 2441
hibernacula
adder **1**: 26
asp **1**: 97
hibernation (including winter sleeping) **19**: 2616
adder **1**: 26
asp **1**: 97
bats **2**: 169
bear **2**: 175, 176
black bear **2**: 222, 223
brown bear **3**: 294
bullfrog **3**: 312
bumblebee **3**: 317
butterflies and moths **3**: 346
chuckwalla **4**: 453
common frog **4**: 510
crossbill **5**: 609
dead man's fingers **5**: 652
dormouse **5**: *700*, 701
edible frog **6**: 743
fire salamander **6**: 831, 832
garden snail **7**: 931–932
garter snake **7**: 934
groundhog **8**: 1075–1076, *1076*, 1077
ground squirrel **8**: 1078, 1080
hamster **8**: 1130
hedgehog **9**: 1173, *1173*
horseshoe bat **9**: 1239
lacewing **10**: 1392
ladybug **10**: 1395
marmot **11**: 1570
natterjack toad **12**: 1720
newt **13**: 1733
pack rat **13**: *1843*
painted turtle **13**: 1853
pipistrelle **14**: 1963, 1964
polar bear **14**: 1999
pond skater **14**: 2009
poorwill **14**: 2013
raccoon dog **15**: 2099, *2099*
racer **15**: 2101–2102
tomb bat **19**: 2703
viviparous lizard **20**: 2837
wood ant **21**: 2964
hiding *see* ambush; burrowing; camouflage; shy; stalking
Hieraaetus
agresii **12**: 1589
fasciatus **12**: 1589
Himalayas **12**: 1669, 1670, 1672
Himantopterus dohertyi **3**: 321
Himantopus **18**: 2540, 2541
Hinney **21**: 2941
Hinnites giganteus **16**: 2248
Hippocamelus **5**: 658; **18**: 2463–2464
hippocampus **13**: 1759
Hippocampus **16**: 2277, 2278, *2279*
Hippodamia **10**: 1396
Hippoglossus **8**: 1121, 1122
Hippolais pallida **20**: *2859*
Hippopotamus (kiboko) **9**: 1192–1195
Pygmy **9**: 1192, *1193*, 1195
Hippopotamus amphibius **9**: 1192
Hippotragus **1**: 63; **13**: *1830*; **16**: 2215, 2216
Hirola **1**: 63
Hirundapus caudactus **19**: 2600
Hirundichthys **7**: 876

Hirundo **2**: 151, 152; **10**: 1428; **12**: 1590
Histioteuthis bonelliana **18**: 2509
Histrio histrio **1**: 51
Hoatzin **9**: 1196–1197
Hobby **9**: 1198–1199
African **9**: 1198
Hog
Giant forest (western forest hog; eastern forest hog; giant hog) **7**: 958–959
Red River *see* Bushpig
Hogchoker **17**: 2444
Hognose snake (hissing sandsnake; blow viper; puff adder; spread adder) **9**: 1202–1203
Eastern **9**: 1202, 1203, *1203*; **17**: *2421*
Southern **9**: 1202, 1203
Western **9**: 1202, *1202*, 1203
Holacanthus
ciliaris **1**: 50
tricolor **1**: 49
Holocentrus rufus **18**: *2512*
hologamy **1**: 42
Holohalaelurus regani **3**: 407, 408
Holothuria **16**: 2271, *2271*
aspus **14**: 1901
forskali **16**: 2272
tubulosa **14**: 1902
Homarus
americanus **4**: 571; **11**: 1475
gammarus **4**: 571; **11**: 1475, 1476
hominids **1**: 70
homobatrachotoxin **1**: 94
Homopus signatus **19**: 2712; **20**: 2795
Homo sapiens **1**: 70–71
honey **2**: 185; **9**: 1210
and honeyguides **9**: 1217–1219
Honey bear *see* Kinkajou
Honeybee **2**: 185; **9**: 1208–1210
Eastern **9**: 1208, 1209
European (hive bee; western honeybee) **9**: 1208–1209
Giant **9**: 1208, 1209
Small **9**: 1208, 1209
Honey buzzard **2**: 209; **9**: 1211–1212
Barred **9**: 1211
Crested **9**: 1211
Eurasian **9**: 1211
Long-tailed **9**: 1211
Honeycreeper **7**: 868; **9**: 1213–1214; **21**: 2985
Green **9**: 1213, *1213*, 1214
Hawaiian **9**: 1213–1214
Purple **9**: *1214*
Red-legged (blue honeycreeper) **9**: 1213
honeydew
and ants **1**: 68, 74; **2**: 243–244; **8**: 1118; **9**: *1225*; **14**: 1894, *1894*; **21**: 2965
from hoppers **9**: 1224, *1225*
from whitefly **21**: 2932
Honeyeater **7**: 868; **9**: 1215–1216; **18**: 2558
Blue-faced **9**: 1216
Brown **9**: 1215
Cardinal **9**: *1216*
Pygmy **9**: 1215
Rufous-banded **9**: 1216
Rufous-throated **9**: 1216
Singing **9**: 1215
Strong-billed **9**: 1215
White-cheeked **9**: *1215*
White-gaped **9**: 1215
Yellow-tufted **9**: 1215, 1216
Honeyguide **9**: 1217–1218
Greater **9**: 1206, 1217–1218
Lesser **9**: *1218*
Lyre-tailed **9**: 1217, 1218

Honeyguide (*continued*)
    Malaysian **9**: 1217
    Orange-rumped **9**: 1217
Hoofed shell **16**: 2248
Hoopoe **9**: 1222–1223
    Madagascar **9**: 1222
hooves **1**: 33
    antelope **1**: 61
    caribou **3**: 391–392
    chamois **3**: 421, *422*
    klipspringer **10**: 1374
    Rocky mountain goat **16**: 2199
Hopkinsia rosacea **16**: *2299*
Hopper **9**: 1224–1225
Hormogaster **6**: 734, 735
Hornbill **9**: 1226–1228
    Abyssinian ground **9**: 1226, *1226*
    African red-billed **9**: *1228*
    Malabar pied **9**: 1227
    Malayan helmeted **9**: 1226
    Red-billed dwarf **9**: 1226
    Rhinoceros **9**: *1227*
    Rufous-necked **9**: 1226
    Silvery-cheeked **9**: 1226–1227
    Southern ground **9**: 1226
    White-crested **9**: 1227
Horned frog **9**: 1229–1230; **18**: 2465
    Asian **9**: *1229*, 1230
    Bell's **9**: 1229
    Colombian **9**: *1230*
    Ornate **9**: 1229, 1230
    Wied's **9**: 1229–1230
Horned lizard (horned toad)
    9: 1231–1232
    Short-horned **9**: *1231*
    Texas **9**: 1231, 1232
Horned pheasant *see* Tragopan
Horned toad *see* Horned lizard
Hornero (*ovenbird*), Rufous (barber)
    13: 1819, 1820, *1820*
Hornet **9**: 1233–1234; **20**: 2824
    European (Eurasian hornet)
        9: 1233, *1233*, 1234; **20**: 2824
    White-faced **9**: 1233, 1234
    see also Wasp, Vespula
Hornet moth **9**: 1234
    Lunar **9**: 1234
horns
    antelope 1: 60
    blackbuck **2**: 226, *226*
    blesbok **2**: 239, *239*
    buffalo **3**: 306–307
    bushbuck **3**: 331
    Cape buffalo **3**: 370, *371*
    eland **6**: *757, 758*
    gaur **7**: 936, 938
    gazelle **7**: *940*
    hartebeest **8**: 1147, 1149
    Indian buffalo **9**: 1286
    klipspringer **10**: 1374
    kudu **10**: *1390*
    markhor **11**: 1559, *1559*, 1560, *1560*
    musk-ox **12**: 1703
    pronghorn antelope **15**: 2049
    rhino **15**: *2152*, 2155
    saiga **16**: 2217, *2218*
    sheep **2**: 200
    takin **19**: 2622
    waterbuck **20**: 2865
    wild goat **21**: 2951, *2952*
    zebu **21**: *3018*, 3019
    see also antlers
Horntail *see* Sawfly
Horse **8**: 1039
    Przewalski's **8**: 1040, *1041*;
        15: 2052–2054
Horsefly (stout; cleg) **6**: *856*;
    9: 1235–1236
Horse mackerel **11**: 1517
Horse mackerel *see* Scad

Horseshoe bat **2**: 168; **9**: 1239–1240;
    16: 2254
    Greater **2**: *168*; **9**: 1239, *1239*, 1240
    Large-eared **9**: 1240, *1240*
    Lesser **9**: 1240
Horse-stinger *see* Dragonfly
Horse of the woods *see* Capercaillie
Houbaropsis bengalensis **3**: 338
Housefly **9**: 1241–1243
    Lesser **9**: 1241, 1242
Hoverfly **6**: *857, 858*; **9**: 1250–1251;
    20: 2826
    Migrant **9**: 1250
Howler monkey **9**: 1252–1254;
    12: 1645
    Black **9**: 1252, *1252*, 1253
    Brown **9**: 1252, 1253
    Mantled **9**: 1252, 1253
    Red **9**: 1252, 1253, *1254*
    Red-handed **9**: 1253
Hubbsiella sardina **8**: 1087, 1088
Hudson Bay Company **2**: 179;
    12: 1705
Huemul
    North Andean **5**: 658
    South Andean **5**: 658
Huia (*extinct*) **21**: *2886*, 2887
Humahuma-nuku-nuka-a-puaa *see*
    Triggerfish, Blackbar
humans
    buffalo association with **3**: 307
    cormorant association with **4**: 538
    and dogfish **5**: 688, *688*
    and dogs **5**: 692
    execution by driver ants **1**: 92
    and multimammate rats **12**: 1694
    see also domestication; hunting (by
        man)
Humblebee *see* Bumblebee
Humboldt, Alexander von **4**: 518;
    10: 1312
Hummingbird **9**: 1255–1257; **12**: 1673
    Bee **9**: 1255
    Giant **9**: 1255; **12**: *1672*, 1673
    Hillstar **12**: 1672, 1673
    Ruby-throated **9**: 1255, 1256, *1256*
    Rufous **9**: 1255, *1255*
    Sword-billed **9**: 1255
    Violet-eared **9**: 1257
hunting (by man)
    addax **1**: 24
    agouti **1**: 33
    antelope 1: 63
    armadillo **1**: 87
    bandicoot **1**: 130
    bat-eared fox **2**: 166
    beaver **2**: 179
    beira antelope **2**: 190
    bison **2**: 215–216, 217; **3**: 307
    blackbuck **2**: 227
    bobcat **2**: 254
    bustard **3**: 339
    caiman **3**: *358*, 359–360, *360*
    capybara **3**: 383
    caribou **3**: 394
    cassowary **3**: 398
    dugong **6**: 725
    edible frog **6**: *744*
    elephant **6**: 770
    elephant seal **6**: 773
    emu **6**: *783*, 784
    fennec fox **6**: 811
    fulmar **7**: 915
    fur seal **7**: 918
    gaper **7**: 927
    garden snail **7**: 932
    godwit **7**: 998
    gorilla **8**: 1026
    gray seal **8**: 1049, 1050
    gray wolf **8**: 1057

groundhog **8**: 1077
guenon **8**: 1092–1093
guinea pig **8**: 1101
hare **8**: 1138
harp seal **8**: 1141
hartebeest **8**: 1149
honey ant **9**: 1205
hooded seal **9**: 1221
king penguin **10**: 1360
leech **10**: 1429
leopard **11**: 1446
manatee **11**: 1535
mandrill **11**: 1538
maned wolf **11**: 1540
moose **12**: 1656
mouflon **12**: 1668
mule deer **12**: 1691
musk deer **12**: 1702
musk-ox **12**: 1705, *1705*
muskrat **12**: 1708
oilbird **13**: 1785, *1785*
opossum **13**: 1794
orangutan **13**: 1799
paca **13**: 1839
pademelon **13**: 1847–1848
pangolin **13**: 1863
peafowl **14**: 1900
pochard **14**: 1993, 1994
polar bear **14**: 2000
porpoise **15**: 2025
pronghorn antelope **15**: 2051
puma **15**: 2070
raccoon **15**: 2098
raven **15**: 2124
red fox **15**: 2132, 2134
rhea **15**: 2150
rhino **15**: *2152*, 2155
ringed seal **15**: 2170
seals **16**: 2289
skunk **17**: 2395
snakes 17: 2422
South American deer **18**: 2464
terrapin **19**: 2666
tinamou **19**: 2692
turkey vulture **20**: 2789
turtle **15**: 2149
turtles **20**: 2799
viscacha **20**: 2836
wallaby **20**: 2847–2848
walrus **20**: 2853
wapiti **20**: 2856
warthog **20**: 2862
wolverine **21**: 2960
wombat **21**: 2963
wood duck **21**: 2972
wood pigeon **21**: 2982
woolly monkey **21**: 2987–2988
see also birds, cage; collecting;
    feathers, collected by man;
    fisheries/fishing; furs, hunted
    for; pets; whaling
hunting in packs/groups
    bushdog **3**: 334
    Cape hunting dog **3**: 373–374
    coyote **4**: 558, 560
    dhole **5**: 671
    dingo **5**: 678, *679*
    gray wolf **8**: 1055–1056, *1057*
    hyena **9**: 1268
    killer whale **10**: 1351–1352
    lion **11**: 1462
    piranha **14**: 1968
    see also helping
Hurricane bird *see* Frigate bird,
    Magnificent
Huso huso **18**: 2556
Hutia **9**: 1263–1264
    Bushy-tailed **9**: 1264
    Cabrera's **9**: 1264
    Cuvier's **9**: 1264

Dwarf **9**: 1264
    Garrido's **9**: 1264
    Large-eared **9**: 1264
    Venezuelan **9**: 1263
Hyaena hyaena **9**: 1267, 1268
Hyas **18**: 2476
Hybopsys gracilis **12**: 1618
hybrids
    Atlantic salmon **1**: 101
    black molly **2**: 234
    canary **3**: 368
    capercaillie **3**: 376
    crocodile **5**: *605*
    crow **5**: 612
    dingo **5**: 677, *677*, 679
    edible crab **6**: 743
    hartebeest **8**: 1149
    muscovy duck **12**: 1700
    peafowl **14**: 1898
    pig-tailed macaque **14**: 1942
    platy **14**: 1984
    Przewalski's horse **15**: 2052
    roach **16**: 2182
    sika **17**: 2375
    stilt **18**: 2542
    swordtail **19**: 2609
    tilapia **19**: 2689–2690
    wapiti **20**: 2854–2855
    wild ass **21**: 2941
    yak **21**: 3000
    zebu **21**: 3018, 3019
Hydra
    oligactis **9**: 1265, 1266
    vulgaris **9**: *1265*, 1266, *1266*
Hydra **9**: 1265–1266
    Brown **9**: 1265, 1266
    Green **9**: 1265, 1266
hydrate mounds **13**: 1773
hydratubas **10**: 1322
Hydrobates pelagicus **18**: 2554, 2555
Hydrochaeris hydrochaeris **3**: 383;
    8: 1040; **9**: 1270
Hydroid *see* Sea fir
Hydromyles **16**: 2270
Hydrophasianus chirurgus **9**: 1289
Hydrophilus
    piceus **20**: 2863, *2863*, 2864
    triangularis **20**: 2863
Hydrophis **20**: 2820
    inornatus **17**: 2309
Hydropotes inermis **4**: 447; **5**: 655
Hydroprogne caspia **19**: 2661
Hydrosaurus **1**: 30
hydrothermal vents **13**: 1773
Hydrurga leptonyx **1**: 82; **10**: 1360;
    11: 1450
Hyena **8**: 1040–1041; **9**: 1267–1269
    Brown **9**: 1267, 1268, 1269
    Spotted (laughing hyena)
        3: 374–375; **9**: 1267–1269, *1267*,
        1269
    Striped **9**: 1267, *1268*, 1269
Hyla **7**: 982, 983, 903; **19**: 2723, 2724,
    2725
Hylobates **1**: 70; **7**: 968, 969
    lar **1**: 71; **7**: 969
    syndactylus **7**: 969
Hylobius abietis **21**: 2903, 2904
Hylochoerus meinertzhageni **7**: 959
Hylocichla mustelina **19**: 2678
Hylophilus **20**: 2831
    decurtatus **20**: 2831
Hylophylax naevioides **1**: 55
Hylotrupes **11**: 1484
    bajulus **11**: 1484
Hymeniacidon sanguinea **16**: 2300
Hymenolaimus malacorhynchos
    19: 2710
Hymenopus coronatus **11**: 1547, 1548
Hyomys goliath **7**: 965

hypermetamorphosis **11**: 1550; **13**: 1783
Hyperolius **15**: 2146, 2147
Hyperoodon
  ampullatus **2**: 170, 171
  planifrons **2**: 170, 171
Hyperoplus lanceolatus **16**: 2228
hyperparasites
  braconid wasps as **3**: 416
  chalcid wasps as **3**: 416
Hyphessobrycon
  copelandi **19**: 2667
  flammeus **19**: 2667
  rodwayi **19**: 2667
Hypnos monopterygium **6**: 764
Hypoatherina temminckii **17**: 2379
Hypogastrura **18**: 2504
Hypomesus pretiosus **17**: 2410
Hypoplectrus **8**: 1081
Hyposmitta corallirostris **20**: 2813
hypostome **19**: 2681
Hypoterus **13**: 1752
Hypsignathus monstrosus **7**: 908
Hypsipetes
  amaurotis **3**: 308
  madagascariensis **3**: 308
Hypsiprymnodon moschatus **15**: 2115
Hypsypops **5**: 639
  rubicunda **5**: 638
Hyrax (coney) **9**: 1270–1271
  Bush (gray hyrax; yellow-spotted
    hyrax) **9**: 1270, 1271
  Cape (dassie) **9**: 1270, *1270*
  Rock **9**: 1270, 1271
  Tree **9**: 1270, 1271
Hysterocarpus traski **18**: 2573
Hystrichopsylla talpae **6**: 852
Hystrix **5**: 595
  cristata **5**: 596

I

Ibex **9**: 1272–1273; **12**: 1672
  Alpine (Nubian ibex; Siberian
    ibex) **9**: 1272, *1272*, 1273, *1273*;
    21: 2951
  East Caucasus (tur) **9**: 1272, 1273;
    21: 2951
  Spanish **9**: 1272, 1273; **21**: 2951
  Walia **9**: 1272, 1273; **21**: 2951
  West Caucasus (tur) **9**: 1272, 1273;
    21: 2951
Ibis **9**: 1274–1276
  Black (red-naped ibis) **9**: 1276;
    21: *2905*
  Glossy **9**: 1274
  Japanese crested **9**: 1274, 1276
  Northern bald (hermit ibis;
    waldrapp) **9**: 1274, 1275, 1276
  Sacred **9**: 1274, 1275, 1276; **18**: *2497*
  Scarlet **9**: 1274, 1275, *1275*, 1276,
    1276; **21**: *2905*
  White **9**: 1274, *1274*; **21**: *2905*
  Wood *see* Stork, Wood
Ibisbill **18**: 2540
Ice fish **9**: 1277–1278
  Blackfin **9**: 1277, *1277*, 1278
  Ocellated **9**: 1277
  Pike **9**: 1277, 1278
Icefish (*glassfish*) **7**: 978
  Japanese **7**: 978
Icerya purchasi **16**: 2246, 2247
Ichthyomyzon **10**: 1404
Ichthyophaga **6**: 837
Ichthyophis **3**: 357
  glutinosus **3**: 356
Icicle fish **7**: 978
Ictalurus
  furcatus **3**: 402
  punctatus **3**: 401–402

Icteria virens **21**: 2985
Icterus galbula **1**: 127; **20**: 2833
Ictonyx striatus **21**: 3021
Idiurus zenkeri **16**: 2251
Idolum diabolicum **11**: 1547, 1548
Iguana **9**: 1282
  iguana **9**: 1281, 1282; **11**: *1469*
Iguana **9**: 1281–1283
  Desert **9**: 1281, 1282
  Green (common iguana) **9**: *1281*,
    1282, 1283; **11**: *1469*
  Land (terrestrial iguana) **9**: 1281,
    1282, 1283, *1283*
  Marine (Galapagos marine
    iguana) **11**: 1555–1556
  Rhinoceros **9**: 1281, 1282
  Tree **9**: 1282
Ilyodrilus **20**: 2769
Impala (pallah) **1**: 61; **9**: 1284–1285
  implantation, delayed **3**: 294;
    11: 1581; **12**: 1616; **16**: 2296;
    21: 2960
imprinting **8**: 1074
Inachis io **20**: 2810–2811
Inachus dorsettensis **18**: 2476
Indicator **9**: 1206, 1217–1218, *1218*
Indigobird **21**: 2938
  Village **21**: 2936
Indostomus paradoxus **20**: 2766
infanticide
  bullhead **3**: 315
  chameleon **3**: *418*
  gerbil **7**: 951
  hake **8**: 1120
  lammergeier **10**: 1401
inflation
  burrowing toad **3**: 326, *326*
  cat shark **3**: 408
  chameleon **3**: 419–420
  chuckwalla **4**: 454
  elephant seal **6**: *771*, 773
  grass snake **8**: *1043*, 1044
  hognose snake **9**: 1203
  hooded seal **9**: 1221
  marine toad **11**: 1558
  moloch **12**: 1630
  porcupine fish **15**: 2022, *2022*
  pufferfish **15**: *2062*, 2063–2064,
    2064
  stifftail **18**: 2537, 2539
  walrus **20**: 2853, *2853*
inguinal gland **4**: 446
Ingwenkala *see* Serval
Inia geoffrensis **5**: 697; **16**: 2176, 2177,
  2178
injury, feigning *see* distraction
  displays
Innkeeper worm *see* Tubeworm
inquilines **7**: 922
instars **2**: 186–187
intermedin **1**: 54
intestine, breathing through **11**: 1473,
  1474
introduced species
  coypu **4**: 561–562
  and desmans **5**: 667
  and dormouse possums **5**: 703
  and flickers **6**: 855
  and fruit pigeons **7**: 912
  garden snail **7**: *930*
  giant snail **7**: 967
  goats **21**: 2953
  and grebes **8**: 1063
  guppy **8**: 1106
  and Hawaiian geese **9**: 1165
  and hutias **9**: 1264
  and Java sparrows **10**: 1316
  and kiwis **10**: 1373
  largemouth bass **2**: 164; **8**: 1063
  and mallee fowl **11**: 1529

marine toad **11**: 1557, 1558
mongoose **3**: 325; **9**: 1165
Nile perch **13**: 1751
  and pademelons **13**: 1848
rabbit **15**: 2108
  and rails **15**: 2108
starling **6**: 855
tahr **19**: 2613
vine pest **20**: 2827
Inyala *see* Nyala
Iphiclides podalirius **18**: 2580
Irediparra gallinacea **9**: 1289
Irena
  cyanogaster **6**: 790
  puella **6**: 790, 791
iridocytes **13**: 1780
irruptions
  crossbill **5**: 609
  nutcracker **13**: 1759
  waxwing **21**: 2892
ischial callosities **7**: 970; **11**: 1541
Ischnomyrmex **9**: 1156, 1157
isinglass **4**: 473; **18**: 2557
Isistius **17**: 2332
Isoodon
  macrourus **1**: *131*
  obesulus **1**: 131
Issoria lathonia **7**: 898
Istiblennius enosimae **2**: 237
Istiophorus
  albicans **16**: 2219
  platypterus **16**: 2219, 2220; **19**: 2605
Isurus oxyrinchus **15**: 2020–2021
ivory **6**: 770; **20**: 2851, 2853
Ixobrychus **2**: 220
  exilis **2**: 220
  minutus **2**: 220
Ixodes **19**: *2681*, 2682
Ixoreus naevius **19**: *2678*
Izak *see* Skaamoog

J

Jabiru **18**: 2551
  mycteria **18**: 2551
Jacana (lily-trotter; lotus-bird; water-
  walker) **9**: 1289–1290
  African **10**: *1398*
  Bronze-winged **9**: *1289*, 1290
  Comb-crested (Australian lotus-
    bird) **9**: 1289, *1290*
  Northern (American jacana)
    9: 1289, 1290
  Pheasant-tailed **9**: 1289
Jacana spinosa **9**: 1289, 1290
Jack **16**: 2242
  Bigeye **13**: *1769*
Jackal **8**: 1041; **9**: 1291–1293
  Black-backed **5**: 692; **9**: 1291–1292,
    1292, 1293, *1293*
  Golden (Asiatic jackal) **5**: 689, 691,
    692; **9**: 1291–1292, *1291*, 1293
  Side-striped **9**: 1291, 1292, 1293
Jackass (*bird*)
  Derwent *see* Butcherbird, Gray
  Tasmanian *see* Butcherbird, Gray
Jackdaw (Eurasian jackdaw; western
  jackdaw) **4**: 452; **5**: 610;
  10: 1300–1301; **18**: 2457
Jack Dempsey **10**: 1302–1303
Jack diver *see* Diving-petrel, Common
Jackie *see* Kookaburra
Jack rabbit **8**: 1138; **10**: 1304–1306;
  18: 2561
  Antelope **10**: 1304, 1305
  Black-tailed (jackass hare)
    10: 1304, 1305, *1305*
  White-sided **10**: 1304, 1305
  White-tailed (plains hare; prairie
    hare) **10**: 1304, *1304*, 1305, 1306

Jacksmelt *see* Silverside
Jacobson's organ **7**: 973; **17**: 2420, *2420*
Jaculus **10**: 1324
Jaeger **10**: 1307–1309
  Long-tailed **10**: 1307, *1307*, 1309
  Parasitic (Arctic skua) **10**: 1307,
    1309
  Pomarine **10**: 1307, 1309
Jaguar (onca; el tigre) **1**: 33; **3**: 406;
  10: 1310–1312
Jaguarondi *see* Jaguarundi
Jaguarundi (jaguarondi; otter cat;
  weasel cat; gato; eyra)
  10: 1313–1314
Janthina **15**: 2073
Java temple bird *see* Sparrow, Java
Jay **5**: 610; **10**: 1317–1319
  Blue **10**: 1317, 1319, *1319*
  Blue (*cuckoo-shrike*) *see* Cuckoo-
    shrike, Black-faced
  Eurasian **10**: 1317, *1317*, 1318,
    1318, 1319; **14**: 1884; **18**: 2457
  Gray (Canada jay) **10**: 1317, 1318,
    1319
  Green **10**: 1318
  Scrub **3**: 426; **10**: 1317, 1319
  Siberian **10**: 1318, 1319; **19**: 2617
  Steller's **10**: 1317, 1319
Jay thrush **1**: 114
Jellyfish **10**: 1320–1322; **13**: *1771*;
  14: 1980; **20**: 2818–2819
  Barrel **16**: 2243
  Blue **16**: 2243
  Box (sea wasp) **10**: 1322;
    20: 2818–2819
  Lion's mane **10**: 1320
  Moon (moon jelly; common
    jellyfish) **10**: 1321; **16**: 2243
Jerboa (jumping rat; kangaroo rat)
  5: 665; **10**: 1323–1324
  Marsupial *see* Marsupial mouse
Jewel fish (African jewel fish; jewel
  cichlid) **10**: 1325–1326
Jewel thrush *see* Pitta
Jewfish **8**: 1082–1083
jinking **18**: 2503
Jird
  Clawed *see* Gerbil, Mongolian
  Midday **7**: 951
John Dory (dory; doree; St. Peter's
  fish; girty) **10**: 1327–1328
Jordanella floridae **19**: 2705
Jordanicus **14**: 1901
  gracilis **14**: 1901, 1902
Jug o'rum *see* Bullfrog
Julbernardia pellegriniana **16**: 2251
Jumping plant louse (lerp) **9**: 1224,
  1225
June beetle *see* Cockchafer
June bug *see* Cockchafer
Jungle babbler **1**: 113
Jungle fowl (wild jungle fowl)
  10: 1329–1330
  Gray **10**: 1329
  Green **10**: 1329, *1329*
  Red **10**: 1329, 1330, *1330*
Junin *see* Grebe, Puna
Jynx
  ruficollis **21**: 2996
  torquilla **21**: 2996, 2997

K

Kagu **18**: 2567
Kaka **10**: 1343
Kallima **10**: 1331, 1332
Kamoshika *see* Serow, Japanese
Kangaroo **10**: 1333–1336
  Eastern gray **10**: 1333, 1334,
    1335–1336, *1336*

Kangaroo (*continued*)
Gray (great gray kangaroo) **10**: 1333, 1334
Rat *see* Rat kangaroo
Red **10**: 1333, *1333*, 1334–1336, *1335*
Tree *see* Tree kangaroo
Western gray **10**: 1333, 1334
Kangaroo mouse (hopping mouse; jumping mouse; jerboa mouse) **10**: 1337–1338
American **10**: 1338
Australian **10**: 1338
New Guinea **10**: 1337, 1338
Kassina **15**: 2146
Katsuwonus pelamis **20**: *2773*
Katydid (long-horned grasshopper; bush cricket) **5**: 598, 600; **8**: 1035, 1037; **10**: 1341–1342
Texas bush **10**: 1341, 1342
Kea **10**: 1343–1344
kelinci kerinci, *see* Rabbit, Sumatran
Kelpfish **2**: 238
keratin **1**: 60
digestion **4**: 470
Kerodon **8**: 1099, 1100
Kestrel (windhover) **1**: 135; **10**: 1345–1347
American **2**: 209; **10**: 1345, *1346*; **18**: 2467
Eurasian **10**: 1345, *1345*, 1346, *1347*
Fox **10**: 1345
Greater **10**: 1345
Mauritius **2**: 210; **10**: 1345
Ketupa **6**: 842, 843; **13**: 1827
Kiang **21**: 2939, 2940, *2941*
Kiboko *see* Hippopotamus
Kilimanjaro, Mount **12**: *1669*
Killdeer **10**: 1348–1349
Killer whale (orca; grampus; great killer whale; blackfish; sea wolf) **1**: 82; **2**: 194; **5**: 695; **6**: 773; **7**: 918; **10**: 1350–1352; **12**: 2211; **21**: 3008
False *see* False killer whale
Killie, Zebra *see* Killifish, Plains
Killifish
Cuban **19**: 2705
Plains (zebra killie) **21**: 3015
killing of siblings *see* siblicide
killing of young *see* infanticide
Kingbird **20**: 2802
Eastern **20**: *2800*, 2802
King crab **4**: 572; **10**: 1353–1354
American **10**: 1353–1354, *1354*
Kingfisher **10**: 1355–1357
Amazon **10**: 1355, *1356*
Belted **16**: 2175, *2175*
Common **10**: 1355, 1356, *1357*
Great brown *see* Kookaburra
Green **10**: 1355
Laughing *see* Kookaburra
Malachite **10**: *1355*
Pied **10**: 1355
Racquet-tailed **10**: 1356
Shoe-billed **10**: 1357
Stork-billed **10**: 1356–1357
Striped **10**: 1357
Yellow sacred **10**: 1355
King of the herrings **13**: 1767, *1767*
Kinglet
Golden-crowned **7**: 999
Ruby-crowned **7**: 999, 1000
King of the salmon **13**: 1768
King snake **10**: 1361–1362
Arizona mountain **10**: 1361, *1361*
California **10**: 1361
Common **4**: 535; **10**: 1362
Desert **10**: 1361
Eastern (chain snake) **10**: 1362

Florida (salt-and-pepper snake) **10**: 1361
Kinkajou (honey bear; nightwalker; potto) **10**: 1363–1364
Kinosternon **20**: 2795
hirtipes murrayi **12**: 1687
subrubrum **12**: 1687, 1688
Kite **2**: 209; **10**: 1367–1369
Black (black-eared kite; large Indian kite; pariah kite; yellow-billed kite) **10**: 1367, 1368–1369
Black-breasted *see* Buzzard, Black-breasted
Black-shouldered (black-winged kite) **10**: 1367–1368, 1369
Brahminy **10**: 1367, 1368–1369, *1369*
Everglade snail (Everglade kite; snail kite; snail hawk) **2**: 209; **6**: 788–789
Red **10**: 1367, *1367*, 1368–1369
White-tailed **2**: 209; **10**: 1367
White-winged **10**: 1368
Kittiwake **10**: 1370–1371
Black-legged **10**: 1370, 1371
Red-legged **10**: 1370
Kiwi **10**: 1372–1373
Brown (common kiwi) **10**: 1372, 1373, *1373*
Great spotted (large kiwi) **10**: 1372
Little spotted (little gray kiwi) **10**: 1372
Klauberina riversiana **13**: 1745, 1746
Klipspringer (cliff-springer) **1**: 61; **10**: 1374–1375
Knife fish **10**: 1376–1377
Banded **10**: 1377
Black ghost **10**: *1376*
Ghost **10**: 1376, 1377
Glass **10**: 1376, 1377
Naked-back **10**: 1376, 1377
Old World *see* Featherback
Sand **10**: 1377
Knifefish, Electric *see* Eel, Electric
Knot **16**: 2233
Koala (koala bear) **10**: 1378–1380
Koayu *see* Sweetfish
Kob **1**: 60, 62; **10**: 1381–1382
Kobus **1**: 62; **10**: 1381, 1382, 1425, 1426; **20**: 2866; **21**: *2907*
Koel **5**: 618, 620, 713
Kogia **18**: 2472
breviceps **18**: 2473; **21**: 2911
simus **18**: 2473; **21**: 2911
Kogmania **21**: 2974
Kokako (wattled crow) **21**: 2886, 2887, *2887*
Komodo dragon (Komodo monitor) **10**: 1383–1384; **11**: 1466; **12**: 1638
Kon Tiki (raft) **12**: 1659; **17**: 2414
Koodoo *see* Kudu
Kookaburra (laughing kingfisher; laughing jackass; jackie; great brown kingfisher; alarmbird) **10**: 1355, 1385–1386
Blue-winged **10**: 1385
Korhaan, Black **3**: *338*
Kouprey (Cambodian forest ox) **3**: 307; **7**: 936, 938
Krait **2**: 265
Sea *see* Sea krait
Krill **4**: 564; **10**: 1387–1388; **16**: 2167
Kuaka *see* Diving-petrel, Common
Kudu (koodoo; grand kudu) **3**: 331; **10**: 1389–1391
Greater **10**: 1389, *1389*, 1390, *1390*
Lesser **10**: 1389, 1390
Kulan **21**: 2939, 2940
Kultarr *see* Marsupial mouse
Kurper *see* Tilapia

L

Laba *see* Paca
labella **9**: 1242
Labia minor **6**: 738, 739
laboratory animals
guinea pig **8**: 1101
guppy **8**: 1106
hamster **8**: 1131
pika **14**: 1944
steppe lemmings **18**: 2529
Labrochromis **4**: 459
Labroides phthirophagus **21**: 2992
Labrus bimaculatus **21**: 2991
Labyrinth fish **13**: 1868
see also Perch, Climbing
lac **16**: 2247
Lacerta **8**: 1068, 1069; **16**: 2232; **20**: 2838
Lacewing **10**: 1392–1393
Brown (*larvae*: aphid lion; aphid wolf) **10**: 1392–1393
Dusty-wing **10**: 1392, 1393
Giant **10**: 1392
Green (golden-eye; stinkfly) **10**: 1392, 1393
Lachesis muta **14**: 1972, 1973
Lactophrys
bicaudalis **20**: *2759*
triqueter **20**: *2758*
Ladybird *see* Ladybug
Ladybug (ladybird; lady beetle) **2**: 187; **10**: 1394–1396
2-spot **10**: 1394
7-spot **10**: 1394, *1394*
10-spot **10**: 1394
22-spot **10**: 1394
Eyed **10**: 1394–1395
Laevicardium
crassum **4**: 488
elatum **4**: 488
substriatum **4**: 488
Lagenorhynchus obscurus **5**: 698
Lagidium **12**: 1672; **20**: 2834
Lagocephalus lagocephalus **15**: 2063
Lagopus **20**: 2778
lagopus **1**: 1084; **15**: 2055
leucurus **15**: 2055, 2056
mutus **15**: 2055
Lagostomus maximus **20**: 2834, 2835
Lagostrophus fasciatus **20**: 2846
Lagothrix
flavicauda **21**: 2987, 2988
lagotricha **21**: 2987, 2988
Lagurus lagurus **18**: 2528, 2529
Lake and pond 10: 1397–1400
Lalage
melanoleuca **5**: 621
nigra **5**: 621
sueurii **5**: 621
Lama
glama **11**: 1470, 1471
guanicoe **11**: 1470, 1471
pacos **11**: 1470, 1471
lamellibranches *see* bivalves
Lammergeier (bearded vulture) **10**: 1401–1402; **19**: 2713; **20**: 2840, 2842
Lamna
ditropis **15**: 2021
nasus **15**: 2021; **17**: 2331
Lampetra **10**: 1403, 1404
planeri **10**: *1403*
Lamprey (nannie nine eyes; stone sucker) **2**: 161; **10**: 1403–1404; **17**: 2329
Brook **10**: 1403, *1403*
River **10**: 1403, 1404, *1404*
Sea **10**: 1403, 1404; **17**: *2332*
Lampris
guttatus **13**: 1790, 1791
immaculatus **13**: 1790, 1791

Lampropeltis
getulus **4**: 535; **10**: 1361, 1362
getulus californiae **10**: 1361
getulus floridana **10**: 1361
getulus getulus **10**: 1361
getulus splendida **10**: 1361
pyromelana **10**: 1361, *1361*
triangulum **10**: 1361
triangulum amaura **10**: 1361
zonata **10**: 1361
Lampshell (arm-footed lampshell) **10**: 1405–1406
Lampyris **6**: 828; **7**: 988
noctiluca **7**: 987
Lancehead, Yellow-jawed **20**: 2817
Lancelet (amphioxus) **10**: 1407–1408
Lancet fish (sea wolf; wolf fish) **10**: 1409
Longnose **10**: 1409
Shortnose **10**: 1409
lancets **18**: 2575, *2577*
Langoustine *see* Spiny lobster
Langur **10**: 1410–1412; **12**: 1644
Brow-ridged **10**: 1411
Delacour's **10**: 1412
Douc **12**: 1645
Golden **10**: 1412
Golden snub-nosed **10**: 1410
Hanuman **10**: 1410, 1411, *1411*; **12**: 1643
Maroon **10**: 1412
Pig-tailed **10**: 1410, 1411
Purple-faced **10**: 1410, 1412
Snub-nosed **10**: 1411
Spectacled (dusty langur) **10**: 1410, 1412
White-bellied **10**: *1412*
White-headed **10**: 1412
Lanius **17**: 2361, 2362
Lantern-eyed fish **18**: 2512
Lantern fish **6**: 836; **10**: 1413–1414; **13**: 1772
Lantern fly **9**: 1225
False **9**: 1225
Lanthanotus borneensis **6**: 733; **12**: 1638
Laothoe populi **9**: 1166
Lapwing **10**: 1415–1416; **16**: 2234
Eurasian (northern lapwing; peewit; green plover) **10**: *1415*, 1416
Red-wattled **10**: 1415, *1416*
Sociable **10**: 1415
larch trees **19**: 2615
Lark
Crested **5**: 593–594
Desert **5**: 594
Thekla **5**: 593
Larosterna inca **19**: 2661
Larus **2**: 230, 231; **9**: 1188, 1189
larvae
ant lion **1**: 64, *64*
blowfly **2**: 241, *242*
burying beetle **3**: 327–328, *327*
caddis fly **3**: 354–355
caecilian **3**: 357
campodeiform **11**: 1549–1550
cicada **4**: 457
clothes moth **4**: 470, *470*
cockchafer **4**: 487
Colorado beetle **4**: 504
crab **4**: 573; **6**: 741–742
crane fly **5**: 588–589, 590
damselfly **5**: 642
diving beetle **5**: 682–683
fire salamander **6**: 832
fluke **7**: 870
fly **6**: 858–859
glowworm **7**: 988, *988*
honeybee **9**: 1210, *1210*

larvae (*continued*)
  hoverfly **9**: 1250–1251
  jellyfish **10**: 1322
  krill **10**: 1387
  lacewing **10**: 1393
  ladybg **10**: 1395
  lampshell **10**: 1406, *1406*
  leptocephalus **4**: 522; **6**: 746–747
  mantis fly **11**: 1549–1550, *1550*
  midge **12**: 1607, 1608, *1608*
  mosquito **12**: 1662, 1664, *1664*
  oil beetle **13**: 1782–1783
  plaice **14**: 1977
  sawfly **16**: 2240–2241, *2241*
  stonefly **18**: 2549, *2549*, 2550
  water beetle **20**: 2864
  see also caterpillars; maggots; nymphs
Lasiorhinus
  krefftii **21**: 2961, 2962
  latifrons **21**: 2961, 2962, *2963*
Lasiurus
  borealis **20**: 2821, 2822, *2822*
  cinereus **20**: 2821, 2822
Lasius
  niger **14**: *1895*
  niger americana **14**: 1895
Lassa fever **12**: 1695
lateral line **6**: 836; **17**: 2332–2333
  cave fish **3**: 409–410
Laterallus ruber **5**: 582
Laterinia laterinia **9**: 1225
Lates niloticus **13**: 1752
Laticauda
  colubrina **17**: *2308*, 2309
  crockeri **17**: 2308
  semifasciata **17**: 2309
Latimeria chalumnae **4**: 497–498
latrines, badger **1**: 121
Latrodectus **2**: 235
  hesperus **2**: 236, *236*
  mactans **2**: 235, *235*, 236
  variolus **2**: 236
Laufschlag **13**: 1803, 1809; **18**: 2501, 2527; **20**: 2867
Laughing jackass *see* Kookaburra
Lazyshark *see* Skaamoog
Leaf fish (Amazon leaf fish; barbeled leaf fish; South American leaf fish) **10**: 1421–1422
  Malaysian **10**: 1421
Leafhopper (dodger) **9**: 1224, 1225
  Rhododendron **9**: *1224*
Leafscraper **13**: 1819
Leatherjacket *see* Crane fly
Lechwe (red lechwe; Kafue lechwe; black lechwe) **1**: 62; **10**: 1425–1427, *1426*; **21**: *2907*
  Nile (Mrs Grey's lechwe) **1**: 62, 63; **10**: 1425, 1426
Leech **10**: 1428–1429
  Horse **10**: 1428
  Medicinal **10**: 1428, *1428*, 1429
legends/myths
  anaconda **1**: 44
  barnacle goose **1**: 141
  basilisk **2**: 157
  climbing perch **4**: 469
  devilfish **5**: 670
  dugong **6**: 724
  fer de lance **6**: 813
  glass lizard **7**: 981
  gray wolf **8**: 1057
  groundhog **8**: 1077
  hedgehog **9**: 1171
  hoopoe **9**: 1223
  John Dory **10**: 1328
  kingfisher **10**: 1357
  manatee **11**: 1535
  moray eel **12**: 1659

nightjar **13**: 1744
python **15**: 2081
roadrunner **16**: 2184
scarab beetle **16**: 2254
shrew **17**: 2359
squid **18**: 2509
swan **18**: 2588
tarantula **19**: 2633
Legless lizard **10**: 1430–1431; **11**: 1466
  Burton's **10**: 1431
  California (Geronimo legless lizard) **10**: 1430, *1430*, *1431*
Lehmannia **17**: 2409
  marginata **17**: 2408
Leicesteria **12**: 1664
Leimadophis epinephelus **20**: 2819
Leiopelma
  archeyi **9**: 1200
  hamiltoni **9**: 1200; **20**: 2869
  hochstetteri **9**: 1201
Leioptilus lienigianus **14**: 1992
Leiothrix, Red-billed (Pekin robin) **1**: 113; **16**: 2190
Leiothrix lutea **16**: 2190
Leipoa ocellata **11**: 1509, 1527, 1528
leishmaniasis **15**: 2156
Leithrix
  argentauris **1**: 113
  lutea **1**: 113
leks (display grounds)
  antelope 1: 62
  bird of paradise **2**: 206–207
  capercaillie **3**: 378
  fruit bat **7**: 908
  grouse **8**: 1086
  manakin **11**: 1532
  prairie chicken **15**: 2036–2037
  ruff **16**: 2213–2214, *2213*
  sandpiper **16**: 2234
  snipe **17**: 2429
Lemming **10**: 1432–1434; **17**: 2438; **20**: 2778
  Bog **10**: 1432, 1433
  Brown **10**: 1433
  Collared **10**: 1432, *1432*, 1433, 1434; **20**: *2777*
  Norwegian **10**: 1432, 1433, 1434
  Steppe **18**: 2528–2529
  Wood **10**: 1432, 1433
Lemmiscus curtatus **18**: 2528, 2529
Lemmus **10**: 1433
  lemmus **10**: 1433
  sibiricus **10**: 1433
Lemniscomys **21**: 3016, 3017
Lemur **10**: 1436
  catta **10**: 1435, 1436
Lemur **10**: 1435–1437
  Black **10**: 1435, 1437
  Broad-nosed gentle **10**: 1435
  Crowned **10**: *1437*
  Dwarf **10**: 1435, 1436, 1437
  Flying *see* Flying lemur
  Fork-crowned **10**: 1435
  Gentle **10**: 1435
  Hairy-eared dwarf **10**: 1435
  Mouse **10**: 1435, 1436, 1437
  Ring-tailed **10**: 1435, *1435*, 1436
  Ruffed **10**: 1435, *1436*, 1437
  Sportive (weasel lemur) **10**: 1435, 1436, 1437
Leontopithecus **19**: 2625, 2626
Leopard (panther) **3**: 406; **8**: 1040; **11**: 1444–1446
  Clouded **3**: *404*
  Snow *see* Snow leopard
Leopard frog (meadow frog; grass frog; shad frog; herring hopper) **11**: 1447–1448
  Northern **11**: 1447
  Southern **11**: 1447

Leopardshark *see* Skaamoog
Leopardus
  pardalis **13**: 1777
  wiedi **13**: 1776
Lepas anatifera **1**: 138, 139
Lepidocephalus guntea **11**: 1474
Lepidochelys
  kempii **20**: 2797; **21**: 2907
  olivacea **20**: 2796
Lepidocolaptes
  affinis **21**: 2968
  angustirostris **21**: 2968
  souleyetii **21**: 2969
Lepidoderma albohirtum **11**: 1557
Lepidosaphes ulmi **16**: 2246, 2247
Lepisma saccharina **2**: 279
Lepisosteus
  oculatus **7**: 928
  osseus **7**: 929
  platostomus **7**: 929
Lepomis
  auritus **18**: 2569
  gibbosus **16**: *2174*; **18**: 2569
  megalotis **18**: 2569
Leporillus
  apicalis **18**: 2535, 2536
  conditor **18**: 2535, *2535*, 2536
leprosy **1**: 87
Leptailurus serval **17**: 2326
Leptasterias groenlandica **18**: 2518
Leptidea sinapis **21**: 2929
Leptinotarsa decemlineata **4**: 505
leptocephali **4**: 522; **6**: 746–747
Leptocircus **18**: 2582
Leptoconops **12**: 1607
Leptodactylus fallax **8**: 1066
Leptomicrurus **4**: 534
Leptomithrax spinulosus **18**: 2475
Leptomyrmex **9**: 1205
Leptonychotes weddelli **1**: 82; **16**: 2210, 2288; **21**: 2899–2900
Leptopelis **15**: 2146, 2147
Leptopterus
  chabert **20**: 2813
  madagascarinus **20**: 2813
Leptoptilos **18**: 2551
  crumeniferus **11**: 1554; **18**: 2551
  dubius **11**: 1553
  javanicus **11**: 1553
Leptosiren paradoxa **11**: 1501, 1502
Lepus **8**: 1136, 1137, 1138; **10**: 1304, 1305; **11**: 1508; **18**: 2561
Lerp *see* Hopper
Leto staceyi **19**: 2603–2604
Leuciscus
  cephalus **12**: 1618; **16**: 2182
  leuciscus **12**: 1618
Leucopsar rothschildi **12**: 1716
Leuresthes tenuis **8**: 1087, 1088, *1088*
Leurognathus marmoratus **11**: 1503
Lialis burtonis **10**: 1431
Liasis **15**: 2080
  childreni **15**: 2079
Libellula quadrimaculata **5**: 710
Lichanura trivirgata **2**: 249
Lichenostomus melanops **9**: 1215, 1216
lichens **1**: 82
Lichmera indistincta **9**: 1215
life span *see* long living; short living
light, animals attracted to **8**: 1072; **11**: 1448
Lightfish *see* Bristlemouth
Lightning bug *see* Firefly
light production **18**: 1771
  Bombay duck **2**: 257
  bristlemouth **2**: 278
  brittle star **2**: 281
  cardinal fish **3**: 390
  deep-sea anglerfish **5**: 653, *653*
  firefly **6**: 827, *827*, 828

glowworm **7**: 987, *987*
hatchet fish **9**: 1161, 1162, *1162*
jellyfish **10**: 1320
krill **10**: 1387
lantern-eyed fish **18**: 2512
lantern fish **10**: 1413–1414, *1413*, *1414*
millipede **12**: 1611
pineapplefish **18**: 2512
pyrosoma **15**: *2078*
sea pen **16**: 2298
springtail **18**: 2505
squid **18**: 2509
tubeworm **20**: 2768
whalefish **21**: 2909, *2909*
Ligia **21**: 2974
  baudiensis **17**: 2398–2399
  oceanica **17**: 2398
Lilliputian fauna **2**: 278
Lily-trotter *see* Jacana
Lima **12**: 1713
Limacina **16**: 2270
  retroversa **16**: 2269–2270
Limanda limanda **5**: 635
Limax **17**: 2409
  maximus **17**: 2409
limbs
  loss (autotomy) **5**: 636, 637, *637*; **6**: 741
  regeneration **6**: 741; **11**: 1477
Limenitis
  archippus **3**: 345; **12**: 1634; **20**: 2810, 2811
  arthemis arthemis **20**: 2810, 2811
  camilla **12**: 1634
Limnaea **7**: 870
Limnodrilus **20**: 2769
Limnogale mergulus **19**: 2656, 2657
Limnophilus **3**: 354
Limosa **7**: 996, 997; **21**: 2906
Limothrips cerealium **19**: 2676
Limpet **11**: 1451–1452
  Antarctic **16**: 2226
  Blue-rayed (peacock's feathers) **11**: 1451
  Common (European limpet; flitter; flither; papshell) **11**: 1451, 1452, *1452*
  Keyhole **11**: 1451, 1452
  Slipper **11**: 1452
  Slit **11**: 1451
  Tortoiseshell **11**: 1451, 1452
Limpkin (crying bird; wailing bird) **11**: 1453–1454
Limulus polyphemus **10**: 1353, 1354
Linckia **18**: 2517
Ling **11**: 1455–1456
  Blue (lesser ling) **11**: 1455, 1456
  Common **11**: 1455–1456, *1455*
  Spanish (Mediterranean ling) **11**: 1455
Lingcod **20**: 2766
Lingula **10**: 1405–1406
Linnaeus, Carl **3**: 357
Linnet **11**: 1457, 1457–1458
  Green *see* Greenfinch
  Pine *see* Siskin, Pine
Linsang (Oriental linsang) **11**: 1459
  Banded **11**: 1459, *1459*
  Spotted **11**: 1459
Lion **3**: 406; **8**: 1040; **11**: 1460–1463; **21**: 2950
  African **3**: 406
  Mountain *see* Puma
Lionfish, Red (turkeyfish; lionfish; zebrafish; dragonfish; butterfly cod) **6**: *835*, 836; **16**: 2260–2261
Lion's mane **10**: 1320
Liothrips vaneeckei **19**: 2677
Liphistius **19**: *2719*

Lipotes vexillifer **5**: 697; **16**: 2177, 2178
Lissemys punctata **17**: 2442
Lithobius forficatus **3**: 412, *412*, 413
Lithodes maia **18**: 2507
Lithrurus umbratilis **18**: 2570
Litocranius walleri **7**: 953
Litoria infrafrenaia **19**: *2724*
Little magpie *see* Magpie lark
Littorina **14**: 1926
    littoralis **14**: 1926, 1927
    littorea **14**: 1926, 1927
    neritoides **14**: 1926, 1927
    saxatilis **14**: 1926, 1927
living in complex societies
    ant **1**: 66–67, 90–92
    bees and wasps 2: 185; 3: 317
    burying beetle **3**: 327
    harvesting ant **9**: 1156–1157, *1156*
    honeybee **9**: 1208–1210
    termite **1**: 66–67; **19**: 2658–2660
    vespula wasp **20**: 2824–2826
    wood ant **21**: 2964–2965
living fossils
    bowfin **2**: 272, *272*, 273
    frilled shark **7**: 896
    king crabs **10**: 1353
    nautilus **12**: 1722
    sea lily **16**: 2280
living in groups
    Adélie penguin **1**: 28–29, *29*
    anchovy **1**: 45–46
    antelope 1: 61
    apes **1**: 73
    avocet **1**: 105–106
    axis deer **1**: 107–108
    babbler **1**: 113–114
    baboon **1**: 115–116
    barbary ape **1**: 136
    bat **2**: 168–169
    beaver **2**: 177
    bee-eater **2**: 181
    beira antelope **2**: 189–190
    beluga whale **2**: 193; **21**: *2914*
    bighorn sheep **2**: 199, *200*
    bishop bird **2**: 213
    bison **2**: 216
    blackbuck **2**: 226–227
    black-headed gull **2**: 231
    blesbok **2**: 239
    bluefish **2**: 245
    broadbill **2**: 282
    brow-antlered deer **2**: 284, *285*
    budgerigar **3**: 302–303, *303*
    buffalo 3: 305–306
    bunting **3**: 319
    burnet moth **3**: 321
    bushbaby **3**: 329
    bushpig **3**: 335
    Canada goose **3**: 365, *367*
    canary **3**: 368
    Cape buffalo **3**: 370–371
    Cape hunting dog **3**: 373
    capuchin monkey **3**: 379–380
    capybara **3**: 382–383
    caribou **3**: 391, 393; **20**: *2778*
    cats 3: 406
    chaffinch **3**: 414
    cheetah **3**: 406, *428*, 429
    chimpanzee **4**: 440–441
    chinchilla **4**: *443*
    Chinese water deer **4**: 446
    coati **4**: 478
    cockatoo **4**: 483–484
    colobus monkey **4**: 501, 502
    coral **4**: 527
    cormorant **4**: 538
    coypu **4**: 562
    crab plover **4**: 568
    dead man's fingers **5**: 651
    dogs 5: 691–692

dolphin **2**: 266, *266*; **4**: 507–508, *508*
dolphins and porpoises 5: 696
dugong **6**: 724
eagle ray **6**: 730
eider **6**: 756
eland **6**: 757
elephant seal **6**: 771
emperor penguin **6**: 780
emu **6**: 782
Everglade snail kite **6**: 789
false killer whale **6**: 797–798
flamingo **6**: 846–847
flying fox **7**: 879, *879*
fruit bat **7**: 908, *908*
gannet **7**: 923
gaur **7**: 936
gibbon **7**: 970
giraffe **7**: 975
godwit **7**: 996–997, *996*
goldfinch **8**: 1016
gorilla **8**: 1025, 1026
grackle **8**: 1032, *1033*, 1034
greenfinch **8**: 1064
greylag goose **8**: 1073
guenon **8**: 1091–1092
haddock **8**: 1113
hammerhead shark **8**: 1127, *1128*
harrier **8**: 1145
hartebeest **8**: 1147, 1148
herring **9**: 1184, *1185*
herring gull **9**: 1188
hippopotamus **9**: 1193
house sparrow **9**: 1247–1248
howler monkey **9**: 1253
hyrax **9**: 1271, *1271*
ibis **9**: 1276, *1276*
Indian buffalo **9**: 1286, *1287*
killer whale **10**: 1350–1351
king penguin **10**: *1360*
krill **10**: 1387
kudu **10**: 1390, *1391*
langur **10**: 1410–1411, 1412
lapwing **10**: 1415
lemur **10**: 1435
lion **3**: 406; **11**: 1460–1461, *1460*
little auk **11**: 1464
locust **11**: 1478–1480
macaque **11**: 1511–1513
macaw **11**: 1515
mackerel **11**: 1517
mangabey **11**: 1543
marine iguana **11**: 1556
marmoset **11**: 1563–1564
martin **12**: 1591
meerkat **12**: *1595*, 1596–1597
minivet **12**: 1614–1615
monkey 12: 1644
mouflon **12**: 1667
mousebird **12**: 1676
mule deer **12**: 1690
narwhal **12**: 1718
night heron **13**: 1739
night monkey **13**: 1748
nilgai **13**: 1753, *1754*
oropendula **13**: 1807
oryx **13**: 1808, *1810*
ostrich **13**: 1814
oystercatcher **13**: *1837*
parakeet **14**: 1879, *1880*
parrotfish **14**: 1888
partridge **14**: 1891, *1891*
patas monkey **14**: 1896
peccary **14**: 1904–1905, *1905*
perch **14**: 1917
phalarope **14**: 1933
pilchard **14**: 1948
pilot whale **14**: 1952
pintail **14**: 1958
pipistrelle **14**: 1964
piranha **14**: 1968

platy **14**: 1983, *1984*
prairie dog **15**: 2038
pratincole **15**: 2041
quail **15**: 2082, 2083
quelea **15**: 2084–2085, *2084*, *2085*,
    2086, *2086*
redpoll **15**: 2137
red-winged blackbird **15**: 2143,
    2143
reedbuck **15**: 2145
rhea **15**: 2150
river dolphin **16**: 2177
Rocky Mountain goat **16**: 2199
rook **16**: 2206, *2207*
rorqual **16**: 2208
sable antelope **16**: 2215, *2216*
saiga **16**: 2217
sand eel **16**: *2227*
sand grouse **16**: 2229
scad **16**: *2243*
sea butterfly **16**: 2269
sea snake **17**: 2309
sea squirt **17**: 2312
sifaka **17**: 2371
silverside **17**: 2379, 2380
skimmer **17**: 2387
smelt **17**: 2411
soft coral **17**: 2440
sperm whale **18**: 2472
spider monkey **18**: 2477
spoonbill **18**: 2497, *2497*, 2498
springbok **18**: 2500
squid **18**: 2508
squirrel monkey **18**: 2513–2514,
    2515
starling **18**: 2521, *2522*
stork **18**: 2552, 2553, *2553*
swallowtail **18**: 2581
takin **19**: 2622
tetra **19**: 2667
tubesnout **20**: 2765
tuna **20**: 2773, *2774*
turnstone **20**: 2791
vampire bat **20**: 2808
viscacha **20**: 2834
walrus **20**: *2851*, 2852
wapiti **20**: 2855–2856
waxwing **21**: 2890–2891
weaver **21**: 2897, *2898*
whales 21: 2913–2914, *2914*
whale shark **21**: *2915*
wild ass **21**: 2939
wild boar **21**: 2942
wood swallow **21**: 2983, *2983*
yellow-eyed penguin
    21: 3003–3004
zebra **21**: 3011
Liza
    aurata **12**: 1692, 1693
    ramada **12**: 1692, 1693
    tricuspidens **12**: 1692, 1693
Lizard
    Agama *see* Agama lizard
    Alligator **3**: 426
    Anole *see* Anole lizard
    Beaded **5**: 665; **7**: 972, 974
    Bearded (bearded dragon; Jew
        lizard) **2**: 172–173
    Centralian blue-tongued **17**: *2389*
    Dragon *see* Agama lizard, Starred
    Fence *see* Fence lizard
    Frilled (frillneck lizard; frilled
        dragon) **7**: 894–895
    Glass *see* Glass lizard
    Green **8**: 1068–1069
    Horned *see* Horned lizard
    Jesus *see* Basilisk, Common
    Legless *see* Legless lizard
    Monitor *see* Monitor lizard
    Night *see* Night lizard

    Ocellated (eyed lizard) **8**: 1068
    Pine *see* Fence lizard, Eastern
    Plateau *see* Fence lizard, Eastern
    Prairie *see* Fence lizard, Eastern
    Rainbow *see* Agama lizard,
        Common
    Ruin (Italian wall lizard) **20**: 2849,
        2850
    Sand **16**: 2231–2232
    Short-horned **9**: *1231*
    Spiny-tailed **1**: 30
    Teiid **19**: 2650
    Viviparous (common lizard)
        20: 2837–2838
    Wall *see* Wall lizard
    Water (sail fin lizard) **1**: 30; **2**: 157
    see also Basilisk; Chameleon;
        Chuckwalla; Gecko; Gila
        monster; Iguana; Komodo
        dragon; Lizards; Moloch; Skink;
        Slowworm; Tuatara
Lizards 11: 1466–1469
Llama **11**: 1470–1472; **12**: 1672
Llama
    glama **12**: 1672
    guanicoe **12**: 1672
    pacos **12**: 1672
llanos **21**: 2906
Loach **11**: 1473–1474
    Clown **11**: *1473*
    Coolie **11**: 1473
    Spiny **11**: 1474
    Stone **11**: 1473, 1474
    Tiger (green tiger botia) **11**: 1473,
        1474
Lobodon carcinophagus **1**: 82; **4**: 565;
    12: 1647; **16**: 2286
Lobster **11**: 1475–1477
    American **4**: 571, 572, *572*;
        11: 1475, 1476, 1477
    European (common lobster)
        4: 571; **11**: 1475, 1476
    Norway (Dublin prawn) **4**: 573;
        11: 1475
    Slipper **4**: 570
    Spiny *see* Spiny lobster
    Squat *see* Squat lobster
    see also Crabs and lobsters
lobtail **18**: 2473
Lobworm *see* lugworm
Lochimas **13**: 1819
Locust **11**: 1478–1480
    Australian plague **11**: 1479
    Desert **11**: 1478, 1479, *1479*, 1480
    Migratory **11**: 1478, *1478*, 1479,
        1480, *1480*
    Red-legged (red locust) **11**: 1478,
        1479, 1480
Locusta
    migratoria **11**: 1479
    migratoria migratorioides **11**: 1478
Locust bird *see* Pratincole, Common
Locustella naevia **20**: 2859
Locust finch **21**: 2888
lodges, beaver **2**: 177, 179
Loligo **18**: 2510
    forbesi **18**: 2508–2509
    peali **18**: 2510
Lonchura
    castaneothorax **11**: 1544, 1545
    cucullata **11**: 1544
    fringilloides **11**: 1544
    nana **11**: 1544
    punctulata **11**: 1544
    striata **11**: 1544
Longclaw **14**: 1965, 1966
    Yellow-throated **14**: 1965
Long-eared bat **20**: 2822
    American **20**: 2821, 2822
    European **2**: *167*; **20**: 2821, 2822

Long-horned beetle (longicorn
    beetle; wood-boring beetle;
    timberman. *larva*: round-headed
    borer) **2**: 188; **11**: 1483–1484
  Asian **11**: 1484
  House **11**: 1484
long living
  albatross **1**: 37
  anemone **1**: 47
  centipede **3**: 413
  flea **6**: 852
  giant snail **7**: 966
  lobster **4**: 572
  moose **12**: 1656
  mud turtle **12**: 1688
  pelican **14**: 1910
  perch **14**: 1918
  rorqual **16**: 2209
  tortoise **19**: 2713
  tuatara **20**: 2764
Longshanks *see* Stilt, Black-winged
Longtail, Doherty's **3**: 321
Lookdown **16**: 2242
Loon **11**: 1485–1487
  Arctic (black-throated diver)
    11: 1485, *1485*; **21**: 2906
  Common (great northern diver)
    11: 1485, 1486, *1486*
  Red-throated (red-throated diver;
    rain goose) **11**: 1485, 1486, 1487,
    1487
  Yellow-billed (white-billed diver)
    11: 1485, 1487
Lophiomys imhausi **13**: 1838
Lophius piscatorius **1**: 51, 52, *52*
Lophocebus **11**: 1541
  albigena **11**: 1541, 1542
  albigena aterrimus **11**: 1542
Lophodytes cucullatus **12**: 1599;
    21: 2972
Lopholithodes **18**: 2507
lophophore **10**: 1405, *1405*, 1406
Lophophorus **14**: 1934
  impeyanus **14**: *1936*
Lophura **14**: 1934
  diardi **14**: *1935*
Lorentzimys nouhuysi **10**: 1337, 1338
Loriculus **11**: 1495
Lorikeet **11**: 1488–1489
  Little **11**: 1488
  Olive-headed (perfect lorikeet)
    11: *1489*
  Purple-crowned (blue lorikeet)
    11: 1488
  Rainbow (rainbow lory; green-
    naped lorikeet; red-breasted
    lorikeet; red-collared lorikeet;
    blue-bellied lorikeet; coconut
    lorikeet) **11**: 1488, *1488*, 1489
  Scaly-breasted **11**: 1488
Loris **11**: 1490–1492
  Pygmy slow (lesser slow loris)
    11: 1490, 1491, 1492
  Slender **11**: 1490, 1491, *1491*, 1492
  Slow **11**: 1490, *1490*, 1491, *1492*
Loris tardidradus **11**: 1490
Lorius
  domicella **11**: 1493, 1494
  garrula **11**: 1493
Lory **11**: 1493–1494
  Dusky **11**: 1493, *1493*
  Purple-naped (purple-capped
    lory) **11**: 1493, 1494, *1494*
  Rainbow *see* Lorikeet, Rainbow
  Red and blue **11**: 1494
  Yellow-backed **11**: 1493
Lota lota **4**: 495; **11**: 1455
Lotus-bird *see* Jacana
Louse
  Graperoot *see* Vine pest

Lousy watchman **16**: 2254
Lovebird **11**: 1495–1496
  Black-cheeked **11**: 1495
  Black-collared (Swindern's
    lovebird) **11**: 1495
  Black-winged (Abyssinian
    lovebird) **11**: 1495, 1496
Lovebird (*budgerigar*) *see* Budgerigar
Lovebird
  Fischer's **11**: *1495*
  Madagascar (gray-headed
    lovebird) **11**: 1495, 1496
  Masked (yellow-collared lovebird)
    11: 1495, 1496, *1496*
  Nyasa (Lilian's lovebird) **11**: 1495
  Red-headed **11**: 1495, 1496
  Rosy-faced (peach-faced lovebird)
    11: 1495, 1496
Lowan *see* Mallee fowl
Loxia
  curvirostra **5**: 607, 608
  leucoptera **5**: 607
  pytyopsittacus **5**: 607
  scotica **5**: 607
Loxocemus bicolor **17**: 2419
Loxodonta africana **6**: 768; **15**: 2113
Lucanus cervus **19**: 2652
Lucastris **9**: 1250
luciferase **6**: 827
luciferin **6**: 827
Lucilia bufonivora **2**: 242
Lugworm (lobworm) **11**: 1497–1498
Lullula arborea **17**: 2396
Lumbricus **6**: 735
  terrestris **6**: 734, 735, 737
Luminodesmus sequoiae **12**: 1611
Lumpfish *see* Lumpsucker
Lumpsucker (lumpfish; sea hen; hen
    fish) **11**: 1499–1500
Lungfish **11**: 1501–1502
  African **11**: 1501, 1502, *1502*
  Australian (Burnett salmon)
    11: 1501
  South American **11**: 1501, *1501*,
    1502
  West African **11**: 1501
lungworm **17**: 2409
Luscinia megarhynchos **13**: 1741;
    18: 2459; **19**: 2678
Luth *see* Turtle, Leatherback
Lutjanus
  analis **17**: *2423*
  apodus **17**: *2423*
  argentimaculatus **17**: *2423*
  griseus **17**: *2423*
  kasmira **17**: *2424*
  sebae **17**: *2423*
  synagris **17**: *2423*
Lutra
  canadensis **16**: 2179, *2179*, 2180
  lutra **16**: 2179, 2180
Lutraria lutraria **7**: *926*, 927
Lutrogale perspicillata **16**: 2180
Luxilus **12**: 1618
  cornutus **12**: 1618
Lycaena
  dispar **4**: 526
  phlaeas **4**: 525–526
Lycalopex **5**: 690
  culpaeus **5**: 689
  griseus **5**: *689*
Lycaon pictus **3**: 374; **5**: 689; **8**: 1041
Lycophotia varia **13**: 1821
Lycosa **11**: 1550; **19**: 2635; **21**: 2956,
    2956, 2957
  barbipes **19**: 2634
  cuneata **19**: 2634
  fabrilis **19**: 2633
  pulverulenta **19**: 2634
  tarentula **2**: 203; **19**: 2633, 2634

Lydekker, Richard **2**: 226
Lyme disease **19**: *2681*, 2682
Lymnaea **14**: 2011
  auricularia **14**: 2010, 2011
  catascopium **14**: 2011
  palustris **14**: 2011
  peregra **14**: 2010
  stagnalis **14**: 2010, 2011
  truncatula **14**: 2010, 2011
Lymnocryptes minimus **17**: 2427
Lynx
  canadensis **3**: 405; **19**: 2616
  lynx **21**: 2906
Lynx **11**: 1506–1508; **12**: 1672;
    19: 2616
  Canada **3**: 405; **11**: 1506, 1507, 1508
  Desert *see* Caracal
  Eurasian (Siberian lynx) **11**: 1506,
    1506, 1507, *1507*; **21**: 2906
  Spanish **11**: 1506, 1507, 1508
Lynx tachete *see* Serval
Lyonogale **19**: 2731
Lyrebird (lyretail; native pheasant)
    11: 1509–1510
  Northern (Albert's lyrebird)
    11: 1509
  Superb **11**: 1509, 1510
Lyretail *see* Lyrebird
Lyristes **4**: 456
  plebejus **4**: 456
Lysandra coridon **2**: 243
Lysiosquilla **11**: *1552*
Lythe *see* Pollack

M

Maasbanker (South African
    mackerel) **17**: 2430
Macaca **1**: 137; **11**: 1511, 1512;
    12: 1643; **14**: 1941, 1942
Macaque **11**: 1511–1513; **12**: 1644
  Assam **11**: 1511, 1512, 1513
  Barbary *see* Ape, Barbary
  Bonnet **11**: 1511–1512, 1513
  Celebes black **11**: 1511, 1513
  Crab-eating (long-tailed macaque)
    11: 1511, 1512, 1513, *1513*;
    14: 1942
  Japanese **11**: 1511, *1511*, 1512, 1513
  Lion-tailed **11**: *1512*, 1513;
    14: 1941, 1942
  Moor **11**: 1511, 1513
  Pig-tailed **14**: 1941–1942
  Rhesus **11**: 1511, 1512, 1513
  Stump-tailed **11**: 1511, 1513
  Tibetan **11**: 1511, 1513
  Toque **11**: 1511, 1512
Macaw **11**: 1514–1516
  Blue-throated **11**: 1516
  Blue and yellow (blue and gold
    macaw) **11**: 1514, *1516*
  Glaucous **11**: 1516
  Hyacinth (hyacinthine macaw)
    11: 1514, *1515*
  Lear's **11**: 1516
  Military (great green macaw)
    11: 1514
  Red-fronted (red-crowned
    macaw) **11**: 1516
  Red and green **11**: 1514
  Scarlet (red and yellow macaw)
    11: 1514, *1514*, 1515
  Spix's (little blue macaw) **11**: 1516
Machaerhamphus alcinus **2**: 210
Mackerel **5**: 694; **11**: 1517–1518
  Atlantic **11**: 1517, 1518, *1518*
  Chub **11**: 1517
  Horse **11**: 1517
  Horse *see* Scad
  Indian **11**: 1517

  Island **11**: 1517
  Japanese Spanish **11**: 1517
  Monterey Spanish **11**: 1517
  Pacific Sierra **11**: 1517
  Pygmy **11**: 1517
  Short **11**: 1517
  Snake *see* Snake mackerel
  South African *see* Maasbanker
  Spanish **11**: 1517
Macolor macularis **18**: *2573*
Macrobrachium **15**: 2043
  carcinus **15**: 2043; **17**: 2364
  rosenbergii **15**: 2043
Macrocheira kaempferi **4**: 571; **18**: 2475
Macroclemys temmincki **17**: 2425, 2426;
    20: 2797
Macrodactyla doreensis **5**: 640
Macroderma gigas **2**: *169*
Macrodipteryx
  longipennis **13**: 1742
  vexillarius **13**: 1742
Macrodontia **11**: 1483, 1484
Macrogalidia musschenbroeki **13**: 1855
Macrogeomys **14**: 1996
Macroglossum stellatarum **9**: 1166
Macrognathus **18**: 2490
  aculeatus **18**: 2490
  aral **18**: 2490
  pancalus **18**: 2490
Macronyx **14**: 1965
  croceus **14**: 1965
Macroperipatus torquatus **14**: 1924
Macropodia rostrata **18**: 2475, 2476
Macropodus
  chinensis **13**: 1868
  concolor **13**: 1868
  opercularis **13**: 1868, 1869
Macropus **10**: 1333, 1334, 1335–1336;
    20: *2846*, 2848
Macrotermes michaelseni **19**: *2658*
Macrotis
  lagotis **1**: 130–131
  leucura **1**: 130
Macrotrema caligans **18**: 2585
Macrovipera schweizeri **17**: 2422
Maculinea
  alcon **2**: 244
  arion **2**: 243–244
Madoqua **1**: 61
  guentheri **5**: 675, 676
  kirkii **5**: 676
  piacentinii **5**: 676
  saltiana **5**: 675, 676
madreporites **18**: 2516
Madtom **3**: 402
maggots
  blowfly **2**: 241, 242, *242*
  housefly **9**: 1241
  "rat-tailed" **9**: 1251
Magicicada **4**: 456
  septemdecim **4**: 456, 457
Magpie **5**: 610; **11**: 1519–1521
  Azure-winged **11**: 1520, *1521*
  Black-billed **11**: 1519, 1520–1521
  Blue **11**: 1520
  Green **11**: 1519
  Little *see* Magpie lark
  Murray *see* Magpie lark
  Sri Lanka (Ceylon magpie)
    11: 1519–1520
  Yellow-billed **11**: 1519, 1520–1521,
    1520
Magpie lark (mudlark; peewit;
    peewee; pugwall; little
    magpie; Murray magpie)
    11: 1522–1523
  New Guinea (torrentlark) **11**: 1522
Maia squinado **17**: 2355; **18**: 2475,
    2476
maidenhair trees **2**: 273

Makaira **11**: 1561
  indica **11**: 1561
  mazara **11**: 1561
  nigricans **11**: 1562
Makira *see* Moorhen, San Cristobel
Mako, Shortfin **15**: 2020–2021, *2021*
Malaclemys terrapin **19**: 2664, 2665
Malacoptila
  mystacalis **15**: *2061*
  panamensis **15**: 2060
Malapterurus
  electricus **6**: 760, 761
  microstoma **6**: 760, 761
malaria **6**: 859; **12**: 1663–1664
Malimbe **21**: 2896
  Red-headed **21**: 2896
Malimbus **21**: 2896
  rubricollis **21**: 2896
Mallard **5**: 717, *717*; **11**: 1524–1526;
  17: 2343
Mallee *see* Chaparral
Mallee fowl (mallee hen; native
  pheasant; lowan; gnow)
  11: 1527–1529
malleoli **18**: 2453
Mallomys rothschildi **7**: 965
Malmignatte **19**: 2634–2635
Mamba **2**: 265; **11**: 1530–1531;
  20: 2817
  Black (blackmouth mamba)
    11: 1530, 1531, *1531*; **12**: 1597;
    15: 2103
  Green **11**: 1530, 1531
man *see* humans
Manacus
  manacus **11**: 1532
  vitellinus **11**: 1532
Manakin **11**: 1532–1533
  Blue-backed **11**: 1532
  Golden-headed **11**: 1532
  Gould's **11**: 1532, *1533*
  Long-tailed **11**: *1532*, 1533
  Red-capped **11**: 1532
  White-bearded **11**: 1532
Manatee **11**: 1534–1535; **21**: 2907
  Amazonian **11**: 1534, 1535
  West African **11**: 1534, 1535
  West Indian **11**: 1534, *1534*, 1535
Mandrill **1**: 115; **11**: 1536–1538;
  12: 1644
Mandrillus
  leucophaeus **11**: 1536, 1537
  sphinx **11**: 1536, 1537; **12**: 1644
Mangabey **11**: 1541–1542
  Agile **11**: 1541, 1542, 1543
  Black-crested **11**: 1541, 1542, *1542*,
    1543
  Crested **11**: 1541
  Crestless **11**: 1541, 1542
  Gray-cheeked **11**: 1541, 1542, 1543
  Red-crowned (collared mangabey)
    11: 1541–1542, *1541*, 1543
  Tana River **11**: 1541, 1542, 1543,
    1543
mangrove swamps **4**: 477, *477*;
  21: 2907
Maniola jurtina **3**: 296–297
Manis **13**: 1862, 1863
manna **16**: 2247
Mannikin **11**: 1544–1545; **21**: 2888
  Bronze **11**: 1544
  Chestnut-breasted (chestnut-
    breasted munia; chestnut-
    breasted finch; chestnut finch;
    barleyfinch; barley sparrow;
    barley bird) **11**: 1544, 1545, *1545*
  Madagascar (bib-finch) **11**: 1544
  Magpie **11**: 1544
  Spotted (spotted munia; nutmeg
    finch; spice finch) **11**: 1544, 1545

Man-o'-war bird *see* Frigate bird,
  Magnificent
Mansonia **12**: 1664
Manta **5**: 668, 669
Mantid *see* Mantis
Mantis (mantid; praying mantis;
  praying mantid) **11**: 1546–1548
  Chinese **11**: 1546
  European **11**: 1546, 1547, *1547*
  Orchid **11**: 1547, 1548
Mantis fly **11**: 1549–1550; **21**: 2957
  Brazilian **11**: 1550
  European **11**: 1550
Mantispa **11**: *1549*
  styriaca **11**: 1549, 1550
Mantispid *see* Mantis fly
Mantis religiosa **11**: 1546, 1547
mantle **4**: 556; **13**: *1832*
Manucode, Trumpeter **20**: 2757
Maoris **21**: 2887
Mara (Patagonian hare) **8**: 1099,
  1100, 1101
Marabou (marabou stork)
  11: 1553–1554; **18**: 2551
Margaronis **13**: 1819
Margarops
  fuscatus **19**: 2671
  fuscus **19**: 2671
Margay cat **3**: 404; **13**: 1776, 1777
Mariana Trench **13**: 1772
Markhor **11**: 1559–1560; **12**: 1672;
  21: 2951
Marlin **11**: 1561–1562; **19**: 2605
  Atlantic blue **11**: 1562
  Atlantic white **11**: 1561
  Black **11**: 1561
  Blue **11**: 1561, *1562*
Marlin-spike *see* Tropic bird
Marmaronetta angustirostris **19**: 2649
Marmessus **8**: 1117, 1118
Marmosa murina **13**: 1792, 1793
Marmoset **11**: 1563–1565; **12**: 1644,
  1645
  Buffy-headed **11**: 1565
  Buffy tufted-ear **11**: 1565
  Emperor **11**: *1565*
  Geoffroy's **11**: 1563, 1565
  Pygmy **11**: 1563, 1564, *1564*
  Silky **11**: 1563, 1565
  Silvery **11**: *1563*
  White **11**: 1565
  White ear-tufted **11**: 1563
  see also Tamarin
Marmot **11**: 1568–1570; **12**: 1672
  Alpine **11**: 1568, *1569*, 1570
  Himalayan (bobak; steppe
    marmot) **11**: 1568, 1569
  Hoary (whistler) **11**: 1568, 1569,
    1570
  Olympic **11**: 1569, *1570*
  Steppe *see* Bobak
  Vancouver Island **11**: 1569
  Yellow-bellied **11**: *1568*
Marmota **8**: 1076; **11**: 1568, *1568*,
  1569; **12**: 1672
Marsupial cat (quoll; native cat)
  11: 1571–1572
  Bronze native cat **11**: 1571
  Eastern native cat (eastern quoll)
    11: 1571, 1572, *1572*
  Geoffroy's native cat **11**: 1571
  Large-spotted native cat **11**: 1571,
    1572
  Little northern native cat
    (northern quoll) **11**: 1571, *1571*,
    1572
  New Guinea native cat **11**: 1571
Marsupial frog **11**: 1573–1574
  Pygmy **11**: 1574, *1574*
Marsupial jerboa *see* Marsupial mouse

Marsupial mole **11**: 1575–1576
Marsupial mouse (marsupial
  jerboa; dunnart; kultarr)
  11: 1577–1578
  Broad-footed **11**: 1577
  Brush-tailed (tuan) **11**: 1577, 1578
  Fat-tailed dunnart **11**: 1577, *1578*
  Flat-skulled *see* Planigale
  Macleay's **11**: 1578
  Short-haired **11**: 1577
Marsupiobdella **10**: 1429
Marten **11**: 1579–1581
  American (American pine marten;
    pine marten) **11**: 1579, 1580;
    19: *2615*, 2616
  Beech (stone marten) **11**: 1579
  Eurasian pine marten **11**: 1579,
    1579, 1580; **19**: 2653
  Yellow-throated **11**: 1579
Martes
  americana **11**: 1579, 1580; **19**: 2616
  flavigula **11**: 1579
  foina **11**: 1579
  martes **11**: 1579; **19**: 2653
  pennanti **11**: 1579; **19**: 2653
  zibellina **11**: 1579
Martin **12**: 1590–1592
  African river **12**: 1590
  Brown-chested **12**: 1590
  Crag **12**: 1590
  House **12**: 1590, *1590*, 1591
  Mascarene **12**: 1590
  Purple **2**: 151; **12**: 1590, 1591, *1591*
  Sand *see* Swallow, Bank
  Tree **12**: 1590
Martineta *see* Tinamou, Crested
Massasauga (swamp rattler; black
  snapper) **15**: 2120
Massoutiera mzabi **8**: 1104
Mastacembelus **18**: 2490
Masticophis
  bilineatus **21**: 2919, 2920
  flagellum **21**: 2919, 2920
  lateralis **15**: 2101
  taeniatus **21**: 2919, 2920
Mastigoproctus giganteus **14**: 1906,
  1907
Mastomys **12**: *1694*, 1695
Mastophora **18**: 2482
Mastotermes **19**: 2658, 2660
Masturus lanceolatus **13**: 1774
Masu *see* Salmon, Cherry
mate guarding
  goby **7**: 994
  greylag goose **8**: 1073–1074
  multimammate rat **12**: 1694
  parakeet **14**: 1880
  pink-footed goose **14**: *1955*, 1956
  secretary bird **17**: 2318
Maticora **4**: 534
mating *see* attracting a mate;
  hermaphrodites; monogamy;
  polygamy; sexual aggression
Mattoral *see* Chaparral
maxillipeds, centipede **3**: 412–413
May beetle *see* Cockchafer
May bug *see* Cockchafer
Mayetiola destructor **6**: 859; **12**: 1606
Mayfly **12**: 1593–1594
Mazama **18**: 2463, 2464
Meadowlark **13**: 1807
  Western **8**: *1039*
meat eating
  adder **1**: 26
  alligator **1**: 39
  anaconda **1**: 43, *43*
  anglerfish **1**: 51–52
  army ant **1**: 91–92
  asp **1**: 97
  bald eagle **1**: 124–125

  barn owl **2**: 149–150
  bear **2**: 176
  binturong **2**: 201, *201*
  black bear **2**: 222
  bullfrog **3**: 312
  burrowing owl **3**: 324
  buzzard **3**: 352
  caiman **3**: 359
  Cape hunting dog **3**: 373, 374
  caracal **3**: 385–386
  cheetah **3**: 427–428
  chimpanzee **4**: 442
  condor **4**: *516*, 517
  cone shell **4**: 519–520
  cowrie **4**: 556
  crocodile **5**: 605, *606*
  crowned eagle **5**: 614–615, *614*
  dragonfly **5**: 710
  ermine **6**: 786
  fennec fox **6**: 809–810
  fer de lance **6**: 812–813, *813*
  goshawk **8**: *1027*, 1028–1029, *1029*
  grass snake **8**: 1043
  gray wolf **8**: 1055–1056, *1057*
  great white shark **8**: 1058,
    1059–1060, *1059*
  gyrfalcon **8**: 1111–1112, *1112*
  harpy eagle **8**: 1143–1144
  hyena **9**: 1267–1269
  jaguar **10**: 1311
  jellyfish **10**: 1321
  kestrel **10**: 1345–1346
  Komodo dragon **10**: 1383, *1384*
  leopard **11**: 1445–1446, *1446*
  leopard seal **11**: 1449–1450
  lion **11**: 1462, *1462*, *1463*
  lungfish **11**: 1501
  lynx **11**: 1507
  mantis **11**: 1546–1547
  mantis shrimp **11**: 1551, *1552*
  marabou **11**: 1553, 1554, *1554*
  marsupial cat **11**: 1572
  martial eagle **12**: 1588, *1589*
  merlin **12**: 1602
  mink **12**: 1616
  monitor lizard **12**: 1640
  peregrine falcon **14**: 1921–1922
  piranha **14**: 1968–1969
  pit viper **14**: 1973
  polar bear **14**: 1998–1999
  polecat **14**: 2002–2003
  puff adder **15**: 2058–2059
  puma **15**: 2068, *2069*, 2070
  python **15**: 2080, *2080*
  raccoon **15**: 2097
  raccoon dog **15**: 2099
  rattlesnake **15**: 2120–2121
  scad **16**: 2242
  sea otter **16**: 2294, *2294*, 2295
  sea slug **16**: 2300
  secretary bird **17**: 2316–2317
  serval **17**: 2325
  shrike **17**: 2361–2363, *2361*
  sidewinder **17**: 2369–2370
  skunk **17**: 2395
  snow leopard **17**: 2433
  solifugid **18**: 2453–2454
  spiders 18: 2482
  starfish **18**: 2517
  tarsier **19**: 2639
  Tasmanian devil **19**: 2642–2643
  tawny owl **19**: 2644–2645
  tiger **19**: 2683–2684
  water spider **20**: 2876
  weasel **21**: 2893–2894
  whelk **21**: 2917–2918
  wildcat **21**: 2946–2947
  wolverine **21**: 2960
  zorilla **21**: 3020
  see also carrion

medicine, animals used for **1**: 40, 94, 96; **11**: 1492; **12**: *1702, 1712*; **15**: 2059, *2152*, 2155; **17**: *2375*, 2405, 2434
medusae **10**: 1320
Meerkat (suricate ; slender-tailed meerkat) **12**: 1595–1597
Megabunus diadema **5**: 637
Megaceryle alcyon **16**: 2175
Megachasma pelagios **17**: 2331
Megachile
    centuncularis **10**: 1419, 1420
    rotundata **10**: 1420
Megaderma lyra **6**: 840
Megadyptes antipodes **14**: 1911; **21**: 3004
Megaladapis **17**: 2373
Megalagrion **5**: 642
    oahuense **5**: 642
Megaloceros **5**: 656
Megalomys **15**: 2156
megalopa larvae **4**: 573; **6**: 742; **17**: 2355; **18**: 2475
Megalops
    atlanticus **19**: *2636, 2636,* 2637
    cyprinoides **19**: *2636,* 2637
Megamuntiacus vuquangensis **12**: 1697
Meganeuramonyi **5**: 711
Meganyctiphanes norvegica **16**: 2167
Megaphobema mesomelas **18**: 2483
Megapodius
    cumingii **11**: 1527
    reinwardt **11**: 1527
Megaptera novaeangliae **5**: 695; **9**: 1259; **16**: 2208; **21**: 2911
Megarhynchus pitangua **20**: 2801
Megarhyssa **16**: 2241
Megascolides **6**: 735
    australis **6**: 734
Megathymus **17**: 2391, 2392
    yuccae **17**: 2392
Megophrys **9**: 1229
    nasuta **9**: *1229,* 1230
Meiacanthus atrodorsalis **2**: 237
Melanargia galathea **21**: 2929
Melanerpes
    erythrocephalus **21**: 2978
    formicivorus **19**: 2654; **21**: *2978,* 2979
melanism **10**: 1310
    industrial **14**: 1916
Melanogrammus aeglefinus **6**: 836; **8**: 1114
melanophores **3**: 420
Melanoptila glabrirostris **3**: 399
Melanosuchus niger **3**: 358
Mesia, Silver-eared **1**: 114
Melanotaenia
    maccullochi **17**: 2380
    trifasciata **17**: *2380*
Melanotis
    caerulescens **12**: 1622
    hypoleucus **12**: 1622
Meleagris **20**: 2784, 2785
Melese laodamia **19**: 2688
Meles meles **1**: 120, 121; **19**: 2653
Melibe **16**: 2300
Melichneutes robustus **9**: 1217
Melichthys niger **20**: 2744
Melierax **8**: 1027
Meliphaga unicolor **9**: 1215
Melitaea cynxia **7**: 898, 899
Melithreptus validirostris **9**: 1215
Mellisuga helenae **9**: 1255
Mellita sexies perforata **16**: *2226*
Mellivora capensis **1**: 120; **9**: 1207, 1217
Meloë proscarabaeus **13**: *1782, 1782,* 1783
Melogale **1**: 120
Melolontha **4**: 486, 487
    melolontha **4**: 486

Melophorus **9**: 1205
    bagoti **9**: *1205, 1205*
Melospiza melodia **4**: 555; **9**: 1247; **18**: 2461
Melursus ursinus **17**: 2404
Menetes berdmorei **13**: 1859, 1860
Menhaden **2**: 245; **12**: 1598
    Atlantic (bonyfish; bugfish; bunker; fatback; mossbunder; pokey; pogy; whitefish) **5**: 694; **9**: 1184; **12**: 1598, *1598*
    Pacific **12**: 1598
Menidia beryllina **17**: 2380
Menura
    alberti **11**: 1509
    novaehollandiae **11**: 1509, 1510
Mephitis
    macroura **17**: 2393, 2394
    mephitis **17**: 2393, 2394
mercaptan **17**: 2395
Merganetta armata **19**: 2708–2710
Merganser **12**: 1599–1601
    Aukland Island **12**: 1600
    Brazilian **5**: 717; **12**: 1600
    Common (goosander) **5**: 717; **12**: 1599, 1600, 1601, *1601*
    Hooded **12**: 1599, 1600, *1600*; **21**: 2972
    Red-breasted **5**: 717; **12**: 1599, *1599,* 1600, 1601
Mergus
    albellus **12**: 1599
    australis **12**: 1600
    merganser **5**: 717; **12**: 1599, 1600
    octosetaceus **5**: 717; **12**: 1600
    serrator **5**: 717; **12**: 1599
Meriones
    meridianus **7**: 951
    unguiculatus **7**: *950*
Merlangius merlangus **21**: 2935
Merlin (pigeon hawk) **12**: 1602–1603
Merluccius
    aungustianus **8**: 1119, 1120
    australis **8**: 1119, 1120
    bilinearis **8**: 1119, 1120
    capensis **8**: 1119, 1120
    gayi **8**: 1119
    merluccius **8**: 1119, 1120
    productus **8**: 1120
mermaids' purses **17**: *2384,* 2385
Merodon equestris **9**: 1251
Merops
    apiaster **2**: 180, 181
    ornatus **2**: 180
    pusillus **2**: *182*
Mesite **12**: 1604–1605
    Brown **12**: 1604–1605
    Subdesert (Bensch's monia) **12**: 1604, *1604, 1605*
    White-breasted **12**: 1604–1605
Mesocapromys **13**: 1263, 1264
    angelcabrerai **9**: 1264
    auritus **9**: 1264
    melanurus **9**: 1264
    nanus **9**: 1264
Mesochaetopterus rickettsi **20**: 2768
Mesocricetus
    auratus **8**: 1130
    neutoni **8**: 1131
Mesoenas **12**: 1604–1605
    benschi **12**: 1605
mesogloea **10**: 1320
mesopelagic zone **13**: 1771
Mesoplodon **2**: 170
    densirostris **2**: *170*
Messor **9**: 1156, 1157, *1157*
Meta **13**: 1801
    segmentata **13**: *1800*

metamorphosis
    backswimmer **1**: 119
    beetle **2**: 187
    bug **2**: 186–187
    butterfly and moth **3**: 346
    complete (holometabolous) **2**: 187; **6**: 858; **9**: 1236; **21**: *2921*
    edible frog **6**: 744
    frogs and toads **7**: 904
    greenhouse frog **8**: *1067*
    hypermetamorphosis **11**: 1550; **13**: 1783
    incomplete (hemimetabolous) **2**: 186–187; **8**: 1037; **12**: 1593–1594
    mayfly **12**: 1593–1594
    salamander **11**: 1505
    see also neoteny
Meterodon platyrhinos **17**: *2421*
Metoecus paradoxus **20**: 2826
Metopidius indicus **9**: 1289
Metridium **1**: 48
    senile **1**: 47
Micrathena **13**: 1801
Micrathene whitneyi **13**: 1827
Microcavia **11**: 1099, 1100
Microcebus **10**: 1436
Microcentrum retinerve **10**: 1341
Microciona **16**: 2299
Microdipodops **10**: 1338
Microgale **19**: 2656, 2657
Microhierax fringillarius **2**: 208
Microlophium **1**: 75
Micromys minutus **9**: 1158, 1159
Micropogon undulatus **5**: 601, 602
Micropotamogale **19**: 2656
    lamottei **13**: 1817, 1818
    ruwenzorii **13**: 1817, 1818
Micropterus salmoides **8**: 1063; **18**: 2569
micropyle **14**: 1947
Microstomum **6**: 848
Microtus **6**: 818
    agrestis **6**: *818,* 819
Micruroides euryxanthus **4**: 534, 535
Micrurus
    fulvius **4**: 534, 535
    nigrocinctus **4**: *535*
Midge (gnat) **6**: 859; **12**: 1606–1608
    Biting (no-seeum; sand fly) **12**: 1606, 1607–1608
    Gall **12**: 1606, 1608
    Meniscus **6**: 859
    Nonbiting (harlequin fly; *larva:* bloodworm) **12**: 1606–1607, *1607,* 1608
    Phantom **12**: 1606
Midshipman **19**: 2698
Midwife toad **7**: 903, 904; **12**: 1609–1610
    Common (bell toad) **12**: 1609–1610
    Iberian **12**: *1609*
migration
    aphid **1**: 74
    Atlantic salmon **1**: 99–100
    Baltimore oriole **1**: 126
    barnacle goose **1**: 140
    barn swallow **2**: 152–153
    beaked whale **2**: 170–171
    bee-eater **2**: 180
    bluefish **2**: 245
    blue whale **2**: 246, 248
    Bombay duck **2**: 257
    brown rat **2**: 298
    bunting **3**: 319
    butterflies and moths 3: 346, *346*
    buzzard **3**: 351
    Canada goose **3**: 365, 367
    caribou **3**: 391, 392, *392,* 394; **20**: 2778, *2778,* 2779

catbird **3**: 399
chaffinch **3**: 414
coalfish **4**: 472
cod **4**: 495
common toad **4**: 512
conch **4**: *514*
corncrake **4**: 541
crabeater seal **4**: 564
crake **5**: 582
crane **5**: *584,* 586, *587*
cuckoo **5**: 617–618
curlew **5**: 626
dolphin **4**: 506–507
dotterel **5**: 704
dragonfly **5**: 709–710
ducks, geese and swans 5: 717
edible crab **6**: 741
eel **6**: 745, 746–747
egret **6**: 750–751
flicker **6**: 853
flycatcher **7**: 872–873
fruit pigeon **7**: 911
gannet **7**: 923
gazelle **7**: 940
godwit **7**: 996–997, *997*
goldcrest **7**: 999
gray whale **8**: 1052–1053, *1052*
green turtle **8**: 1070
hake **8**: 1119
hammerhead shark **8**: 1127
harp seal **8**: 1139, 1140–1141
hatchet fish **9**: 1162
hawkmoth **9**: 1166
honey buzzard **9**: 1211
hummingbird **9**: 1255–1256
humpback whale **9**: 1258, 1259–1260, *1260*
jaeger **10**: 1308
killdeer **10**: 1348
lamprey **10**: 1403
lantern fish **10**: 1413
lapwing **10**: 1416
lemming **10**: 1432–1433, *1434*
lobster **4**: 573
locust **11**: 1479
martin **12**: 1590–1591
merganser **12**: 1600
millipede **12**: 1613
monarch **3**: *346*; **12**: 1632, *1634*
mule deer **12**: 1689–1690
narwhal **12**: 1717
nightingale **13**: 1741
osprey **13**: 1811
owlet moth **13**: 1821
Pacific salmon **13**: 1840, *1841,* 1842
pilchard **14**: 1948, *1948,* 1949
pilot whale **14**: 1953
pintail **14**: 1959
plover **14**: 1990
pollack **14**: 2005
porpoise **15**: 2024–2025
redpoll **15**: 2138
redstart **15**: 2139
red-winged blackbird **15**: 2143, *2143*
right whale **16**: 2167
robin, American **16**: 2188
Rocky Mountain goat **16**: 2199
roller **16**: 2204
rorqual **16**: 2208, *2208*
ruff **16**: 2213
sailfish **16**: 2220
sand grouse **16**: 2229
sandpiper **16**: 2233
seabirds 16: 2264–2265
sea hare **16**: *2275*
seed snipe **17**: 2319
shearwater **17**: 2335, *2336*
snake mackerel **17**: 2414, 2415
songbirds 18: 2456, 2458–2459

migration (*continued*)
 springbok **18**: 2501
 squid **18**: *2510*
 stargazer **18**: 2520
 stork **18**: 2553
 sturgeon **18**: 2556
 swift **19**: 2598
 swordfish **19**: 2606
 tanager **18**: *2456*; **19**: 2628
 tarpon **19**: 2636
 tern **19**: 2661–2663
 thresher shark **19**: 2674
 to the taiga **19**: 2617
 tuna **20**: 2773
 turnstone **20**: 2791
 turtledove **20**: 2793–2794
 vanessa butterflies **20**: 2812
 vireo **20**: 2832, 2833
 walrus **20**: 2851
 warbler **20**: 2857–2858
 waxwing **21**: 2890, 2892
 whales 21: 2912, 2913
 whiting **21**: 2934, *2935*
 wildebeest **21**: 2948, *2949, 2950*
 wood pigeon **21**: 2982
 wood warbler **21**: 2985
Miliaria calandra **3**: 320
milk, Indian buffalo **9**: 1287
Milk snake (king snake) **4**: 535
 Common **10**: 1361
 Louisiana (false coral snake)
  10: 1361
Miller *see* Clothes moth, Common
Miller's-thumb *see* Bullhead,
  European
Millions fish *see* Guppy
Millipede **12**: 1611–1613
 Pill **12**: 1612
 Sierra luminous **12**: 1611
 Snake **12**: 1612
 Sucking **12**: 1612
milt **1**: 88
Milvus
 migrans **10**: 1367, 1368
 milvus **10**: 1367
Mimas tiliae **9**: 1166
mimicry, by
 blue butterflies **2**: 244
 butcherbird **3**: 340
 butterflies and moths 2: 244; 3: 345
 cockatoo **4**: 484
 copper butterflies **4**: 525
 coral snake **4**: 535
 cuckoo **5**: 619–620
 egg-eating snake **6**: 749
 flies **6**: *857*, 858
 hawkmoth **9**: *1167, 1167*, 1168
 hoverfly **6**: *857*; **9**: 1250, *1251*
 jay **10**: 1318
 kallima butterflies **10**: 1331, *1332*
 katydid **10**: *1341*
 leaf fish **10**: 1421, *1421*
 long-horned beetle **11**: 1484, *1484*
 mantis **11**: 1548
 mockingbird **12**: *1622*, 1624
 mynah **12**: 1716
 parrot **14**: 1883
 sawfly **16**: *2240*
 songbirds 18: 2458
 starling **18**: 2523
 swallowtail **18**: 2582
 tiger moth **19**: 2688
 whydah **21**: 2938
mimicry, of
 bees and wasps 2: 184
 hornet **9**: 1234
Mimus
 gilvus **12**: *1624*
 gundlachii **12**: 1622
 polyglottos **12**: 1622, 1623

Miner (*bird*) **13**: 1819
Miner's cat *see* Ringtail
Minivet **12**: 1614–1615
 Ashy **12**: 1614–1615
 Fiery **12**: 1614, 1615
 Mountain **12**: 1614
 Scarlet (flamed minivet; orange
  minivet) **12**: 1614–1615, *1614*
 Small **12**: *1615*
Mink **12**: 1616–1617
 American **12**: 1616, *1616*, 1617
 European **12**: 1616, 1617
Minnow **12**: 1618–1619
 Eurasian **12**: 1618, *1618*, 1619
 Sheepshead **19**: 2705
 see also Carp
Minyobates **1**: 94
miracadia **7**: 870
Miracinonyx inexpectatus **3**: 429
Mirounga
 angustirostris **6**: 771, 772; **16**: 2285
 leonina **6**: 771, 772
Misgolas rapax **19**: 2720
Misgurnus
 anguillicaudatus **11**: 1474
 fossilis **11**: 1473, 1474
missing links
 earless monitor **6**: *732*, 733
 olm **13**: 1789
Mississippi River **16**: 2176
Mistletoe bird (mistletoe
  flowerpecker) **7**: 868, *868*
Misumena vatia **5**: 580, 581, *581*
Mite **3**: 318; **12**: 1620–1621
 Freshwater **12**: 1621, *1621*
 Gall **12**: 1621
 Harvest (harvest bug) **12**: 1620
 Itch **12**: 1620, 1621
 Spider **12**: 1621
 Vine *see* Vine pest
Mithan *see* Gaur, Domestic
Mitu
 mitu **5**: 623
 mitu mitu **5**: 625
 mitu tuberosa **5**: *625*
Mniotilta varia **21**: 2986
Mobula **5**: 668
 diabolis **5**: 668, 669
Moccasin
 Highland *see* Copperhead
 Water (cottonmouth; gapper;
  trapjaw) **14**: 1972, 1973
Mockingbird **12**: 1622–1624
 Bahama **12**: 1622
 Blue **12**: 1622
 Blue and white **12**: 1622
 Galapagos **12**: 1622, 1623–1624
 Northern **12**: 1622, *1622*, 1623,
  1624
 Tropical **12**: *1624*
Moco *see* Cavy, Rock
Modiolaria **12**: 1713
Modiolus modiolus **12**: 1711
Moenkenhausia oligolepsis **19**: 2667
Moho
 bishopi **9**: 1215
 braccatus **9**: 1215
 nobilis **9**: 1215
Mola
 mola **13**: 1774, 1775
 ramsayi **13**: 1774
Mole **12**: 1625–1627
 Eastern **12**: 1625
 European (common mole)
  12: 1625, 1626, *1626*, 1627
 Golden *see* Golden mole
 Kurdistan **12**: 1627
 Marsupial *see* Marsupial mole
 Mediterranean **12**: 1625
 Star-nosed **12**: 1625, 1627

Mole rat (bamboo rat; sand puppy)
  12: 1628–1629
 Common **12**: *1629*
 Indian *see* Bandicoot rat, Lesser
 Naked **12**: *1628*, 1629
 Palestine **12**: 1628–1629
 South American *see* Tuco-tuco
Molly
 Amazon **2**: 234, *234*
 Black **2**: 233–234
 Sail-bearer **2**: 233, *233*, 234
 Sail-fin **2**: 233, 234
 Wedge-face **2**: 233, 234
Moloch (thorny devil; mountain
  devil) **5**: *664*, 665; **12**: 1630–1631
Moloch horridus **5**: 665; **12**: 1631
Molothrus
 ater **3**: 400; **4**: 553, 554; **15**: 2141;
  18: 2459
 badius **4**: 553
 bonariensis **4**: 553
 rufoaxillaris **4**: 553
molting
 bird-eating spider **2**: 204
 bug **2**: 186–187
 cockroach **4**: 494
 crabs and lobsters 4: 572
 crayfish **5**: 592
 daddy longlegs **5**: 637
 lobster **11**: 1477
 mayfly **12**: 1594
 prawn **15**: 2043–2044
 woodlouse **21**: 2973
Molva
 dypterygia **11**: 1455
 macrophthalma **11**: 1455
 molva **11**: 1455, 1456
Momotus momota **12**: 1665, 1666
Monachus
 monachus **12**: 1647, 1648;
  16: 2289
 schauinslandi **12**: 1647, 1648;
  16: 2286
 tropicalis **12**: 1647, 1648; **16**: 2286
Monarch (milkweed butterfly;
  common tiger butterfly) **3**: 345,
  346; **12**: 1632–1634; **20**: 2810
Monasa **15**: 2060
 atra **15**: 2061
 morpheus **15**: 2060–2061
Mongoose **3**: 325; **6**: 814;
  12: 1635–1637
 Banded **12**: 1635, *1637*
 Crab-eating **12**: 1635, 1637
 Dwarf **12**: 1635, 1637
 Egyptian **12**: 1635
 Lesser Indian **12**: 1637
 Malagasy broad-striped
  12: 1635
 Malagasy ring-tailed **12**: *1636*,
  1637
 Water **12**: 1636–1637
 Yellow **12**: 1595, *1635*, 1636
Monia, Bensch's *see* Mesite,
  Subdesert
Monitor lizard (monitor; goanna)
  9: 1281; **12**: 1638–1640
 Asian water **12**: 1638, 1639, 1640
 Bornean rough-necked **12**: 1639
 Desert **12**: 1639
 Earless **6**: 732–733; **12**: 1638
 Indian **12**: *1638*, 1639, 1640
 Komodo monitor *see* Komodo
  dragon
 Lace **12**: 1639
 Nile **12**: 1638, 1639, 1640
 Savanna **12**: *1639*
 Short-tailed **12**: 1638
Monkey
 Allen's swamp **12**: 1643

 Blue (gentle monkey) **8**: 1091, 1092,
  1093; **12**: 1643
 Capuchin *see* Capuchin monkey
 Colobus *see* Colobus monkey
 De Brazza **8**: 1091, 1093; **12**: 1643
 Diana **8**: 1091, *1091*, 1092, 1093;
  12: 1643
 Green (callithrix monkey) **8**: 1091,
  1092, 1093; **12**: 1643
 Howler *see* Howler monkey
 Mustached **8**: 1091, 1093
 Night *see* Night monkey
 Owl *see* Night monkey
 Owl-faced **8**: 1091
 Patas (red guenon; hussar
  monkey; military monkey)
  14: 1896–1897
 Proboscis **10**: 1410, 1411, 1412;
  12: 1643, *1643*, 1644
 Putty-nosed **8**: 1091
 Red-tailed **8**: 1091, 1092, 1093,
  1093; **12**: 1643
 Ring-tail *see* Capuchin monkey
 Squirrel *see* Squirrel monkey
 Vervet **8**: 1091, *1092*; **12**: 1643
 Widow **19**: 2696, 2697
 Woolly *see* Woolly monkey
 see also Ape, Barbary; Baboon;
  Guenon; Langur; Macaque;
  Mandrill; Mangabey;
  Marmoset; Monkeys; Saki;
  Tamarin; Titi; Uakari
Monkeys 12: 1641–1645
Monkfish (angler; common
  goosefish) **1**: 51, 52, *52*
Monkfish (*shark species*; angelshark;
  fiddle fish; angelfish) **12**: 1646;
  17: 2329–2330, 2385
Monklet **15**: 2060
Monk seal **12**: 1647–1648
 Caribbean **12**: 1647–1648; **16**: 2286,
  2289
 Hawaiian **12**: 1647–1648, *1647,
  1648*; **16**: 2286, 2289
 Mediterranean **12**: 1647, 1648;
  16: 2289
Mono *see* Squirrel monkey
Monocentris japonica **18**: 2512
Monocirrhus **10**: 1422
 polyacanthus **10**: 1421
Monodon monoceros **12**: 1718; **21**: 2911
monogamy
 accentor **1**: 23
 beaver **2**: 179
 greylag goose **8**: 1073
 partridge **14**: 1893
 songbirds **18**: 2457
 see also birds, mating for life
Monomorium **9**: 1157
monophagy **21**: 3005
Monopterus albus **18**: 2585
monsoons **12**: 1671
montane biomes **12**: 1671–1672
Moonfish, Spotted *see* Opah,
  Common
Moorhen **12**: 1649–1651
 Common (common gallinule; blue
  rail; red-billed mud hen; Florida
  gallinule) **12**: 1649, 1650
 Dusky **12**: 1649
 Gough's **12**: 1649; **15**: 2108
 Lesser **12**: 1649
 Samoan **12**: 1649
 San Cristobel (makira) **12**: 1649
Moorish idol **12**: 1652–1653
Moose (elk) **5**: 655; **12**: 1654–1656;
  19: *2617*
Mopoke *see* Frogmouth
Moray eel **12**: 1657–1659
 Fine-spotted **12**: *1659*

Green **12**: 1657–1658, *1658*, 1659
Leaf-nosed (ribbon eel) **12**: 1658
Leopard **12**: 1657, 1658
Zebra **12**: *1657*, 1658
Mordacia **10**: 1403, 1404
Morelia **15**: 2080
  amethistina **15**: 2079, 2080, *2114*
  spilota **15**: 2079
Morningbird **21**: 2923
Morone **2**: 164; **8**: 1082
Morpho **12**: 1660–1661
  Common **12**: 1661
morrillo gland **3**: 382
Morus
  bassana **7**: 923, 924; **16**: *2268*
  capensis **7**: 923
  serrator **7**: 923
Moschus **5**: 655
  berezovskii **12**: 1701, 1702
  chrysogaster **12**: 1701, 1702
  moschiferus **12**: 1701, 1702
mososaurs **12**: 1640
Mosquito **6**: 859; **12**: 1662–1664
  Common **12**: *1662*
Mosquito hawk *see* Crane fly
Mossbunder *see* Menhaden, Atlantic
Motacilla **4**: 499; **20**: 2843, *2843*
Moth
  Angle shades **13**: *1821*, 1822
  Atlas **1**: 102; **3**: *343*
  Bent-wing **19**: 2603–2604
  Boisduval's carpenter **7**: 992
  Brown-tail **19**: 2611
  Burnet *see* Burnet moth
  Clearwing **3**: 343–344
  Clothes *see* Clothes moth
  Common emperor **3**: 346
  Cotton **13**: 1821, 1822
  Ermine **19**: 2686
  Feather-winged **14**: 1991
  Footman **19**: 2686
  Ghost **19**: 2603, *2603*, 2604
  Goat **7**: 991–992
  Gypsy **3**: 346; **19**: 2611
  Hercules **3**: 344
  Hornet *see* Hornet moth
  Kitten **15**: 2076
  Leopard **7**: 991
  Lobster **15**: 2077
  Madagascan sunset **20**: 2805, 2806
  Maize *see* Corn borer
  Many-plume **14**: 1991, *1991*
  Mediterranean brocade **3**: 346
  Owlet (Noctuid moth; *larvae*:
    cutworm; army worm) **3**: 344;
    13: 1821–1822
  Peppered **19**: 1915–1916
  Plume *see* Plume moth
  Processionary *see* Processionary
    moth
  Prominent **15**: 2076
  Puss **3**: 345; **15**: 2076–2077
  Reed leopard **7**: 992
  Silk *see* Silk moth
  Swift *see* Swift moth
  Tapestry **4**: 470
  Tiger *see* Tiger moth
  Urania **20**: 2805–2806
  Vaporer **3**: 346
  Wax **3**: 318
  Yucca **21**: 3005–3006
  Zodiac **20**: 2805, 2806
  see also Bollworm; Butterflies and
    moths; Corn borer
Mother Carey's chicken *see* Storm
  petrel
Mother Shipton **13**: 1821
Motmot **12**: 1665–1666
  Blue-crowned (clock bird)
    12: 1665–1666, *1665*

Rufous **12**: 1665
  Turquoise-browed **12**: 1665, *1666*
Mouflon **2**: 198; **12**: 1667–1668
Mountain 12: 1669–1673
Mountain beaver (sewellel; whistler)
  12: 1674–1675
Mountain cat *see* Ringtail
Mountain chicken (*frog*) **8**: 1066
Mountain devil *see* Moloch
Mountain lion *see* Puma
Mourning cloak (Camberwell
  beauty) **20**: 2810, *2811*
Mouse
  Bank *see* Vole, Bank
  Brush **5**: 660
  Cactus **5**: 660
  California **3**: 426; **5**: 660
  Deer *see* Deer mouse
  Flying *see* Scaly-tailed squirrel
  Four-striped grass **21**: *3016*
  Gapper's *see* Red-backed vole,
    Southern
  Harvest *see* Harvest mouse
  Hopping *see* Kangaroo mouse
  House **9**: 1244–1246
  Jerboa *see* Kangaroo mouse
  Jumping *see* Kangaroo mouse
  Kangaroo *see* Kangaroo mouse
  Marsupial *see* Marsupial mouse
  Multimammate *see*
  Multimammate rat
  Oldfield **5**: 660
  Pocket **5**: 665; **10**: 1338
  Red *see* Vole, Bank
  Red-backed *see* Red-backed vole
  Spiny *see* Spiny mouse
  Striped grass *see* Zebra mouse
  Vesper *see* Deer mouse
  White-footed **5**: 660; **19**: 2682
  Wood **21**: 2975–2976
  Yellow-necked **21**: 2975, 2976, *2976*
  Zebra *see* Zebra mouse
Mousebird **12**: 1676–1677
  Blue-naped **12**: 1676
  Red-faced **12**: 1676, *1676*
  Speckled **12**: 1676
  White-backed **12**: 1676, 1677, *1677*
  White-headed **12**: 1676
Mouse-hare *see* Pika
mouth-brooders
  fish **3**: 389–390; **4**: 459; **6**: 836;
    19: 2690
  mouth-brooding frog
    12: 1678–1679
Mud eel *see* Siren
Mudfish, American *see* Bowfin
mudflats **4**: 476–477
Mud hen, Red-billed *see* Moorhen,
  Common
Mudlark *see* Magpie lark
Mudminnow **12**: 1680–1681
  Central **12**: 1680, 1681
  Eastern **12**: 1680, 1681
  European **12**: 1680, 1681
  Olympic **12**: 1680, 1681
Mudpuppy (water dog)
  12: 1682–1683
  Louisiana water dog **12**: 1682
Mudskipper **7**: 993; **12**: 1684–1686
Mud turtle **12**: 1687–1688; **20**: 2795
  Big Bend **12**: 1687
  Common (Florida mud turtle;
    eastern mud turtle) **12**: 1687,
    1687, 1688
Mugger *see* Crocodile, Marsh
Mugil cephalus **12**: 1692, 1693
Mule **21**: 2941
Mule-killer **20**: 2815
Mulgara **11**: 1577, 1578
  Large **11**: 1578

Mullet **5**: 694; **12**: 1692–1693
  Flathead **12**: 1692, 1693
  Golden gray **12**: 1692, 1693
  Striped **12**: 1692, 1693
  Striped red (surmullet) **7**: 989,
    990
  Thicklip gray **12**: 1692, *1692*, 1693
  Thinlip gray **12**: 1692, 1693
Mulloidichthys **7**: 990
  martinicus **7**: 990
Mullus **7**: 990
  auratus **7**: 990
  surmuletus **7**: 989, 990
Multimammate rat (multimammate
  mouse) **12**: 1694–1695
multiple fission **1**: 42
Mungos mungo **12**: 1635
Munia **21**: 2888
  Chestnut-breasted *see* Mannikin,
    Chestnut-breasted
  Red *see* Avadavat
  Spotted *see* Mannikin, Spotted
  White-rumped (striated munia)
    11: 1544
Muntiacus **5**: 658; **12**: 1696, 1697
Muntjac (barking deer; rib-faced
  deer) **5**: 655; **12**: 1696–1698
  Borneo **12**: 1696, 1697
  Chinese (Reeve's muntjac)
    12: 1696, 1697, 1698
  Fea's **5**: 658; **12**: 1696–1697
  Giant **12**: 1697
  Hairy-fronted **12**: 1696
  Indian **12**: 1696, *1696*, 1697, *1697*
Murexia **11**: 1577
Murray magpie *see* Magpie lark
Murre *see* Guillemot
Murrelet, Xantus' **16**: 2264–2265
Musang *see* Palm civet, Common
Mus avellanarius **19**: 2653
Musca domestica **9**: 1241, 1242
Muscardinus avellanarius **5**: 699, 700,
  702
Muscicapa striata **7**: 872–873
Muscivora forficata **20**: 2800
musk **1**: 120; **4**: 461; **12**: 1702, *1702*,
  1703, 1710; **21**: 2894–2895
Musk deer **5**: 655; **12**: 1701–1702
  Dwarf **12**: 1701, 1702
  Mountain (forest musk deer)
    12: 1701, 1702
  Siberian **12**: 1701, *1701*, 1702, *1702*
Muskellunge **14**: 1945, 1946
musk glands **4**: 446; **5**: 667; **12**: 1703;
  14: 1905; **18**: 2539
Musk-ox (musk sheep) **12**: 1703–1705
Musk rat *see* Shrew, Musk
Muskrat (musquash) **12**: 1706–1708;
  19: 2616
  Round-tailed (Florida water rat)
    12: 1706
Musk turtle **12**: 1687; **20**: 2795
  Common (stinkpot; stinking-jim)
    12: 1688, *1688*
  Keel-backed **12**: 1687
Mus musculus **9**: 1245
Musophaga
  porphyreolopha **20**: 2780
  rossae **20**: 2780, 2781
  violacea **20**: 2780
musquash **12**: 1706
Mussel **12**: 1711–1713; **14**: 1982
  Australian **12**: 1711
  California **12**: 1711
  Common (blue mussel; edible
    mussel) **12**: 1711, 1712
  Fan **12**: 1711
  Green-lipped **12**: *1712*
  Horse **12**: 1711
  Swan **12**: 1711

Mussel-digger *see* Whale, Gray
Musselpecker *see* Oystercatcher
Mussel worm *see* Ragworm
Mussurana **6**: 814
Mustela
  erminea **6**: 786; **14**: 2003
  eversmanni **2**: 228; **14**: 2001
  lutreola **12**: 1616, 1617
  nigripes **2**: 229
  nivalis **14**: 2003; **21**: 2893–2895
  putorius **2**: 228; **14**: 2001, 2002
  putorius furo **14**: 2001
  vison **12**: 1616, 1617
mutations, fruit fly **7**: 910
Mutilla
  europaea **20**: 2815
  lilliputiana **20**: 2815
Muttonbird *see* Shearwater, Short-
  tailed
Muttonbird, New Zealand *see*
  Shearwater, Sooty
mutualism (symbiosis)
  bitterling **2**: 218–219
  clam **4**: 463
  copper butterfly **4**: 525
  coral **4**: 529
  damselfish **5**: 638, 640, *640*
  flowerpecker **7**: 868
  hermit crab **9**: 1178, *1178*
  leafcutter ant **10**: 1417
  mycorrhizae **19**: 2616
  nutcracker **13**: 1758
  oxpecker **13**: 1830, *1830*
  yucca moth **21**: 3005, *3005*, 3006
  see also commensalism;
    zooxanthellae
Mutum *see* Curassow, Razor-billed
Mya **7**: 926
  arenaria **7**: 927
  binghami **7**: 927
  truncata **7**: 927
mycorrhizae **19**: 2616, 2655
Mycteria **18**: 2551
  americana **18**: 2552
  ibis **18**: 2552
  leucocephala **18**: *2553*
Myctophum punctatum **10**: 1414
Mydaus javanensis **1**: 120
Mygale *see* Spider, Bird-eating
Myiopsitta monachus **14**: 1878
Mylagra azureocapilla **2**: 283
Myliobatis
  aquila **6**: 729, 730
  californica **6**: 729, 730
  freminvilli **6**: 729, 730
Mylopharodon conocephalus **12**: 1618
Mynah **12**: 1714–1716
  Bank (Bengal mynah) **12**: 1714,
    1716
  Brahminy **12**: *1716*
  Chinese crested **12**: 1714
  Common (Indian mynah) **7**: 912;
    12: 1714–1715, *1714*, 1716
  Golden-crested **12**: 1714
  Indian hill (greater hill mynah)
    12: 1714
  Papuan **12**: 1714
  Rothschild's (Bali mynah; Bali
    starling; Rothschild's starling;
    Rothschild's grackle) **12**: *1715*
Myocastor coypus **4**: 562; **5**: 667
Myomimus **5**: 699
Myoprocta pratti **1**: 32
Myopus **10**: 1433
Myotis
  adversus **6**: 840
  evotis **2**: 168
  mystacinus **20**: 2822, 2823
Myotis, Long-eared **2**: 168
Myripristis **18**: 2511

Myrmecobius fasciatus **1**: 129
myrmecochory **9**: 1157
Myrmecocystus **9**: 1205
  mexicanus **9**: 1205
Myrmeleon
  formicarius **1**: 64
  obsoletus **1**: 64
Myrmeleotettix **8**: 1036
  maculatus **8**: 1037
Myrmotherula
  axillaris **1**: 55
  brachyura **1**: 55
Mysateles **9**: 1263, 1264
  garridoi **9**: 1264
mysis larvae **11**: 1477
Mysis relicta **13**: 1796
myths *see* legends/myths
Mytilus
  californianus **12**: 1711
  edulis **12**: 1711, 1712
  obscurus **12**: 1711
myxamatosis **15**: 2093
Myxas **14**: 2011
  glutinosa **14**: 2010, 2011
Myzomela cardinalis **9**: *1216*
Myzostoma cirriferum **6**: 805
Myzus persicae **1**: 74

N

Naemorhedus
  baileyi **8**: 1022
  caudatus **8**: 1022
  goral **8**: 1022
nagana **20**: 2760, 2761
Naja **2**: 265; **4**: 480–482; **20**: 2818
Naked snake *see* Caecilian
Naked-tailed rat, Giant **7**: 964, *964*, 965
Nandinia binotata **13**: 1855, 1856
Nandus **10**: 1422
  oxyrhynchus **10**: 1421
Nannie nine eyes *see* Lamprey
Nannopterum harrisi **4**: 536
Napoleon fish *see* Wrasse, Humphead
Narceus **12**: 1611
Narcine tasmaniensis **6**: 764
Narwhal (narwhale) **12**: 1717–1719; **21**: 2911
Nasalis **10**: 1411
  larvata **12**: 1643
Naso **18**: 2576
  tuberosus **18**: 2576
Nassarius glans **21**: *2917*
Nasua **4**: 478, *478*, 479, *479*
Nasuella olivacea **4**: 478, 479
Nasutitermes **1**: 68
  triodiae **1**: 68
Native Americans
  and the bison **2**: 215–216, 217
  and chuckwallas **4**: 454
  and scale insects **16**: 2247
  and smelts **17**: 2410
  and wapiti **20**: 2856
Native cat *see* Marsupial cat
Native pheasant *see* Lyrebird; Mallee fowl
Native porcupine *see* Spiny anteater
Natrix **20**: 2874
  maura **20**: 2874, 2875
  natrix **8**: 1043; **9**: 1203; **15**: 2103; **17**: *2420*; **20**: 2874, 2875
  tesselata **20**: 2874, 2875
natural selection **5**: 648; **14**: 1915–1916
Naucrates ductor **14**: *1951*; **16**: 2242
nauplius larvae **1**: 139; **2**: 276; **6**: 792; **10**: 1387; **14**: 1981
Nautilus **12**: 1722, 1723

Nautilus (pearly nautilus) **12**: 1722–1723
Nautilus, Paper (argonaut) **12**: 1722; **13**: 1779
navigation
  bear **2**: 176
  parrotfish **14**: 1888–1889
Neanura **18**: 2505
Necrophorus
  americanus **3**: 328
  investigator **3**: 327
  vespilloides **3**: 327
Necrosyrtes monachus **20**: 2841
nectar-eaters
  bee **2**: 185; **9**: 1209
  flowerpecker **7**: 868
  fruit bat **7**: 907
  honeycreeper **9**: 1213
  honeyeater **9**: 1215, *1215*
  hummingbird **9**: 1257
  lorikeet **11**: 1488–1489
  moth **9**: 1167–1168
  sugarbird **18**: 2558, *2558*
  sunbird **18**: 2565, *2565*
Nectarinia
  amethystina **18**: *2565*
  asiatica **18**: *2458*, 2566
Nectonema **9**: 1237
Necturus **12**: 1682
  maculosus **12**: 1683
  maculosus louisianensis **12**: 1682
  maculosus maculosus **12**: 1682
Needle bug *see* Water scorpion
Needlefish (garfish) **12**: 1724–1725
  Atlantic **12**: 1725
  California **12**: *1725*
Needletail
  Brown-backed **18**: 2484, *2484*
  White-throated **19**: 2600
Negaprion brevirostris **17**: 2333
nekton **13**: 1770
nematocysts (stinging cells) **1**: 47–48; **9**: 1265, *1266*; **10**: 1322; **15**: 2029, 2030–2031, *2031*; **20**: 2818
Nematus compressicornis **16**: 2240
Nemesia **19**: 2719
Nemorhaedus
  crispus **17**: 2324
  sumatraensis **17**: 2324
  swinhoei **17**: 2324
Nemoura **18**: 2549
Nene *see* Goose, Hawaiian
Neoceratodus forsteri **11**: 1501
Neofelis nebulosa **3**: *404*
Neofiber alleni **12**: 1706
Neomys
  anomalus **20**: 2872
  fodiens **20**: 2872
Neomysis integer **13**: *1795*, 1796
Neopagetopsis **9**: 1278
Neophasia menapia **21**: 2928
Neophema
  elegans **14**: *1880*
  pulchella **14**: 1878–1879
Neophoca cinerea **16**: 2282, 2283, 2288
Neophocaena phocaenides **15**: 2024, 2025
Neophron percnopterus **20**: 2840–2841
Neoseps reynoldsi **17**: 2390
neoteny
  axolotl **1**: 109, 110, *110*
  emu **6**: 782
  hellbender **9**: 1175
  midge **12**: 1608
  mudpuppy **12**: 1683
  olm **13**: 1789
  siren **17**: 2382
  see also metamorphosis
Neotermis militaris **1**: 68
Neotoma **13**: 1843, 1844, *1844*

Neotragus
  batesi **18**: 2571, 2572
  moschatus **1**: 61; **18**: 2571, 2572
  pygmaeus **1**: 60; **18**: 2571, 2572
Nepa cinerea **20**: 2871, *2871*
Nephila **13**: 1800–1801
Nephrops norvegicus **4**: 573; **11**: 1475
Nephruros **11**: 1468
Neptunea antiqua **21**: 2917
Nereis
  fucata **15**: 2104
  limnicola **15**: 2105
  longissima **15**: 2104
  virens **15**: 2104
Nerodia **20**: 2874, 2875, *2875*
Nesiarchus nasutus **17**: 2414
Nesocites micromegas **14**: 1937
Nesokia indica **1**: 132
Nesolagus netscheri **18**: 2561
Nesomimus **12**: 1622
Nesoryzomys **15**: 2157
Nestor
  meridionalis **10**: 1343
  notabilis **10**: 1344
nests
  American robin **16**: 2190
  avadavat **1**: 103–104
  bald eagle **1**: 125
  Baltimore oriole **1**: 127
  barnacle goose **1**: 140
  bee and wasp **2**: 184, 185; **3**: 317–318
  broadbill **2**: 282–283
  bubble **1**: 89; **6**: 821–822, 836; **13**: 1868–1869; **18**: 2585
  budgerigar **3**: 303–304
  bunting **3**: 320
  cock **21**: 2888–2889
  cuckoo-shrike **5**: 622
  curlew **5**: *626*, 628
  flicker **6**: *854*, 855
  flycatcher **7**: 874
  gallinule **7**: 920, *920*
  gannet **7**: 924, *924*
  gliding frog **7**: 982
  golden eagle **7**: 1003
  hamerkop **8**: 1124–1125, *1125*
  harrier **8**: 1145, *1146*
  hornbill **9**: 1228
  jackdaw **10**: 1300, *1301*
  kittiwake **10**: 1370, 1371
  magpie lark **11**: 1523
  mallee fowl **11**: 1527–1529
  mannikin **11**: 1545
  martial eagle **12**: 1588–1589
  merlin **12**: 1602–1603, *1603*
  muskrat **12**: 1706
  mynah **12**: 1716
  orangutan **13**: 1797–1798, *1798*
  oropendula **13**: *1805*, 1806–1807, *1807*
  osprey **13**: 1812–1813, *1813*
  ovenbird **13**: 1819, *1819*, 1820, *1820*
  pack rat **13**: 1843
  palmchat **13**: 1854
  paper wasp **13**: 1864, *1864*, 1865, *1865*
  seabird 16: 2268
  songbird 18: 2458
  spinetail swift **18**: 2484–2485
  stick-nest rat **18**: 2535
  stork **18**: *2551*, 2552
  swift **19**: 2600, *2600*
  swiftlet **19**: 2601
  tailorbird **19**: 2618, 2619, *2619*
  thornbill **19**: 2669–2670, *2669*
  tree squirrel **20**: 2742
  trogon **20**: 2747
  vespula wasp **20**: 2824–2825
  vireo **20**: *2832*, 2833

weaver **21**: *2897*, 2898
  woodpecker **21**: 2979
Netta **14**: 1993
  erythrophthalma **14**: 1995
  rufina **14**: *1993*
Nettapus **5**: 714; **21**: 2927
  coromandelianus **5**: 714
neuromasts **6**: 836; **17**: 2333
Neuroterus **7**: 922
  quercus-baccarum **7**: 921, 922, *922*
Newmaster rubiginosa **16**: *2280*
Newt **13**: 1732–1734
  California **13**: 1733, 1734; **19**: *2653*
  Crested (warty newt) **13**: 1732, 1733, *1733*, 1734
  Palmate **13**: 1732, *1732*, 1733
  Siberian **19**: 2616
  Smooth **13**: 1732, 1733, 1734, *1734*
Nezara viridula **17**: 2344
Nicaragua, Lake **21**: 3007
nictitating membrane **2**: 197; **5**: 680; **21**: *2997*
nidifugous species **4**: 569; **5**: 717
Night adder **6**: 749; **13**: 1735–1736
  Common (rhombic night adder) **13**: 1735, 1736, *1736*
  Lichtenstein's **13**: 1735
  Snouted **13**: 1735
  Velvety (green night adder) **13**: 1735, *1735*
Nightcrawler *see* Earthworm
Nighthawk **13**: 1742
  Common **13**: *1744*; **14**: 2013
Night heron **13**: 1737–1739
  Black-crowned **13**: 1737–1738, *1738*, 1739, *1739*
  Japanese **13**: 1737
  Malaysian **13**: 1737
  Rufous (Nankeen night heron) **13**: 1737, 1739
  White-backed **13**: 1737
  White-eared **13**: 1737
  Yellow-crowned **13**: 1737, *1737*
Nightingale **13**: 1740–1741; **18**: 2459; **19**: 2678
Nightjar **13**: 1742–1744
  Egyptian **13**: 1742
  Eurasian (fern owl; churn owl; eve-jarr; dorhawk; goatsucker) **13**: 1742, *1742*, 1743–1744; **14**: 2013
  Jungle **13**: 1742
  Large-tailed **13**: 1742, *1743*
  Long-tailed **13**: 1742
  Pennant-winged **13**: 1742
  Red-necked **13**: 1742
  Spotted **13**: 1742; **14**: 2013
  Standard-winged **13**: 1742
  White-throated **13**: 1742
  Wood *see* Potoo
Night lizard **13**: 1745–1746
  Arizona **13**: 1745, 1746
  Cuban **13**: 1745
  Desert **13**: 1745, 1746
  Granite **13**: 1745, *1745*
  Island **13**: 1745, 1746
Night monkey (douroucouli; owl monkey) **12**: 1642, 1644, 1645; **13**: 1747–1748
Nightwalker *see* Kinkajou
Nile fish (aba) **10**: 1377, *1377*; **13**: 1749–1750
Nile River **16**: 2176
Nilgai (blue bull; blue buck) **1**: 60; **13**: 1753–1754
Nilghai *see* Nilgai
Nilssonia formosa **17**: 2443
Nine-eyes *see* Butterfish
Nipponia nippon **9**: 1274
Nithoptera priamus **15**: 2110

Noctilio leporinus **2**: 167; **6**: 840, *840*, 841
Noctua pronuba **13**: 1822
Noctule **20**: 2822, 2823, *2823*
   Slender-winged **20**: 2822
nocturnal
   agouti **1**: 32
   anteater **1**: 58, *58*
   armadillo **1**: 85
   aye-aye **1**: 111
   badger **1**: 120–121
   bandicoot **1**: 130
   barn owl **2**: 149
   bat-eared fox **2**: 165
   bats **2**: 169
   big-headed turtle **2**: 197
   binturong **2**: 201, *201*
   bird-eating spider **2**: 203
   boa constrictor **2**: 250
   bobcat **2**: 253, *253*
   bristletail **2**: 279
   bullhead **3**: 314, 315
   bushbaby **3**: 330
   bushpig **3**: 335
   butterfly fish **3**: 348
   caracal **3**: 386
   cardinal fish **3**: 390
   centipede **3**: 411
   chinchilla **4**: 444
   civet **4**: *460*
   common toad **4**: 512, *512*
   coral **4**: 529
   crab-eating fox **4**: 566, *566*
   crab plover **4**: 568
   crested pocupine **5**: 596, *596*
   diving-petrel **5**: 684
   egg-eating snake **6**: 748
   fish owl **6**: 842
   flying squirrel **7**: 884, *885*
   frogmouth **7**: 900
   gecko **7**: 945
   genet **7**: 948, *948*
   hamster **8**: 1129
   horseshoe bat **9**: 1239
   mudpuppy **12**: 1683
   muntjac **12**: 1696
   night heron **13**: 1737
   night monkey **13**: 1747
   ocelot **13**: 1776
   oilbird **13**: 1784
   otter shrew **13**: 1818
   owlet-nightjar **13**: 1823
   paca **13**: 1838, *1838*
   phalanger **14**: 1929, 1930
   poorwill **14**: 2012, *2013*
   Port Jackson shark **15**: 2027
   potoo **15**: 2032
   raccoon dog **15**: 2099
   red fox **15**: 2134, *2134*
   scaly-tailed squirrel **16**: *2251*
   slater **17**: 2398, *2398*
   solifugid **18**: 2453
   spadefoot toad **18**: 2465
   springhare **18**: 2502, *2503*
   stone curlew **18**: 2546
   tarsier **19**: 2639–2640
   Tasmanian devil **19**: 2641
   tawny owl **19**: 2644, *2645*
   vesper bat **20**: 2823
   wild boar **21**: 2942
Noddy **13**: 1755–1756
   Black **13**: 1755, 1756
   Blue **13**: 1755
   Brown **13**: 1755, 1756
   Gray **13**: 1755
   Lesser **13**: 1755, *1756*
Nomadacris septemfasciata **11**: 1478, 1479
nomadism
   flamingo **6**: 844

fruit bat **7**: 908
Gouldian finch **8**: 1030–1031
pelican **14**: 1910
plains-wanderer **14**: 1978
polar bear **14**: 1998
pronghorn antelope **15**: 2050
spoonbill **18**: *2498*
swan **18**: *2588*
waxwing **21**: 2891
Nomeus albula **15**: 2031
Nonnula **15**: 2060
Noodlefish **7**: 978
No-seeeum *see* Midge, Biting
Notemigonus crysoleucas **16**: 2182
Notharchus
   macrorhynchus **15**: 2061
   pectoralis **15**: 2061
Nothoprocta ornata **19**: 2691
Notiomystis cincta **9**: 1215
notochord **4**: 497–498; **17**: 2313
Notomys **10**: 1338
   alexis **10**: *1337*
Notonecta
   glauca **1**: 118, 119
   hoffmanni **1**: 118
   insulata **1**: 118, 119
   undulata **1**: 119
Notopterus **6**: 803, 804
Notornis mantelli **19**: 2621
Notoryctes typhlops **11**: 1576
Novumbra hubbsi **12**: 1680, 1681
Nucifraga
   caryocatactes **13**: 1757, 1758
   caryocatactes caryocatactes **13**: 1757
   columbiana **13**: 1757
nuclei, paramecium **14**: 1881, 1882, *1882*
Nuculus
   sulcata **13**: 1764
   tenuis **13**: 1763, 1764
Nudibranch *see* Sea slug
Numbat *see* Anteater, Banded
Numbfish *see* Electric ray
Numenius **5**: 626, 627, 628
Numida meleagris **8**: 1097
Nunbird **15**: 2060
   Black **15**: 2061
   White-fronted **15**: 2060–2061
Nunlet **15**: 2060
Nurse hound *see* Dogfish, Greater spotted
Nutcracker **13**: 1757–1759
   Clark's **13**: 1757, *1757*, 1759, *1759*
   European (slender-billed nutcracker; thick-billed nutcracker) **13**: 1757, 1758, *1758*, 1759
Nuthatch **13**: 1760–1762
   Coral-billed **20**: 2813
   Corsican **13**: 1762
   Eurasian **13**: 1760–1761, *1760*
   Red-breasted **13**: *1761*, 1762
   White-breasted **13**: *1762*
Nutshell **13**: 1763–1764
   Brown **13**: 1764
   Thin **13**: 1763, 1764
Nyala (inyala) **3**: 331; **13**: 1765–1766
   Mountain **13**: 1765, 1766
Nyasa, Lake **10**: 1398
Nyctalemon **20**: 2805
   patroclus **20**: 2805, *2806*
Nyctalus noctula **20**: 2822
Nyctea scandiaca **13**: 1827; **17**: 2438; **20**: 2778
Nyctereutes **5**: 689
   procyonoides **15**: 2100
Nyctibius
   grandis **15**: 2033
   griseus **15**: 2032
   jamaicensis **15**: 2032

Nycticebus
   coucang **11**: 1490, 1491
   pygmaeus **11**: 1490, 1491
   tardigradus **11**: 1490, 1491
Nycticorax
   caledonicus **13**: 1737
   nycticorax **13**: 1737, 1738
   violacus **13**: 1737
Nycticryphes semicollaris **13**: 1849
Nyctiphrynus ocellatus **14**: 2013
Nylgau *see* Nilgai
Nymphalis antiopa **20**: 2810, 2811
Nymphicus hollandicus **4**: 483
Nymphon gracile **17**: *2310*, 2311
nymphs
   backswimmer **1**: 119
   bug **2**: 186
   dragonfly **5**: 710, 711
   earwig **6**: 740
   mayfly **11**: 1593–1594, *1594*
Nystalus maculatus **15**: *2060*

O

oak apples **7**: 921
Oarfish **13**: 1767–1768
   King of the herrings **13**: 1767, *1767*
Obelia **16**: 2273
Ocean 13: 1769–1773
   deep scattering layers **9**: 1162
   floor **13**: 1772–1773
Oceanites
   hornbyi **18**: 2554
   oceanicus **18**: 2554
   tethys **18**: 2554
Oceanodroma leucorhoa **18**: *2554*
Ocean perch **16**: 2260, 2261
ocean trenches **13**: 1772
ocelli **18**: 2505
Ocelot **13**: 1776–1778
Ochotona **14**: 1944
   pallasi **14**: 1944
   princeps **14**: 1944
Ochrogaster lunifer **15**: 2048
Octopus **13**: 1779, *1779*, 1780
Octopus **13**: 1779–1781
   Blue-ringed **13**: 1781; **20**: 2819–2820, *2819*
   Common **13**: 1779, 1780, 1781, *1781*
   Lesser (curly octopus) **13**: 1779
   Pacific **13**: 1779
Ocypode cursor **4**: 573
Ocyurus chrysurus **17**: 2424
Odobenus rosmarus **7**: 927; **16**: 2285; **20**: 2852
Odocoileus **3**: 426; **5**: 656; **12**: 1689, 1690; **19**: 2653, 2682
Odontesthes regia **17**: 2380
Odontodactylus scyllarus **11**: *1551*
Odontophorus gujanensis **15**: 2082
odor *see* attracting a mate with scent; scent marking; smell (for defense)
Oedistoma pygmaeum **9**: 1215
Oilbird (guácharo; diablotin) **13**: 1784–1785
Oilfish *see* Escolar
oil spills
   and guillemots **8**: 1096
   and penguins **14**: 1914
   and puffins **15**: 2067
   and razorbills **15**: 2126
   and sea otters **16**: 2296
oil-spitting, fulmar **7**: 913, 915
Okapi **7**: 975; **13**: 1786–1787
Okapia johnstoni **13**: 1787
Okavango Delta **16**: 2176; **21**: 2907, *2907*
Okeechobee, Lake **21**: 2907

Old maid *see* Gaper, Sand
Oldsquaw *see* Duck, Long-tailed
oligophagy **21**: 3005
Olingo **10**: 1363
Olm **13**: 1788–1789
ommatidia **6**: 857; **18**: 2505
Ommatophoca rossi **16**: 2211
Omocestus **8**: 1036
   viridulus **8**: 1037
Onager **21**: 2939, 2940, *2940*
Onca see Jaguar
Oncomeris flavicornis **17**: 2344
Oncorhynchus **13**: 1840, 1841; **20**: 2751, 2752
Ondatra zibethicus **5**: 667; **12**: 1706–1708; **19**: 2616
Oniscus **21**: 2974
   asellus **21**: 2973
Onuxodon **14**: 1901
Onychiurus **18**: 2505
Onychogalea fraenata **20**: 2846
Onychorhynchus mexicanus **20**: 2800
Onychoteuthis banksi **18**: 2510
Oo
   Bishop's **9**: 1215
   Great (oo) **9**: 1215
   Kauai (ooaa) **9**: 1215
Ooaa *see* Oo, Kauai
Opah **13**: 1790–1791
   Common (spotted moonfish) **13**: 1790, *1790*, 1791
   Southern **13**: 1790, 1791
Openbill **18**: 2553
Opheodrys aestivus **19**: 2733
Ophiocoma **2**: 159
Ophiodon elongatus **20**: 2766
Ophion **9**: 1279
Ophiophagus **2**: 265
   hannah **17**: 2420
ophiopluteus larvae **2**: 281
Ophiothrix **2**: 159, 281
Ophisaurus **6**: 733; **7**: 980, 981; **10**: 1430
Ophisternon candidum **18**: 2585
Ophiura **2**: 281
Opifex **12**: 1664
Opisthocomus hoazin **9**: 1197
opisthoglyphs **13**: 1736
Opossum **13**: 1792–1794
   Common mouse (murine opossum) **13**: 1792, 1793, *1794*
   Dormouse *see* Possum, Dormouse
   Pygmy *see* Possum, Dormouse
   Virginia (common opossum) **13**: 1792, *1792*, 1793, *1793*
   Water (yapok) **13**: 1792, 1793
Oppel, Michael **3**: 357
Opsanus tau **19**: 2698, 2699
Orange tip, Desert **21**: 2928
Orangutan **1**: 70, 71, 72, 73, *73*; **13**: 1797–1799
Orb spider (orb-web spider) **13**: 1800–1802; **18**: 2480, *2481*, 2482
   Araneus **13**: *1801*
Orca *see* Killer whale
Orcaella brevirostris **21**: 2911
Orchopeas howardi **6**: *850*
Orcinus orca **1**: 82; **5**: 695; **10**: 1351; **16**: 2211
Oreamnos americanus **16**: 2200
Oreaster occidentalis **18**: 2516
Orectochilus **21**: 2921
   villosus **21**: 2921, 2922
Orectolobus **21**: 2954, 2955
Oreochromis
   alcalicus grahami **19**: 2689
   mossambicus **19**: 2689–2690
Oreoica gutturalis **2**: 191, 192
Oreoscoptes montanus **19**: 2671

Oreotragus oreotragus **1**: 61; **10**: 1375
Oreotrochilus **12**: 1672
Organ-grinder *see* Wallaby, Nail-tailed
Oribi **13**: 1803–1804
Oriole
    Baltimore **1**: 126–127; **20**: 2833
    Black **8**: 1012
    Black-headed **8**: 1013
    Golden **8**: 1012–1013
    Maroon **8**: 1012
Oriolia bernieri **20**: 2813
Oriolus **8**: 1012, 1013
Ornithodoros **19**: 2682
Ornithoptera **2**: 212; **20**: 2806
    alexandrae **3**: 344; **18**: 2582
Ornithorhynchus anatinus **14**: 1986
Oropendula **13**: 1805–1807
    Chestnut-headed **13**: *1805, 1807*
    Montezuma **13**: 1806
    Russet-backed **13**: *1806*
Ortalis vetula **20**: 2757
Orthogeomys **14**: 1997
Orthopsittaca **11**: 1514
Orthotomus **19**: 2618, 2619
Ortygospiza locustella **21**: 2888
Orycteropus afer **1**: 17
Oryctolagus cuniculus **15**: 2092
Oryx **1**: 62, 63; **5**: 665; **13**: 1808, 1809
Oryx **1**: 62; **5**: 665; **10**: 1382;
    13: 1808–1810
    Arabian **1**: 63; **13**: 1808, 1809, *1809,*
        1810
    Beisa *see* Gemsbok
    Scimitar **1**: 63; **13**: 1808, 1809, 1810
    Tufted *see* Gemsbok
Oryzomys **15**: *2157*
    alfaroi **15**: 2156
    antillarum **15**: 2156
    bauri **15**: *2156*
    couesi **15**: 2156
    melanotis **15**: 2156
    palustris **15**: 2157
Oryzorictes **19**: 2656, 2657
Osbornictis piscivora **7**: 946, 947
Oscines **3**: 340
Osmerus eperlanus **17**: 2410
osmeterium **2**: 212; **18**: 2580, 2582
osmoregulation **5**: 606
Osmylus fulvicephalus **10**: 1392
Osprey (fish hawk) **1**: 82, 124;
    2: 209–210; **13**: 1811–1813;
    21: 2906
Osteolaemus
    tetraspis **5**: 603
    tetraspis osborni **5**: 603
Ostracion meleagris **20**: 2758
ostracitoxin **20**: 2758
Ostrea edulis **13**: 1832, 1833
Ostrich **13**: 1814–1816
Ostrinia nubilalis **4**: 540
Otaria flavescens **16**: 2282, 2283
Otis tarda **3**: 337, 338
Otocyon **5**: 689
    megalotis **2**: 166; **5**: 665
Otolemur **3**: 330
    argentatus **3**: 329
    crassicaudatus **3**: 329, 330
Otter
    Canadian (river otter) **16**: 2179,
        2179, 2180
    Clawless **16**: 2179
    European (common otter)
        16: 2179, 2180, 2181
    Giant Brazilian **16**: 2179, 2180
    River **16**: 2179–2181
    Sea **16**: 2179, 2294–2296
    Small-clawed **16**: 2179
    Smooth-coated **16**: 2179, 2180
Otter cat *see* Jaguarundi

Otter-shell, Common **7**: 927
Otter shrew **13**: 1817–1818; **19**: 2656
    Giant **13**: 1817–1818, *1817*
    Mount Nimba **13**: 1817, 1818
    Ruwenzori **13**: 1817, 1818
Otus **13**: 1827, 1828, *1828*, 1829
Ounce *see* Snow leopard
Ourebia ourebia **13**: 1804
outback **8**: 1038
Ouzel
    American water *see* Dipper,
        American
    Ring **19**: 2680
Ovenbird **13**: 1819–1820; **18**: 2455
    Rufous hornero **13**: 1819, 1820,
        1820
    True (hornero) **13**: *1819*
Ovenbird (*wood warbler*) **13**: 1819;
    21: 2985, 2986
Ovibos moschatus **12**: 1703–1705
ovigers **17**: 2310
ovipositor **2**: 183; **16**: 2240, 2278;
    19: 2676; **20**: 2826
Ovis **2**: 198, 199; **12**: 1668
ovoviviparity
    in caecilians **3**: 357
    in fish **1**: 46; **5**: 670; **7**: 897; **8**: 1103,
        1107; **18**: 2544
    in lizards **10**: 1431; **11**: 1468
    in scorpions **16**: 2259
    in sharks **17**: 2333; **19**: 2675;
        21: 2916
    slowworm **17**: 2407
    in snakes **1**: 27, 44, 98; **7**: 935;
        15: 2059, *2059*
Owl
    Barn **2**: 148–150; **3**: 323; **13**: 1825,
        1826, 1829
    Barred **13**: 1825
    Boreal **13**: 1829
    Burrowing **3**: 323–325; **13**: 1828
    Eagle **13**: 1825, 1828, *1829*
    Elf **13**: *1826*, 1827, 1828
    Fish *see* Fish owl
    Galapagos **5**: 650
    Great gray **13**: *1825*, 1827
    Hume's desert **13**: 1827, 1828
    Least pygmy **13**: 1825
    Little **13**: *1827*
    Long-eared **13**: 1826
    Masked **18**: 2536
    Pygmy **13**: 1829
    Screech *see* Screech owl
    Short-eared **7**: 893; **13**: 1828
    Snowy **13**: 1827–1828, 1829;
        17: 2437–2439; **20**: 2778
    Tawny (brown owl) **13**: 1827;
        19: 2644–2646
    see also Owls; Scops-owl
Owlet, Forest **13**: 1826, 1829
Owlet-nightjar **13**: 1823–1824
    Australian **13**: 1823, 1824, *1824*
    Mountain **13**: *1823*
Owls 13: 1825–1829
Ox, Cambodian forest *see* Kouprey
oxbow lakes **16**: 2175
Oxpecker (tick bird) **13**: 1830–1831;
    18: 2457
    Red-billed **13**: 1830, *1831*
    Yellow-billed **13**: 1830
Oxybelis aeneus **19**: 2733
oxyluciferin **6**: 827
Oxyporhamphus **7**: 876
Oxyura **18**: 2537, 2538
Oxyuranus
    microlepidotus **5**: 665
    scutellatus **20**: *2817*
Oyster **13**: 1832–1834
    American **13**: 1832, 1834
    European flat **13**: 1832–1833, *1834*

    Japanese **13**: 1832
    Pearl **13**: 1832, 1834
    Portuguese **13**: 1832
    Saddle **13**: 1832
    Thorny **13**: 1832
Oystercatcher (musselpecker)
    13: 1835–1837
    American **13**: 1835, *1835*, 1837
    American black **13**: 1835
    Black **13**: 1835
    Canary Islands **13**: 1835
    Eurasian **13**: 1835, 1836, 1837, *1837*
    Magellanic **13**: 1835, *1836*
    Pied **13**: 1835
    Sooty **13**: 1835
ozone layer, depletion **1**: 82
Ozotoceros bezoarticus **18**: 2463

P

Paca (spotted cavy; laba; agouti)
    13: 1838–1839
    False (pacarana) **13**: 1838
    Lowland **13**: 1838, *1838*, 1839
    Mountain **13**: 1838, 1839
Pacarana *see* Paca, False
Pachycephala **21**: 2923, 2924
Pachycheles **18**: 2507
pachyderms **6**: 767
Pachygrapsus crassipes **17**: 2353
Pachyphantes **21**: 2896
Pachyptila **15**: 2045, *2045*, 2046
Pachyramphus
    niger **4**: 543
    polychopterus **4**: 543
Pacifastacus leniusculus **5**: 592
Pack rat (trade rat; wood rat)
    13: 1843–1844
    Bushy-tailed wood rat **13**: 1843,
        1843, 1844
    Desert wood rat **13**: 1844, *1844*
    Dusky-footed wood rat **13**: 1843,
        1844
    Eastern (Florida pack rat) **13**: 1843,
        1844
packs, hunting in *see* hunting in
    packs
Padda oryzivora **10**: 1316; **11**: 1544
Paddlefish **13**: 1845–1846
    Mississippi (spoonbill sturgeon)
        13: 1845, *1845*, 1846, *1846*
Paddy *see* Sheathbill, Snowy
Pademelon (Tasmanian pademelon)
    13: 1847–1848
    Calaby's (alpine wallaby) **13**: 1847,
        1848
    Dusky **13**: 1847, 1848
    New Guinea **13**: 1847, 1848
    Red-bellied (rufous wallaby)
        13: 1847, *1847*, 1848
    Red-legged **13**: 1847, 1848
    Red-necked **13**: 1847, 1848
    Short-tailed *see* Quokka
Padochela hemphilli **18**: 2476
Pagetopsis **9**: 1278
Pagodroma nivea **17**: 2436
Pagophilus groenlandica **8**: 1140
Paguma larvata **13**: 1855
Paguropsis typica **9**: 1178
Pagurus
    bernhardus **9**: 1176, 1177
    megistas **9**: *1176*
    prideauxi **9**: 1178
Pahmi *see* Badger, Ferret
Paille-en-queue *see* Tropic bird
Painted lady *see* Butterfly, Thistle
Painted snipe **13**: 1849–1850
    Old World **13**: 1849, *1849*, 1850
    South American **13**: 1849
Painted thrush *see* Pitta

Palaemon **15**: 2043, 2044
Paleosuchus
    palpebrosus **3**: 358
    trigonatus **3**: 358
Palinurus **4**: 573; **11**: 1475
    argus **11**: *1475*
    elephas **11**: 1476, *1476*
    interruptus **11**: 1475
Pallah *see* Impala
Palmchat **13**: 1854
Palm civet **13**: 1855–1856
    African (two-spotted civet)
        13: 1855, 1856
    Banded **13**: 1855
    Borneo banded **13**: 1855
    Common (musang; toddy cat)
        13: 1855, 1856, *1856*
    Malay banded **13**: 1855, *1855*, 1856
    Masked **13**: 1855, 1856
    Owston's **13**: 1855
    Small-toothed **13**: 1855
    Sulawesi **13**: 1855
Palmer worm **19**: 2687
Palm squirrel **13**: 1859–1860
    African **13**: 1859, 1860
    Asian (striped palm squirrel)
        13: 1859, 1860
    Berdmore's **13**: 1859, 1860
    Biafran Bight **13**: 1859
    Five-striped **13**: *1860*
    Three-striped **13**: *1859*
    Western **13**: 1859
Palolo worm **13**: 1861
Palophus titan **18**: 2530
Palpares **1**: 64, 65
palps **2**: 236; **13**: 1764, 1833; **18**: 2481
pampas **8**: 1038
Pan
    paniscus **1**: 70, 71; **4**: 439–440
    troglodytes **1**: 71; **4**: 439, 440
Panaxia dominula **19**: 2687
Panda
    Giant (beishung) **2**: 174, 176;
        7: 960–963
    Red **10**: 1363
    Red (lesser panda; cat-bear; fox-cat; fire-fox) **15**: 2135–2136
Pandaka pygmaea **7**: 993
Pandinus imperator **16**: 2257
Pandion haliaetus **13**: 1812; **21**: 2906
Pangio kuhlii **11**: 1473
Pangolin (scaly anteater)
    13: 1862–1863
    Cape **13**: 1862, *1862*, 1863
    Chinese **13**: 1862
    Giant **13**: 1862
    Indian **13**: 1862, 1863
    Long-tailed tree **13**: 1862, 1863
    Malayan **13**: 1862, 1863, *1863*
    Small-scaled tree **13**: 1862
Pan-pan-panella *see* Bellbird,
    Crested
Pantanal **21**: 2906–2907, *2908*
Panther *see* Leopard; Puma
Panthera **3**: 404, 405, 406; **8**: 1040;
    10: 1311, **11**: 1445, 1461, 1559;
    12: 1672; **17**: 2433; **19**: 2683–2685;
    21: 2907
Pantholops hodgsoni **16**: 2217, *2218*
Pantodon buchholzi **3**: 347–348
Panulirus pencillatus **4**: 573
Panurus biarmicus **14**: 1886, 1887;
    18: 2458; **19**: 2693
Panyptila sancti-hieronymi **19**: 2600
Papagomys armandvillei **7**: 965
Papilio **3**: 344; **18**: 2580, *2580*, 2581,
    2581, 2582
papillae, oral **14**: 1924
Papio **1**: 115, 116; **11**: 1536; **12**: 1644
Pappogeomys **14**: 1997

Papshell *see* Limpet, Common
Papyrocranus 6: 803, 804
Parabuteo unicinctus 3: 426
Paracanthurus hepatus 18: 2577
Parachanna 17: 2412
Paracheirodon
    axelrodi 19: 2667
    innesi 19: 2667, 2668
Paradipus 10: 1324
Paradisaea 2: 205, 206–207
Paradise fish 13: 1868–1869
    Black 13: 1868
    Round-tailed 13: 1868
    Spike-tailed 13: 1868
Paradise-whydah 21: 2936, 2937,
    2938
Paradoxornis
    flavirostris 14: 1886
    heudei 14: 1886
    poliotis 14: 1886
Paradoxurus 13: 1856
    hermaphroditus 13: 1855, 1856
Paraexocoetus 7: 876
Paragordius 9: 1237
Parahyaena brunnea 9: 1267, 1268
Parakeet 14: 1878–1880
    American (conure) 14: 1878
    Bat *see* Parrot, Hanging
    Blossom-headed 14: 1878
    Carolina (*extinct*) 14: 1880
    Crested 14: 1879
    Eastern slaty-headed 14: *1878*
    Elegant grass 14: *1880*
    Grass 14: 1878–1879
    Gray-breasted 14: 1878, 1880
    Green 14: 1878
    Ground 14: 1879
    Lineolated 14: 1880
    Long-tailed ground (ground
        parrot; swamp parrot)
        14: 1879–1880, 1885
    Rose-ringed (ring-necked
        parakeet) 14: 1878, 1879, *1879*
    Short-tailed (night parrot) 14: 1880
    Slender-billed 14: 1878
    Turquoisine grass 14: 1878
Parambassis ranga 7: 978–979
Paramecium 14: 1881–1882
Paramecium
    bursaria 14: 1881, 1882
    caudatum 14: 1881, *1881, 1882, 1882*
Parapagurus pilosimanus 9: 1178
Paraplesiops 5: 668
    meleagris 5: 668
parapodia 15: 2104
Parascyllium variolatum 3: *408*
parasites
    of ants 10: 1418
    of bumblebee 3: 318
    of gall wasps 7: 922
    of parasites *see* hyperparasites
    of weavers 21: 2898
    see also inquilines;
        kleptoparasitism; robbery
parasites, animals as
    basket star 2: 159
    blue butterfly 2: 243–244
    catfish 3: 402
    chalcid wasp 3: 416, 417; 21: 2928,
        2933
    cowbird 1: 127; 3: 400; 4: 553, *553*,
        554–555; 15: 2143; 18: 2459, 2461
    cuckoo 5: 617, 619–620, *619, 620*
    deep-sea anglerfish 5: 653–654
    flea 6: 850, *850*, 851–852
    flies 6: 859
    fluke 7: 870, *870*
    honeyguide 9: 1218
    horsehair worm 9: 1237, 1238,
        1238

    ichneumon wasp 9: 1279–1280,
        1279; 19: 2611
    lamprey 10: 1403, *1404*; 17: *2332*
    leech 10: 1428, *1428*
    mantis fly 11: 1549, *1550*
    mites 7: 988; 12: 1620, *1620*
    oil beetle 2: 188
    pearlfish 14: 1902
    stifftail 18: 2539
    tachina fly 17: 2345; 19: 2610–2611
    tick 19: 2681
    velvet ant 20: 2815, 2816, *2816*
    whydah 21: 2889, 2937, 2938
parasitoids 9: 1280
Parategeticula 21: 3005
Paraxerus capapi 20: *2741*
Parchment-tube worm *see*
    Tubeworm
Pardosa 21: 2957
Parenteroxenos dogieli 16: 2272
Parioglossus 7: 995
Parnassian *see* Butterfly, Apollo
Parnassius
    apollo 1: 76
    autocrator 1: 76
Parophasma galinieri 3: 399
parr 1: 99
Parrot 14: 1883–1885
    African gray 14: 1883, 1885
    Amazon *see* Amazon
    Black-headed 15: *2112*
    Canary *see* Budgerigar
    Eclectus 14: 1883
    Ground *see* Parakeet, Long-tailed
        ground
    Hanging (bat parakeet) 11: 1495
    Night *see* Parakeet, Short-tailed
    Swamp *see* Parakeet, Long-tailed
        ground
Parrotbill (crow tit) 1: 113;
        14: 1886–1887
    Bearded 18: 2458
    Black-throated 14: 1886
    Blyth's 14: 1886
    Gould's (black-breasted parrotbill)
        14: 1886
    Reed bed 14: 1886
Parrotfish 4: 532; 6: 836;
        14: 1888–1890
    Bullethead 14: *1888*, 1889
    Bumphead 14: 1889
    Large 14: 1889
    Princess 14: 1888, 1889
    Rainbow 14: 1890
    Redfin 14: 1890
    Redlip 14: *1889*
    Rusty 14: *1890*
    Striped 14: 1888, 1889
Parrotlet, Green-rumped 14: 1885
parthenogenesis
    aphid 1: *74, 75*
    brine shrimp 2: 276
    chalcid wasp 3: 416
    flowerpot snake 17: 2421
    stick insect 18: 2532
    teiid lizard 19: 2650
Parthenolecanium corni 16: *2246*
Parthenope investigatoris
    18: 2475–2476
Partridge 14: 1891–1893
    Bamboo 14: 1891
    Black *see* Francolin, Black
    Gray (Hungarian partridge)
        14: 1891, 1892, *1892*,
        1893, *1893*
    Red-legged 14: 1891, 1892, 1893
    Rock 14: *1891*
    Tree 14: 1891
    Wood 14: 1891
Parula americana 21: 2986

Parupeneus 7: 990
    multifasciatus 7: 990
Parus 19: 2617, 2653, *2654*, 2693,
    2694, 2695
Passer 7: 999; 9: 1247, 1248, 1249;
    10: 1316; 18: 2457
Passerculus princeps 19: 2649
Passerina ciris 3: 319
Pasteurella pestis 6: 850
Patagona gigas 9: 1255; 12: 1673
Patagonian hare *see* Mara
Patella vulgata 11: 1451, 1452
Patina pellucida 11: 1451
Pauxi pauxi 5: 623
Pavo
    cristatus 14: 1898, 1899
    muticus 14: 1898
PCBs 5: 698
Peacock (*bird*) *see* Peafowl, Indian
Peacock (*butterfly*) 20: 2810–2811
Peacock eel 18: 2490
Peacock's feathers *see* Limpet, Blue-
    rayed
Peacock worm 6: 802
Peafowl 14: 1898–1900
    Congo 14: 1898
    Green 14: 1898
    Indian (common peafowl; blue
        peafowl; peacock (*male*); peahen
        (*female*)) 14: 1898, *1898*, 1899,
        1899, 1900
Peahen *see* Peafowl, Indian
Pearlfish 14: 1901–1902; 16: 2272
pearl organs 8: 1019
pearls 13: 1834
peatlands 21: 2906
Pecari tajacu 14: 1903, 1904
Peccary 14: 1903–1905
    Chacoan 14: 1903, 1905
    Collared 14: 1903, *1903*, 1904, *1904*,
        1905, *1905*
    White-lipped 14: 1903, 1904, 1905
Pecten
    latiauritus 16: 2248
    maximus 16: 2248, 2249
Pectinator, Speke's 8: 1104
Pectinator spekei 8: 1104
Pectinophora gossypiella 2: 256
Pedetes capensis 18: 2502–2503
pedicellariae 16: 2226; 18: 2516
Pedionomus torquatus 14: 1979
Pedipalpi (whip scorpion;
    vinegaroon; tailless whip
    scorpion; whip spider; short-
    tailed whip spider)
    14: 1906–1907
pedipalps 6: 799; 12: 1620; 14: 1906,
    1907; 16: 2257; 17: 2310; 18: 2453;
    21: 2956
pedogenesis 12: 1608; 17: 2313
pedorrera *see* Tody, Cuban
Peep 16: 2233
Peeper, Spring 19: 2723, 2724
Peewee *see* Magpie lark
Peewee, Water *see* Phoebe, Eastern
Peewit *see* Lapwing, Eurasian;
    Magpie lark
Pegasus volitans 16: 2290
Pegusa lascaris 17: 2444
Pekan *see* Fisher
Pelagia noctiluca 10: 1320
pelagic species 4: 475
Pelagothuria natatrix 16: 2271
Pelamis platurus 17: 2309
Pelargopsis capensis 10: 1356–1357
Pelecanoides 5: 684, 685
Pelecanus 14: 1908, 1909, *1909*, 1910;
    16: 2265
Pelican 14: 1908–1910; 16: 2264, 2265,
    2267, 2268

    American white 14: 1908
    Australian 14: *1909*
    Brown (Chilean pelican; Peruvian
        pelican) 14: 1908, *1908*, 1909;
        16: 2265
    Dalmatian 14: 1908
    Great white 14: 1910, *1910*
    Pink-backed 14: 1908
    White 14: 1909
Peliolepisma 2: 279
pellets, owl 2: 149
Pelobates 18: 2466
    fuscus 18: 2466
Pelochelys bibroni 17: 2443
Pelomedusa subrufa 9: 1191; 20: *2799*
Peloscolex 20: 2769
Peludo *see* Armadillo, Hairy
Pempheris
    mangula 17: *2441*
    oualensis 17: *2441*
Peneus
    duorarum 17: 2364
    setiferus 17: 2364
Penguin
    Adélie 1: 28–29, *80*; 14: 1912, 1914;
        21: 3003, 3004
    Chinstrap 14: 1912
    Emperor 1: 82; 6: 779–781;
        14: 1911, 1912, 1913, *1914*
    Erect-crested 14: 1914
    Fjordland 14: 1914
    Flying *see* Diving-petrel, Common
    Galapagos 14: 1911, *1913*, 1914
    Gentoo 14: *1911*
    Jackass 14: 1911, 1913, 1914
    King 10: 1358–1360; 14: 1912, 1913,
        1914
    Little blue (fairy penguin)
        14: 1911, 1913
    Rockhopper 14: 1913
    Snares Island 14: 1914
    White-flippered 14: 1911
    Yellow-eyed 14: 1911, 1914;
        21: 3003–3004
    see also Penguins
Penguins 1: 82; 14: 1911–1914
Pennatula 16: *2298*
pentacrinule larvae 6: 806
Pentatoma rufipes 17: 2345
Peradorcas concinna 16: 2198
Perameles
    bougainville 1: 130
    gunnii 1: *130*
    nasuta 1: 131
Perca
    flavescens 14: 1917, 1918
    fluviatilis 14: 1917, 1918; 16: 2175
Perch 6: 834–835; 14: 1917–1918
    Blue 10: 1421
    Climbing (climbing fish; climbing
        gourami; labyrinth fish)
        4: 468–469
    European 14: 1917, *1917*, 1918,
        1918; 16: 2175
    Nile 13: 1751–1752
    Ocean 16: 2260, 2261
    Sacramento 18: 2569
    Shiner (yellow-banded seaperch)
        18: 2574
    Tule 18: 2573
    Yellow 14: 1917
Percopsis
    omiscomaycus 20: 2754, 2755
    transmontana 20: 2754
Perdix perdix 14: 1891, 1892
Peregrine *see* Falcon, Peregrine
Perentie 12: 1638, 1639, *1640*
pereon 17: 2398
pereonites 17: 2398
perfumes 4: 461

Periclemenes
  amethysteus **17**: *2365*
  imperator **17**: *2366*
Pericrocotus **5**: 621; **12**: 1614, *1615*
Perillus bioculatus **4**: 504
Periophthalmodon **12**: 1684
  schlosseri **12**: 1684, 1685
Periophthalmus **12**: 1684–1685
  chrysospilos **12**: 1684, 1685
  kalolo **12**: 1684–1685
  koelreuteri **12**: 1685
Peripatoides novaezealandiae
  14: 1925
Peripatus **14**: 1924–1925
Peripatus **14**: 1924
  novaezealandiae **14**: *1925*
Periplanta **4**: 493
  americana **4**: 493
Perisoreus
  canadensis **10**: 1317
  infaustus **10**: 1318; **19**: 2617
Perissocephalus tricolor **4**: 543
perissodactyls **19**: 2630
Periwinkle **14**: 1926–1928
  Common (edible periwinkle)
    14: 1926, *1926*, *1927*
  Flat **14**: 1926, 1927, *1927*, 1928
  Rough **14**: 1926, 1927, 1928, *1928*
  Small **14**: 1926–1927, 1928
Perla **18**: 2549
permafrost **19**: 2614; **20**: 2776, 2779
Perna canaliculus **12**: *1712*
Pernis **9**: 1211
  apivorus **2**: 209; **9**: 1211, 1212
  celebensis **9**: 1211
  ptilorhynchus **9**: 1211
Perodicticus potto **15**: 2035
Perognathus **10**: 1338
Perolicula **15**: 2082
Peromyscus **3**: 426; **5**: 659,660, 661;
    19: 2682
Peroryctes
  broadbeuti **1**: 130, 131
  eremiana **1**: 130
pest control, biological *see* biological
    pest control
pesticides
  causing mite pests **12**: 1621
  and golden eagles **7**: 1003
  and locusts **11**: 1480
  and owls **2**: 148
  and peregrine falcons
    14: 1921–1922
  and sparrow hawks **18**: 2468
  see also DDT
pests
  agouti **1**: 33
  aphid **1**: 74, *75*
  armadillo **1**: 87
  bandicoot rat **1**: 133
  barbary ape **1**: 136
  barnacle **1**: 138
  blowfly **2**: 241
  bollworm **2**: 255–256
  brown rat **3**: 298, *299*
  brush tail possum **3**: 301
  bushpig **3**: 336
  capybara **3**: 383
  caracal **3**: 385
  cardinal **3**: 387
  click beetle **4**: 467
  clothes moth **4**: 470, *470*, 471
  cockatoo **4**: 485
  cockroach **4**: 494
  Colorado beetle **4**: 504, 505
  corn borer **4**: 539–540
  cottontail **4**: *545*
  coyote **4**: 558, 559–560
  coypu **4**: 563
  earwig **6**: 739

emu **6**: 782
flies 6: 859
flowerpecker **7**: 868–869
giant snail **7**: 966, *966*, 967
grackle **8**: 1033
gray seal **8**: 1050
groundhog **8**: 1076
honey badger **9**: 1206
hopper **9**: 1224
house sparrow **9**: 1248
jack rabbit **10**: 1305
Java sparrow **10**: *1315*
locust **11**: 1478
marmot **11**: *1568*
millipede **12**: 1611–1612
mole rat **12**: 1628
mousebird **12**: 1676
multimammate rat **12**: 1694
musk shrew **12**: 1709
oil beetle **13**: 1783
owlet moth **13**: 1822
oystercatcher **13**: 1837
parakeet **14**: 1880
plume moth **14**: 1991
polecat **14**: 2003
quelea **15**: 2084, 2085, *2085*, *2086*
raccoon **15**: 2097
rice rat **15**: 2156
sawfly **16**: 2241
scale insect **16**: 2246, *2246*
shieldbug **17**: 2344
ship rat (black rat) **17**: 2348, *2348*
shipworm **17**: 2349, *2349*, 2350
springtail **18**: 2504
tenrec **19**: 2656
termite **1**: 68; **19**: 2660
weevil **21**: 2904
white butterflies **21**: 2928, *2929*
whitefly **21**: 2932, 2933
wood pigeon **21**: 2980–2981
petasma **10**: 1387
Petaurus **7**: 985
Petrel **16**: 2268
  Snow (gray-faced petrel; long-
    winged petrel) **17**: 2435–2436
  Storm *see* Storm petrel
Petrobius maritimus **18**: 2505
Petrochelidon nigricans **12**: 1590
Petrodromus tetradactylus **6**: 775
Petrogale
  brachyotis **16**: 2197
  penicillata **16**: 2197, 2198
  xanthopus **16**: 2197, 2198, *2198*
Petroica **16**: 2190
Petrolisthes **4**: *571*; **18**: 2507
Petromyzon **10**: 1404
  marinus **10**: 1403, 1404; **17**: *2332*
Petrosia **18**: *2495*
pets
  alligator **1**: 40
  chinchilla **4**: 445
  gecko **7**: *943*
  gerbil **7**: 950
  gibbon **7**: 969–970
  guinea pig **8**: 1101
  hamster **8**: 1129
  kinkajou **10**: 1364
  marmoset **11**: 1565
  ocelot **13**: 1778
  orangutan **13**: 1799
  snakehead **17**: *2412*
  uakari **20**: 2804
  woolly monkey **21**: 2988
  zebra mouse **21**: 3017
  zorilla **21**: 3021
  see also birds, cage
Pezoporus wallicus **14**: 1879–1880
Phacochoerus
  aethiopicus **20**: 2861
  africanus **20**: 2861

Phaethon
  aethereus **20**: 2748
  lepturus **20**: 2748
  rubricauda **20**: 2748, 2749
Phagocata **6**: 849
phagocytosis **1**: 41
Phainopepla, Crested **21**: 2890
Phainopepla nitens **21**: 2890
Phalacrocorax **4**: 536, 537, 538
Phalaenoptilus nuttalli **14**: 2013
Phalanger **5**: 630; **14**: *1930*
Phalanger **14**: 1929–1930
  Dormouse *see* Possum, Dormouse
  Lesser flying *see* Gliding possum,
    Lesser
Phalangium opilio **5**: 636
Phalarope **14**: 1931–1933
  Red (gray phalarope) **14**: 1931,
    1931, 1932–1933
  Red-necked (northern phalarope)
    14: 1931, 1932–1933, *1932*
  Wilson's **14**: 1931, 1932–1933, *1933*
Phalaropus
  fulicarius **14**: 1931, 1932
  lobatus **14**: 1931
Phaner **10**: 1436
Pharomachrus mocinno **15**: 2088
Pharus
  legumen **15**: 2127
  pellucidus **15**: 2127, *2128*
Phascogale **11**: 1577
  calwa **11**: 1578
Phascolarctos cinereus **10**: 1379
Phasianus **14**: 1934
  colchicus **14**: 1934, 1935
  versicolor **14**: 1934
Pheasant **14**: 1934–1936
  Eared **14**: 1934–1935
  Golden **14**: 1935
  Green **14**: 1934
  Horned *see* Tragopan
  Lady Amherst's **14**: 1935
  Long-tailed **14**: 1935
  Monal **14**: 1934, *1936*
  Native *see* Lyrebird; Mallee fowl
  Ring-necked **14**: 1934, *1934*,
    1935–1936
  Ruffed **14**: 1935
  Silver **14**: 1934
Phedina borbonica **12**: 1590
Pheidole **9**: 1156, 1157, *1225*
Phelsuma madagascariensis **7**: *943*
Phenacogramma interruptus **19**: 2667
Phengodes **7**: 988
pheromones
  alarm **2**: 183
  from queen wasps **13**: 1864
  tamarin **19**: 2626
  see also attracting a mate with
    scent; scent marking
Phibalura flavirostris **4**: *544*
Phigys **11**: 1493
Philaenus spumarius **9**: 1224
Philanthus **9**: 1262
Philemon **9**: 1215
  buceroides **9**: 1215
  corniculatus **9**: 1215
Philesturnus
  carunculatus **21**: 2886
  carunculatus carunculatus **21**: 2886
  carunculatus rufusater **21**: 2886
Philohela minor **21**: 2966
Philomachus pugnax **16**: 2213
Philothamnus semivariegatus **19**: 2733
Philydor **13**: 1819
Phlogophora meticulosa **13**: 1822
Phoca
  hispida **1**: 82; **16**: 2285
  sibirica **10**: 1398
  vitulina **8**: 1133; **12**: 1647

Phocaena **15**: 2026
  phocaena **15**: 2024, 2025
  sinus **15**: 2024, 2025
  spinipinnis **15**: 2024, 2025
Phocarctos hookeri **16**: 2282, 2283
Phocine Distemper Virus (PDV)
  8: 1134
Phocoenoides dalli **5**: 695; **15**: 2024,
    2025
Phodopus sungorus **8**: 1130
Phoebe
  Black **20**: 2801
  Eastern (water peewee) **20**: 2800
Phoenicopterus **6**: 844, 845
Phoenicurus
  moussieri **15**: 2139
  phoenicurus **15**: 2139, 2140
Pholadidea loscombiana **14**: 1939,
    1940
Pholas dactylus **14**: 1939, 1940
Pholcus phalangioides **18**: 2482
Pholis gunnellus **3**: 342
Phoneutria **20**: 2820
Phonygammus keraudreni **20**: 2757
phoresy (phoresia) **6**: 800
Photinus **6**: 828
  pyralis **6**: 827
  scintillans **6**: 828
photosynthesis **19**: 2654
Photuris **6**: 828
  pennsylvanica **6**: 828
Phoxinus
  erythrogaster **12**: 1618–1619
  phoxinus **12**: 1618, 1619
Phrixothrix **7**: 988
Phronima **15**: *2078*; **16**: 2224
Phryganea **3**: 354, 355
Phrynocephalus **1**: 30
Phrynosoma
  cornutum **9**: 1231, 1232
  douglassi **9**: *1231*
Phtheirichthys lineatus **15**: 2148
Phylidonris nigra **9**: *1215*
Phyllastrephus **3**: 308
Phyllirhoe **16**: 2300
Phyllobates **1**: 94; **20**: 2819
  terribilis **1**: 93
Phyllomedusa **19**: 2725
  burmeisteri **19**: *2723*
Phyllopertha **4**: 487
  horticola **4**: 487
Phyllophryne didactyla **1**: 93
Phylloscopus
  collybita **20**: 2857
  sibilatrix **21**: 2985
  trochilus **20**: 2857
Phyllotoma aceris **16**: 2240
Phylloxera, Grape *see* Vine pest
Phylloxera vitifoliae **20**: 2827, 2828
Phyllurus cornutus **7**: 945
Physalia **20**: 2819
  physalis **14**: 1980–1981; **15**: 2029,
    2030; **20**: 2798
  utricularis **15**: *2029*
Physeter
  catodon **18**: 2510
  macrocephalus **18**: 2472; **21**: 2911
Physopoda *see* Thrips
phytoplankton **4**: 475; **10**: 1388, 1400;
    13: 1770; **14**: 1980, 1981–1982
  Antarctic **1**: 81
Pica
  nuttalli **11**: 1519, 1520
  pica **11**: 1519, 1520
Pichi *see* Armadillo, Pygmy
Pichiciago *see* Armadillo, Lesser fairy
Pickerel
  Chain **14**: 1945, 1946
  Redfin (grass pickerel) **14**: 1945,
    1946

Picoides **21**: 2977
  major **21**: 2979
  villosus **21**: 2978
Picromerus bidens **17**: 2344
Picuda *see* Barracuda, Great
Piculet **14**: 1937–1938
  Antillean **14**: 1937
  Olivaceous **14**: 1937, 1938, *1938*
  Rufous **14**: 1937
  Speckled **14**: *1937*, 1938
Picumnus
  innominatus **14**: 1938
  olivaceus **14**: 1937
Picus **21**: 2977
  viridis **21**: 2965, 2977
Piddock **14**: 1939–1940
  Common **14**: 1939, *1939*, 1940, *1940*
  Great **14**: 1939, 1940
  Little **14**: 1939, 1940
  Paper (American rock borer) **14**: 1939, 1940
  White **14**: 1939–1940
Pieris
  brassicae **21**: 2928, *2929*
  rapae **3**: 416; **21**: 2928
Pig, Wild *see* Boar, Wild
Pigeon
  Domestic *see* Dove, Rock
  Fancy *see* Dove, Rock
  Feral *see* Dove, Rock
  Fruit *see* Fruit pigeon
  Green *see* Green pigeon
  Imperial **7**: 911
  Mauritius blue **7**: 912
  New Zealand (New Zealand fruit pigeon; native pigeon) **7**: 912
  Passenger (*extinct*) **3**: 328
  Racing *see* Dove, Rock
  Topknot **7**: 911
  Torres Strait (nutmeg pigeon) **7**: 911
  Wood (ring dove) **21**: 2980–2982
pigeon's milk **7**: 911–912; **13**: 1858; **16**: 2196; **20**: *2794*; **21**: 2982, *2982*
Pig-tail *see* Macaque, Pig-tailed
Pika (rock rabbit; mouse-hare; rock coney; calling hare; piping hare; whistling hare) **14**: 1943–1944
  Pallas' **14**: 1944
  Rocky Mountain **14**: 1944
Pike **14**: 1945–1947
  Black-spotted **14**: 1945, 1946
  Northern **14**: 1945, 1946, *1946*
  Sea *see* Snoek
Pikeperch *see* Walleye
Pilchard (sardine) **14**: 1948–1949
  Australian **14**: 1948
  European **14**: 1948, 1949
  Japanese **14**: 1948
  South American **14**: 1948, *1948*, 1949
  Southern African **14**: 1948; **17**: 2430
Pile worm *see* Ragworm
Piliocolobus **4**: 501, 502
  badius **4**: 502
  oustaleti **4**: 502
Pill bug *see* Wood louse
Pilot fish **14**: 1950–1951; **16**: 2242
Pilot whale **5**: 695
  False *see* False killer whale
  Long-finned **14**: 1952, 1953, *1954*
  Short-finned **14**: 1952, *1952*, 1953, 1954
Pimephales promelas **12**: 1618
Pinaroloxias **5**: 649
Pincerbug *see* Earwig
Pinctada **13**: 1834
Pineapplefish **18**: 2512

Pinecone fish **18**: 2512
Pine marten *see* Marten
Pinguinus impennis **14**: 1911; **16**: 2268
Pinna fragilis **12**: 1711
pinocytosis **1**: 41
Pintail **14**: 1957–1959; **17**: 2343
  Northern **14**: 1957, *1957*, 1958, 1959, *1959*
  White-cheeked (Bahama pintail) **14**: 1957, *1958*
Pionites melanocephala **15**: 2112
Piophila **18**: 2505
Pipa pipa **18**: 2579
Pipefish **14**: 1960–1961; **20**: 2766
  Bay **14**: 1961
  Broadnosed **14**: 1961
  Dusky **14**: 1960, 1961
  Ghost **20**: 2766
  Greater **14**: *1961*
  Multibar **14**: *1960*
  Network **14**: 1960
  Ringed **14**: 1960
Pipile pipile **5**: 623
Piping hare *see* Pika
Pipistrelle **14**: 1962–1964
  Common **14**: 1962, *1962*, 1963, 1964, *1964*
  Eastern **14**: 1962, 1964
  Soprano **14**: 1962
  Western **14**: 1962, *1963*, 1964
Pipistrellus
  hesperus **14**: 1962
  pipistrellus **14**: 1962, 1963
  pygmaeus **14**: 1962
  subflavus **14**: 1962
Pipit **14**: 1965–1966
  American **14**: 1966
  Bogota **14**: 1965
  Golden **14**: 1965
  Meadow **5**: 620; **14**: 1965, 1966
  Red-throated **14**: 1965
  Rock **14**: 1965, *1966*
  Sprague's **14**: 1965
  Tree **14**: 1965, 1966
  Yellow-breasted **14**: 1965
Pipra
  erythrocephala **11**: 1532
  mentalis **11**: 1532
Piquero *see* Booby, Peruvian
Piranga
  flava **19**: 2627
  ludoviciana **19**: 2627
  olivacea **18**: 2457; **19**: 2627
  rubra **19**: 2627
Piranha (piraya) **14**: 1967–1969
  Blackspot **14**: 1969
  Red (common piranha; Natterer's piranha) **14**: 1968, *1968*, 1969
  San Francisco **14**: 1967, 1969
  Speckled **14**: 1969
Pirate-perch **20**: 2755
Piraya *see* Piranha
Pisaura **21**: 2957
  mirabilis **18**: 2483
Pithecia **16**: 2221, 2222, *2222*
Pithecophaga jefferyi **15**: 2111
Pitohui **20**: 2819; **21**: 2923
pit organs, of snakes **14**: 1973–1974, *1974*; **15**: *2081*
pits, of ant lion larvae **1**: 64, *64*, 65
Pitta **14**: 1970, *1970*, 1971
Pitta (painted thrush; ground thrush; jewel thrush) **14**: 1970–1971
  African **14**: 1970, 1971
  Banded **14**: *1970*
  Gurney's **14**: 1970
  Indian (Bengal pitta; blue-winged pitta) **14**: 1970, 1971, *1971*
  Noisy **14**: 1970
  Steere's **14**: 1970

Pit viper **13**: 1736; **14**: 1972–1974; **17**: 2420
  Eyelash **14**: *1972*, *1973*
  Himalayan **14**: 1973, 1974
  Japanese *see* Habu
  Wagler's **14**: 1973
  White-lipped **14**: *1974*
Pizonyx vivesi **2**: 167; **6**: 840; **20**: 2822, 2823
placenta, armadillo **1**: 87
Plagiodontia **9**: 1263, 1264
  aedium **9**: 1264
Plagiolepis **9**: 1205
  trimeni **9**: 1205
Plagionotus **11**: 1484
  detritus **11**: *1484*
plague, bubonic **17**: 2346–2347
Plaice **14**: 1975–1977
Plains-wanderer (collared plains-wanderer; turkey quail) **14**: 1978–1979
Planigale **11**: 1577
  ingrami **11**: 1577, 1578
Planigale (flat-skulled marsupial mouse) **11**: 1577
  Ingram's **11**: 1577, 1578
Plankton **13**: 1770; **14**: 1980–1982
  blooms **13**: 1770; **14**: 1981–1982
  see also phytoplankton; zooplankton
Planorbis **7**: 870
Plantain-eater, Violet **20**: 2780
plantigrade stance **2**: 174
Plant louse *see* Aphid
plants
  chaparral **3**: 423, *423*, 424–425
  desert **5**: 663, 664–665
  polar **1**: 82
planula larvae **10**: 1322; **17**: 2441
Plasmodium **12**: 1663
plastron **9**: 1169; **13**: 1851; **17**: 2425; **19**: 2713; **20**: 2795
Platalea **18**: 2496, 2497
Platambus **5**: 683
Platanista
  gangetica **5**: 697; **16**: 2177, 2178
  minor **5**: 697; **16**: 2177
Platichthys flesus **6**: 860, 861
Platy **14**: 1983–1984
  Black **14**: 1983
  Blue (blue moon platy; blue coral platy) **14**: 1983
  Golden **14**: 1983
  Hi-fin **14**: *1983*
  Red **14**: 1983
  Southern **14**: 1983, 1984
  Spotted **14**: 1984
  Variable (variegated platy) **14**: 1983
  Wagtail **14**: 1983
Platyarthrus **21**: 2974
  hoffmannseggi **21**: 2974
Platynereis megalops **15**: 2105
Platyptilia
  calodactyla **14**: 1992
  carduidactyla **14**: 1991
  gonodactyla **14**: 1991
Platypus (watermole; duckmole; duckbill) **14**: 1985–1987
Platyrhina **8**: 1103
Platyrhinoidis **8**: 1103
Platyspiza **5**: 649
Platysternon megacephalum **2**: 197
play
  dolphin **4**: 508
  ermine **6**: 785
  polecat **14**: *2002*
  river otter **16**: *2180*, 2181
  sea lion **16**: 2284
  sperm whale **18**: 2473

Plebejus
  argus **2**: 243
  icarioides missionensis **2**: 244
Plecoglossus
  altivelis altivelis **19**: 2597
  altivelis ryukyuensis **19**: 2597
Plecotus
  auritus **2**: 167; **20**: 2821, 2822
  macrotis **20**: 2821, 2822
  townsendii **2**: 168
Plectronemia **3**: 354
Plectrophenax
  hyperboreus **3**: 319
  nivalis **3**: 319, 320
Plectropterus gambiensis **5**: 717
Plegadis falcinellus **9**: 1274
Plethodon **11**: 1504
  cinereus **11**: 1503
  glutinosus **11**: 1504
Pleuronectes platessus **14**: 1976
Pliny **7**: 992
Ploceus **21**: 2896, 2897, *2897*
Ploughbird, Wattled **21**: 2923
Plover **14**: 1988–1990
  Black-bellied **14**: 1988, *1989*
  Blacksmith **10**: 1415
  Crab **4**: 568–569
  Egyptian (crocodile bird) **4**: 551, 552
  Golden *see* Golden plover
  Green *see* Lapwing, Eurasian
  Little ringed **14**: 1988
  Mountain **14**: 1989
  Ringed **14**: 1988, *1988*, 1990
  Sand **14**: 1988–1989
  Semipalmated **14**: 1988
  Snowy (Kentish plover) **14**: 1988–1989
  Swallow *see* Pratincole
plumage
  eclipse **14**: 1957, 1993; **21**: 2896–2897
  see also feathers
Plumaria setacea **16**: 2273
Plume moth **14**: 1991–1992
  Artichoke **14**: 1991
  Grape **14**: 1991, 1992
  Triangle **14**: 1991, 1992
  White **14**: *1992*
Plusiotus gloriosa **16**: *2253*
pluteus larvae **17**: 2315
Pluvialis **14**: 1988, 1989, *1989*
Pluvianus aegyptus **4**: 551, 552
pneumatophore **15**: *2030*
pneumostome **7**: 930
Pochard **14**: 1993–1995
  Common (European pochard) **14**: 1993, 1994, 1995, *1995*
  Ferruginous **14**: 1993
  Madagascar **5**: 717
  Red-crested **14**: *1993*
  Southern **14**: 1995
Pocket gopher **8**: 1040; **14**: 1996–1997
  Botta's **14**: *1996*
  Northern **14**: *1997*
Podarcis
  hispanica **20**: 2849
  muralis **20**: 2849, 2850
  sicula **20**: 2849, 2850
Podargus strigoides **7**: 901
Podica senegalensis **6**: 823
Podiceps **8**: 1061, 1063
Podilymbus
  gigas **8**: 1063; **19**: 2649
  podiceps **8**: 1061, 1062
Podisus **17**: 2344
  maculiventris **4**: 504
Podley *see* Pollack
Podlok *see* Pollack

Podocnemis expansa **9**: 1191
Podura **18**: 2505
  aquatica **18**: 2504–2505
Poecilia **2**: 234; **8**: 1107; **12**: 1664
Poecilictis libyca **21**: 3020
Poecilogale albinucha **21**: 3020
Pogona **5**: 665
  barbata **2**: 172
  vitticeps **2**: 173
Pogonias chromis **5**: 601, 602
Pogonichthys macrolepidotus **12**: 1618
Pogonomyrmex **9**: 1156, 1157
Pogy *see* Menhaden, Atlantic
poisons/venom **20**: 2817–2820
  adder **1**: 26, 27
  anemone **1**: 47
  arrow-poison frog **1**: 93, 94; **7**: 905
  asp **1**: 98
  bee **2**: 183
  bee-eater immunity to **2**: 182
  black widow **2**: 235
  boomslang **2**: 264, 265
  bumblebee **3**: 318
  burnet moth **3**: 322
  butterflies and moths 3: 345
  centipede **3**: 412–413
  cobra **1**: 98; **4**: 480–481, *481*
  common toad **4**: 512, 513
  cone shell **4**: 519, 520
  coral snake **4**: 534
  dogfish **5**: 688
  eagle ray **6**: 731
  false scorpion **6**: 799, *799*, *800*
  fer de lance **6**: 812, 814
  fire salamander **6**: 831–832
  gila monster **7**: 972, *973*, 974
  jellyfish **10**: 1322
  marine toad **11**: 1557–1558
  midwife toad **12**: 1609
  millipede **12**: 1612–1613
  monarch **12**: 1633
  moray eel **12**: 1659
  newt **13**: 1734
  night adder **13**: 1735, *1735*, 1736
  octopus **13**: 1781; **20**: *2819*
  platypus **14**: 1985
  Portuguese man-of-war
    15: 2030–2031
  puff adder **15**: 2058
  pufferfish **15**: 2064
  rabbitfish **15**: *2095*
  rattlesnake **15**: 2120, *2121*
  scorpion **16**: 2257–2258, *2257*;
    20: 2818
  scorpionfish **16**: 2260, 2261;
    20: *2820*
  sea snake **17**: 2308
  solifugid **18**: 2454
  spider crab **5**: 580
  spiders 18: 2482
  squid **18**: 2510
  stargazer **18**: 2519
  stingray **18**: 2543–2544
  stonefish **18**: 2547–2548, *2548*
  swallowtail **18**: 2581, 2582
  tarantula **19**: 2633, 2634, 2635
  trunkfish **20**: 2758–2759
  water shrew **20**: 2872
  weever **21**: 2901–2902
  see also Venomous animals;
    warning coloration
Pokey *see* Menhaden, Atlantic
polar regions **1**: 79–82
Polecat (fitchew; foumart; foul-
  marten) **14**: 2001–2003
  European **2**: 228; **14**: 2001, 2002,
    2003
  Marbled **14**: 2001
  Steppe **2**: 228; **14**: 2001
  Striped *see* Zorilla

Polemaetus bellicosus **12**: 1589
Polihierax semitorquatus **2**: *209*
Polistes **2**: 185; **20**: 2824
  annularis **13**: 1865
Pollachius
  pollachius **14**: 2004–2005
  virens **4**: 473
Pollack (lythe; pollock; podlok;
  podley) **14**: 2004–2005
  Black *see* Coalfish
pollen basket **10**: 1419
Pollenia rudis **2**: 242
Pollicepes pollicepes **1**: 139
pollination
  by bumblebees **3**: 316
  by bushbabies **3**: 330
  by fig wasps **3**: 417
  by flowerpeckers **7**: 868
  by flying foxes **7**: 879
  by hawkmoths **9**: 1168
  by honeyeaters **9**: 1216
  by hummingbirds **9**: 1257
  by leafcutter bees **10**: 1420
  by lorikeets **11**: 1488
  by songbirds **18**: 2456–2457
  by yucca moths **21**: 3005–3006
Pollock *see* Pollack
pollution
  and Atlantic salmon **1**: 101
  and beluga whales **2**: 194
    coastal **4**: 477
  on coral reefs **4**: 533
  deep-sea **13**: 1773
  detection by flies **3**: 355; **6**: 859;
    18: 2550
  and dolphins **5**: 698
  and grayling **8**: 1048
  and harbor seals **8**: 1134
  and killer whales **10**: 1352
  and penguins **14**: 1914
  and the peppered moth **14**: 1916
  in wetlands **21**: 2908
  see also oil spills
Polyboroides **17**: 2373
Polycarpa aureta **17**: *2312*
Polycentrus **10**: 1422
polyembryony, chalcid wasp
    3: 416–417
Polyergus rufescens **1**: 69
polyestrus **9**: 1194
polygamy (*including* polygyny;
    polyandry)
  accentor **1**: 23
  Baltimore oriole **1**: 127
  bishop bird **2**: 214
  black widow **2**: 236
  bunting **3**: 320
  button quail **3**: 350
  common agama **1**: 31
  deer 5: 656–657
  flycatcher **7**: 874
  harrier **8**: 1146
  lion **11**: 1463
  ostrich **13**: 1814–1815
  painted snipe **13**: 1849
  rabbit **15**: 2092
  red-winged blackbird
    15: 2142
  rhea **15**: 2150
  songbirds 18: 2457–2458
  starling **18**: 2523
  turkey **20**: 2785
Polygonia **3**: 345; **20**: 2811, 2812
Polyipnus **9**: 1162
polymorphism
  ant **9**: 1156
  reed frog **15**: 2146
Polyodon spathula **13**: 1845, 1846
Polyprion **2**: 164
  americanus **8**: 1082

polyps
  coral **4**: 527, 529, *529*; **17**: 2440
  dead man's fingers **5**: 651
  jellyfish **10**: 1320, 1322
  Portuguese man-of war **15**: 2030;
    20: 2819
  sea fir **16**: 2273
  sea pen **16**: 2297, *2297*, 2298
Polypterus **2**: 196
  birchir **2**: 196
  ornatipinnis **2**: 195
Polysticta stelleri **6**: 754
Pomacanthus
  imperator **1**: *50*
  semicirculatus **1**: 49
  striatus **1**: 49
Pomacentrus coeruleus **5**: 638
Pomatoceros **6**: 801
  triqueter **6**: 801
Pomatomus saltatrix **2**: 245
Pomatostomus isidorei **1**: *114*
Pompadour fish **14**: 2006–2007
Pompano, Florida **16**: 2242
Pomponia imperatoria **4**: 455, 456
Pond *see* Lake and pond
Pond skater (water strider) **2**: 188;
    14: 2008–2009
Pond snail **14**: 2010–2011
  Dwarf **14**: 2010, 2011
  Ear **14**: 2010, 2011
  Great **14**: 2010, *2010*, 2011
  Marsh **14**: 2011
  North American **14**: 2011
  Wandering **14**: 2010
Pongo
  pygmaeus **1**: 70
  pygmaeus abelii **13**: 1797, 1798
  pygmaeus pygmaeus **13**: 1797, 1798
Pontoporia blainvillei **5**: 697; **16**: 2177,
    2178
Poorwill **14**: 2012–2013
  Common **14**: 2012, *2012*, 2013
  Ocellated **14**: *2013*
population cycles **6**: 820; **10**: 1309,
    1433, *1433*; **15**: 2157; **17**: 2438
population growth
  elephant overpopulation
    6: 768–769
  fruit fly **7**: 910
  lemming **10**: 1433–1434
  prairie dog **15**: 2040
  see also irruptions
Porania pulvillus **18**: 2518
Porbeagle **15**: 2020–2021; **17**: 2331
Porcellana longicornis **18**: 2507
Porcellio **21**: 2974
  scaber **21**: 2973, 2974
Porcupine (*New World*) *see* Tree
    porcupine
Porcupine (*Old World*)
  Brush-tailed **5**: 595
  Crested (African crested
    porcupine; African porcupine)
    5: 595–597
  Long-tailed **5**: 595
Porcupine
  Native *see* Spiny anteater
  Spiny *see* Spiny anteater
Porcupine fish **15**: 2022–2023
  Long-spine (spiny puffer) **15**: *2022*
Porichthys **19**: 2698
Porina **19**: 2603
Porites **4**: 528
Porphyrio
  martinica **7**: 919, 920
  porphyrio **7**: 919
Porpita **16**: 2300
Porpoise **15**: 2024–2026
  Black *see* Dolphin, Bottlenose
  Burmeister's **15**: 2024, 2025

  Common (bottlenose dolphin)
    2: 266–268; **4**: 507; **15**: 2024
  Common (harbor porpoise;
    puffing pig) **15**: 2024, 2025,
    2025, *2026*
  Dall's (True's porpoise) **5**: 695,
    698; **15**: *2024*, 2025, 2026
  Finless **15**: 2024, 2025
  Gray *see* Dolphin, Bottlenose
  Spectacled **15**: 2024, 2025
  White-bellied *see* Dolphin,
    Common
  see also Dolphins and porpoises
Portia **18**: 2482
Portuguese man-of-war
    15: 2029–2031; **16**: 2300; **20**: 2819
Porzana
  carolina **5**: 582, 583; **15**: *2106*
  palmeri **15**: 2108
  porzana **5**: 582
  pusilla **5**: 582
Possum
  Brush tail (vulpine possum;
    foxlike possum) **3**: 300–301;
    14: 1929
  Dormouse (dormouse phalanger;
    dormouse opossum; pygmy
    possum; pygmy opossum)
    5: 702–703
  Gliding *see* Gliding possum
  Mountain pygmy possum
    (burramys) **5**: 703
  Scaly-tailed **14**: 1929–1930, *1929*
Potamochoerus porcus **3**: 336
Potamogale **19**: 2656
  velox **13**: 1817, 1818
Potato bass *see* Grouper, Potato
Potato bug *see* Beetle, Colorado
Potato cod *see* Grouper, Potato
Potoo (wood nightjar) **15**: 2032–2033
  Gray (common potoo) **15**: 2032
  Great **15**: 2032, *2032*, 2033, *2033*
  Northern **15**: 2032, 2033
Potoroo *see* Rat kangaroo,
    Long-nosed
Potos flavus **10**: 1364
Potoyunco *see* Diving-petrel,
    Peruvian
Potto (*African*; Bosman's potto)
    11: 1490; **15**: 2034–2035
  Golden (angwantibo; golden
    angwantibo; Calabar
    angwantibo) **8**: 1014–1015;
    11: 1490
Potto (*American*) *see* Kinkajou
Pouched rat, African giant (Gambian
    pouched rat; African giant rat;
    hamster rat) **7**: 964, 965, *965*
Prairie chicken **15**: 2036–2037
  Greater **15**: 2036, 2037, *2037*
  Lesser **15**: 2036, *2036*, 2037
Prairie dog **8**: 1040; **15**: 2038–2040
  Black-tailed **15**: 2038, 2039, *2039*
  Gunnison's **15**: *2038*
  Utah **15**: 2040
  White-tailed **15**: 2039
Prairie runner *see* Coachwhip
prairies **8**: 1038
Pratincole (swallow plover)
    15: 2041–2042
  Australian **15**: 2041, *2042*
  Black-winged **15**: 2041
  Common (collared pratincole;
    red-winged pratincole; locust
    bird) **15**: 2041, *2041*, 2042
Prawn **15**: 2043–2044
  African **15**: 2043
  Burrowing **18**: 2507
  Dublin *see* Lobster, Norway
  Giant freshwater **15**: 2043

Malaysian **15**: 2043
 Painted harlequin **15**: *2044*
Prawn killer *see* Shrimp, Mantis
precocial young, agouti **1**: 32
predators, dive-bombed by gulls
 2: 232
preening, other birds (allopreening)
 1: 104, 114
Premnas **5**: 639, 640
 biaculeatus **5**: *640*
Prenolepis imparis **9**: 1204
Presbytis **10**: 1411
 entellus **12**: 1643
 melalophus **10**: *1410*
 thomasi **10**: *1412*
prezoea larvae **4**: 573; **6**: 741
prickets **6**: 796
Priodontes giganteus **1**: 85
Prion **15**: 2045–2046; **16**: 2268
 Antarctic **15**: 2045
 Broad-billed **15**: 2045, 2046, *2046*
 Fairy **15**: 2045, *2045*
 Fulmar **15**: 2045
 Salvin's **15**: 2045, 2046
 Thin-billed **15**: 2045
Prionace glauca **17**: 2331
Prionodon
 linsang **11**: 1459
 pardicolor **11**: 1459
Prionodura newtoniana **2**: 271
Prionotus
 carolinus **8**: 1109
 scitulus **8**: 1109
Prionus **11**: 1484
Pristella, Albino *see* Tetra, X-ray
Pristella maxillaris **21**: 2998, 2999
Pristis
 pectinata **16**: 2238, 2239
 perotteti **16**: 2238
Pristolepis **10**: 1422
 fasciata **10**: 1421
Prosciger aterrimus **4**: 483, 484
Procapra **7**: 939
Procapromys geayi **9**: 1263
Procavia **9**: 1270, 1271
 capensis **9**: 1271
Procelsterna cerulea **13**: 1755
Processionary moth (procession
 moth; processional moth)
 15: 2047–2048
 Australian **15**: 2048
 Oak **15**: 2047, 2048
 Pine **15**: 2047, *2047*, 2048, *2048*
 Pistachio **15**: 2048
Procnias
 alba **4**: 543
 averano **4**: 543, 544
 tricarunculata **2**: 191; **4**: 543
Procolobus verus **4**: 501, 502
Procyon **15**: 2096, 2097
Prognathochromis **4**: 459
Progne
 subis **2**: 151; **12**: 1590, 1591
 tapera **12**: 1590
Prognichthys **7**: 876
prolactin **14**: 1933
Promerops
 cafer **18**: 2558, 2559
 gurneyi **18**: 2558
Pronghorn *see* Antelope, Pronghorn
pronking **18**: 2499
 see also stotting
Propithecus **17**: 2372
propolis **9**: 1209
Propyrrhura **11**: 1514
prosoma **18**: 2453
Prosthemadera novaeseelandiae **9**: 1215
Proteles cristatus **1**: 19
proteroglyphs **13**: 1736
Proteus anguinus **13**: 1789

Protonotaria citrea **21**: 2986
protonymphons **17**: 2311
protoplasm, of amoeba **1**: 41, 42
Protopterus
 aethiopicus **11**: 1501, *1502*
 annectens **11**: 1501
Protoxerus **13**: 1860
 aubinnii **13**: 1859
 stangeri **13**: 1859
Prunella **1**: 22
Psammomys obesus **7**: 950, 951
Psammoperca **13**: 1752
Psarisomus dalhousiae **2**: 282
Psarocolius
 angustifrons **13**: *1806*
 montezuma **13**: 1805, 1806
 wagleri **13**: *1805*
Psephurus gladius **13**: 1845
Pseudacris **19**: 2723
 nigrita **19**: 2724
Pseudanthias **8**: 1081
Pseudemydura umbrina **9**: 1191;
 20: 2797–2798
Pseudemys **13**: 1851
Pseudeos **11**: 1493
 fuscata **11**: 1493
Pseudibis papillosa **9**: 1276
Pseudis paradoxa **14**: 1877
Pseudobalistes flavimarginatus **20**: *2744*
Pseudobranchus
 axanthus **17**: 2382
 striatus **17**: 2382
Pseudocalyptomena graueri **2**: 283
Pseudocarcinus gigas **4**: 571
Pseudoceros **6**: 849
 ferrugineus **6**: *849*
Pseudochaenichthys **9**: 1278
Pseudochilidon **12**: 1590
 eurystomina **12**: 1590
 sirintarae **12**: 1590
Pseudococcus **10**: 1396
Pseudocreobotra wahlbergi **11**: 1548
Pseudogyps **20**: 2841
Pseudois najaur **12**: 1672
pseudopodia **1**: 41, *41*, 42
Pseudopontia paradoxa **21**: 2929
pseudopupae **13**: 1783
Pseudorca see False killer whale
Pseudorca crassidens **6**: 798
Pseudoscorpion *see* False scorpion
Pseudoseisura lophotes **13**: 1820
Pseudosphromenus dayi **13**: 1868
pseudotracheae **9**: 1242
Pseudotriakis microdon **3**: 407
Pseudupeneus **7**: 990
 grandisquamis **7**: 990
Psittacula
 cyanocephala **14**: 1878
 himalayana finschii **14**: *1878*
 krameri **14**: 1878, 1879
Psittacus erithacus **14**: 1883
Psolus diomediae **16**: 2271
Psophia
 crepitans **20**: 2756, 2757
 leucoptera **20**: 2756, 2757
 viridis **20**: 2756, 2757
Ptarmigan **15**: 2055–2057; **20**: 2778
 Rock **15**: 2055, *2055*, 2056, *2056*
 White-tailed **15**: 2056
 Willow *see* Grouse, Willow
Pteridophora alberti **2**: 205
Pterocles **16**: 2230
Pterocnemia pennata **15**: 2151
Pterodroma cahow **16**: 2268
Pteroglossus torquatus **19**: 2714, 2715
Pterognatha gigas **11**: 1484
Pterois volitans **16**: 2260
Pteromalus **3**: 416; **21**: 2928–2929
Pteromylaeus **6**: 730
Pteronarcys californica **18**: 2549

Pteronura brasiliensis **16**: 2179, 2180
Pterophorus
 pentadactyla **14**: *1992*
 periscelidactylus **14**: 1991, 1992
Pterophylla camellifolia **10**: 1341
Pterophyllum **4**: 459
 altum **16**: 2244, 2245
 leopoldi **16**: 2244, 2245
 scalare **4**: 458; **16**: 2244, 2245
Pteropod *see* Sea butterfly
Pteroptyx malaccae **6**: *828*
Pteropus **7**: 907; **15**: 2110
 giganteus **2**: 167; **7**: *877*, 878, 879
 poliocephalus **7**: 878, 879
 vampyrus **2**: 167; **7**: *877*, 878
Ptilocercus lowi **19**: 2730, 2731
Ptilonorhynchus violaceus **2**: 270
Ptiloris **2**: 207
 magnificus **2**: 205
Ptyas mucosus **15**: 2117; **17**: 2422
Ptychocheilus **12**: 1618
Ptychozoon **7**: 945
Pudu
 mephistophiles **5**: 655; **18**: 2463, 2464
 pudu **18**: 2463, 2464
Pudu **5**: 655; **18**: 2463
 Northern Andean **18**: 2464
 Southern Andean **18**: 2464, *2464*
Puffback **17**: 2361
Puffbird (silly John) **15**: 2060–2061
 Black-breasted **15**: 2061
 Mustached **15**: *2061*
 Spot-backed **15**: *2060*
 White-necked **15**: 2060, 2061
 White-whiskered **15**: 2060, 2061
Puffer *see* Pufferfish
Pufferfish (puffer; balloonfish;
 swellfish; globefish; blowfish)
 4: 532; **6**: 833; **15**: 2062–2064
 Freshwater **15**: 2063, *2063*
 Oceanic **15**: 2063
 Spiny puffer *see* Porcupine fish,
 Long-spine
 Spotted **15**: *2062*
 Star **15**: 2063
Puffin **15**: 2065–2067; **16**: 2268
 Atlantic (common puffin) **15**: 2065,
 2066, *2066*, 2067, *2067*; **16**: *2228*,
 2264
 Horned **15**: 2065
 Tufted **15**: 2065, *2065*
Puffing pig *see* Porpoise, Common
Puffinus **16**: 2264; **17**: 2334, 2335, 2336
Pugettia producta **18**: 2475
puggles **18**: 2488
Pugwall *see* Magpie lark
Puku **3**: 1381, 1382
Pulex irritans **6**: 850, 851
Puma
 concolor **15**: 2069
 concolor coryi **15**: 2070
Puma (cougar; mountain lion;
 panther) **15**: 2068–2070; **19**: 2653,
 2729; **21**: 2907
Pumpkinseed *see* Sunfish
 (freshwater), Common
Pungitius pungitius pungitius **18**: 2533
Puntius eugrammus **21**: 3015
pupae
 ant lion **1**: 65
 bee and wasp 2: 184
 blowfly **2**: 241–242
 burnet moth **3**: 322
 plume moth **14**: 1992
pupation, false **21**: 2933
Pupfish, Desert *see* 2705
Purple emperor **15**: 2071–2072
 Lesser **15**: *2071*
Purple sea snail *see* Sea snail,
 Purple

Pusa
 caspica **16**: 2169, 2170
 hispida **14**: 1998; **16**: 2169, 2170
 sibirica **16**: 2169, 2170
Pycnodonta **13**: 1832
Pycnogonum **17**: 2311
 stearnsi **17**: 2311
Pycnonotus
 barbatus **3**: 309
 cafer **3**: 309
 leucogenys **3**: 308
Pygathrix **10**: 1411; **12**: 1645
Pygocentrus **14**: 1967, 1968, 1969
Pygoplites diacanthus **6**: *833*
Pygoscelis **14**: 1913
 adeliae **1**: 29, 80; **14**: 1912
 antarctica **14**: 1912
 papua **14**: *1911*
Pylocheles **9**: 1176
Pylodictis olivaris **3**: 401
Pylopagurus **9**: 1176
Pylylia **21**: 2938
Pyrocephalus rubinus **20**: 2800, 2801
Pyrophorus **6**: 828
 noctilucus **6**: 827
Pyrosema **15**: 2078, *2078*
 atlanticum **15**: 2078
Pyrosomella **15**: 2078
Pyrostremma **15**: 2078
Pyrrhocorax
 graculus **4**: 451; **12**: 1673
 pyrrhocorax **4**: 451, 452; **12**: 1673
Pyrrhopyge **17**: 2391
Pyrrhula pyrrhula **3**: 311
Python **15**: 2079, 2080; **17**: 2419, 2422
Python **15**: 2079–2081; **17**: 2418, 2419,
 2420, 2422
 African (African rock python)
 15: 2079, 2080, *2080*
 Amethystine (Australian scrub
 python) **15**: 2079, 2080, *2114*
 Ball (royal python) **15**: 2079, 2080,
 2081
 Black-headed **15**: 2079
 Carpet (diamond python) **15**: 2079
 Children's **15**: 2079
 Dwarf **15**: 2079
 Green tree (green python)
 15: 2079, *2079*, 2080
 Indian **15**: 2079, 2081
 Reticulated **15**: 2079, 2080, *2081*;
 17: 2419
Pytilia melba **21**: 2938
Pyxipoma **21**: 2989

Q

Q fever **19**: 2682
Quagga *see* Zebra, Plains
Quahog *see* Clam, Hard
Quail **15**: 2082–2083
 Button *see* Button quail
 Common **15**: 2082, 2083, *2083*
 Crested **15**: 2083
 Gambel's **15**: *2082*
 Lark **3**: 349
 Little bustard *see* Button quail,
 Common
 Marbled wood **15**: 2082
 Turkey *see* Plains-wanderer
Quay louse *see* Slater
Quay lowder *see* Slater
Quelea **15**: 2084
 cardinalis **15**: 2085
 erythrops **15**: 2085
 quelea **15**: 2085; **21**: 2897
Quelea **15**: 2084–2086
 Cardinal **15**: 2085
 Red-billed (Sudan dioch) **15**: *2084*,
 2085, *2085*, *2086*; **21**: 2897

Quelea (*continued*)
Red-headed **15**: 2085
questing **19**: 2681
Quetzal **15**: 2087–2088
Resplendent (resplendent trogon)
15: 2087, *2088*
Quiscalus **8**: 1032, 1033
Quokka (short-tailed pademelon)
15: 2089–2090
Quoll *see* Marsupial cat

R

Rabbit (Old World rabbit; domestic
rabbit) **15**: 2091–2093
Brush **3**: 426; **4**: 546
Jack *see* Jack rabbit
Rock *see* Pika
Snowshoe **18**: 2561
Sumatran (Sumatran hare; short-
eared rabbit; Sumatran short-
eared rabbit; Kerinci rabbit;
kelinci kerinci) **18**: 2560–2561
Swamp (canecutter) **4**: 545, 546
see also Cottontail
Rabbitfish **4**: 532; **15**: 2094–2095
Masked **15**: *2095*
Rabbit-rat, White-footed **18**: 2536
rabies **20**: 2809
Raccoon **15**: 2096–2098
Bahama **15**: 2096, 2097
Barbados **15**: 2096
Common (North American
raccoon) **15**: 2096, 2097
Cozumel Island **15**: 2096
Crab-eating **15**: 2096, 2097
Guadeloupe **15**: 2096
Tres Marías **15**: 2096
Raccoon dog (ussuri dog) **5**: 689;
15: 2099–2100
Raccoon fox *see* Ringtail
Racer **15**: 2101–2103
Antiguan **15**: *2103*
Black **15**: 2101, *2101*, 2102
Blue **15**: 2102, *2102*
Cedar *see* Whip snake, Striped
Eastern yellow-bellied **15**: *2102*
Red *see* Coachwhip
Striped **15**: 2101
Racerunner **19**: 2650
Prairie **19**: 2650
Six-lined **19**: 2650
radula **1**: 20; **4**: 519, 520; **14**: 1926,
2011; **17**: 2409
Rag, King **15**: 2104
Ragworm (mussel worm; clam
worm; pile worm)
15: 2104–2105
Rail **15**: 2106–2108
African **15**: 2106
Blue *see* Moorhen, Common
Clapper **15**: 2106, *2107*
Flightless **15**: 2108
King **15**: 2106
Land *see* Corncrake
Laysan (*extinct*) **15**: 2108
Lewin's **15**: 2106
Sora **15**: *2106*
Sora *see* Sora
Virginia **15**: 2106
Wake Island (*extinct*) **15**: 2108
Water **15**: 2106, 2107, *2108*
Rainbow bird *see* Bee-eater, Rainbow
Rainbow fish *see* Guppy
Rainbowfish (silverside) **17**: 2379
Banded **17**: *2380*
MacCulloch's **17**: 2380
Rain forest 8: 1093; **12**: 1642–1643;
15: 2109–2114, 2114
Rain goose *see* Loon, Red-throated

rain shadows **5**: 663; **12**: 1671
Raja
binoculata **17**: 2383
clavata **17**: 2330, 2383
rakers *see* gill rakers
Rallus **15**: 2106, 2107
Ramphastos **19**: 2714, 2716
dicolorus **19**: *2716*
sulfuratos **19**: 2714
toco **19**: *2715*
vitellinus **19**: *2714*
Ramphocelus passerinii **19**: 2629
Ramphocinclus brachyurus **19**: 2671
Ramphotyphlops braminus **17**: 2421
Ramsar Convention **21**: 2908
Ramshorn (trumpet snail; wheel
snail) **14**: 2010
Rana **3**: 313; **4**: 511; **6**: 743, 744; **7**: 903;
11: 1447, 1448
Ranatra linearis **20**: *2870*, 2871
Rangifer tarandus **3**: 391, 392; **5**: 655;
20: 2778
Ranzania laevis **13**: 1774
Raphicerus
campestris **18**: 2527
melanotis **18**: 2527
sharpei **18**: 2527
Rastrelliger
brachyosoma **11**: 1517
feughni **11**: 1517
kanapurta **11**: 1517
Rat
Bamboo *see* Mole rat
Bandicoot *see* Bandicoot rat
Brown (common rat; Norway rat)
3: 298–299; **20**: 2764
Canepiece **15**: 2156
Cloud **7**: 965
Giant *see* Giant rat
Indian Mole *see* Bandicoot rat,
Lesser
Jumping *see* Jerboa
Kangaroo (*American*) **5**: 665;
10: 1339–1340
Kangaroo (*Old World*) *see* Jerboa
Maned **13**: 1838
Marmoset **11**: 1566–1567
Mole *see* Mole rat
Multimammate *see*
Multimammate rat
Musk *see* Shrew, Musk
Naked-tailed *see* Naked-tailed rat
Pack *see* Pack rat
Pest *see* Bandicoot rat, Short-tailed
Polynesian **20**: 2764
Rice *see* Rice rat
Sand *see* Gerbil
Ship (black rat) **3**: 299;
17: 2346–2348; **20**: 2764
Stick-nest *see* Stick-nest rat
Tree *see* Tree rat
Water *see* Water rat
Wood *see* Pack rat
Ratel *see* Badger, Honey
Rat kangaroo **15**: 2115–2116
Desert **15**: 2115, 2116
Long-nosed (potoroo) **15**: 2116
Musky **15**: 2115, *2115*
Rufous **15**: 2115, 2116
Short-nosed **15**: 2115, 2116
Rat snake (black rat snake; mountain
black rat snake; yellow rat
snake; gray rat snake; Everglades
rat snake; Texas rat snake;
chicken snake; pilot snake; striped
house snake) **15**: 2117–2118
Great Plains *see* Snake, Corn
Green **15**: 2117, 2118
Indian **15**: 2117; **17**: 2422
Rosy *see* Snake, Corn

Rattler
Hardwood *see* Snake, Fox
Swamp *see* Massasauga
see also Rattlesnake
Rattlesnake (rattler) **3**: 426;
15: 2119–2121; **17**: 2422
Aruba Island **15**: 2120
Eastern diamondback **15**: 2119,
2120
Horned *see* Sidewinder
Mohave **15**: *2120*
New Mexico ridge-nosed
17: 2422
Prairie **15**: *2119*, 2120, 2121
Pygmy (ground rattler) **15**: 2119,
2120
Red diamond **15**: 2120, 2121
Tiger **15**: 2120
Timber (banded rattlesnake)
15: 2120
Western diamondback (coontail
rattler) **3**: *426*; **15**: 2120, *2121*
Rattus
exulans **20**: 2764
norvegicus **3**: 298–299; **20**: 2764
rattus **17**: 2347; **20**: 2764
Raven **5**: 610; **15**: 2122–2124
Australian **15**: 2122
Chihuahua **15**: 2122
Common **15**: 2122, 2123; **18**: 2455,
2457
Fan-tailed **15**: 2122
Forest **15**: 2122
Little **15**: 2122
Thick-billed **15**: 2122
White-necked **15**: 2122
Ray **17**: 2329–2330, 2332
Bullnose **6**: 729, 730
California bat **6**: 729, 730, 731
Cownose **6**: 729
Devil *see* Devilfish
Eagle *see* Eagle ray
Electric *see* Electric ray
Fiddle *see* Guitarfish
Giant reef *see* Stingray, Black-
spotted
Manta *see* Devilfish
Round **17**: 2333
Shovelnose *see* Guitarfish
Thornback **17**: 2330
Torpedo *see* Electric ray
Razor
Eggshell **15**: 2127
Transparent **15**: 2127, *2128*
Razorbill (razorbilled auk)
15: 2125–2126
Razorfish **17**: 2367, *2367*, 2368
Pearly **21**: 2991
Razor shell (spoutfish) **15**: 2127–2128
Pod (jackknife clam) **15**: 2127,
2127, *2128*
see also Razor
rectum, used as gills **17**: 2442
Recurvirostra
americana **1**: 105
andina **1**: 105
avosetta **1**: 105, 106
novaehollandiae **1**: 105
Red back *see* Black widow
Red-backed vole (red vole; red-
backed mouse) **1**: 134; **15**: 2129–2131
Northern **15**: 2129
Southern (Gapper's mouse)
15: 2129, 2130
Western **15**: 2129
Red bird *see* Cardinal
Redcoat **18**: 2511, *2511*, 2512
redds **18**: 2569; **20**: 2753
Redfish (scorpionfish), Norway
16: 2260, 2261

Redfish (*sea trout*) *see* Croaker,
Meager
Redhead **14**: 1993, 1994–1995, *1994*
rediae **7**: 870
Red mark *see* Black widow
Redpoll **15**: 2137–2138
Arctic (hoary redpoll)
15: 2137–2138
Greenland **15**: 2137
Lesser **15**: 2137
Mealy **15**: 2137, *2137*, 2138
Redshank **16**: 2233
Redstart (*Old World*) **15**: 2139–2140
Black **15**: 2139
Common **15**: 2139, 2140, *2140*;
19: 2678
Moussier's **15**: 2139
red tides **14**: 1982
Redunca **1**: 61
arundinum **15**: 2144, 2145
fulvorufula **15**: 2144, 2145
redunca **15**: 2144, 2145
Reedbuck **1**: 61; **15**: 2144–2145
Bohor **15**: 2144, 2145
Common (southern reedbuck)
15: 2144, *2144*, 2145
Mountain **15**: 2144, 2145, *2145*
Reed frog (sedge frog) **15**: 2146–2147
Brown and white spiny **15**: 2146,
2147
Five-lined **15**: 2146
Green **15**: 2146
Painted (marbled reed frog)
15: 2146, *2146*, 2147, *2147*
Small golden spiny **15**: 2146, 2147
Reedling, Bearded (bearded tit)
14: 1886, *1886*, 1887, *1887*
Reed warbler **20**: 2859
Blyth's **20**: 2859
Great **5**: 620
reefs *see* coral reefs
refection *see* coprophagy
Regalecus glesne **13**: *1767*, *1767*, 1769
regeneration
earthworm **6**: 736
fanworm **6**: 802
featherstar **6**: 805
flatworm **6**: 849
limb **6**: 741; **11**: 1477
sea pen **16**: 2297
starfish **18**: 2517
tubeworm **20**: 2768
Regulus
calendula **7**: 999
goodfellowi **7**: 999
ignicapillus **7**: 999
regulus **7**: 999, 1000; **18**: 2455
satrapa **7**: 999
teneriffae **7**: 999
Reindeer *see* Caribou
Reinhardtius
evermanni **8**: 1121
hippoglossoides **8**: 1121
Reithrodontomys **9**: 1159
humulis **9**: 1158
megalotis **9**: 1158
raviventris **9**: 1158
relapsing fever **19**: 2682
Remora
australis **15**: 2148
brachyptera **15**: 2148
osteochir **15**: 2148
remora **15**: 2148, *2149*
Remora **15**: 2148–2149
Common **15**: *2149*
Spearfish **15**: 2148
Striped *see* Sharksucker, Live
Remorina albescens **15**: 2148
Renilla **16**: 2298
repletes **9**: 1204–1205

Repomucenus **5**: 707
resilin **6**: 852
rete mirabile **11**: 1490
Reticulitermes **1**: 68
retinaculum **17**: 2392; **19**: 2603
Rhabdamia **3**: 390
Rhabdomys, pumilio **21**: 3016, *3017*
Rhacophorus
   nigropalmatus **7**: 982, 983
   reinwardtii **7**: 982, 983
Rhagium **11**: 1484
Rhagodes nigracinctus **18**: 2454
Rhantus **5**: 683
Rhea **15**: 2150–2151; **19**: 2692
   Greater (common rhea) **15**: 2150, 2151
   Lesser (Darwin's rhea) **15**: 2150, *2150*, 2151, *2151*
Rhea americana **15**: 2151
Rhina **8**: 1103
   ancylostoma **8**: *1102*
rhinarium **3**: 292
Rhincodon typhus **6**: 834; **17**: 2331; **21**: *2915*, 2916
Rhinecanthus aculeatus **20**: 2744
Rhinobatos **8**: 1103
   lentiginosus **8**: 1103
   rhinobatos **8**: 1103, *1103*
Rhinoceros
   sondaicus **15**: 2152, 2153
   unicornis **15**: 2152, 2153
Rhinoceros **15**: 2152–2155
   Black (browse rhinoceros) **15**: 2152, *2152*, 2153, 2154, 2155
   Greater Indian (Asian great one-horned rhinoceros) **15**: 2152, 2153, *2153*, 2154, 2155
   Javan **15**: 2152, 2153, 2154–2155
   Sumatran **15**: 2152, 2153, *2154*, 2155
   White (grass rhinoceros) **15**: 2152, 2153, 2154, 2155, *2155*
Rhinoderma
   darwinii **12**: 1678–1679
   rufum **12**: 1678, 1679
Rhinolophus
   ferrum **2**: 168
   ferrumequinum **9**: 1239, 1240
   hipposideros **9**: 1240
   philippiensis **9**: *1240*
Rhinomuraena quaesita **12**: 1658
Rhinophrynus dorsalis **3**: 326
Rhinopias aphanes **16**: 2261
Rhinopithecus **10**: 1411
Rhinoplax virgil **9**: 1226
Rhinopoma
   hardwickei **19**: 2703
   microphyllum **19**: 2703
   muscatellum **19**: 2703
Rhinoptilus
   africanus **4**: *552*
   bitorquatus **4**: 551
Rhipidura leucophrys **20**: 2843
Rhizostoma **10**: 1320, 1321
Rhodalia cardinalis **10**: 1396
Rhodeus sericeus **2**: 219
Rhodites rosae **7**: 921
Rhodonessa caryophyllacea **5**: 717
Rhodostethia rosea **16**: *2266*
Rhombobomys opimus **7**: 949
Rhopalodina **16**: 2271
Rhyacophila **16**: 2175
Rhynchdermus **6**: 849
   terrestris **6**: 848
Rhynchobatos **8**: 1103
   djiddensis **8**: 1102, 1103
Rhynchocyon
   chrysopygus **6**: 774, 775
   cirnei **6**: 774
Rhynochetus jubatus **18**: 2567

Rhyssa persuasoria **9**: 1279, *1279*, 1280
Ribbonfish **13**: 1768
   Whiptail (king of the salmon) **13**: 1768
Ricebird *see* Sparrow, Java
Rice rat **15**: 2156–2157
   Brazilian **15**: 2157
   Galapagos **15**: 2157
   Giant (*extinct*) **15**: 2156
rictal bristles **20**: 2801
Ridley turtle
   Atlantic **20**: 2797; **21**: 2907
   Pacific (olive turtle) **20**: *2796*
Riflebird **2**: 207
   Prince Albert (magnificent riflebird) **2**: 205
Rifleman (titipounamu; thumbie) **16**: 2164–2165
Rift Valley **12**: 1671–1672
Right whale **16**: 2166–2168, 2209; **21**: 2913, 2914
   Northern (black right whale) **16**: 2166, 2167; **21**: 2910
   Pygmy **16**: 2166, 2167, 2168; **21**: 2910
   Southern (black right whale) **16**: 2166, 2167, *2167*, 2168; **21**: 2910
rinderpest **3**: 307, 370–371; **7**: 958; **13**: 1830; **21**: 2948
Ringed worm **6**: 805
ring formation (defense)
   musk-ox **12**: 1705
   Przewalski's horse **15**: 2054
   wildebeest **21**: 2950
ring species **9**: 1189
Ringtail (band-tailed cat; cacomistle; cat-squirrel; coon cat; miner's cat; mountain cat; raccoon fox; ring-tailed cat; tepemixtle) **16**: 2171–2172
   Central American (guayonoche) **16**: 2171, 2172
Ring-tailed cat *see* Ringtail
Riparia riparia **2**: 151; **6**: 850; **12**: 1590
Rissa
   brevirostris **10**: 1370
   tridactyla **10**: 1370, 1371
ritualization **12**: 1601
River dolphin **16**: 2177–2178
   Amazon (boto) **5**: 697; **16**: 2176, 2177, 2178
   Chinese (baji) **5**: 697; **16**: 2177, *2177*, 2178
River and stream 16: 2173–2176
Rivulus **19**: 2704
Roach (*New World*) **16**: 2182
Roach (*Old World*) **16**: 2182–2183
Roadrunner **3**: *424*, 426; **16**: 2184–2185
   Greater (chaparral cock) **16**: 2184, *2184*, 2185, *2185*
   Lesser **16**: 2184
robbery by
   bald eagle **1**: 124
   black-headed gull **2**: 231
   crow **5**: 613
   death's head hawkmoth **9**: 1210
   frigate bird **2**: 263; **7**: 891–892
   grackle **8**: 1034
   hawkmoth **9**: 1168
   jaeger **10**: 1309
   magpie **11**: *1519*, 1520
   sheathbill **17**: 2338
   spotted hyena **3**: 374–375
Robin
   American (Canada robin; northern robin; robin redbreast) **4**: 555; **15**: 2141; **16**: 2188–2190; **19**: 2678; **20**: 2833

   European (robin redbreast) **16**: 2191–2193; **19**: 2678
   Indian **16**: 2190
   Pekin *see* Leiothrix, Red-billed
   see also Tody, Jamaican
Robin chat, Red-capped **6**: 776
robin's pincusion **7**: 921
Roccus
   chrysops **8**: 1082
   saxatilis **8**: 1082
Rock beauty **1**: 49
Rock borer, American *see* Piddock, Little
Rock cat *see* Caracal
Rock coney *see* Pika
Rock eel *see* Dogfish
Rockfowl **1**: 113
   Gray-necked **1**: 114
   White-necked **1**: 114
Rock rabbit *see* Pika
Rock salmon *see* Dogfish
Rock wallaby **16**: 2197–2198
   Brush-tailed **16**: *2197*, 2198
   Little **16**: 2198
   Ring-tailed (yellow-footed rock wallaby) **16**: 2197, 2198, *2198*
   Short-eared **16**: 2197
Rocky Mountains **3**: 423; **12**: 1669–1670, 1672
Rocky Mountain spotted fever **19**: 2682
Roguefish, Bearded **16**: 2260
Rollandia microptera **8**: 1063
Roller **16**: 2204–2205
   Abyssinian (Senegal roller) **16**: 2204
   Broad-billed **16**: 2204
   Eurasian **16**: 2204
   Indian **16**: *2205*
   Lilac-breasted **16**: 2204, *2205*
Rollulus **14**: 1891
Roly-poly *see* Wood louse
Romans
   and dormice **5**: 701
   and electric eels **6**: 766
   and flamingos **6**: 847
   and gazelles **7**: 942
   and goatfish **7**: 990
   and snakes **15**: 2118
Roncador stearnsi **5**: 602
Rondeletia loricata **21**: 2909
Rook **4**: 452; **5**: 610; **16**: 2206–2207
rookeries, Adélie penguin **1**: 28
rook-worm **4**: 487
Roosevelt, Theodore **2**: 223
roosting, aerial **19**: 2600
Rorqual **16**: 2167–2168, 2208–2209; **21**: 2910–2911
   Great northern *see* Whale, Blue
   Lesser *see* Whale, Minke
Rosefinch, Scarlet **4**: 499
Rosefish, Blackbelly (bluemouth; bluethroat) **16**: 2260, 2261
Rostanga rufescens **16**: 2299–2300
Rostratula benghalensis **13**: 1849, *1849*, 1850
Rostrhamus
   hamatus **6**: 788
   sociabilis **2**: 209; **6**: 789
rostrum **18**: 2475; **20**: 2871; **21**: 2903, *2903*
Round-headed borer *see* Long-horned beetle
Rousettus **2**: 167
   aegyptiacus **7**: *906*
Rudd **16**: 2182
Ruff (reeve) **16**: 2212–2214
Ruffe **14**: 1917, 1918
Rupicapra rupicapra **3**: 422

Rupicola
   peruviana **4**: 491, 543
   peruviana aurea **4**: 491
   peruviana sanguinolenta **4**: 491
   rupicola **4**: 491, 543
Rutilus **6**: 2182, 2183
Ruvettus pretiosus **17**: 2415
Ruwenzorornis johnstoni **20**: 2780
Rynchops
   albicollis **17**: 2387
   flavirostris **17**: 2387
   nigra **17**: 2387

S

Sabella **6**: 801
   pavonina **6**: 802
   spallanzanii **6**: 801
   starte indica **6**: *802*
Sable **11**: 1579, 1581
Saddleback (*bird*; tieke) **21**: 2886–2887
Saddleback (*seal*) *see* Seal, Harp
Saddlebill **18**: 2551
Sagittarius serpentarius **2**: 210; **17**: 2316–2318
Saguinus **12**: 1645, *1645*; **19**: 2624, 2625
Sahara Desert **5**: 663
Saiga **1**: 61; **16**: 2217–2218
Saiga
   tatarica **1**: 61; **16**: 2217, 2218
   tatarica mongolica **16**: 2217
Sailfish **16**: 2219–2220; **19**: 2605
   Atlantic **16**: 2219
   Indian and Pacific Ocean **16**: 2219, 2220
Saimiri **18**: 2513, 2514
St. James' shell *see* Scallop
St. Peter's fish *see* John Dory
Saithe *see* Coalfish
Sakesphorus canadensis **1**: 55
Saki **16**: 2221–2222
   Bald-faced **16**: 2222
   Bearded **16**: 2221, 2222
   Black **16**: 2221, 2222
   Equatorial **16**: 2222
   Red-bearded (monk's saki) **16**: 2221, 2222
   White **16**: 2222
   White-faced **16**: 2221, *2221*, 2222
   White-nosed **16**: 2221, 2222
Salamander
   Arboreal **11**: 1503–1504, 1505
   California slender **11**: 1503
   Dusky **11**: 1503, 1505
   Eastern tiger **11**: *1504*
   Fire (spotted salamander) **6**: 831–832
   Four-toed **11**: 1503
   Giant *see* Giant salamander
   Hellbender *see* Hellbender
   Jefferson **11**: *1505*
   Long-tailed **11**: 1503, *1503*
   Lungless **11**: 1503–1505
   Ocoee **11**: 1503
   Pygmy **11**: 1503
   Red-backed **11**: 1503
   Shovel-nosed **11**: 1503
   Slimy **11**: 1504, 1505
   Tiger **1**: 109, 110
   Yellow-blotched **11**: 1505
Salamandra salamandra **6**: 832
Salamandrella keyserlingii **19**: 2616
salamandrin **6**: 831
Salangichthys microdon **7**: 978
Salanx
   cavieri **7**: 978
   chinensis **7**: 978
saliva, licking **13**: 1844

Salmo **1**: 100; **16**: 2175, 2176; **20**: 2751, 2752
Salmon **6**: 836
    Atlantic **1**: 99–101; **16**: 2176
    Burnett *see* Lungfish, Australian
    Cherry (masu) **13**: 1840
    Chinook (tyee salmon; quinnat salmon; king salmon; spring salmon; Sacramento salmon; Columbia River salmon) **13**: 1840, *1840*, 1842
    Chum (keta salmon; dog salmon) **13**: 1840, 1841, 1842
    Coho (silver salmon) **13**: 1840, 1842
    Pacific **13**: 1840–1842
    Pink (humpback salmon) **13**: 1840, 1841, 1842
    Rock *see* Dogfish
    Sockeye (red salmon; blueback salmon) **13**: 1840, *1841*, 1842, *1842*
    Zebra **21**: 3015
salmonella **3**: 299
Salp **16**: 2223–2224
    Giant **16**: 2224
Salpa **16**: *2223*
    maximus **16**: 2224
Salpinctes obsoletus **21**: 2994
Salpingotus **10**: 1324
salt
    and brine shrimp **2**: 277
    and the flounder **6**: 861
    and green turtles **8**: 1070
    and penguins **14**: 1913
    and tree porcupines **19**: 2728
salt glands **14**: 1932
salt licks **7**: 937
salt marshes **4**: 477
Salvelinus
    fontinalis **20**: 2751
    malma **20**: 2751
    namaycush **20**: 2751
Sand dollar (cake urchin) **16**: 2225–2226
    Common **16**: 2225, 2226
Sand eel **16**: 2227–2228
    Great **16**: 2228
    Pacific (Pacific sand lance) **16**: 2228
    Small **16**: 2227
    Smooth **16**: 2227
Sandelia **4**: 468
Sanderling **16**: 2233
Sandfish **17**: 2390
Sand grouse **5**: 665; **16**: 2229–2230
    Double-banded **16**: *2230*
    Namaqua **16**: *2229*, 2230
    Pallas's **16**: 2229, 2230
    Pintailed **16**: 2230
    Spotted **16**: 2229–2230
Sand lance **16**: 2227
    Pacific *see* Sand eel, Pacific
Sandpiper **16**: 2233–2234
    Buff-breasted **16**: 2233, 2234
    Common **16**: 2233, *2233*, 2234
    Curlew **16**: *2234*
    Gray-rumped (gray-tailed tattler) **16**: 2233
    Green **16**: 2233, 2234
    Pectoral **20**: *2779*, 2833
    Purple **16**: 2233, 2234
    Solitary **16**: 2234
    Terek **16**: 2233
    Upland **16**: 2234
    Wood **16**: 2234; **21**: 2906
Sand puppy *see* Mole rat
Sandroller *see* Trout-perch, Blunt-nosed
Sand smelt *see* Silverside

Sandsnake, Hissing *see* Hognose snake
Sapajou *see* Capuchin monkey
sap feeding
    aphid **1**: 74
    cicada **4**: 457
    hopper **9**: 1224
    sapsucker **16**: 2235–2236, 2237, *2237*
    tachina fly **19**: *2611*
    thrips **19**: 2676
    whitefly **21**: 2932
saprophytes **15**: 2112
Sapsucker **16**: 2235–2237
    Red-breasted **16**: 2235
    Red-naped **16**: 2235
    Williamson's **16**: 2235, *2236*
    Yellow-bellied **16**: 2235, *2235*, 2236, *2237*
Sarcogyps calvus **20**: 2841
Sarcophilus harrisii **5**: 677; **19**: 2642
Sarcophyton trocheilophorum **17**: 2441
Sarcorhamphus papa **20**: 2787, 2841
Sarda sarda **2**: 258
Sardina pilchardus **14**: 1948, 1949
Sardine *see* Pilchard
Sardinops **14**: 1948, 1949; **17**: 2430
sargassum **1**: 51; **7**: 875
Sargocentron
    rubrum **18**: 2511, 2512
    spiniferum **18**: 2511, 2512
    violaceum **18**: 2512
Sarken *see* Snow leopard
Sarotherodon
    galilaeus **19**: 2689, *2689*
    galilaeus galilaeus **19**: 2690
Sasia abnormis **14**: 1937
satellite tracking, of sharks **2**: 162
Satinbird *see* Bowerbird
Sauromalus
    obesus **4**: 454
    varius **4**: 454, *454*
Saury **12**: 1724
Saury pike (skipper) **17**: 2368
savannas **8**: 1038–1039, 1041
Sawfish **16**: 2238–2239
    Knifetooth **16**: 2238
    Largetooth **16**: 2238
    Smalltooth **16**: 2238, 2239, *2239*
Sawfly (wood wasp; horntail) **9**: 1279, 1280; **16**: 2240–2241
    Black-and-yellow horntail **16**: 2240
    European spruce **16**: 2241
    Jumping-disc **16**: 2240–2241
    Palisade **16**: 2240
    Pear (*larva*: slugworm) **16**: 2240
Saxicoloides fulicata **16**: 2190
Sayornis
    fusca **20**: 2800
    nigricans **20**: 2801
Scad (horse mackerel) **16**: 2242–2243
    Atlantic horse mackerel **16**: 2243
    Bigeye (akule) **16**: *2243*
Scaeochlamys lividus **16**: 2248
Scalare (angelfish) **4**: 458, 459; **16**: 2244–2245
    Deep angelfish **16**: 2244, 2245
    Lesser angelfish (freshwater angelfish) **16**: 2244, 2245
Scale insect **16**: 2246–2247
    Cottony-cushion **10**: 1396; **16**: 2246, 2247
    Elm scale **16**: 2246, 2247
    Heath scale **16**: 2246, 2247
    Mussel scale **16**: 2246, 2247
    Pine leaf scale **16**: 2246, 2247
Scale worm **16**: 2292
Scallop **16**: 2248–2249
    Doughboy **16**: *2249*

    Great (St. James' shell; coquille St. Jacques) **16**: 2248, 2249
    Purple-hinged **16**: 2248
    Queen **16**: *2248*, 2249
    Small (southern scallop) **16**: 2248
Scalytail *see* Scaly-tailed squirrel
    Scaly-tailed squirrel (anomalure; scalytail; scaly-tailed flying squirrel; flying mouse) **16**: 2250–2251
    Beecroft's **16**: 2250, 2251
    Flightless (Cameroon scaly-tailed squirrel) **16**: 2250, *2250*, 2251
    Pel's **16**: 2250, *2251*
    Pygmy **16**: 2251
Scaphiopus **18**: 2466
    bombifrons **18**: 2466
    couchii **18**: 2466
    hammondi **18**: 2466
    holbrooki **18**: *2465*
    multiplicatus **18**: 2465
Scarabaeus sacer **16**: 2252
Scarab beetle **16**: 2252–2254
    Green **16**: *2252*
    Silver striped **16**: *2253*
    see also Cockchafer
Scardinius erythrophthalmus **16**: 2182
Scartelaos **12**: 1684, 1685
Scarus **14**: 1888, *1888*, 1889–1890, *1889, 1890*
Scat (argusfish) **16**: 2255–2256
    Spotted **16**: 2255, *2255*, 2256
    Tiger **16**: 2255
Scatophagus **16**: 2256
    argus **16**: 2256
    rubifrons **16**: 2255
    tetracanthus **16**: 2255, 2256
Scatty **16**: 2255
Scaup **14**: 1993
    Greater (bluebill; broadbill) **14**: 1995
    New Zealand **14**: 1993, 1995
scavenging
    basket star **2**: 158
    edible crab **6**: 741
    hyena **9**: 1269, *1269*
    marabou **11**: 1553
    mussel **12**: 1712
    opossum **13**: 1793
    red fox **15**: 2134
    sheathbill **17**: 2337
    slater **17**: 2398, *2398*
    spider crab **18**: 2475, *2476*
    toadfish **19**: 2699
    vespula wasp **20**: 2826
    vultures **20**: **2839**, *2839*, **2840**
    wood louse **21**: 2973
    see also carrion
Sceliphron **9**: 1262
Sceloporus **6**: 807
    occidentalis **6**: 807, 808
    undulatus **6**: 807, 808
    undulatus elongatus **6**: *807*
Scenopinus fenestralis **4**: 471
scent marking
    aardwolf **1**: 19
    Asiatic black bear **1**: 95
    badger **1**: 120
    brown rat **3**: 298
    bushbaby **3**: 329
    bushdog **3**: 334
    capybara **3**: 382
    cats 3: 406
    civet **4**: 461
    hartebeest **8**: 1147
    house mouse **9**: 1245
    jaguar **10**: 1310
    lemur **10**: 1435
    marten **11**: 1580
    muntjac **12**: 1697

    musk shrew **12**: 1710
    nyala **13**: 1765
    oribi **13**: 1803
    peccary **14**: 1905
    polecat **14**: 2002
    potto **15**: 2034
    rabbit **15**: *2093*
    red panda **15**: 2135
    rhino **15**: 2154, 2155
    roe deer **16**: 2203
    sloth **17**: 2402
    titi **19**: 2696
    water vole **21**: 2885
    weasel **21**: 2894–2895
    wildcat **21**: 2945–2946
Schetba rufa **20**: 2813
Schistocerca gregaria **11**: 1478, 1479
Schistometopum thomense **3**: 356
Schistosoma **7**: 871
schistosomiasis *see* bilharzia
Schizothaerus nuttallii **7**: 927
Schoblia **21**: 2974
schools *see* living in groups
Sciaena **5**: 602
    aquila **5**: 601
Scimitar babbler **1**: 113
Scincus
    mitranus **17**: *2390*
    scincus **17**: *2390*
Sciurus **20**: 2740
    carolinensis **5**: 700; **20**: 2740, 2741
    vulgaris **4**: 436; **20**: 2740, 2741, *2742*
Sclerurus **13**: 1819
Scolopax rusticola **21**: 2966, 2967
Scolopendra **3**: 411
    gigantea **3**: 411
    morsitans **3**: 411, 412
Scomber
    japonicus **11**: 1517
    scombrus **11**: 1517, 1518
Scomberesox **17**: 2368
Scomberomorus **11**: 1517
Scophthalmus
    maximus **20**: 2782, *2782*, 2783
    maximus maestica **20**: 2782
    rhombus **20**: 2782
Scops-owl
    Grand Comoro **13**: 1829
    Rain forest **13**: 1827
    Seychelles **13**: 1829
    Sokoke **13**: *1828*
    Torotoroka **13**: 1827
Scopus umbretta **8**: 1124; **18**: 2551
Scorpaena **20**: 2818
    brasiliensis **16**: 2260
    guttata **16**: 2260
    scrofa **16**: 2260
Scorpaenopsis diabolus **16**: *2260*
Scorpion **16**: 2257–2259; **20**: 2818, 2820
    Asian forest **16**: 2258
    Book *see* False scorpion
    Emperor **16**: 2257
    False *see* False scorpion
    Whip *see* Pedipalpi
    Wind *see* Solifugid
Scorpionfish **16**: 2260–2261; **20**: 2818, 2820
    California (sculpin) **16**: 2260, 2261
    Devil **16**: *2260*
    Orange **16**: 2260
    Weedy **16**: *2261*
Scotopelia **6**: 842; **13**: 1827
    bouvieri **6**: 842
    peli **6**: 842, 843
    ussheri **6**: 842
Scotornis climacurus **13**: 1742
Scourfish *see* Escolar
Scrawled fish **20**: 2743

Screamer **5**: 716; **16**: 2262–2263
   Horned **16**: 2262
   Northern **16**: 2262, 2263
   Southern **16**: 2262–2263, *2262*
Screech owl **13**: 1827
   Eastern **13**: 1828
   Western **13**: 1828
Scrub fowl
   Orange-footed (Reinwardt's
     scrub fowl) **11**: 1527
   Philippine (tabon) **11**: 1527
Scudderia texensis **10**: 1341
Sculpin (*bullhead*) *see* Bullhead,
   European
Sculpin (*scorpionfish*) *see*
   Scorpionfish, California
Scutigera forceps **3**: 411–412
Scutisorex somereni **9**: 1182
Scyliorhinus
   canicula **5**: 686, 687
   stellaris **5**: 686
Scyllaea **16**: 2300
Scyllarides **4**: 570
Scypha compressa **15**: 2075
scyphistomas **10**: 1322
Scythebill **21**: 2968
Sea anemone *see* Anemone
Sea bass **2**: 163, *163*
   Giant (wreckfish) **2**: 163, 164
Seabirds 16: 2264–2268
Sea biscuit **16**: 2226
Sea blubber **10**: 1320
Sea butterfly (pteropod)
   16: 2269–2270
Sea canary *see* Whale, Beluga
Sea cow *see* Dugong
Sea cucumber **14**: 1901–1902, *1901*,
   1981; **16**: 2271–2272
Sea dragon *see* Sea moth, Longtail;
   Weever
Sea fingers *see* Dead man's fingers
Sea fir (hydroid) **16**: 2273–2274
   Naked **16**: 2273
Sea fox *see* Thresher shark
Sea hare **16**: 2275–2276
   California **16**: *2276*
Sea hen *see* Lumpsucker
Sea horse **16**: 2277–2279; **20**: 2766
   Dwarf **16**: 2277
   Longsnout (longnose sea horse)
     16: 2278
   Pygmy **16**: 2277, *2278*
   White's **16**: 2277
   Yellow (Pacific sea horse) **16**: *2277*
Sea krait **17**: 2308
   Giant banded (wide-faced sea
     krait; amphibious sea snake)
     17: *2308*
   Rennell Island **17**: 2308
Seal **1**: 82
   Baikal **10**: 1398; **16**: 2169, 2170
   Bearded **1**: *82*; **16**: 2286
   Caspian **16**: 2169, 2170, *2170*
   Crabeater (white seal) **1**: 82;
     4: 564–565; **12**: 1647; **16**: 2286
   Eared **16**: 2285, *2287*
   Elephant *see* Elephant seal
   Fur *see* Fur seal
   Gray (horse head seal)
     8: 1049–1051; **16**: 2289
   Harbor (common seal) **4**: *475*;
     8: 1049, 1132–1134; **12**: 1647
   Harp (Greenland seal;
     saddleback)
     8: 1139–1141; **16**: *2285*
   Hooded (bladdernose)
     9: 1219–1221
   Leopard **1**: 29, 82; **4**: 565; **10**: 1360;
     11: 1449–1450
   Monk *see* Monk seal

   Ringed **1**: 82; **14**: 1998;
     16: 2169–2170, 2285, 2286
   Ross (big-eyed seal) **16**: 2210–2211
   True (earless) **16**: 2285, 2286
   Weddell **1**: *82*; **16**: 2288;
     21: 2899–2900
   see also Sea lion; Seals; Walrus
Sea lemon **16**: 2299, 2300
Sea lily (stone lily) **16**: 2280–2281
Sea lion **16**: 2282–2284, 2285, 2287,
   2289
   Australian **16**: 2282, 2283, *2283*,
     2287, 2288
   California **16**: 2282, *2282*, 2283,
     2284
   Hooker's **16**: 2282, 2283
   Southern **16**: 2282, 2283
   Steller's **16**: 2283, 2284, *2284*, 2289
Seals 16: 2285–2289
Sea moth **16**: 2290–2291
   Dragon (short dragonfish)
     16: 2291, *2291*
   Hawaiian **16**: 2290
   Longtail (batfish; sea dragon;
     dragonfish) **16**: 2290, *2290*
Sea mouse **16**: 2292–2293
Sea pansy **16**: 2298
Sea peach **17**: 2313
Sea pen **16**: 2297–2298
Seaperch *see* Surfperch
Sea pheasant (corkwing) **21**: 2991
Sea pike *see* Snoek
Sea robin *see* Gurnard
sea serpents, myth **12**: 1659
Sea-slater (shore-roach; quay louse;
   quay lowder) **17**: 2398–2399
Sea slug (nudibranch) **16**: 2299–2301
   Rose nudibranch **16**: *2299*
Sea snail (*fish*) **11**: 1499
   Common **11**: 1499
Sea snail (*mollusk*) **16**: 2272
   Purple (violet sea snail; violet
     snail; floating shell of the high
     seas) **15**: 2073
Sea snake **7**: 935; **17**: 2308–2309;
   20: *2818*, 2820
   Amphibious *see* Sea krait, Giant
     banded
   Olive **17**: 2309
   Yellow-bellied **17**: 2308
Sea spider **17**: 2310–2311
   Giant deep-sea **17**: *2311*
Sea squirt (tunicate) **4**: 556; **14**: *1980*;
   15: 2078; **17**: 2312–2313
   Lightbulb **17**: 2313
Sea star *see* Starfish
Sea swallow *see* Tern
Sea trout (*Cynoscion*) *see* Croaker
Sea urchin (sea hedgehog)
   17: 2314–2315
   Black-spined **17**: 2315
   Edible **17**: *2314*, 2315
   Long-spined **17**: 2315
   Pencil-spined **17**: *2315*
sea wash balls **21**: 2918
Sea wasp *see* Jellyfish, Box
seaweed **4**: 476
Sea wolf *see* Killer whale; Lancet
   fish
Sebastes
   marinus **16**: 2260
   viviparus **16**: 2260
Secretary bird **2**: 210; **17**: 2316–2318
secretive *see* shy
sedentary lifestyle *see*
   stationary/sedentary lifestyle
Sedge fly *see* Caddis fly
Seed snipe **17**: 2319–2320
   Gray-breasted (D'Orbigny's seed
     snipe) **17**: 2319

   Least (pygmy seed snipe;
     Patagonian seed snipe)
     17: 2319, *2319*, 2320, *2320*
   Rufous-bellied (Gay's seed snipe)
     17: 2319
   White-bellied **17**: 2319
Sehirus dubius **17**: 2344–2345
Seiurus **21**: 2985
   aurocapillus **13**: 1819; **21**: 2985
Seladang *see* Gaur
Selar crumenophthalmus **16**: *2243*
Selasphorus rufus **9**: 1255
Selene vomer **16**: 2242
Selevina betpakdalaensis **5**: 699
self-anointing **19**: 2657
Semibalanus balanoides **1**: 138, *139*
semidigitigrade walking **6**: 774
Semiotus affinis **4**: 466
Semotilus corporalis **12**: 1619
Sepia officinalis **5**: 632
Sepioteuthis sepioidea **18**: *2509*
Serengeti **21**: *2949*, 2950
Sergeant major **5**: 638, 639
Sericulus
   bakeri **2**: 271
   chrysocephalus **2**: 271
Seriema (cariama) **17**: 2321–2322
   Burmeister's **17**: 2321
   Red-legged (crested seriema; red-
     legged cariama; crested
     cariama) **17**: 2321–2322, *2321*,
     2322
Serin **4**: 499; **8**: 1016
Serinus
   canaria **3**: 369
   dermus **8**: 1016
   serinus **4**: 499
Seriola
   dumerili **16**: 2242
   lalandi **16**: 2242
Serow **17**: 2323–2324; **19**: 2623
   Japanese (kamoshika) **17**: 2323,
     2324, *2324*
   Mainland (maned serow)
     17: 2323, 2324
   Taiwanese **17**: 2323, 2324
Serpulorbis
   gigas **21**: 2990
   novaehollandiae **21**: 2990
   squamigerus **21**: 2990
Serrasalmus **14**: 1967, *1967*, 1969
   spilopleura **14**: 1969
Sertularella speciosa **16**: 2274
Sertularia **16**: 2274
   abietena **16**: 2274
Serval (servaline cat; lynx tachete;
   aner; ingwenkala) **3**: 403;
   17: 2325–2326
Servaline cat *see* Serval
Sesarma meinerti **4**: *573*
Sesia apiformis **9**: 1234
Setaceous Hebrew character
   13: 1821
Setifer
   setosus **19**: 2656
   spinosus **19**: 2656, 2657
Setonix brachyurus **15**: 2090
sets, badger **1**: 121
Severn Wildfowl Trust **9**: 1165
Sewellel *see* Mountain beaver
sex change
   fanworm **6**: 801
   grouper **8**: 1083
   limpet **11**: 1452
   oyster **13**: 1834
   purple sea snail **15**: 2073
   shipworm **17**: *2350*
   swordtail **19**: 2608–2609
sex determination *see* temperature,
   · affecting sex

sexual activity or display *see*
   attracting a mate with color;
   attracting a mate with scent;
   attracting a mate with sound;
   sexual aggression
sexual aggression
   bighorn sheep **2**: 199
   chimpanzee **4**: 441
   chinchilla **4**: 444
   eland **6**: 758–759
   fallow deer **6**: 796
   fighting fish **6**: 822
   gaur **7**: 938
   giraffe **7**: 977
   haddock **8**: 1114
   hamster **8**: 1131
   oryx **13**: 1809
   swordtail **19**: 2608
sexual difference
   antwren **1**: 55
   bird of paradise **2**: 206
   bullfrog **3**: 312, *313*
   copper butterfly **4**: 526
   deep-sea anglerfish **5**: 653
   dolphin **4**: 506
   dragonet **5**: 706
   earwig **6**: 738, *738*
   eider **6**: 754, *755*
   elephant seal **6**: 771
   flicker **6**: 855
   fur seal **7**: 916
   gharial **7**: *955*, 956
   golden oriole **8**: 1012, *1013*
   green lizard **8**: 1068
   guppy **8**: 1106
   lion **11**: 1460
   mallard **11**: 1524
   mandrill **11**: 1536
   parakeet **14**: *1879*
   parrot **14**: *1883*
   patas monkey **14**: 1896
   peafowl **14**: 1898
   pheasant **14**: 1934, 1935
   pintail **14**: *1959*
   quelea **15**: 2084
   songbirds 18: 2457
   Surinam toad **18**: 2578
   terrapin **19**: 2664
   tooth-carp **19**: 2704, *2704*
   wattlebird **21**: 2886, 2887
   whistler **21**: 2923
   wrasse **21**: 2991
sexual dimorphism *see* sexual
   difference
Shad **17**: 2327–2328
   Allis **17**: 2327, *2327*, 2328
   American (common shad; white
     shad) **9**: 1184; **17**: 2327, 2328
   Gizzard **17**: 2328
   Green-tailed (hardhead shad;
     yellow-tailed shad) **17**: 2328
   Hilsa **17**: 2327
   Reeve's **17**: 2327
   Skipjack **17**: 2328
   Twaite (goureen) **17**: 2327
Shag **4**: 537
   Spotted **16**: *2265*
Shanny **2**: 237, 238
Shapu **2**: 198
Shark **4**: 532; **13**: 1770, 1772
   Basking **2**: 160–162; **17**: 2331
   Blacktip **17**: 2332
   Blue **11**: 1562; **13**: *1770*; **17**: 2329,
     2331, 2333
   Bonnethead **8**: 1126, 1127
   Bramble *see* Dogfish, Alligator
   Bull **17**: 2332
   Bullhead *see* Bullhead shark
   Caribbean reef **15**: *2148*; **17**: 2329,
     2332

Shark (*continued*)
Carpet *see* Wobbegong
Cat *see* Cat shark
Cookiecutter **17**: 2332
Fox *see* Thresher shark
Frilled (frill shark) **7**: 896–897;
17: 2330
Great white (maneater shark;
great blue shark; white pointer;
white shark; white death)
8: 1058–1060; **17**: 2331, 2332
Hammerhead *see* Hammerhead
shark
Horn **15**: 2027
Lemon **17**: 2333
Longtail *see* Thresher shark
Mako **8**: 1059; **11**: 1562
Mallethead **8**: 1127
Megamouth **17**: 2331
Port Jackson **15**: 2027–2028
Requiem **17**: 2331, 2332
Salmon **15**: 2021
Sand tiger **17**: 2331, 2332
Saw **16**: 2238
Scoophead **8**: 1127
Shovelnose *see* Guitarfish
Swell **3**: 407, *407*, 408
Thresher *see* Thresher shark
Tiger **17**: 2332
Whale **6**: 834; **17**: 2331;
21: 2915–2916
Whitetip **14**: *1950*
Winged *see* Hammerhead shark,
Winged
Zambezi (bull shark; shovelnose
gray shark; slipway gray shark;
van Rooyen's shark; cub shark;
ground shark; whaler shark;
Lake Nicaragua shark) **17**: 2329;
21: 3007–3008
Zebra **21**: 3015
see also Devilfish; Dogfish; Eagle
ray; Electric ray; Guitarfish;
Lamprey; Monkfish; Porbeagle;
Sawfish; Sharks; Skate;
Stingray; Wobbegong
Sharks 17: 2329–2333
warm-blooded **15**: 2021
Sharksucker
Live (striped remora) **15**: 2148,
2148, 2149
Whitefin **15**: 2148
Shearwater **17**: 2334–2336
Christmas **17**: 2336
Cory's **17**: 2334
Fluttering **17**: 2334
Great **17**: 2334, *2334*, 2335
Heinroth's **17**: 2334
Little **17**: 2334
Manx **17**: 2335, *2335*
Short-tailed (muttonbird) **17**: 2335
Sooty (somber shearwater;
ghostbird; New Zealand
muttonbird) **17**: 2334, 2335
Tailed **16**: 2264
Sheatfish *see* Catfish, European
Sheathbill **17**: 2337–2338
Lesser **17**: 2337
Snowy (American sheathbill; pale-
faced sheathbill; paddy)
17: 2337, *2337*, 2338
Sheba, Queen of **3**: 364
Sheep
Bighorn *see* Bighorn sheep
Blue **12**: 1672
Dall **2**: 198; **12**: *1673*
Marco Polo (argali) **2**: 198
Musk *see* Musk-ox
Sheldgoose **5**: 714
Shelduck **17**: 2341–2343

Common (shelduck) **17**: *2341*,
2342, *2342*
Crested **5**: 717
Paradise (paradise duck; painted
duck; rangitata goose) **17**: 2341
Ruddy **17**: 2341–2342, 2343, *2343*
shellfish poisoning **14**: 1982
shells
clam **4**: 462, *462*
cockle **4**: 488, *488, 489*
conch **4**: 514, 515, *515*
cowrie **4**: 556, *556*, 557
cuttlefish **5**: 631
elephant tusk shell **6**: *777*
garden snail **7**: 930
green turtle **8**: 1070
lampshell **10**: 1405
limpet **11**: 1451
mussel **12**: 1711
nautilus **12**: *1722, 1723*
nutshell **13**: 1763, *1763*
oyster **13**: 1832
painted turtle **13**: 1851
Papuan turtle **13**: 1866
periwinkle **14**: 1926
piddock **14**: 1939
pond snail **14**: 2010
turtles 20: 2795, 2796
worm shell **21**: 2989
Shieldbug (stinkbug) **17**: 2344–2345
Shikra **18**: 2467
Shiner (*insect*) *see* Cockroach,
German
Shiner (*minnow*) **12**: 1618
Common **12**: 1618
Shiner (*surfperch*) **18**: 2574
Redfin **18**: 2570
Shingleback **17**: 2390
Shipworm **13**: 1772; **17**: 2349–2350
shoals *see* living in groups
Shore-roach *see* Slater
short living
glowworm **7**: 987
mayfly **12**: 1593, *1593*
sawfly **16**: 2240
water shrew **20**: 2872
Shrew **17**: 2358–2360
Dwarf (Savi's pygmy shrew)
12: 1709
Elephant *see* Elephant shrew
Eurasian common **17**: *2358*, 2359
European pygmy **17**: 2358, 2359,
2359
Hero (armored shrew)
9: 1182–1183
Least short-tailed (lesser-short-
tailed shrew; little short-tailed
shrew) **17**: 2358
Masked (common shrew) **17**: 2358,
2359
Musk (musk rat; money shrew;
house shrew) **12**: 1709–1710;
17: 2358
North American pygmy **17**: 2359
Otter *see* Otter shrew
Pygmy **17**: 2358
Red-toothed **17**: 2358
Short-tailed **17**: 2358, 2359, *2360*
Smoky **17**: 2358, 2359
Tree *see* Tree shrew
Water *see* Tenrec, Web-footed;
Water shrew
White-toothed **12**: 1710; **17**: 2358
Shrike **17**: 2361–2363; **18**: 2456–2458
Loggerhead **17**: 2361, 2362
Northern (great gray shrike)
17: 2361, *2361*
Pied *see* Triller, Pygmy
Red-backed **17**: 2361, *2363*
Woodchat **17**: *2362*

Shrike-thrush **21**: 2923
Whistling **21**: 2923
Shrimp **14**: *1982*; **17**: 2364–2366
Brine **2**: 276–277
Cleaner **17**: 2364
Common European (brown
shrimp; edible shrimp)
17: 2364–2365
Emperor **17**: *2366*
Fairy **6**: 792–793
Freshwater *see* Freshwater
shrimp
Mantis (split-thumb; squilla;
prawn killer; shrimp mammy;
nurse shrimp) **11**: 1550,
1551–1552
Opossum **13**: 1795–1796
Oriental **15**: 2043
Pistol **17**: 2364, 2366
Red **17**: *2364*
Red-handed **15**: 2043
Sewing **17**: 2364
Shrimpfish **17**: 2367–2368; **20**: 2766
Grooved **17**: 2368
Smooth **17**: 2368
Speckled **17**: 2368
Shrimp mammy *see* Shrimp, Mantis
Shuffle-wing *see* Accentor, Hedge
Shufflewing *see* Cuckoo-shrike,
Black-faced
shy
agouti **1**: 32
boat-billed heron **2**: 252
broadbill **2**: 283
bulbul **3**: 310
bullhead **3**: 314–315
bushdog **3**: 333, 334
bushpig **3**: 335
button quail **3**: 349
colobus monkey **4**: 501
corncrake **4**: 541, *541*
cottontail **4**: 545–546
coucal **4**: 549
crake **5**: 583
dibatag **5**: 674
dik-dik **5**: 676
finfoot **6**: 823
francolin **7**: 890
golden oriole **8**: 1013
goshawk **8**: 1027–1028
hero shrew **9**: 1182
linnet **11**: 1458
nightingale **13**: 1740
pitta **14**: 1970
wildcat **21**: *2946*
zebra mouse **21**: 3016
Shy eye *see* Skaamoog
Sialia sialis **18**: 2457
Siamang **7**: 969, 970, 971
siblicide
crowned eagle **5**: 616
electric eel **6**: 763
siblings, killing of *see* siblicide
Sicklebill **9**: 1255
Sidewinder (horned rattlesnake;
horned rattler) **17**: 2369–2370
Sierra *see* Snoek
Sifaka **17**: 2371–2373
Diadem **17**: 2372
Tattersall's (golden-crowned
sifaka) **17**: 2371, 2372
Verreaux's **17**: 2371, *2371*, 2372,
2372, 2373
Siganus **15**: 2094, 2095, *2095*
Sigmoceros lichensteini **8**: 1148
sign stimuli **18**: 2534
Sika **17**: 2374–2375
Siliquaria **21**: 2989
silk
orb spider **13**: *1801*, 1802, *1802*

silk moth **17**: 2376, *2377*, 2378
water beetle **20**: 2864
water spider **20**: 2876
Silk moth (*larva*: silkworm) **3**: *345*;
17: 2376–2378
Eri **17**: 2377
Indian **17**: 2377
Mulberry **17**: *2376*
Silkworm *see* Silk moth
Silly John *see* Puffbird
Silurus see Catfish, European
Silurus glanis **3**: 402; **6**: 760
Silverbill **11**: 1544
Silvereye (a waxbill species) **21**: 2888
Silver-eye (a white-eye species) **21**: 2930
Silverfish **2**: 279
Silverside (jacksmelt; sand smelt)
17: 2379–2380
Inland (tidewater silverside)
17: 2379, 2380
Peruvian **17**: 2380
Samoan (whitebait) **17**: 2379
Silver Tommy *see* Topshell, Painted
Silver Y **13**: 1821
Singa **13**: 1801
singing *see* attracting a mate with
sound; songs
Singsing *see* Waterbuck
Sinobdella **18**: 2490
Siphamia **3**: 390
versicolor **3**: 389
Siphonogorgia **17**: *2440*
siphuncle **12**: 1723
Sipyloidea sipylus **18**: 2532
Siren
intermedia **17**: 2382
lacertina **17**: 2382
Siren (*amphibian*; mud eel)
17: 2381–2382
Greater **17**: 2381, 2382
Lesser **17**: *2381*, 2382, *2382*
Northern dwarf **17**: 2381, 2382
Southern dwarf **17**: 2381, 2382
Siren (*mammal*) *see* Dugong
Sirex **9**: 1279
gigas **16**: 2240
Sirpus **4**: 571
Siskin **3**: 368; **8**: 1016
Pine (pine linnet) **11**: 1457
Sistrurus **3**: 426; **15**: 2119
catenatus **15**: 2120
miliarius **15**: 2120
Sitatunga **1**: *63*; **3**: 331; **20**: 2760
Sitophilus granarius **21**: 2904
Sitta **13**: 1760, 1761, 1762, *1762*
size, body **21**: 2930
Skaamhaai *see* Skaamoog
Skaamoog (shy eye; skaamhaai; izak;
lazyshark; leopardshark;
tigershark) **3**: 407, 408
Skate **17**: 2329–2330, 2333, 2383–2385
Common (blue skate; grey skate)
17: 2383, 2384, 2385
Deep-sea **17**: 2383
Thornback **17**: *2383*
Skimmer **16**: 2266; **17**: 2386–2388
African **17**: 2386
Black **17**: 2386, 2387–2388, *2387*,
2388
Indian **17**: 2386
skin
breathing through **6**: 745; **11**: 1505;
13: 1867
volar **21**: 2987
Skink **17**: 2389–2390
Bermuda rock **17**: 2390
Burrowing **11**: *1467*
Five-lined **17**: 2390
Florida sand **17**: 2390
Giant **17**: 2389, 2390

Great Plains **17**: 2389, 2390
Kalahari burrowing **17**: 2390
Keel-bearing **17**: 2389
Many-lined **17**: 2389, 2390
Sand **17**: 2389, *2390*
Snake-eyed **17**: 2389
Skipjack *see* Click beetle
Skipper (*butterfly*) **17**: 2391–2392
  Dingy **17**: 2391
  Essex **17**: *2391*, 2392
  Giant **17**: 2392
  Great yucca **17**: 2392
  Lulworth **17**: *2392*
  Regent **17**: 2392
Skipper (*fish; saury pike*) **17**: 2368
Skua **1**: 29; **10**: 1307, 1308; **21**: 3004
  Arctic *see* Jaeger, Parasitic
  Chilean **10**: 1307, 1309
  Great (bonxie; Falkland skua)
    10: 1307, *1308*, *1309*
  South Polar **10**: 1307–1308, 1309
Skunk **3**: 318; **17**: 2393–2395
  Hog-nosed **17**: 2393–2394, 2395
  Hooded **17**: 2393, 2394
  Spotted **17**: 2393, 2394, 2395
  Striped **17**: 2393, 2394, *2394*, 2395, *2395*
Skylark (northern skylark)
    17: 2396–2397; **18**: 2459
  Small (oriental skylark) **17**: 2396
Slater (shore-roach; quay louse; quay
    lowder) **17**: 2398–2399
  see also Wood louse
slave-making ants **1**: 69
Sleeper *see* Dormouse, Hazel
sleeping sickness (trypanosomiasis)
    6: 859; **20**: 2760
Slider **13**: 1851; **20**: 2796
Slipper shell *see* Limpet, Slipper
Sloth **17**: 2400–2402
  Three-toed **15**: *2113*; **17**: 2400, 2401, 2402
  Two-toed **17**: 2400, 2401, 2402
slow moving
  beluga whale **2**: 193
  boa constrictor **2**: 250
  chameleon **3**: 418
  cone shell **4**: 519
  dragonet **5**: 706
  dugong **6**: 724
  fire salamander **6**: 831
  gila monster **7**: 972
  giraffe **7**: 975–976
  loris **11**: 1491
  manatee **11**: 1534
  mantis **11**: 1547
  porcupine fish **15**: 2022
  potto **15**: 2034
  pufferfish **15**: 2063
  right whale **16**: 2167, 2168
  sloth **17**: 2400
  sunfish **13**: 1775
  tortoise **19**: 2711–2712
  trunkfish **20**: 2758
Slowworm (blind worm) **10**: 1430;
    17: 2406–2407
  Blue-spotted **17**: 2406
Slug **17**: 2408–2409
  Banana **17**: 2408
  Gray field **17**: 2408
  Great gray **17**: 2409
  Keelback **17**: 2408, 2409
  Roundback **17**: 2408, 2409
  Shelled **17**: 2408
  Tree **17**: 2408
Slugworm **16**: 2240
smell (for defense)
  ermine **6**: 785
  fire-bellied toad **6**: 825
  fulmar **7**: 913, 915

grass snake **8**: 1044
palm civet **13**: 1856
parrotfish **14**: 1890
rat snake **15**: 2118
skunk **17**: 2393, 2395, *2395*
water snake **20**: 2874
whirligig beetle **21**: 2922
zorilla **21**: 3020, 3021
see also chemical defense
smell (scent) *see* attracting a mate
    with scent; scent marking
Smelt **17**: 2410–2411
  European **17**: 2410, *2410*
  Longfin **17**: 2410
  Sand *see* Silverside
  Surf **17**: 2410
Smerinthus ocellata **9**: 1168
Smew **12**: 1599, 1600, 1601
Sminthillus limbatus **8**: 1066
Sminthopsis
  crassicaudata **11**: 1577
  douglasi **11**: 1578
Sminthurides **18**: 2505
Sminthurus viridis **18**: 2504
Smithorius rufolateralis **2**: 282
smolt, salmon **1**: 101
Snail
  Bladder **14**: 2010
  Garden (brown garden snail;
    common snail) **7**: 930–932
  Giant **7**: 966–967
  Glutinous **14**: 2010, 2011
  Pond *see* Pond snail
  Purple sea *see* Sea snail, Purple
  River **14**: 2010
  Trumpet *see* Ramshorn
  Tube *see* Worm shell
  Valve **14**: 2010
  Violet *see* Sea snail, Purple
  Violet sea *see* Sea snail, Purple
  Wheel *see* Ramshorn
Snailfish **11**: 1499
Snake
  Aesculapian **15**: 2117, 2118
  Brown vine **19**: 2733
  Chain **10**: 1362
  Chicken *see* Rat snake
  Coral *see* Coral snake
  Corn (Great Plains rat snake; rosy
    rat snake) **15**: 2117, 2118
  Dice **20**: 2874, 2875
  Egg-eating *see* Egg-eating snake
  Elephant's trunk **20**: 2874, 2875
  False coral *see* Milk snake,
    Louisiana
  File **17**: 2419; **20**: 2874, 2875
  Four-lined **15**: 2117
  Fox (hardwood rattler) **15**: 2117,
    2118
  Garter *see* Garter snake
  Glossy **17**: 2370
  Grass (water snake; ringed snake)
    8: 1042–1044; **9**: 1203; **15**: 2103;
    17: *2420*; **20**: 2874, 2875
  Hognose *see* Hognose snake
  King *see* King snake
  Ladder **15**: 2117
  Leopard **15**: 2117
  Mexican burrowing **17**: 2419
  Milk *see* Milk snake
  Naked *see* Caecilian
  Pilot *see* Rat snake
  Rat *see* Rat snake
  Ribbon *see* Garter snake
  Rough green (vine snake) **19**: 2733
  Salt-and-pepper *see* King snake,
    Florida
  Sea *see* Sea snake
  Spotted bush **19**: 2733
  Striped house *see* Rat snake

Thread (flowerpot snake) **17**: 2421
Tree *see* Tree snake
Viperine **20**: 2874, *2874*, 2875
Water *see* Snake, Grass; Water
    snake
Whip *see* Whip snake
Yellow-and-green *see* Whip snake,
    Dark green
see also Adder; Anaconda; Asp;
    Boa constrictor; Boomslang;
    Cobra; Fer-de-lance; Mamba;
    Python; Racer; Rattlesnake;
    Sidewinder; Snakes; Viper
Snakebird *see* Anhinga, American;
    Wryneck
snake-charming **4**: 482
Snake-eagle, Brown *see* Eagle, Brown
    harrier
Snakehead **17**: 2412–2413
  Giant (Indian snakehead; great
    snakehead) **17**: 2412, 2413
Snake mackerel **17**: 2414–2415
Snake-necked turtle **17**: 2416–2417;
    20: 2795
  Common (eastern snake-necked
    turtle; long-necked tortoise)
    17: 2416, *2416*, 2417
  Giant *see* Turtle, Broad-shelled
    river
Snakes 17: 2418–2422
snakes
  fangs **2**: 264; **13**: 1736; **17**: 2420,
    2421
  venomous **2**: 265; **20**: *2818*, 2820
Snake-weasel **21**: 3020
Snapper (*fish*) **4**: 532; **17**: 2423–2424
  Bluestripe **17**: *2424*
  Emperor red **17**: 2423
  Gray **17**: *2423*
  Lane **17**: 2423
  Mangrove red **17**: 2423
  Mutton **17**: 2423
  Yellow **17**: *2423*
  Yellowtail **17**: 2424
Snapper (*snake*), Black *see*
    Massasauga
Snapper (*turtle*) *see* Snapping turtle
Snapping turtle **9**: 1190; **12**: 1688;
    17: 2425–2426
  Alligator (alligator snapper)
    17: 2425, 2426; **20**: 2797, 2798
  Common (snapping turtle;
    snapper) **17**: 2425, 2426, *2426*
Snipe **17**: 2427–2429
  Chatham Island **17**: 2429
  Common (Wilson's snipe; fantail
    snipe) **17**: 2427, 2428, *2428*, 2429
  Great **17**: 2427, 2429
  Jack **17**: 2427
  Magellan **17**: *2427*
  Noble **17**: 2429
  Painted *see* Painted snipe
  Pintail **17**: 2429
  Seed *see* Seed snipe
Snipefish **20**: 2766
Snoek (barracouta; sierra; couta; sea
    pike; snook) **17**: 2414, 2430–2431
Snook **17**: 2430
Snow-finch **9**: 1247
Snow leopard (ounce; xue bao;
    barfani chita; sarken) **3**: 404,
    406; **12**: *1670*, 1672;
    17: 2432–2434
Soapfish **8**: 1081
sociable *see* living in groups
Softshell *see* Soft-shelled turtle
Soft-shelled turtle (pancake turtle;
    flapjack turtle; softshell)
    17: 2442–2443; **20**: 2796
  Asian giant **17**: 2442, 2443

Black **17**: 2443; **20**: 2798
Burmese peacock softshell
    17: 2443
Chinese **17**: 2442
Florida **17**: 2442, 2443, *2443*
Smooth **17**: 2442, 2443
Spiny **17**: 2442, *2442*, 2443
Soldierfish **4**: 532
Sole **6**: 833; **17**: 2444–2445
  Common (Dover sole) **17**: 2444,
    2445, *2445*
  Deepwater **17**: 2444
  Naked **17**: 2444
  Sand **17**: 2444
Solea solea **17**: 2445
Solenodon
  arredondoi **18**: 2452
  cubanus **18**: 2452
  marcanoi **18**: 2452
  pardoxus **18**: 2452, *2452*
Solenodon **18**: 2452
  Haitian **18**: *2452*
solenoglyphs **13**: 1736
Solifugid (false spider; sun spider;
    wind scorpion; camel spider;
    solipugid) **18**: 2453–2454
  Giant **18**: *2453*
Solipugid *see* Solifugid
Somateria
  mollissima **6**: 754, 755
  spectabilis **6**: 754
sonar **14**: 1953; **16**: 2283
Song babbler **1**: 113
Songbirds 18: 2455–2459
songs
  canary **3**: 368
  cardinal **3**: 387, 388
  chaffinch **3**: 415, *415*
  European robin **16**: 2191–2192
  katydid **10**: 1341–1342, *1342*
  nightingale **13**: 1740, 1741
  oropendula **13**: 1805
  pipit **14**: 1965
  rorqual **16**: *2209*
  skylark **17**: 2396, 2397, *2397*
  songbird 18: 2457
  song sparrow **18**: 2460
  thrasher **19**: 2672–2673
  thrush **19**: 2678–2679
  vireo **20**: 2831–2832
  whistler **21**: 2923
  wren **21**: 2995
Sora (sora rail; Carolina crake) **5**: 583
Sorex **17**: 2358, 2359; **20**: 2872, 2873
sounds *see* attracting a mate with
    sound; ultrasonic sounds;
    warning sounds
soup
  bird's nest **19**: 2601
  sharkfin **2**: 162; **17**: 2333
Sow bug *see* Wood louse
Spadefoot toad **18**: 2465–2466
  Common (garlic toad) **18**: 2465,
    2466
  Couch's **18**: 2466
  Eastern (Hunter's spadefoot toad)
    18: 2465
  Hammond's **18**: 2466
  Mexican **18**: 2465
  Plains **18**: 2466
  Western **7**: 904
Sparisoma **14**: 1888
Sparrow
  Barley *see* Mannikin, Chestnut-
    breasted
  Chestnut **9**: 1249
  Chipping **18**: 2460
  Desert **9**: 1247
  Grasshopper **18**: 2460
  Hedge *see* Accentor, Hedge

Sparrow (*continued*)
House 7: 999; 9: 1247–1249;
18: 2457, 2459
Ipswich 19: 2649
Java (Java finch; Java temple bird;
ricebird) 10: 1315–1316; 11: 1544
Lark 18: 2460
Song 4: 555; 18: 2460–2461
Spanish 9: 1247, 1249
Tree 9: 1247
Sparrow hawk 18: 2467–2469
Levant 18: 2467
Spathius exarator 4: 471
spats 4: 489
Spearfish
Longbill 11: 1561
Shortbill 11: 1561
speciation, sympatric 5: 650
species 21: 2924
cryptic (sibling species) 20: 2859
speculum 14: 1957; 19: 2708
speed *see* fast moving; slow moving
Speleomantes 11: 1503, 1504
Speothos venaticus 3: 334; 17: 2357
spermaceti 18: 2473, 2474
spermatathecae 6: 737
spermatophores 1: 109; 5: 600; 6: 831;
10: 1387, 1428; 12: 1620–1621;
13: 1734, 1781, 1788; 14: 1907;
18: 2505, 2510; 20: 2769–2770
Spermophilus 8: 1040, 1078, 1079
Sperm whale 13: 1772, *1772*;
18: 2472–2474, 2510; 21: 2911,
2913–2914
Dwarf 18: 2473; 21: 2911
Pygmy 18: 2472, 2473; 21: 2911
Sphaeramia, nematoptera 3: *390*
Sphaerodactylus 11: 1466
Sphecia bembeciformis 9: 1234
Sphecotheres
viridis 8: 1012
viridis flaviventris 8: 1012
Spheniscus 14: 1914
demesus 14: 1911
mendiculus 14: 1911
Sphenodon
guntheri 20: 2763, 2764
punctatus 20: 2763, 2764
Sphiggurus 19: 2729
Sphinx ligustri 9: 1166
Sphyraena 2: 155
Sphyrapicus
thyroideus 16: 2235
varius 16: 2235, 2236
Sphyrna 8: 1126, 1127
Spider
Bird-eating (goliath tarantula;
mygale) 2: 203–204; 19: *2634*
Bolas 13: 1801–1802; 18: 2482
Burrowing 18: 2480
Camel *see* Solifugid
Cellar (daddy longlegs) 5: 636;
18: 2482
Crab 5: 580–581; 18: 2480, 2482
Cross *see* Garden spider,
European
Diadem *see* Garden spider,
European
False *see* Solifugid
Fishing 18: 2482
Funnel web *see* Funnel-web spider
Garden *see* Garden spider
Hourglass *see* Black widow
Jumping 18: 2480, *2480*, 2481, 2482
Lynx 18: 2481
Mothercare 18: 2483
Net-casting 18: 2482
Nursery-web 18: 2483; 21: 2957
Orb *see* Orb spider
Orb-web *see* Orb spider

Purse-web 19: 2719, 2720
Red back *see* Black widow
Sheetweb (tunnel-web spider;
funnel-web spider)
17: 2339–2340
Shoe button *see* Black widow
Sun *see* Solifugid
Trapdoor *see* Trapdoor spider
Water (European water spider)
18: 2481–2482; 20: 2876–2877
Whip *see* Pedipalpi
White-tailed 20: 2818
Wolf 18: *2482*, 2483; 21: 2956–2957
see also False scorpion; Scorpion;
Spiders; Tarantula
Spider crab (slender-legged spider
crab; long-legged spider crab;
decorator crab) 4: 572–573;
17: 2355; 18: 2475–2476
Giant 4: 571; 18: 2475
Long-beaked 18: 2475, 2476
Spiny (thornback crab) 18: 2475
Spider monkey 12: 1644;
18: 2477–2479
Black-handed 12: *1641*
Common 18: 2477, 2478
Woolly 18: 2477, 2478
Spiders 18: 2480–2483
Spilocuscus
maculatus 5: 629–630
rufoniger 5: 630
Spilogale
gracilis 17: 2393, 2394
putorius 17: 2393
pygmaea 17: 2393, 2394
Spinachia spinachia 18: 2533
Spinebill 9: 1215
Spinefoot
Dusky 15: 2095
Little 15: 2094
Marbled 15: 2095
Orange-spotted 15: 2094
Streaked 15: 2094, 2095
Vermiculated 15: 2094, 2095
Spinetail (*an ovenbird species*;
castlebuilder) 13: 1819
Pale-breasted 13: 1819
Spinetail swift 18: 2484–2485
Batlike spinetail 18: 2484
Boehm's spinetail 18: 2484
spinnerets 18: 2482
Spinus pinus 11: 1457
Spiny anteater (long-beaked
echidna; short-beaked echidna;
long-nosed echidna; short-
nosed echidna; native
porcupine; spiny porcupine)
18: 2486–2488
Spiny eel 18: 2489–2490
Barred 18: 2490
Blind 18: 2490
Lesser (elephant trunk fish)
18: 2490
One-striped 18: 2490
Tiretrack 18: 2490
Spiny lobster (crawfish; langoustine)
4: 571–572, 573; 11: 1475, *1476*
California 11: 1475
Common 4: 573
West Indian 11: *1475*
Spiny mouse 18: 2491–2492
Cairo 18: 2491
Cape 18: 2491
Golden 18: 2491
Spiny porcupine *see* Spiny anteater
spiracles, backswimmer 1: 118
Spirinchus thaleichthys 17: 2410
Spirorbis 6: 801
borealis 6: 801
pagenstecheri 6: 801

Spittlebug *see* Hopper
Spiza americana 18: 2458
Spizella passerina 18: 2460
Splittail, Sacramento 12: 1618
Split-thumb *see* Shrimp, Mantis
Spodoptera littoralis 3: 346
Spondylus 13: 1832
Sponge 18: 2493–2495
Barrel 18: *2495*
Bath 18: 2493, 2494, 2495
Horny 18: 2493
Purse 15: 2074–2075
Red Sea 18: *2493*
Sponge fly (spongilla fly) 10: 1392,
1393
Spongia 18: 2493
officinalis 18: 2494
Spoonbill 18: 2496–2498
African 18: 2497, *2498*
Eurasian 18: 2496, *2497*, 2498
Lesser 18: 2496–2497
Roseate 18: 2496, *2496*, 2497;
21: 2907
Royal (black-billed spoonbill)
18: 2496–2497
Yellow-billed (yellow-legged
spoonbill) 18: 2496–2497
Spotfin *see* Bowfin
Spoutfish *see* Razor shell
Springbock *see* Springbok
Springbok (springbuck; springbock)
18: 2499–2501
Springbuck *see* Springbok
Springhare (jumping hare)
18: 2502–2503
Springtail 18: 2504–2505
Sprite *see* Damselfly
Squalodon 16: 2178
Squalus
acanthias 5: 686, 687; 17: 2333
suckleyi 5: 686
Squatina 12: 1646; 16: 2244; 17: 2385
argentina 12: 1646
californica 12: 1646
Squat lobster 4: 572; 18: 2506–2507
Common 18: 2506
Squawfish 12: 1618
Squid 18: 2508–2510
European 18: 2509
Flying 18: 2510
Giant 13: 1772; 18: 2473–2474,
2509–2510
Longfin 18: 2509
Pop-eyed 18: *2509*
Squilla *see* Shrimp, Mantis
Squilla
desmaresti 11: 1552
empusa 11: 1551, 1552
mantis 11: 1551, 1552
Squirrel
Douglas 4: 436, 437
Eurasian red 4: 436
Flying *see* Flying squirrel; Gliding
possum, Lesser
Gray 20: 2740, *2740*, 2741
Ground *see* Ground squirrel
Mearn's 4: 436, 437
Oil-palm (African giant squirrel)
13: 1859, 1860
Palm *see* Palm squirrel
Red 20: 2740, 2741, *2742*
Red (pine squirrel) 4: 436, 437,
438
Rock *see* Ground squirrel
Scaly-tailed *see* Scaly-tailed
squirrel
Tree *see* Tree squirrel
Squirrelfish 4: 532; 18: 2511–2512
Longspine 18: *2512*
Sabre 18: 2511, 2512

Squirrel monkey (titi; mono)
12: 1644; 18: 2513–2515
Gothic 18: 2513, *2513*, 2515
Roman 18: 2513, 2515
stabilimenta 13: 1801; 18: 2482
Stagemaker *see* Catbird,
Tooth-billed
Stagmomantis carolina 11: 1546
stalking
bittern 2: 221
cheetah 3: 428
cuttlefish 5: 633
darter 5: 647
gar 7: 928
John Dory 10: 1327–1328, *1327*
lion 11: 1462
marsupial cat 11: 1572
night heron 13: 1738
ocelot 13: 1778
puma 15: 2069–2070
stamping behavior 11: 1451
Star *see* Starfish
Starfish 18: 2516–2518
Blood sea star (bloody Henry)
18: *2517*, 2518
Burrowing star (sand star)
18: 2517, 2518
Crown-of-thorns 18: 2517, 2518,
2518
Cushion star (starlet) 18: 2516,
2518
Scarlet nubbly 18: *2516*
Stargazer 18: 2519–2520
Atlantic 18: 2520
Mediterranean 18: 2520
Northern 18: 2519
Starling 18: 2521–2523
Bali *see* Mynah, Rothschild's
Common (European starling)
6: 855; 18: 2521–2523, *2521*, *2522*,
2523
Glossy 18: 2521
Rose-colored 18: 2521
Rothschild's *see* Mynah,
Rothschild's
Wattled 18: 2521
stationary/sedentary lifestyle
anemone 1: 47
barnacle 1: 138–139
basket star 2: 158
dead man's fingers 5: 651
fanworm 6: 801
mussel 12: 1711
oyster 13: 1832, 1833, *1833*, *1834*
sea lily 16: 2280
sea squirt 17: 2312
sponge 18: 2493, 2494
worm shell 21: 2989
statocysts 13: 1795
Stauropus fagi 15: 2077
Steamer duck 5: 716; 18: 2524–2525
Flightless (Falkland Island
flightless steamer duck; logger
duck) 18: 2524, *2524*, 2525, *2525*
Flying 18: 2524–2525
Magellanic flightless 18: 2524,
2525
Steamfly *see* Cockroach,
German
Steatornis caripensis 13: 1785
Steenbok (steinbuck; steinbok;
steenbuck) 18: 2526–2527
Steenbuck *see* Steenbok
Steephead *see* Whale, Northern
bottlenose
Steganopus tricolor 14: 1931
Stegostoma fasciatum 21: 3015
Steinbok *see* Steenbok
Steinbuck *see* Steenbok
Steinernema carpocapsae 4: 540

Stenella
  attenuata **5**: 698
  coeruleoalba **5**: 698
  frontalis **5**: 698
  longirostris **5**: 698
Stenobothrus **8**: 1036
  lineatus **8**: 1037
Stenophylax **3**: 354
Stenoptilia pneumonanthes
  14: 1991–1992
Stenorhynchus seticornis **18**: *2475*
Stephanoaetus coronatus **5**: 615; **8**: 1093
Stephopoma **21**: 2989
steppes **8**: 1038
Stercorarius
  longicaudus **10**: 1307
  parasiticus **10**: 1307, 1308
  pomarinus **10**: 1307
Stereolepis gigas **2**: 164
Sterlet **18**: 2556
Sterna **16**: 2264, 2265, 2268;
  19: 2661, 2662, *2663*
Sternoptyx **9**: 1162
  diaphana **9**: 1162
Sternotherus **20**: 2795
  carinatus **12**: 1687
  odoratus **12**: 1687, 1688
Stethophysma grossum **8**: 1036
Sthenoteuthis **18**: 2508
Stichodactyla haddoni **5**: 640
Stichopus
  chloronotus **14**: 1901
  regalis **14**: 1902
Stick insect **18**: 2530–2532
  Giant prickly **18**: 2531, *2531*
  Laboratory **18**: 2530, 2531–2532
  Madagascar **18**: 2532
  Prickly **18**: 2530, 2532
  Smooth **18**: 2530, 2531, 2532
  Water *see* Water stick insect
Stickleback **6**: 836; **18**: 2533–2534
  Brook **18**: 2533
  Four-spined **18**: 2533
  Nine-spined **18**: 2533
  Sea (15-spined stickleback)
    18: 2533–2534
  Three-spined **18**: 2533, 2534
Stick-nest rat **18**: 2535–2536
  Greater **18**: *2535*
Stictocarbo punctatus **16**: *2265*
Stictonetta naevosa **5**: 714
Stifftail **18**: 2537–2539
Stigmella ridiculosa **3**: 344
Stilt **18**: 2540–2542
  Banded **18**: 2540, 2541, 2542
  Black **18**: 2540, 2542
  Black-necked **18**: 2540, 2541, *2542*
  Black-winged (pied stilt; white-
    headed stilt; longshanks;
    stiltbird) **18**: 2540, *2540*, 2541,
    2541
Stiltbird *see* Stilt, Black-winged
Stiltia isabella **15**: 2041
Stingaree *see* Stingray, Southern
Stingfish *see* Weever
stinging
  anemone **1**: 47–48
  bees and wasps 2: 183
  bumblebee **3**: 318
  dead man's fingers **5**: 652
  hydra **9**: 1265, *1266*
  jellyfish **10**: 1322
  Portuguese man-of-war **15**: 2029,
    2030–2031, *2031*
  scorpion **16**: *2258*, 2259
  scorpionfish **16**: 2261
  sea pen **16**: *2297*
  stingray **17**: *2331*; **18**: 2543–2544
  stonefish **18**: 2547–2548
  velvet ant **20**: 2815

vespula wasp **20**: 2826
  weever **21**: 2901–2902
Stingray **17**: 2330; **18**: 2543–2544
  Black-spotted (giant reef ray)
    18: *2544*
  Blue-spotted **17**: *2331*; **18**: *2543*
  Southern (clam-cracker; stingaree)
    18: 2544
Stinkbug *see* Shieldbug
Stinking-jim *see* Musk turtle,
  Common
Stinkpot *see* Musk turtle, Common
Stint **16**: 2233
Stitchbird **9**: 1215
Stizostedion vitreum **14**: 1917, 1918
Stoat *see* Ermine
Stockfish **8**: 1119, 1120
Stoichactis **1**: 47, 48
Stomoxys calcitrans **9**: 1242–1243
Stone curlew (thick-knee)
  18: 2545–2546
  Beach **18**: 2545
  Bush thick-knee **18**: 2545, 2546
  Double-striped thick-knee **18**: 2545
  Eurasian **18**: 2545, *2545*, 2546
  Great thick-knee **18**: 2545
  South American thick-knee
    18: 2545–2546
  Water thick-knee (water dikkop)
    18: 2545, 2546
Stonefish **16**: 2260; **18**: 2547–2548;
  20: 2818
  Estuarine **18**: 2547, *2548*
Stonefly **18**: 2549–2550
Stone lily *see* Sea lily
Stoneroller, Central **12**: 1618
Stone sucker *see* Lamprey
Stork **18**: 2551–2553
  Adjutant **11**: 1553; **18**: 2551
  African wood **18**: 2552
  Black **18**: 2551, 2552, *2552*, 2553
  Black-necked **18**: 2551
  Jabiru **18**: 2551
  Lesser adjutant **11**: 1553
  Maguari **18**: 2551, 2553
  Marabou *see* Marabou
  Openbill **18**: 2551
  Painted **18**: 2552, *2553*
  Saddlebill **18**: 2551
  Shoebill (whale-headed stork; abu
    markub) **8**: 1123; **17**: 2351–2352;
    18: 2551
  White **18**: 2551, *2551*, 2552, *2552*,
    2553
  Wood (wood ibis) **18**: 2551, 2552
Stormfly *see* Thrips
Storm petrel (Mother Carey's
  chicken) **16**: 2266; **18**: 2554–2555
  European **18**: 2554, 2555
  Galapagos (wedge-rumped storm
    petrel) **18**: 2554, 2555
  Hornby's **18**: 2554
  Leach's **18**: *2554*
  Wilson's **18**: 2554, 2555, *2555*
stotting **1**: 61; **7**: 942; **13**: 1804
  see also pronking
Stout *see* Horsefly
strandings
  porpoise **15**: *2025*
  whale **6**: 797, 798; **14**: 1952
Strangalia **11**: 1484
Stream-creeper **13**: 1819
Streptopelia
  decaocto **4**: 500
  senegalensis **13**: 1858
  turtur **4**: 499; **20**: 2794
stridulation **3**: 315; **5**: 600; **6**: 817;
  7: 880; **8**: 1036, 1037; **9**: 1168,
  1168; **11**: 1483; **16**: 2254, 2258;
  20: 2815

Strigamia maritima **3**: 411, 412
Striped weasel **21**: *3021*
  Libyan **21**: 3020
Strix
  aluco **13**: 1827; **19**: 2645
  butleri **13**: 1827
  nebulosa **13**: 1827
  varia **13**: 1825
Stromb shell *see* Conch
Strombus **4**: 514–515
  gigas **4**: 514, 515
  goliath **4**: 514, 515
Strongylocentrotus **17**: 2315
Strongylura
  exilis **12**: *1725*
  marina **12**: 1725
Struthidea cinerea **11**: 1522
Strutio
  camelus **13**: 1815
  camelus syriacus **13**: 1814
Strymonidia **8**: 1118
Sturgeon **18**: 2556–2557
  Atlantic **18**: 2556, 2557, *2557*
  Russian (beluga) **18**: 2556
  Spoonbill *see* Paddlefish,
    Mississippi
  Swordbill *see* Swordfish, Chinese
  White **18**: 2556, 2557
Sturmia scutellata **19**: 2611
Sturnella neglecta **8**: *1039*
Sturnus
  pagodarum **12**: *1716*
  roseus **18**: 2521
  vulgaris **6**: 855; **18**: 2521, 2522
Stylatula elongata **16**: 2297
style **13**: 1833
Stylodipus **10**: 1324
Subcoccinella vigintiquatuorpunctata
  10: 1396
  Sucker, Marlin **15**: 2148
Suckerfish
  Slender **15**: 2148
  White **15**: 2148
Suez Canal, and rabbitfish **15**: 2095
Sugarbird **18**: 2558–2559
  Blue **18**: 2558
  Cape **18**: 2558, 2559
  Gurney's **18**: 2558
Suillotazus marchei **1**: 120
Sula **2**: 261, 262; **16**: 2264
Sulphur butterfly (yellow sulphur
  butterfly) **21**: 2929
  Orange sulphur **21**: 2929
Sunbird **7**: 868; **18**: 2456, *2458*,
  2565–2566
  Black **18**: 2565
  Collared **18**: *2566*
  Purple **18**: *2458*, 2566
  Pygmy **18**: 2455, 2565, 2566
Suncus **17**: 2358
  etruscus **12**: 1709
  murinus **12**: 1710; **17**: 2358
Sundarbans **21**: 2907
Sunfish (*freshwater*) **2**: 163, 164;
  18: 2569–2570
  Banded **18**: *2570*
  Blue spotted **18**: 2569
  Common (pumpkinseed) **16**: *2174*;
    18: 2569, *2569*, 2570
  Long-eared **18**: 2569
  Red-breast **18**: 2569
Sunfish (*marine*)
  Ocean **13**: 1774–1775
  Sharptail **13**: 1774
  Slender **13**: 1774
Sungrebe *see* Finfoot, American
Suni **1**: 61; **18**: 2571–2572
Sun spider *see* Solifugid
Surfbird **20**: 2790, 2791
Surffish *see* Surfperch

Surfperch (seaperch; surffish)
  18: 2573–2574
  Midnight **18**: *2573*
  Pink **18**: 2573
  Yellow-banded seaperch (shiner)
    18: 2574
Surgeon, Atlantic Ocean **18**: 2577
Surgeonfish **4**: 532; **18**: 2575–2577
  Five-banded **18**: 2575–2576
  Palette **18**: 2577
Suricata suricatta **12**: 1596
Suricate *see* Meerkat
Surmullet *see* Mullet, Striped red
Suslik *see* Ground squirrel
Sus scrofa **19**: 2653; **21**: 2907, 2943
Sutorectus tentaculatus **21**: 2955
Swallow
  Bank (sand martin) **2**: 151;
    12: 1590, *1592*
  Barn (swallow; chimney swallow;
    house swallow) **2**: 151–153
  Cliff **2**: 151
  Golden **2**: 151
  Mosque **2**: 151
  Red-rumped **2**: 151
  Rufous-chested **2**: 151
  Sea *see* Tern
  Tree **2**: 151
  Wood *see* Wood swallow
Swallower, Black **13**: *1773*
Swallow plover *see* Pratincole
Swallow-shrike *see* Wood swallow
Swallowtail **18**: 2580–2582
  African **18**: 2582
  Asian **18**: 2582
  Birdwing *see* Birdwing
  Citrus (orchard swallowtail)
    18: 2581
  Common **18**: 2580, 2581
  European **3**: *344*
  Five-bar **18**: *2582*
  Giant **18**: 2581
  Giant African **18**: 2581, 2582
  Kite **18**: 2582
  Oriental **18**: 2582
  Royal **18**: 2582
  Ruby-spotted **18**: 2581
  Scarce **18**: 2580
  Tiger **18**: *2580*, 2581
Swallow-tanager **18**: 2583–2584
Swallow-wing **15**: 2060
Swamp cuckoo *see* Coucal, Pheasant
Swamp eel **18**: 2585
Swamphen, Purple *see* Gallinule,
  Purple
swamps **21**: 2905, 2908
Swan **18**: 2586–2589
  Black **5**: *714*; **18**: 2587, *2588*
  Black-necked **18**: 2587
  Coscoroba **18**: 2586; **21**: 2925, 2927
  Mute **5**: 715; **18**: 2586, *2586*, 2588,
    2589
  Trumpeter **5**: 714, 715, 717;
    18: 2587, *2587*, 2588; **20**: 2757
  Tundra (Bewick's swan; whistling
    swan) **5**: 717; **18**: 2586–2587;
    20: 2779
  Whooper **5**: 714; **18**: 2586, 2588,
    2589
  see also Ducks, geese and swans
Sweeper
  Black-margin **17**: *2441*
  Copper **17**: *2441*
Sweetfish (sweetmouth; ayu; koayu)
  19: 2596–2597
Sweetmouth *see* Sweetfish
Swellfish *see* Pufferfish
Swift (*bird*) **19**: 2598–2600
  African white-rumped **19**: 2598
  Alpine **19**: 2598

Swift (*bird; continued*)
  Asian white-rumped **19**: 2598
  Batlike (Boehm's spinetail)
    18: 2484
  Black **19**: 2598
  Chimney **18**: 2484, 2485, *2485*;
    19: 2598
  Common **19**: 2598, *2598*, 2599,
    2600
  Crested **19**: 2598
  Edible-nest **19**: 2601
  Great dusky **19**: *2599*
  Pallid **19**: 2598
  Palm (*New World*) **19**: 2598, 2600
  Palm (*Old World*) **19**: 2598, 2600
  Scissor-tailed **19**: 2600
  Spinetail *see* Spinetail swift
  True **19**: 2598
  Vaux's **18**: 2484, 2485; **19**: 2598
  White-throated **19**: 2598, 2600
Swift (*lizard*) *see* Fence lizard
Swiftlet **19**: 2601–2602
  Black-nest **19**: 2601
  Cave **19**: *2601*
  Glossy (white-bellied swiftlet)
    19: 2601–2602
  White-nest **19**: 2601
Swift moth **19**: 2603–2604
  Common **19**: 2603, *2604*
  Gold **19**: 2603
  Map-winged **19**: 2604
swim-bladder
  for breathing **6**: 804, *804*; **10**: 1376;
    12: 1680; **14**: 1917
  for buoyancy control **4**: 473
    helping to make sounds **1**: 88;
      5: 601, *601*; **8**: 1109, *1109*;
      19: 2698
  paired **2**: 195
  transmitting sound waves **3**: 395;
    18: 2511
Swordfish (broadbill) **19**: 2605–2607
Swordfish, Chinese (*a paddlefish
  species*; swordbill sturgeon)
  13: 1845
Swordtail **14**: 1984; **19**: 2608–2609
Sygnathus
  acus **14**: *1961*
  floridae **14**: 1960
  leptorhynchus **14**: 1961
  typhle **14**: 1961
Sylvia
  atricapilla **20**: 2857
  communis **18**: 2459
  undata **20**: 2857, 2858
Sylvicapra **6**: 727
  grimmia **6**: 726, 727
Sylvilagus **3**: 426; **4**: 546
Sylviorthorhynchus desmursii **13**: 1819
symbiosis *see* mutualism
symmetry, radial **6**: 806
Sympetrum
  flaveolum **5**: 709
  fonscolombei **5**: 709
  striolatum **5**: 710
Symphasis varia **11**: 1550
Symphodus melops **21**: 2991
Symphysodon **4**: 459
  aequifasciatus **14**: 2006
  discus **14**: 2006, 2007
Synallaxis **13**: 1819
  subpudica **13**: 1819
Synanceia **5**: 668
  horrida **16**: 2260; **18**: 2547
  verrucosa **18**: 2547, 2548
Synaptomys **10**: 1433; **20**: 2778
Syncerus caffer **3**: 371
Sypheotides indica **3**: 338
syrinx **3**: 340; **14**: 1884; **18**: 2455, 2458
Syrmaticus **14**: 1935

Syrphus **9**: 1251
  balteus **9**: 1250
  ribesii **6**: *857*, 858
Syrrhaptes paradoxus **16**: 2229

T

Tabanus **9**: 1235, 1236, *1236*
  distinguendus **6**: *856*
  sudeticus **9**: 1235
Tabon *see* Scrub fowl, Philippine
Tachornis squamata **19**: 2598
Tachpleus **10**: 1354
Tachuris rubigastra **20**: 2800
Tachybaptus ruficollis **8**: 1062
Tachycineta
  bicolor **2**: 151
  euchrysea **2**: 151
Tachyeres
  brachypterus **5**: 716; **18**: 2524, 2525
  leucocephalus **5**: 716; **18**: 2524
  patachonicus **5**: 716; **18**: 2524
  pteneres **5**: 716; **18**: 2524
Tachyglossus **18**: 2486
  aculeatus **18**: 2487
Tadorna
  ferruginea **17**: 2341
  tadorna **17**: 2342
  variegata **17**: 2341
Tadorna cristata **5**: 717
tadpoles **4**: 465, 511; **9**: 1230; **11**: 1574;
  12: 1610, 1679, 1721;
  18: 2578–2579
  fastest developing **18**: 2466
  larger than the frog **14**: 1876–1877
Taeniopygia **21**: *3012*
  guttata **21**: 3013
Taeniura
  lymma **18**: *2543*
  melanospilos **18**: *2544*
Tahr **19**: 2612–2613
  Arabian **19**: 2612, 2613
  Himalayan **19**: 2612, 2613, *2613*
  Nilgiri **19**: 2612, *2612*, 2613
Taiga 19: 2614–2617
tail
  loss (autotomy), by lizards **4**: *453*,
    454; **7**: 945, 981, *981*
  prehensile **5**: 629; **9**: 1254; **10**: 1363,
    1363; **15**: *2079*; **16**: *2277*;
    18: 2477–2478; **21**: 2987
Tailorbird **19**: 2618–2619
  Ashy **19**: 2618
  Ashy-naped **19**: 2618, 2619
  Australian **19**: 2618
  Black-necked **19**: 2618
  Common **19**: 2618, 2619, *2619*
Taipan **5**: 665; **20**: *2817*
Takahe **19**: 2620–2621
Takin **19**: 2622–2623
Talpa
  caeca **12**: 1625
  europaea **12**: 1625, 1626
  micrura **12**: 1625
  streeti **12**: 1627
Tamandua
  mexicana **1**: 57
  tetradactyla **1**: 57
Tamandua
  Northern **1**: 57, 58, 59, *59*
  Southern **1**: 57, 58, 59
Tamarau **3**: 305, 307; **9**: 1286–1287
Tamarin (marmoset—*incorrectly*)
  12: 1644, 1645; **19**: 2624–2626
  Black-and-gold lion **19**: 2625
  Black-chested mustached **19**: 2624
  Black-faced golden lion **19**: 2625
  Black-mantle **19**: 2624
  Brown-headed (saddle-back
    tamarin) **19**: 2624

  Emperor **12**: 1645; **19**: 2624, *2625*
  Geoffroy's **19**: 2625
  Golden lion **19**: *2624*, 2625
  Golden-rumped lion **19**: 2625
  Lion **19**: 2625, 2626
  Long-tusked **19**: 2625
  Mottle-faced **19**: 2625
  Pied **19**: 2625
  Pinché (cotton-top tamarin)
    12: *1642*; **19**: 2625, *2626*
  Red-chested mustached
    19: 2624
  Red-handed (black tamarin; midas
    tamarin) **19**: 2624, 2626
  Saddleback **12**: *1645*
  White-footed (silvery-brown bare-
    face tamarin) **19**: 2625
Tamiasciurus
  douglasii **4**: 436, 437
  hudsonicus **4**: 436, 437
  mearnsi **4**: 436, 437
Tamias striatus **4**: 448, 449; **19**: 2653
Tanager **18**: 2456, 2458; **19**: 2627–2629
  Azure-winged **19**: 2627
  Blue-gray **19**: 2627, *2627*, 2628,
    2629
  Blue-winged mountain **19**: 2628
  Gray-headed **19**: 2629
  Hepatic **19**: 2627
  Magpie **19**: 2627
  Masked **19**: 2627, 2629
  Red-rumped paradise **19**: 2627
  Scarlet **18**: *2456*, 2457; **19**: 2627,
    2628, 2629
  Scarlet-rumped black **19**: 2629
  Speckled **19**: *2629*
  Summer **19**: 2627
  Western **19**: 2627
Tang
  Blue **18**: 2576, *2576*
  Yellow **18**: 2575, *2577*
Tanganyika, Lake **10**: 1398
Tangara **19**: 2627
  chilensis **19**: 2627
  larvata **19**: 2629
  nigrocincta **19**: 2627
Tanysiptera galatea **10**: 1356
Tanytarsus **12**: 1607
Tapera naevia **5**: 619
tapetum lucidum **3**: *330*, 403;
  21: 3007–3008
Tapeworm, Common **6**: 852
Taphozous **19**: 2703
  mauritianus **19**: *2703*
Tapir **19**: 2630–2632
  Baird's **19**: 2630, 2631, *2631*, 2632
  Malayan **19**: 2630, *2630*, 2631, 2632
  Mountain (Andean tapir) **19**: 2630,
    2631, 2632
  South American (Brazilian tapir;
    lowland tapir) **19**: 2630, 2631,
    2632
Tapirus
  bairdii **19**: 2630, 2631
  indicus **19**: 2630, 2631
  pinchaque **19**: 2630, 2631
  terrestris **19**: 2630, 2631
tarantism **19**: 2634
Tarantula **2**: 203; **19**: 2633–2635;
  21: 2956
  Goliath *see* Spider, Bird-eating
  Mexican superb **19**: *2633*
Taricha **20**: 2819
  torosa **13**: 1733, 1734; **19**: *2653*
Tarpan (*extinct*) **15**: 2052
Tarpon **19**: 2636–2637
Tarsier **19**: 2638–2640
  Horsfield's **19**: 2638, 2639, *2639*
  Philippine **19**: 2638, *2638*, 2639,
    2640

  Spectral (Celebes tarsier) **19**: 2638,
    2639–2640
Tarsius
  bancanus **19**: 2638, 2639
  dianae **19**: 2638, 2639
  pumilus **19**: 2638, 2639
  spectrum **19**: 2638, 2639
  syrichta **19**: 2638, 2639
Tasmacetus sheperdi **2**: 170
Tasmania **19**: 2643
Tasmanian devil (devil) **5**: 677;
  11: 1571; **19**: 2641–2643
Tasmanian wolf (thylacine; *extinct*)
  5: 677
Tatera indica **7**: 950
Tattler
  Gray-tailed *see* Sandpiper, Gray-
    rumped
  Wandering **16**: 2233
Tauraco
  corythaix **20**: *2781*
  hartlaubi **20**: 2780
  macrorhynchus **20**: 2780
  ruspolii **20**: 2780
  schuttii **20**: 2781
Taurotragus **7**: 942
  derbianus **1**: 60; **6**: 757, 758
  oryx **6**: 757, 758
Taxidea taxus **1**: 120
Tayassu pecari **14**: 1903, 1904
Tchagra **17**: 2361
Teal **5**: 715; **19**: 2647–2649
  Andaman **19**: 2648
  Aukland Islands **19**: 2647
  Baikal **19**: 2647, 2649
  Blue-winged **19**: *2647*
  Brown (Auckland teal) **19**: 2649
  Campbell Island **19**: 2647, 2649
  Cape **19**: 2647
  Chestnut **19**: 2647
  Gray **19**: 2647, 2648
  Green-winged (teal) **19**: 2647,
    2648, *2648*
  Hottentot **19**: 2647
  Laysan **19**: 2648, 2649
  Madagascar **19**: 2649
  Marbled **19**: 2649
  Speckled (sharp-winged teal)
    19: 2647, *2649*
teddy bears **2**: 223
teeth
  aardvark **1**: 17
  buffalo **3**: 305
  camel **3**: 361
  capybara **3**: 382
  carnivore **2**: 202
  chickaree **4**: 438
  Chinese water deer **4**: *446*
  coypu **4**: 561
  dogs 5: 690
  eagle ray **6**: 731
  elephant **6**: 769
  flying lemur **7**: 882
  frilled shark **7**: 896–897
  guitarfish **8**: 1103
  insectivore **2**: 202
  mole rat **12**: *1629*
  palm civet **18**: 1855–1856
  pharyngeal (throat) **3**: 395; **5**: 638;
    6: 860; **12**: 1618; **14**: 1889, 1976,
    1977; **16**: *2182*
  pike **14**: *1946*, 1947
  piranha **14**: 1967–1968, *1967*
  porcupine **5**: 595–596
  rock wallaby **16**: 2198
  rodent **4**: 438
  seal **4**: 564
  shark 17: 2331–2332
  tree squirrel **20**: 2741, 2742
  see also carnassials; tusks

Tegenaria **17**: 2339
  domestica **17**: 2339, 2340
Tegeticula **21**: 3005
  maculata **21**: 3005
  synthetica **21**: 3005
  yuccasella **21**: 3005, *3005*, 3006
Tegu **19**: 2650, *2650*
Teinopalpus imperialis **18**: 2582
Temperate forest 19: 2651–2655
temperature
  affecting color **1**: 53, *53*
  affecting eel vertebrae **6**: 747
  affecting sex **1**: 40; **9**: 1170; **17**: 2426
temperature control
  Arctic fox **1**: 83
  bat-eared fox **2**: 165, *165*
  bearded lizard **2**: 173
  beaver **2**: 177
  birdwing butterfly **2**: 211
  buffalo 3: 305
  bumblebee **3**: 318
  camel **3**: 363
  caribou **3**: 391
  chuckwalla **4**: 453
  crocodile **5**: 603–604, *603*, *604*
  dolphin **2**: 267
  elephant **6**: 767, 768
  emperor penguin **6**: 781, *781*
  fence lizard **6**: 808
  fennec fox **2**: 165; **6**: 809, *811*
  ground squirrel **8**: 1080
  hedgehog **9**: 1173
  jack rabbit **10**: *1305*, 1306
  jerboa **10**: 1323
  leatherback turtle **10**: 1423
  legless lizard **10**: 1430
  lizards 11: 1468–1469
  marine iguana **11**: 1556
  meerkat **12**: 1596, *1597*
  morpho **12**: *1661*
  oil beetle **13**: 1782
  oryx **13**: 1810
  painted turtle **13**: *1851*, 1853
  poorwill **14**: 2012
  ptarmigan **15**: 2057
  python **15**: 2081
  quokka **15**: *2089*, 2090
  sheathbill **17**: *2337*
  sidewinder **17**: 2369, *2370*
  slowworm **17**: 2406
  termite **19**: 2660
  turtles 20: 2798–2799
  velvet ant **20**: 2816
  see also torpor
Tench (doctor fish) **12**: 1619
Tenodera aridifolia sinensis **11**: 1546
Tenrec **19**: 2656–2657
  Common **19**: 2656–2657, 2657
  Hedgehog **19**: 2656, 2657
  Large hedgehog **19**: 2657
  Lesser hedgehog (small hedgehog
    tenrec) **19**: 2657, *2657*
  Long-tailed (shrewlike tenrec)
    19: 2656, 2657
  Rice **19**: 2656, 2657
  Streaked **19**: 2656, *2656*, 2657
  Web-footed (water shrew)
    19: 2656, 2657
Tenrec ecaudatus **19**: 2656, 2657
Tenualosa
  ilisha **17**: 2327
  reevesi **17**: 2327
tenuis larvae **14**: 1902
Tepemixtle *see* Ringtail
Teraba major **1**: 56
Terapon, Jarbua (zebra fish; tiger
  fish) **21**: 3015
Terapon jarbua **21**: 3015
Teratornis incredibilis **4**: 516
Terebratella **10**: 1406

Terebratulina **10**: 1406
Teredo navalis **17**: 2350, *2350*
termitaria **1**: 67, 68
Termite (white ant) **19**: 2658–2660
  Black-mound **19**: 2658–2659
  Compass (magnetic termite)
    19: 2660
  see also Ants and termites
termite-feeders
  aardvark **1**: 16, *17*
  aardwolf **1**: 18–19
  anteater **1**: 59
  armadillo **1**: 86
  banded anteater **1**: 128
  bat-eared fox **2**: 166
  pangolin **13**: 1862, *1862*
  sloth bear **17**: 2403–2404
  spiny anteater **18**: 2487
termites, puffbirds breeding in nests
  of **15**: 2061
Tern (sea swallow) **16**: 2265;
  19: 2661–2663
  Arctic **16**: 2265; **19**: 2661–2662,
    2662, 2663
  Black **19**: 2661
  Caspian **19**: 2661
  Common **19**: 2661, *2661*, 2663
  Forster's **16**: 2264; **19**: 2661,
    2662–2663
  Gull-billed **19**: 2661
  Inca **19**: 2661
  Roseate **16**: 2268; **19**: 2663
  Royal **19**: *2663*
  Sooty **19**: 2661, 2663
  Whiskered **19**: 2661
  White-winged black **19**: 2661
Terpsiphone mutata **7**: *873*
Terrapene **20**: 2796
Terrapin **19**: 2664–2666; **20**: 2796
  Basking *see* Turtle, Painted
  Cape **20**: *2799*
  Diamondback **19**: 2664, *2664*,
    2665–2666, *2665*, *2666*
  Marsh *see* Turtle, African
    helmeted
  Painted *see* Turtle, Painted
territoriality *see* calling; display
  (territorial); fighting for
  territory; scent marking
Tersina viridis **18**: 2584
Testacella **17**: 2409
Testudo
  graeca **19**: 2712
  hermanni **19**: 2713
Tethys **16**: 2300
  californica **16**: 2276
Tethyum pyriforme **17**: 2313
Tetra **19**: 2667–2668
  Black **19**: 2667
  Buenos Aires **19**: 2667
  Cardinal **19**: 2667, *2668*
  Congo **19**: 2667, 2668
  Copeland's **19**: 2667
  Glass **19**: 2667
  Glowlight **19**: 2667, *2667*
  Neon **19**: 2667, 2668
  Rio (flame tetra) **19**: 2667
  Rodway's **19**: 2667
  X-ray (water goldfinch; albino
    pristella; pristella tetra)
    21: 2998–2999
Tetracanthagyna plagiata **5**: 711
Tetracerus quadricornis **1**: 60; **13**: 1753
Tetragonopterus **19**: 2667
Tetramorium caespitum **1**: 69
Tetrao
  parvirostris **3**: 376, 377
  tetrix **8**: 1084
  urogallus **3**: 377
Tetraodon palembangensis **15**: *2063*

Tetrapturus **11**: 1561
  albidus **11**: 1561
  angustirostris **11**: 1561
  audax **11**: 1561
  pfluegeri **11**: 1561
Tetrarope barbata **16**: 2260
Tetrix **8**: 1036
tetrodotoxin (TTX) **13**: 1734; **15**: 2064
Thalassoica antartica **17**: 2435
Thalassoma bifasciatum **21**: 2991, 2992
Thalassornis leuconotus **18**: 2538
Thaleichthys pacificus **17**: 2410, 2411
Thamin *see* Deer, Brow-antlered
Thamnophilus doliatus **1**: 55
Thamnophis **7**: 933, 934
  elegans **7**: 934
  marcianus **7**: *935*
  radix **7**: 934
  sirtalis **7**: 934
  sirtalis tetrataenia **17**: 2422
thanatosis *see* death, feigning
Thaumetopoea
  pityocampa **15**: 2047, 2048
  processionea **15**: 2047, 2048
  solitaria **15**: 2048
Thea vigintiduopunctata **10**: 1394
Thecla **8**: 1118
  betulae **8**: *1117*
Theophila religiosae **17**: 2377
Theraphosa leblondi **2**: 204
Theridion sisyphium **18**: 2483
Thermobia domestica **2**: 279
thermoregulation *see* temperature
  control
Theropithecus gelada **1**: 115, 116;
  12: 1643
Thickhead *see* Whistler
Thick-knee *see* Stone curlew
Thinocorus
  orbignyianus **17**: 2319
  rumicivorus **17**: 2319, 2320
Thistle *see* Butterfly, Thistle
Thomisus onustus **5**: 580, 581
Thomomys **8**: 1040; **14**: 1997
  bottae **14**: *1996*
  talpoides **14**: *1997*
Thoracocharax **9**: 1161
Thornbill (tit; Australian warbler;
  wren-warbler) **19**: 2669–2670;
  20: 2857
  Brown **19**: 2669
  Slaty-backed **19**: 2669
  Tasmanian **19**: 2669
  Western **19**: 2669, 2670
  Yellow-rumped **19**: 2669, 2670,
    2670
Thornbug **9**: 1225
Thorny devil *see* Moloch
Thrasher **19**: 2671–2673
  Bendire's **19**: 2672, 2673
  Brown **18**: 2458; **19**: 2671,
    2672–2673
  California **3**: 426; **19**: 2671, 2673
  Crissal **19**: 2673
  Curve-billed **19**: *2671*, *2672*
  Le Conte's **19**: 2673
  Long-billed **19**: 2673
  Pearly-eyed **19**: *2671*, *2672*
  Sage **19**: 2671
  Scaly-breasted **19**: 2671, 2672
  White-breasted **19**: 2671
Thraupis
  cyanoptera **19**: 2627
  episcopus **19**: 2628
  virens **19**: 2627
threat displays
  bearded lizard **2**: 172, *172*
  big-headed turtle **2**: 197
  black molly **2**: 233
  bullhead **3**: 315

chameleon **3**: 419–420
hairstreak butterflies **8**: 1117–1118
mandrill **11**: 1538
mantis shrimp **11**: 1552
pit viper **14**: *1973*
puss moth **15**: 2077, *2077*
skunk **17**: 2395, *2395*
snapping turtle **17**: *2425*
snowy owl **17**: 2439
swallowtail **18**: 2582
swift moth **19**: 2604
see also display (territorial)
Thresher shark (fox shark; sea fox;
  longtail shark) **19**: 2674–2675
  Bigeye thresher **17**: 2333; **19**: 2675
  Common thresher (thintail
    thresher) **19**: 2674, 2675
  Pelagic thresher **19**: 2675
Threskiornis aethiopicus **9**: 1274;
  18: *2497*
Thrips
  flavus **19**: 2676
  tabaci **19**: 2676
Thrips (thunderbug; stormfly;
  bladderfoot; fringewing)
  19: 2676–2677
  Grain (corn thrips) **19**: 2676
  Greenhouse **19**: 2676
  Honeysuckle **19**: 2676
  Lily **19**: *2677*
  Onion **19**: 2676
  Western flower **19**: 2677
throat-teeth *see* teeth, pharyngeal
Thrush **18**: 2458; **19**: 2617, 2678–2680
  Gray-cheeked **19**: 2678
  Ground *see* Pitta
  Hermit **19**: 2678
  Jewel *see* Pitta
  Mistle **19**: 2678–2680
  Painted *see* Pitta
  Song **19**: 2678, 2679, *2679*, 2680
  Varied **19**: *2678*
  Wood **19**: 2678
Thryomanes bewickii **21**: 2993
Thrysites atun **17**: 2431
Thryssa, New Guinea **1**: 45
Thryssa scratchleyi **1**: 45
Thumbie *see* Rifleman
Thunderbird *see* Whistler, Golden
Thunderbug *see* Thrips
Thunnus **19**: 2605
  alalunga **20**: 2773, 2774
  albacares **20**: 2773, 2774
  thynnus **20**: 2773, 2774
Thylogale
  billardierii **13**: 1847, 1848
  browni **13**: 1847, 1848
  brunii **13**: 1847, 1848
  calabyi **13**: 1847, 1848
  stigmatica **13**: 1847, 1848
  thetis **13**: 1847, 1848
Thymallus
  arcticus **8**: 1047
  arcticus baicalensis **8**: 1047
  arcticus grubii **8**: 1047
  arcticus pallasi **8**: 1047
  thymallus **8**: 1047, 1048
Thymelicus
  lineola **17**: 2391, 2392
  octeon **17**: 2392
thyroxine **1**: 110
Thysania agrippina **3**: 344; **13**: 1821,
  1822
Thysanoessa inermis **16**: 2167
Tibicen **4**: 456
  canicularis **4**: 456
Tichodroma muraria **19**: 2721
Tick **19**: 2681–2682
  Common dog **19**: 2682
  Deer **19**: *2681*

Tick *(continued)*
　　Lone-star **19**: 2682
Tick bird *see* Oxpecker
Tieke *see* Saddleback
Tiger **3**: 405, *405*, 406; **19**: 2683–2685;
　　21: 2907
Tiger fish *see* Terapon, Jarbua
Tiger moth **19**: 2686–2688
　　Cream-spot **19**: 2686, *2686*, 2687
　　Garden **19**: 2686, 2687, *2687*
　　Jersey **19**: 2687
　　Scarlet **19**: 2687
Tigershark *see* Skaamoog
Tilapia **4**: 459
　　mariae **19**: *2690*
　　zillii **19**: 2689
Tilapia (kurper; freshwater bream)
　　19: 2689–2690
　　Mango **19**: 2690
　　Mozambique **19**: 2689
　　Red-belly **19**: 2689
Tiliqua multifasciata **17**: *2389*
Timberman *see* Long-horned beetle
Tinamou **8**: 1040; **19**: 2691–2692
　　Crested (martineta) **19**: 2691
　　Great **19**: 2691
　　Little **19**: 2691
　　Ornate **19**: 2691–2692
　　Variegated **19**: 2691, 2692
Tinamus major **19**: 2692
Tinbergen, Nikolaas **18**: 2534
Tinca tinca **12**: 1619
Tineola bisselliella **4**: 470, 471
Tipula oleracea **5**: 589
Tit *(thornbill) see* Thornbill
Tit (titmouse; chickadee) **19**: 2653,
　　2693–2695
　　Bearded *see* Reedling, Bearded
　　Black-capped chickadee **19**: 2653,
　　　　2693
　　Blue **19**: 2693, *2693*, 2694, 2695
　　Bush **19**: 2693
　　Chestnut-backed chickadee
　　　　19: *2695*
　　Crow *see* Parrotbill
　　Great **19**: *2654*, 2693, *2694*, 2695
　　Long-tailed **19**: 2693, 2694, 2695
　　Penduline **19**: 2693, 2695
　　Siberian **19**: 2617
　　True **19**: 2695
　　Willow **19**: 2695
Tit babbler **1**: 113
Titi **19**: 2696–2697
　　Dusky **19**: 2696, *2696*
　　Gray **19**: 2696, 2697
　　Masked **19**: 2696, 2697, *2697*
　　see also Squirrel monkey
Titipou *see* Rifleman
Titipounamu *see* Rifleman
Titmouse *see* Tit
Tityra **4**: 543–544
　　Masked **4**: 544
Tityra semifasciata **4**: 544
Tmetothylacus tenellus **14**: 1965
Toad **7**: 903, *905*; **20**: 2819
　　African clawed *see* Frog, Clawed
　　Bell *see* Frog, American tailed;
　　　　Midwife toad, Common
　　Burrowing (Mexican burrowing
　　　　toad; digger toad) **3**: 326
　　Common **4**: 510, 512–513
　　Darwin's *see* Frog, Mouth-
　　　　brooding
　　Fire-bellied (firebelly) **6**: 825–826;
　　　　7: 903; **20**: 2819
　　Garlic *see* Spadefoot toad,
　　　　Common
　　Giant (cane toad; giant toad)
　　　　11: 1557–1558
　　Green **12**: 1720

Horned *see* Horned lizard
Midwife *see* Midwife toad
Natterjack **12**: 1720–1721
Spadefoot *see* Spadefoot toad
Surinam (tongueless frog)
　　18: 2578–2579
Yellow-bellied (variegated fire-
　　toad) **6**: 825
see also Frogs and toads
Toadfish **19**: 2698–2699
　　Oyster **19**: 2698, 2699
Tockus
　　camurus **9**: 1226
　　erythrorhynchus **9**: 1228
Toddy cat *see* Palm civet, Common
Todus
　　angustirostris **19**: 2700
　　mexicanus **19**: 2700
　　multicolor **19**: 2700, 2701
　　subulatus **19**: 2700
　　todus **16**: 2190; **19**: 2700
Tody **19**: 2700–2701
　　Broad-billed **19**: 2700
　　Cuban (pedorrera) **19**: 2700, *2700*,
　　　　2701
　　Jamaican (robin) **16**: 2190; **19**: 2700
　　Narrow-billed **19**: 2700
　　Puerto Rican **19**: 2700, 2701, *2701*
Toe-biter *see* Water scorpion
Tolypeutes **1**: 85
　　matacus **1**: 85
Tomato fruitworm *see* Bollworm,
　　American
Tomato Grub *see* Bollworm,
　　American
Tomistoma schlegelii **5**: 603; **7**: 954
Tomocerus longicornis **18**: 2504
tool using
　　apes 1: 72
　　buzzard **3**: 353
　　capuchin monkey **3**: 380
　　chimpanzee **4**: 442, *442*
　　Darwin's finches **5**: 649–650, *650*
　　sea otter **16**: 2295, 2296
　　vultures 20: *2842*
Tooth-carp **19**: 2704–2705
Topi **21**: 2950
Topshell **19**: 2706–2707
　　Brick-red **19**: 2706
　　Gray **19**: 2706
　　Painted (painted topsnail; silver
　　　　Tommy) **19**: 2706, *2706*, 2707,
　　　　2707
Topsnail, Painted *see* Topshell,
　　Painted
Torgos tracheliotus **20**: 2842
Tornierella **15**: 2146
Torpedo see Electric ray
Torpedo
　　fuscomaculata **6**: *764*
　　marmorata **6**: 765
　　nobiliana **6**: 764, 765; **17**: 2332
　　torpedo **6**: 766
Torpedo fish *see* Electric ray
torpor
　　chipmunk **4**: 449
　　common frog **4**: 510
　　dormouse possum **5**: 702
　　hummingbird **9**: 1257
　　lemur **10**: 1436
　　poorwill **14**: 2013
　　swift **19**: 2598, 2600
　　water-holding frog **20**: 2868
Torrentlark *see* Magpie lark, New
　　Guinea
Tortoise **19**: 2711–2712; **20**: 2796
　　Aldabra **20**: 2799
　　Aldabra giant **19**: 2712;
　　　　20: 2799
　　Desert **19**: 2711, 2712; **20**: 2799

Galapagos giant (Indefatigable
　　Island tortoise; Santa Cruz
　　tortoise) **19**: *2711*, 2712, 2713;
　　20: 2799
Gopher **19**: 2711, 2713; **20**: 2795
Greek (spur-thighed tortoise)
　　19: 2712, *2712*
Hermann's **19**: 2713
Leopard **19**: 2712; **20**: *2798*
Long-necked *see* Snake-necked
　　turtle, Common
Marion's **19**: 2713
Speckled Cape (speckled tortoise)
　　19: 2712; **20**: 2795
Texas **19**: 2711, 2712
Water *see* Turtle, African helmeted
　　tortoiseshell **9**: 1169, *1169*, 1170
Small **20**: 2812, *2812*
Totoaba **5**: 601, 602
Totoaba macdonaldi **5**: 601, 602
Toucan **19**: 2714–2716
　　Channel-billed **19**: *2714*
　　Rainbow-billed **19**: 2714, 2715
　　Red-breasted **19**: *2716*
　　Toco **19**: *2715*
Toucanet **19**: 2714
Touraco *see* Turaco
tourism
　　and coral reefs **4**: 533
　　and gray whales **8**: 1052
　　and llamas **11**: 1472
　　and loggerhead turtles **11**: 1482
toxins *see* poisons
Toxostoma
　　bendirei **19**: 2672
　　dorsale **19**: 2673
　　lecontei **19**: 2673
　　longirostre **19**: 2671
　　redivivum **3**: 426; **19**: 2671, 2673
　　rufum **18**: 2458; **19**: 2671, 2672
Toxotes
　　chatereus **1**: 77
　　jaculator **1**: 78
Trachinotus carolinus **16**: 2242
Trachinus **20**: 2818
　　araneus **21**: 2902
　　armatus **21**: 2902
　　draco **21**: 2901, 2902
　　lineolatus **21**: 2902
　　pellegrini **21**: 2902
　　radiatus **21**: 2902
Trachipterus
　　arcticus **13**: 1768
　　trachypterus **13**: 1768
Trachops cirrhosus **2**: 167
Trachurus **16**: 2242
　　trachurus **16**: 2243; **17**: 2430
Trachycardium
　　egmontianum **4**: 488
　　muricatum **4**: 488
　　quadragenarium **4**: 488
Trachydosaurus rugosus **17**: 2390
Trachypithecus **10**: 1411
Tragelaphus **3**: 331; **13**: 1765
　　angasi **3**: 331; **13**: 1765, 1766
　　bongo **10**: 1390
　　buxtoni **13**: 1765, 1766
　　eurycerus **1**: 63; **10**: 1389–1390
　　imberbis **3**: 331; **10**: 1390
　　scriptus **1**: 63; **3**: 332
　　spekei **1**: 63; **3**: 331; **20**: 2760
　　strepsiceros **10**: 1390
Tragopan
　　blythi **19**: 2717, 2718
　　caboti **19**: 2717
Tragopan (horned pheasant)
　　19: 2717–2718
　　Blyth's **19**: 2717, 2718
　　Cabot's **19**: 2717, 2718
　　Satyr **19**: 2717

Temminck's **19**: 2717, *2717*, *2718*
Western **19**: 2717, 2718
Tragopan
　　melanocephalus **19**: 2717
　　satura **19**: 2717
　　temmincki **19**: 2717
transport, "hitching" a ride *see*
　　phoresy
Trapdoor spider **2**: 203; **18**: 2480;
　　19: 2719–2720
　　Adelaide **19**: 2720
　　Brown **19**: 2720
　　California **19**: 2720
　　Malaysian **19**: *2719*
Trapjaw *see* Moccasin, Water
Tratratratra *(extinct)* **17**: 2373
Treecreeper (spotted creeper)
　　19: 2721–2722
　　Australian **19**: 2721
　　Common **19**: 2721
　　Eurasian **19**: 2721, *2721*, 2722, *2722*
　　Short-toed **19**: 2721
　　Spotted **19**: 2721, 2722
Tree frog **7**: 903; **19**: 2723–2725
　　Brazilian **7**: 983
　　Canyon **19**: 2724
　　Cuban **19**: 2723
　　Giant **19**: *2724*
　　Golden (bell tree frog) **19**: 2724
　　Red-eyed **7**: *903*
　　Rohde's **19**: 2725
　　Southern **19**: *2725*
Treehopper **9**: 1225, *1225*
　　Buffalo **9**: 1225
Tree kangaroo **19**: 2726–2727
　　Bennett's **19**: 2726, *2726*
　　Goodfellow's **19**: 2726
　　Lumholtz's **19**: 2726, *2727*
　　Matschie's **19**: 2726
　　Scott's **19**: 2727
Tree porcupine **19**: 2728–2729
　　Amazon basin porcupines
　　　　19: 2729
　　Amazonian porcupine **19**: 2729
　　Brazilian **19**: 2728, 2729, *2729*
　　Canadian (North American
　　　　porcupine) **19**: 2728, *2728*, 2729
　　Prehensile-tailed porcupines
　　　　19: 2729
Tree rat
　　Giant **7**: 965
　　New Guinea giant **7**: 965
Tree rat *(squirrel) see* Tree squirrel
Treerunner **13**: 1819
Tree shrew **19**: 2730–2731
　　Belanger's **19**: *2730*
　　Long-nosed **19**: *2731*
　　Mindanao **19**: 2730
　　Pen-tailed **19**: 2730, 2731
　　Smoothtailed **19**: 2730
Tree snake **19**: 2732–2733
　　Common **19**: 2733
　　Golden **19**: 2732, 2733
　　Paradise **19**: 2732, *2733*
Tree squirrel (tree rat)
　　20: 2740–2742
　　South African **20**: *2741*
Tremarctos **2**: 174
　　ornatus **12**: 1672; **18**: 2471
Trembler
　　Brown **19**: 2671
　　Gray **19**: 2671
Trialeurodes vaporariorum **3**: 416;
　　21: 2932, *2932*, 2933
Trichechus
　　inunguis **11**: 1534, 1535
　　manatus **11**: 1534, 1535; **21**: 2907
　　senegalensis **11**: 1534, 1535
trichinosis **21**: 2943
Trichiurus nitens **17**: *2415*

trichobothria **18**: 2481
trichocysts **14**: 1881
Trichoglossus
  chlorolepidotus **11**: 1488
  euteles **11**: *1489*
  haematodus **11**: 1488, 1489
Trichoniscus **21**: 2974
Trichosurus
  arnhemensis **3**: 300, 301
  caninus **3**: 300, 301
  vulpecula **3**: 300, 301
Trichys
  fasciculata **5**: 595
  lipura **19**: 2729
Tridacna gigas **4**: 462, 463
Triggerfish **4**: 532; **20**: 2743–2744
  Black **20**: 2744
  Blackbar (humahuma-nuku-nuka-
    a-puaa) **20**: 2744
  Clown **20**: 2743, 2744
  Gray **20**: 2744
  Orange-lined **20**: 2744
  Yellow-margined **20**: *2744*
Trigla lyra **8**: 1109
Trigonoceps **20**: 2841
Triller
  Black-and-white **5**: 621
  Black-breasted **5**: 621
  Pied **5**: 621
  Pygmy (pied shrike) **5**: 621
  White-winged **5**: 621
Trimeresurus
  albolabris **14**: *1974*
  flavoviridis **14**: 1973
  wagleri **14**: 1973
Trimma **7**: 994
Trimmatom nanus **6**: 834; **7**: 993
Trinectes maculatus **17**: 2444
Tringa **16**: 2233, 2234; **21**: 2906
Trionyx **17**: 2442
Tritonia **5**: 652
  hombergi **5**: 652; **16**: 2300
Triton shell, Giant **18**: 2518
Triturus **13**: 1732
  cristatus **13**: 1732, 1733
  helveticus **13**: 1732, 1733
  vulgaris **13**: 1732, 1733
trochophore larvae **1**: 20;
  19: 2706
Trochus niloticus **19**: 2707
Troglodytes **21**: *2993*
  aedon **21**: 2993
  troglodytes **7**: 999; **21**: 2993, 2994
Trogon
  elegans **20**: 2745, 2746, *2746*
  massena **20**: 2745
  mexicanus **20**: 2746–2747
  rufus **20**: 2747
  violaceus **20**: 2747
  viridis **20**: *2747*
Trogon **20**: 2745–2747
  Blue-throated **20**: 2747
  Coppery-tailed **20**: *2746*
  Elegant **20**: 2745, 2746
  Gartered **20**: 2747
  Mountain **20**: 2746–2747
  Narina **20**: 2745
  Resplendent *see* Quetzal,
    Resplendent
  Slaty-tailed **20**: *2745*
  White-tailed **20**: *2747*
Trogonoptera **2**: 212
Troides
  allotai **2**: 212
  brookiana **2**: 211
  helena **2**: 211
  paradisea **2**: 211
  victoriae **2**: 211
trophallaxis **19**: 2658
Tropheus duboisi **10**: *1400*

Tropic bird (paille-en-queue; marlin-
  spike; bo'sun bird; boatswain
  bird) **16**: 2265, 2268;
  20: 2748–2750
  Red-billed **20**: 2748, *2749*, *2750*
  Red-tailed **20**: 2748, 2749
  White-tailed **20**: 2748, *2748*
Tropicranis albocristatus **9**: 1227
Tropidacris latriellei **8**: 1035
Trout **20**: 2751–2753
  Brook **20**: 2751
  Cut-throat **20**: 2751, 2752
  Dolly Varden **20**: 2751
  European (brown trout; Aral
    trout; sea trout; lake trout-
    European; Amu-Darya trout)
    20: 2751, 2752, 2753
  Golden **20**: 2751
  Lake (*North American*) **20**: 2751
  Rainbow **20**: 2751, *2751*, 2752, *2753*
  Sea (*Cynoscion*) *see* Croaker
Trout-perch **20**: 2754–2755
  Blunt-nosed (sandroller) **20**: 2754
True lover's knot **13**: 1821
Trumpeter (trumpet bird)
  20: 2756–2757
  Gray-winged **20**: 2756, *2756*, 2757
  Green-winged (dark-winged
    trumpeter) **20**: 2756, 2757, *2757*
  Pale-winged **20**: 2756, 2757
Trumpetfish **20**: 2766
Trumpet shell *see* Conch
Trunkfish (boxfish; cofferfish;
  dingleberry) **20**: 2758–2759
  Four-sided blue **20**: 2758
  Smooth **20**: *2758*
  Spotted **20**: *2759*
Trygonorhina **8**: 1103
Tryngites subruficollis **16**: 2233,
  2234
Trypanosoma
  brucei **20**: 2760
  brucei brucei **20**: 2761
  brucei congolense **20**: 2761
  brucei gambiense **20**: 2760
  rhodesiense **20**: 2760
trypanosomiasis *see* sleeping
  sickness
Tsaine *see* Banteng
Tuan *see* Marsupial mouse, Brush-
  tailed
Tuatara **20**: 2762–2764
  Gunther's (Brother's Island
    tuatara) **20**: 2763
tube feet **2**: 280; **16**: 2226, 2271;
  17: 2314; **18**: 2516
Tube mollusk *see* Worm shell
Tubenose **20**: 2766
tuberculosis **1**: 122; **3**: 307
Tube snail *see* Worm shell
Tubesnout **20**: 2765–2766
Tubeworm (parchment-tube worm;
  innkeeper worm) **13**: 1773;
  20: 2767–2768
Tubifex
  costatus **20**: 2769
  tubifex **20**: 2769, 2770
Tubifex worm (bloodworm)
  20: 2769–2770
Tubularia
  bellis **16**: 2274
  indivisa **16**: 2274
  larynx **16**: 2274
Tuco-tuco (South American mole
  rat; tucu-tucu) **20**: 2771–2772
Tucu-tucu *see* Tuco-tuco
Tuea pellionella **4**: 470
Tui **9**: 1215
tularemia **9**: 1235; **19**: 2682
Tumblebug **2**: 188

Tuna (tunny) **1**: 46; **19**: 2605;
  20: 2773–2775
  Northern bluefin **20**: 2773,
    2774–2775, *2775*
  Skipjack **20**: 2773, 2774
  Yellowfin (yellow-finned
    albacore) **20**: 2773, 2774
Tundra 20: 2776–2779
Tunga penetrans **6**: 850, 851
Tunicate *see* Sea squirt
Tunny *see* Tuna
Tupaia
  belangeri **19**: *2730*
  glis **19**: 2730
  tana **19**: 2731
Tupinambis nigripunctatus **19**: *2650*
Tur *see* Ibex, East Caucasus; Ibex,
  West Caucasus
Turaco (touraco) **20**: 2780–2781
  Black-billed (Congo turaco)
    20: 2781
  Giant (great blue turaco) **20**: 2780
  Hartlaub's **20**: 2780, 2781
  Knysna **20**: *2781*
  Prince Ruspoli's **20**: 2780
  Purple-crested **20**: 2780, *2780*
  Ross's **20**: 2780, 2781
  Ruwenzori **20**: 2780
  Yellow-billed **20**: 2780
Turbot **20**: 2782–2783
  Mediterranean (Black Sea turbot)
    20: 2782
Turdoides caudatus **1**: 114
Turdus
  merula **2**: 225; **19**: 2678
  migratorius **4**: 555; **15**: 2141;
    16: 2289; **19**: 2678; **20**: 2833
  philomelos **19**: 2678, 2679
  pilaris **18**: *2459*
  torquatus **19**: *2680*
  viscivorus **19**: 2678
Turkey **20**: 2784–2786
  Ocellated **20**: 2784, 2786, *2786*
  Wild (Merriam's wild turkey; Rio
    Grande turkey; Gould's wild
    turkey) **20**: 2784, 2785, *2785*
Turkeyfish *see* Lionfish, Red
Turkey quail *see* Plains-wanderer
Turnix
  melanogaster **3**: 350
  sylvatica **3**: 350
Turnstone (ruddy turnstone)
  20: 2790–2792
  Black **20**: 2790, 2791, 2792
Turquoisine **14**: 1878
Tursiops truncatus **2**: 267
Turtle
  African helmeted (water tortoise;
    marsh terrapin) **9**: 1190–1191,
    1191
  Amazon river (arrau) **9**: 1190,
    1190, 1191
  Big-headed **2**: 197
  Box **19**: 2711; **20**: 2796
  Broad-shelled river (Murray River
    turtle; giant snake-necked
    turtle) **17**: 2417
  Flapjack *see* Soft-shelled turtle
  Flapshelled *see* Flapshelled turtle
  Green (edible turtle) **8**: 1070–1072;
    20: 2797, 2798
  Hawksbill **9**: 1169–1170; **20**: 2798
  Hidden-necked **9**: 1190–1191;
    20: 2795, 2796
  Leatherback (leathery turtle; luth)
    10: 1423–1424; **20**: 2795, 2796,
    2798, 2799
  Loggerhead **11**: 1481–1482;
    15: 2030–2031; **20**: 2797, 2798
  Long-headed **17**: 2442

  Mud *see* Mud turtle
  Murray **17**: *2417*
  Musk *see* Musk turtle
  Olive *see* Ridley turtle, Pacific
  Painted (painted terrapin; basking
    turtle; basking terrapin; eastern
    painted turtle; western painted
    turtle; southern painted turtle;
    Midland painted turtle)
    13: 1851–1853
  Pancake *see* Soft-shelled turtle
  Papuan (pitted-shell turtle; pig-
    nosed turtle; hog-nosed turtle;
    Fly River turtle) **13**: 1866–1867;
    20: 2796
  Ridley *see* Ridley turtle
  Side-necked **20**: 2795, 2796, *2799*
  Snake-necked *see* Snake-necked
    turtle
  Snapping *see* Snapping turtle
  Soft-shelled *see* Soft-shelled turtle
  Western swamp **9**: 1191; **20**: 2797
  see also Terrapin; Tortoise; Turtles
Turtledove **20**: 2793–2794
Turtle dove, Collared *see* Dove,
  Collared
Turtles 20: 2795–2799
tusks
  bushpig **3**: 335
  elephant **6**: 767
  hippopotamus **9**: 1192
  narwhal **12**: *1717*, *1718*, 1719
  walrus **20**: 2851, 2852–2853, *2852*
  warthog **20**: 2860, *2860*
Tusk shell, Elephant *see* Elephant
  tusk shell
Twinspot **21**: 2938
Twite **8**: 1016; **11**: 1457
Tylos **21**: 2974
tymbals **4**: 456
tympanal organs **13**: 1821, 1822
Tympanuchus
  cupido **15**: 2037
  pallidicinctus **15**: 2037
  phasianellus **8**: 1084
Typhlacontias gracilis **17**: 2390
Typhlogobius californiensis **7**: 995
Typhlonectes **3**: 357
Typhologarra widdowsoni **3**: 410
typhus **3**: 299
Tyrannus **20**: 2802
  tyrannus **20**: 2802
Tyrant, Many-colored **20**: 2800
Tyto
  alba **2**: 148–150; **13**: 1825
  novaehollandiae **18**: 2536

U

Uacari *see* Uakari
Uakari (uacari) **12**: 1645;
  20: 2803–2804
  Bald (white uakari) **20**: 2803–2804,
    2803, *2804*
  Black-headed **20**: 2803, 2804
Uca **4**: 573; **6**: 815
  annulipes **6**: 815, 816
  paradussumieri **6**: 817
  pugilator **6**: *817*
  tangeri **6**: 817
  uruguayensis **6**: 816
ultrasonic sounds
  brown rat **3**: 298
  dolphin **2**: 268
  flying squirrel **7**: 885
  house mouse **9**: 1246
ultraviolet light
  detection **10**: 1347; **11**: 1467
  and webs of spiders **13**: 1801;
    18: 2482

Umbonia crassicornis **9**: 1225
Umbra
  krameri **12**: 1680, 1681
  limi **12**: 1680, 1681
  pygmaea **12**: 1680, 1681
Umbrellabird **4**: 543
Umbrina, roncador **5**: *601*, 602
Underwing
  Beautiful yellow **13**: 1821, 1822, *1822*
  Large yellow **13**: 1822
ungulates **1**: 33; **5**: 655
Unicorn fish, Humpnose **18**: 2576
unicorns **2**: 227
United States of America, national emblem **1**: 123, 125, *125*
Upucerthia **13**: 1819
Upupa
  epops **9**: 1223
  marginata **9**: 1222
upwellings **4**: 475
Uraeginthus angolensis **21**: *2889*
Uraeotyphlus **3**: 357
Urania **20**: 2805
  leilus **20**: 2805, *2805*
  sloanus **20**: 2806
Urania, Sloane's **20**: 2806
Uranomys **7**: 964
  sapientis **7**: 964
Uranoscopus **18**: 2520
  scaber **18**: 2520
urban/suburban habitats
  black-headed gulls **2**: *230*, 232
  bulbuls **3**: 309
  coyotes **4**: 558
  herring gulls **9**: 1188
  racoons **15**: 2096–2097
  red fox **15**: 2133
  see also birds, garden
Urchin (*echinoderm*)
  Cake *see* Sand dollar
  Common heart **17**: 2314
  Keyhole **16**: *2226*
  Pea **16**: 2226
  see also Sea urchin
Urchin (*mammal*) *see* Hedgehog
Uria aalge **8**: 1095
Urial **2**: 198
Urocissa **11**: 1520
Urocyon **5**: 689
  cinereoargenteus **3**: 426; **8**: 1045, 1046
  littoralis **8**: 1045, 1046
Urogale **19**: 2731
  everetti **19**: 2730
Urolophus halleri **17**: 2333
Uromastix **1**: 30
Uromys caudimaculatus **7**: 965
Urosalpinx cinerea **21**: 2917
Ursavus elmensis **2**: 174
Ursus **1**: 82, 96, *175*; **2**: 223; **3**: 293, 295; **12**: 1672; **14**: 1999; **19**: 2616, 2617, 2653
Ussuri dog *see* Raccoon dog

V

valuable *see* commercially important
Vampire bat **2**: 167; **20**: 2807–2809
  Common **20**: 2807, 2808, *2808*, 2809, *2809*
  False **6**: 840
  Hairy-legged **20**: 2808, 2809
  White-winged **20**: 2807–2808, 2809
Vandellia cirrhosa **3**: 402
Vanellus
  armatus **10**: 1415
  gregarius **10**: 1415
  indicus **10**: 1415
  vanellus **10**: 1415, 1416; **16**: 2234

Vanessa
  atalanta **20**: 2810, 2811
  canace **20**: 2810
  cardui **20**: 2812
  indica **20**: 2810
Vanga curvirostris **20**: 2813
Vanga (vanga-shrike) **12**: 1605; **20**: 2813–2814
  Bernier's **20**: 2813
  Blue **20**: 2813
  Chabert's **20**: 2813, 2814, *2814*
  Coral-billed **20**: 2813
  Helmet (helmetbird) **20**: 2813
  Hook-billed **20**: 2813
  Pollen's **20**: 2813
  Red-shouldered **20**: 2813, 2814
  Red-tailed **20**: 2813
  Rufous **20**: 2813
  Sickle-billed **20**: 2813, *2813*, 2814
  Van Dam's **20**: 2813
Vaquita **15**: 2024, 2025
Varanus
  albigularis **12**: *1639*
  bengalensis **12**: 1639
  brevicaudata **12**: 1638
  giganteus **12**: 1638, 1639
  griseus **12**: 1639
  komodoensis **10**: 1384; **11**: 1466; **12**: 1638
  niloticus **12**: 1638, 1639
  rudicollis **12**: 1639
  salvator **12**: 1638, 1639
  varius **12**: 1639
Varecia **10**: 1436
  variegata **10**: *1436*
Veery **19**: 2678
vegetarians *see* browsing; grazing
Velella **15**: 2073; **16**: 2300
veliger larvae **1**: 20–21; **13**: 1834; **15**: 2073; **16**: 2270; **17**: 2349; **19**: 2706; **21**: 2989
Velvet ant **20**: 2815–2816
  Large **20**: 2815
Velvet worm *see* Peripatus
Venomous animals 20: 2817–2820
  see also poisons/venom
Verasper variegatus **8**: 1122
Vermetid *see* Worm shell
Vermetus cristatus **21**: 2990
Vermicularia knorri **21**: *2990*
Vermivora
  chysoptera **21**: 2985
  pinus **21**: 2985
vertebrae
  amphicoelous **9**: 1200
  of hero shrews **9**: 1183
Vesitarses see Thrips
Vespa crabro **9**: 1234; **20**: 2824
Vespula **20**: 2824
  austriaca **20**: 2824
  germanica **9**: 1233; **20**: 2824
  maculata **9**: 1233, 1234
  norvegica **20**: 2824
  rufa **20**: 2824
  squamosa **20**: 2824
  sylvestris **20**: 2824
  vulgaris **9**: 1233; **20**: 2824
vestigial organs, boa constrictor **2**: 251
vexillifer larvae **14**: 1902
Viceroy **12**: 1634; **20**: 2810, 2811, 2812
Vicugna vicugna **11**: 1470, 1471; **12**: 1672; **20**: 2772
Vicuña **11**: 1470, *1471*, 1472; **12**: 1672
Vidua
  chalybeata **21**: 2936
  macroura **21**: 2889, 2936, 2937
  paradisaea **21**: 2936
  regia **21**: *2938*

Vinegaroon *see* Pedipalpi
Vine louse *see* Vine pest
Vine mite *see* Vine pest
Vine pest (vine mite; vine louse; graperoot louse; grape phylloxera) **20**: 2827–2828
Viper **13**: 1736; **17**: 2418, 2420; **20**: 2817
  Aspic *see* Asp
  Blow *see* Hognose snake
  Brazilian **17**: 2421
  Brown saw-scaled **6**: 749
  Gaboon **15**: 2058–2059
  Horned **17**: *2419*
  June *see* Asp
  Lataste's **1**: 97; **21**: 2919
  Long-nosed **1**: 97
  Lowland **6**: 749
  Pit *see* Pit viper
  Rhinoceros **15**: 2058; **17**: *2422*
Vipera
  ammodytes **1**: 97
  aspis **1**: 98
  berus **1**: 27; **7**: 935
  latastei **1**: 97
Viperfish **10**: 1413; **20**: 2829–2830
  Dana **20**: 2829
  Sloane's **20**: 2829
Vireo
  flavifrons **20**: 2831
  gilvus **20**: 2831
  gracilirostris **20**: 2831
  griseus **20**: 2831–2832
  olivaceus **20**: 2831, 2832
  philadelphicus **20**: 2831
  solitaricius **20**: 2832
Vireo **20**: 2831–2833
  Blue-headed **20**: 2832
  Brown-capped **20**: 2831
  Noronha **20**: 2831
  Red-eyed **20**: 2831, 2832, 2833, *2833*
  Warbling **20**: 2831
  White-eyed **20**: 2831–2832
  Yellow-throated **20**: 2831
Vireosa **7**: 995
Viscacha **20**: 2834–2836
  Mountain (mountain chinchilla) **12**: 1672; **20**: 2834, 2835–2836, *2835, 2836*
  Plains **20**: 2834–2835, 2836
Viverra
  civettina **4**: 461
  zibetha **4**: 460
viviparity **17**: 2333, 2421
vocalization *see* calling
Vole **10**: 1347
  Bank (wood vole; bank mouse; red mouse) **1**: 134–135; **15**: 2129
  Field (short-tailed vole) **6**: 818–820
  Red *see* Red-backed vole
  Red-backed *see* Red-backed vole
  Sagebrush **18**: 2528, 2529, *2529*
  Water *see* Water vole
Volucella **9**: 1251; **20**: 2826
  bombylans **6**: 858; **9**: 1251
Vombatus ursinus **21**: 2961, 2962
Vormela pergusna **14**: 2001
Vulpes **5**: 690
  bengalensis **5**: *690*
  pallida **5**: 689
  velox **4**: 560; **5**: *692*
  velox macrotis **5**: *691*
  velox mutica **4**: 560
  vulpes **5**: 689; **15**: 2133
  vulpes necator **15**: 2133
Vulture
  Bearded *see* Lammergeier
  Black **20**: 2787, 2840, *2840*, 2841

  Cinereous (Eurasian black vulture) **20**: 2840
  Egyptian **20**: 2840–2841, 2842, *2842*
  Greater yellow-headed **4**: 518; **20**: 2787
  Griffon **10**: 1401; **20**: 2841, 2842
  Hooded **20**: 2841, 2842
  Indian black **20**: 2841
  King **20**: 2787, 2841, *2841*
  Lappet-faced **20**: 2842
  Lesser yellow-headed (yellow-headed vulture) **4**: 518; **20**: 2787
  Palm nut **20**: 2841, 2842
  Turkey **4**: 518; **20**: 2787–2789, 2840, 2841
  White-backed **10**: 1401; **20**: *2839*, 2841, 2842
  Yellow-headed **20**: 2841
  see also Condor; Lammergeier; Vultures
Vultures 20: 2839–2842
Vultur gryphus **4**: 516, 517; **12**: 1673; **20**: 2841

W

Wagtail **20**: 2843–2845
  African pied **20**: 2844
  Blue-headed **20**: *2843*, 2844
  Cape **20**: 2843
  Forest **20**: 2843
  Gray **4**: 499; **20**: 2843, 2844, 2845
  Madagascar **20**: 2843
  Mountain **20**: 2843
  Pied (white wagtail) **20**: 2843, 2844, 2845, *2845*
  Willie **20**: 2843
  Yellow **20**: 2843, *2843*, 2844, 2845
Wailing bird *see* Limpkin
Waldrapp *see* Ibis, Northern bald
Wallaby **20**: 2846–2848
  Alpine *see* Pademelon, Calaby's
  Banded hare **20**: 2846, 2848
  Bennett's **20**: 2848
  Bridled nail-tailed **20**: *2848*
  Hare **20**: 2846, 2847–2848
  Nail-tailed (organ-grinder) **20**: 2846
  New Guinea forest **20**: 2846
  Parma **20**: *2846*
  Rock *see* Rock wallaby
  Rufous *see* Pademelon, Red-bellied
  Scrub **20**: 2848
Wallace, Alfred Russel **10**: 1331; **11**: 1494
Wallaroo, Common (euro) **10**: 1333
Wallcreeper **19**: 2721
Walleye (pikeperch) **14**: 1917, 1918
Wall lizard **20**: 2849–2850
  Common **20**: 2849, *2849*, 2850
  Iberian **20**: 2849, 2850
  Italian (ruin lizard) **20**: 2849, 2850
Walrus **16**: 2285, 2286, 2287, 2288, 2289, *2289*; **20**: 2851–2853
Wapiti (elk—*incorrectly;* red deer) **5**: *655*, 656, 657; **17**: 2375; **19**: 2653; **20**: 2854–2856
Warbler, Australian *see* Thornbill
Warbler (*New World*) *see* Wood warbler
Warbler (*Old World*) **18**: 2456; **20**: 2857–2859
  Arctic **20**: 2857
  Cetti's **20**: 2857
  Dartford **20**: 2857, 2858
  Grasshopper **20**: 2859
  Marsh **18**: 2458
  Olivaceous **20**: *2859*

Reed *see* Reed warbler
Rüppell's **20**: 2857
Sedge **18**: 2459; **20**: *2858*, 2859
Willow **20**: 2857, 2859
Warbler-finch **5**: 648, 649, *649*
warning coloration **20**: 2819–2820
    arrow-poison frog **1**: 93, *93*, 94
    bee and wasp 2: 184
    birdwing butterfly **2**: 212
    burnet moth **3**: *321*, 322
    butterflies and moths 3: 345
    caecilian **3**: 356
    centipede **3**: 413
    coral snake **4**: 534–535, *535*
    fire-bellied toad **6**: 825, *825*
    gila monster **7**: 972, *974*
    honey badger **9**: 1206
    ladybug **10**: *1394*
    monarch **12**: 1633, *1633*
    sawfly **16**: 2241
    shieldbug **17**: 2344, *2344*, 2345
    trunkfish **20**: *2758*
    see also mimicry; poisons; threat displays
warning sounds
    baboon **1**: 117
    beaver **2**: 179
    blackbird **2**: 224
    boa constrictor **2**: 251
    chaffinch **3**: 415
    dik-dik **5**: 675
    eland **6**: 757
    flying gurnard **7**: 880, *881*
    hawkmoth caterpillar **9**: *1168*
    lobster **11**: 1475
    long-horned beetle **11**: 1483
    marmot **11**: 1570
    meerkat **12**: 1597
    muntjac **12**: 1698
    prairie dog **15**: *2039*, 2040
    rabbit **15**: 2093
    roe deer **16**: 2202
    snipe **17**: 2428
    song sparrow **18**: 2461
    squirrelfish **18**: 2511
Warthog **20**: 2860–2862
Wasp **18**: 2482–2483
    Chalcid **3**: 416–417; **4**: 471
    Common **20**: 2824, 2826
    Cuckoo **2**: 185; **13**: 1865; **20**: 2824
    Fig **3**: 417
    Gall **2**: 183; **7**: 921–922
    German **20**: 2824, *2825*, *2826*
    Hunting (digger wasp) **9**: 1261–1262; **21**: 2957
    Ichneumon (ichneumon fly) **2**: 183; **9**: 1279–1280
    Mason **9**: 1262
    Mud-dauber **9**: 1262
    Norwegian **20**: 2824
    Paper **2**: 185; **13**: 1864–1866; **20**: 2824
    parasitic **1**: 75; **2**: 183; **3**: 416, 417; **4**: 504–505; **20**: 2815, 2816, *2816*; **21**: 2928, 2933, 2957
    Potter **9**: 1262, *1262*
    Red **20**: 2824
    social and solitary types **9**: 1261
    Spider-hunting **2**: 185; **9**: *1261*, 1262
    Tree **20**: 2824
    Vespula (yellowjacket; hornet— *incorrectly*) **20**: 2824–2826
    Wood *see* Sawfly
    see also Bees and wasps; Velvet ant
water, running over **2**: 156, *156*
Water beetle (water scavenger beetle) **20**: 2863–2864

Giant water scavenger beetle **20**: 2863
    Silver **20**: 2863, *2863*, 2864
    see also Whirligig beetle
Water boatman *see* Backswimmer
Waterbuck (common waterbuck; defassa; singsing) **1**: 62; **20**: 2865–2867
Water cavy *see* Capybara
Water dikkop **18**: 2545, 2546
Water dog *see* Mudpuppy
Water flea *see* Daphnia
Water goldfinch *see* Tetra, X-ray
Watermole *see* Platypus
Water pig *see* Capybara
Water rat
    Florida *see* Muskrat), Round-tailed
    Giant **7**: 965
Water scorpion (toe-biter; needle bug) **20**: 2870–2871
Water shrew **20**: 2872–2873
    American **20**: 2872, 2873
    European **20**: 2872, *2873*
    Mediterranean **20**: 2872
    Pacific **20**: 2872
    see also Tenrec, Web-footed
Water snake **20**: 2874–2875
    Banded (Florida water snake; broad-banded water snake; Gulf salt marsh water snake; mangrove water snake) **20**: 2875
    Green **20**: 2875
    Northern (common water snake) **20**: 2874, 2875
    Red-bellied (copperbelly; yellow-bellied water snake; plain-bellied water snake; blotched water snake) **20**: *2875*
Water stick insect **20**: 2870, *2870*, 2871
Water strider *see* Pond skater
Waterthrush **21**: 2985
Water tiger *see* Diving beetle
water vascular system **18**: 2516
Water vole **21**: 2884–2885
    Southwestern **21**: 2884, 2885
Water-walker *see* Jacana
Wattlebird **21**: 2886–2887
Waxbill **21**: 2888–2889, 2938
    Blue *see* Cordon-bleu
    Common (St. Helena waxbill) **21**: 2888
    Gray (red-eared waxbill; black-rumped waxbill; common waxbill) **21**: 2888
    Orange-cheeked **21**: *2888*
    Strawberry *see* Avadavat
    Zebra (orange-breasted waxbill) **21**: 2888, 2889
wax-eating **9**: 1217
Wax insect, Chinese **16**: 2247
Waxwing **21**: 2890–2892
    Bohemian **21**: 2890, *2890*, 2891, 2892, *2892*
    Cedar **21**: 2890, *2891*
    Japanese **21**: 2890
Weakfish **5**: 602
Weasel **21**: 2893–2895
    Least (cane weasel; common weasel) **14**: 2003; **21**: 2893–2895
    Short-tailed *see* Ermine
    Striped *see* Striped weasel
    White-naped **21**: 3020
Weasel cat *see* Jaguarundi
Weatherfish **11**: 1473, 1474
    Chinese **11**: 1474
Weaver **21**: 2896–2898
    Baya **21**: 2896
    Cape **21**: 2896

Masked **21**: 2896, 2898
    Southern masked **21**: *2897*
    Spectacled **21**: 2897
    Village (black-headed weaver) **21**: 2896, 2898
Webnerian mechanism **3**: 395
webs
    orb spider **13**: 1800–1801, *1800*, *1801*; **18**: *2481*
    processionary moth **15**: 2047, *2047*
    sheetweb **17**: 2339, *2339*, *2340*
Weever (weeverfish; stingfish; sea dragon) **17**: 2445; **20**: 2818; **21**: 2901–2902
    Cape Verde **21**: 2902
    Greater **21**: 2901, 2902
    Guinean **21**: 2902
    Lesser **21**: 2901, *2901*, 2902, *2902*
    Sailfin **21**: 2902
    Spotted **21**: 2902
    Striped **21**: 2902
Weevil **21**: 2903–2904
    Cotton boll **21**: 2904
    Figwort **21**: 2903, 2904
    Gorse **21**: 2903, 2904
    Grain **21**: 2904
    Large pine **21**: 2903, 2904
    Nut **21**: 2903, *2903*, 2904, *2904*
Weil's disease **3**: 299
Weka **15**: 2108
Wels *see* Catfish, European
Werewolf *see* Aardwolf
Whale
    Baleen **21**: 2910, 2911, 2913, 2914
    Beaked *see* Beaked whale
    Beluga (belukha; white whale; sea canary) **2**: 193–194; **21**: 2911, 2914, *2914*
    Blue (sulfur-bottom whale; great northern rorqual) **2**: 246–248; **13**: 1770; **14**: 1954; **16**: 2208, 2209; **21**: 2910–2911, 2913, 2914
    Bowhead (Arctic whale; Greenland whale) **16**: 2166, 2167, 2168; **21**: 2910, 2913
    Bryde's **16**: 2208, 2209; **21**: 2911, 2913
    Fin **16**: 2208, 2209; **21**: 2914
    Gray (California gray whale; scrag whale; mussel-digger; devilfish) **8**: 1052–1053; **21**: 2910, *2911*, 2913
    Humpback (humpbacked whale) **9**: 1258–1260; **13**: 1770; **16**: 2208, 2209, *2209*; **21**: *2910*, 2911, *2913*, 2914
    Killer *see* Killer whale
    Minke (lesser rorqual; pike whale; little piked whale) **16**: 2208, *2208*, 2209; **21**: *2912*, 2914
    Northern bottlenose (North Atlantic bottlenose whale; bottlehead; steephead) **2**: 171
    Pilot *see* Pilot whale
    Right *see* Right whale
    Sei **16**: 2208, 2209; **21**: 2914
    Southern bottlenose (Antarctic bottlenose whale) **2**: 171
    Sperm *see* Sperm whale
    Toothed **21**: 2910, 2911, 2914
    see also Narwhal; Rorqual; Whales
Whalebird **15**: 2045
Whalefish **21**: 2909
    Redmouth **21**: 2909
    Velvet **21**: 2909
Whales 21: 2910–2914
    see also whaling
Whalesucker **15**: 2148

whaling **2**: 248; **8**: 1052; **9**: 1260; **12**: 1719; **14**: 1954; **16**: 2168, 2208; **18**: 2474; **21**: 2910, 2914
Wheat-stem borer **16**: 2241
Whelk (drill) **7**: 927; **21**: 2917–2918
    Channeled **21**: 2917, 2918
    Common (white whelk; buccinum; buckie) **21**: 2917, 2918, *2918*
    Red (almond whelk) **21**: 2917
    Red **21**: 2917
    Rough (whelk-tingle; oyster drill) **21**: 2917
Whelk-tingle *see* Whelk, Rough
Wherry-man *see* Backswimmer
Whimbrel *see* Curlew, Hudsonian
Whipbird *see* Whistler, Golden
Whippoorwill **14**: 2012, 2013
Whip scorpion *see* Pedipalpi
Whip snake (whipsnake) **17**: 2421; **21**: 2919–2920
    Dark green (western whip snake; yellow-and-green snake) **21**: 2919, *2919*, 2920
    Green **21**: 2920
    Sonora **21**: 2919, 2920
    Striped (ornate whip snake; cedar racer) **21**: 2919, 2920
    Yellow-faced **21**: 2919, 2920
Whip spider *see* Pedipalpi
Whiptail **19**: 2650
    New Mexico **19**: 2650
    Western **19**: 2650
Whirligig beetle **21**: 2921–2922
Whirligig *see* Water beetle
whiskers, cat **3**: 403–404
Whistle pig *see* Groundhog
Whistler *see* Marmot, Hoary; Mountain beaver
Whistler (thickhead) **21**: 2923–2924
    Bare-throated **21**: *2923*, *2924*
    Golden (cutthroat; thunderbird; whipbird) **21**: 2923–2924
    Olive **21**: 2923
    Rufous **21**: 2923, 2924
    White-breasted **21**: 2923
Whistling duck (tree duck) **5**: 714, 716; **21**: 2925–2927
    Black-bellied (red-billed whistling duck; red-billed tree duck; cornfield duck) **21**: 2926, 2927, *2927*
    Black-billed **21**: 2926, 2927
    Fulvous **21**: 2925, *2926*
    Lesser **21**: 2925
    Plumed **21**: 2925
    Wandering **21**: 2925
    White-faced **21**: 2925, *2925*, *2927*
Whistling hare *see* Pika
Whistling Jack *see* Butcherbird, Gray
Whitebait *see* Silverside, Samoan
White butterfly **21**: 2928–2929
    Green-veined white **21**: *2928*
    Large white **21**: 2928
    Marbled white **21**: 2929
    Margined white **21**: 2928–2929
    Mustard white **21**: 2928
    Pine white **21**: 2928
    Small white (European cabbage butterfly) **9**: 1280; **21**: 2928
    West Virginia white **21**: 2928, 2929
    Wood white **21**: 2929
White death *see* Shark, Great white
White-eye **14**: 1993; **21**: 2930–2931
    African yellow (green white-eye; yellow white-eye) **21**: 2931
    Javan **21**: *2930*
    Oriental **21**: *2931*
    Yellow-spectacled (Wallace's white-eye) **21**: 2930

Whitefish *see* Menhaden, Atlantic
Whitefly **3**: 416; **21**: 2932–2933
  Cabbage **21**: 2932, 2933
  Citrus **21**: 2932, 2933
  Greenhouse **21**: 2932, *2932*, 2933
White, Gilbert **3**: 337
White pointer *see* Shark, Great white
Whitethroat, Common **18**: 2459
Whiting **4**: 495; **21**: 2934–2935
Whydah **21**: 2936–2938
  Pin-tailed (pin-tailed widowbird) **21**: 2889, 2936, *2936*, 2937, *2937*
  Queen (shaft-tailed whydah) **21**: *2938*
Widowbird, Pin-tailed *see* Whydah, Pin-tailed
Widow bird, Yellow-shouldered *see* Bishop Bird, Burton's black-and-yellow
Wild ass **21**: 2939–2941
  African (burro) **21**: 2939
  Asiatic **8**: 1040; **21**: 2939, *2939*, *2941*
Wildcat (African wildcat) **3**: *403*; **21**: 2945–2947
Wildebeest (gnu) **1**: 61; **8**: 1040; **21**: 2948–2950
  Black (white-tailed wildebeest; white-tailed gnu) **16**: 2216; **21**: 2948, 2949
  Blue (white-bearded wildebeest; brindled gnu) **21**: 2948, *2948*, 2949, *2949*, 2950
Wilsonia pusilla **21**: *2985*
Windhover *see* Kestrel
Wind scorpion *see* Solifugid
wings, hoatzin claws on **9**: 1197, *1197*
Wiretail, Des Mur's **13**: 1819
Wireworm *see* Click beetle
Wisent *see* Bison, European
Wobbegong (carpet shark) **17**: 2331, *2333*; **21**: 2954–2955
  Cobbler **21**: 2955
  Japanese **21**: 2955
  Northern **21**: 2955
  Ornate **21**: 2955
  Spotted **21**: 2954, 2955
  Tassled **17**: 2331; **21**: *2954*, 2955
Wolf **3**: 393
  Ethiopian **5**: 692
  Gray (timber wolf; tundra wolf; plains wolf) **5**: 689, 691–692; **8**: 1054–1057; **19**: 2653; **21**: 2906
  Maned **11**: 1539–1540
  Red **5**: 692; **8**: 1054, 1057
  Red (*dhole*) *see* Dhole
Wolf fish *see* Lancet fish
Wolverine (glutton) **19**: 2729; **21**: 2958–2960
Woma **15**: 2079
Wombat **21**: 2961–2963
  Common (coarse-haired wombat) **21**: *2961*, 2962, 2963
  Northern hairy-nosed **21**: 2961, 2962, 2963
  Southern hairy-nosed **21**: 2961, 2962, *2962*, 2963
wood
  digestion **4**: 493, 494; **11**: 1483
  goatfish eating **7**: 992
  and shipworm **17**: 2349–2350, *2349*
  and termites **19**: 2660
Wood ant **1**: 68; **21**: 2964–2965
  Common **21**: 2964–2965
  Hairy (meadow ant) **21**: 2964, 2965
  Red **21**: 2965
Woodchuck *see* Groundhog
Woodcock **17**: 2427; **21**: 2966–2967
  American **21**: 2966, *2966*
  Eurasian **21**: 2966, 2967

Woodcreeper **19**: 2721; **21**: 2968–2969
  Barred **21**: 2968
  Buff-throated **21**: *2968*, 2969
  Great rufous **21**: 2968
  Narrow-billed **21**: 2968
  Spotted-crowned **21**: 2968
  Streak-headed **21**: 2969
  Tawny-winged **21**: 2969
  Wedge-billed **21**: *2969*
Woodlark **17**: 2396
Wood louse (pill bug; sow bug; slater; roly-poly) **21**: 2973–2974
  Pill **21**: 2973
Woodpecker **19**: 2653–2654; **21**: 2977–2979
  Acorn **19**: 2654; **21**: *2978*, 2979
  Downy **19**: 2654
  Great spotted **19**: 2654; **21**: *2979*
  Green **21**: 2965, 2977, *2977*, 2978–2979
  Ground **21**: 2979
  Hairy **21**: 2978
  Imperial (*extinct*) **21**: 2977
  Ivory-billed (*extinct*) **21**: 2977
  Lesser spotted **19**: 2617
  Pied (spotted woodpecker) **21**: 2977
  Pileated **21**: 2979
  Red-headed **21**: 2978, 2979
  Rufous **21**: 2979
Woodpecker-finch **5**: 648, 649–650, *650*; **16**: 2296
Wood rat *see* Pack rat
Wood swallow (swallow-shrike) **21**: 2983–2984
  Ashy **21**: 2983, 2984
  Black-faced **21**: *2983*
  Dusky **21**: 2983
  Little **21**: 2983, 2984
  Masked **21**: 2983
  Papuan **21**: 2984
  White-backed **21**: 2984
  White-breasted **21**: 2983, *2984*
  White-browed **21**: 2984
Wood warbler (*New World warbler*) **18**: 2458; **19**: 2617; **21**: 2985–2986
  Black-and-white warbler **21**: 2986
  Black-poll warbler **21**: 2985
  Black-throated blue warbler **21**: 2985
  Blue-winged warbler **21**: 2985
  Chestnut-sided warbler **21**: 2985
  Golden-winged warbler **21**: 2985
  Kirtland's warbler **21**: 2985
  Myrtle warbler **21**: 2986
  Parula warbler **21**: 2986
  Prothonotary warbler **21**: 2986
  Yellow warbler **4**: *553*; **21**: 2985
Wood wasp *see* Sawfly
Woolly monkey **21**: 2987–2988
  Hendee's (yellow-tailed woolly monkey) **21**: 2987, 2988
  Humboldt's **21**: 2987, 2988
World War I, use of blowfly maggots **2**: 242
Worldwide Fund for Nature **2**: 162
Worm, Horsehair (hair worm; Gordian worm) **9**: 1237–1238
Worm shell (tube mollusk; tube snail; vermetid) **21**: 2989–2990
  Scaled **21**: 2990
wounds, and blowfly maggots **2**: 242
Woylie *see* Bettong, Brush-tailed
Wrasse **4**: 533; **6**: 836; **21**: 2991–2992
  Ballan **21**: 2991
  Broomtail **6**: *836*
  Corkwing **21**: *2992*
  Cuckoo **21**: 2991
  Humphead (Napoleon fish) **21**: *2991*
  Mediterranean rainbow **21**: 2991

Wreckfish *see* Bass, Stone; Sea bass, Giant
Wren **21**: 2993–2995
  Bewick's **21**: 2993
  Bush (South Island bush wren) **16**: 2164
  Cactus **21**: 2993–2994, *2995*
  Common European **7**: 999
  Giant **21**: 2993
  House **21**: 2993, 2995
  Marsh (long-billed marsh wren) **21**: *2994*, 2995
  Rock *see* Rifleman
  Rock (*true wren*) **21**: 2994
  Short-billed marsh wren (sedge wren) **21**: 2994
  South Island *see* Rifleman
  Winter (wren; European wren; Jenny wren) **21**: 2993, *2993*, 2994, 2995
  Zapata **21**: 2994
Wren babbler **1**: 113
Wrentit **1**: 113; **3**: 426
Wren-warbler *see* Thornbill
Wrybill **14**: 1989, 1990
Wryneck (snakebird) **21**: 2996–2997
  Eurasian **21**: 2996, *2996*, 2997
  Red-breasted **21**: 2996
Wuchereria **12**: 1664
Wyulda squamicaudata **14**: 1929, 1930

X

Xanthocephalus xanthocephalus **2**: 224
Xanthopan morgani predicta **9**: 1168
xanthophores **3**: 420
Xantusia
  arizonae **13**: 1745, 1746
  henshawi **13**: 1745
  vigilis **13**: 1745, 1746
Xenia **17**: 2440, 2441
Xenicus
  gilviventris **16**: 2164
  longipes **16**: 2164
  lyalli **16**: 2165
Xenomystus **6**: 803, 804
  nigre **6**: *804*
Xenopirostris
  damii **20**: 2813
  polleni **20**: 2813
Xenops, Plain **13**: 1819
Xenops minutus **13**: 1819
Xenopsylla cheopis **6**: 850
Xenopus laevis **4**: 465; **7**: 903
Xenus cinereus **16**: 2233
Xestia c-nigrum **13**: 1821
Xiphias gladius **19**: 2606
Xiphocolaptes major **21**: 2968
Xiphophorus
  helleri **14**: 1984; **19**: 2609
  maculatus **14**: 1983, 1984
  variatus **14**: 1983
Xiphorhynchus guttatus **21**: 2968, 2969
Xixuthrus heros **11**: 1483
X-organ **17**: 2355
Xue bao *see* Snow leopard
Xyleutes cinereus **7**: 992
Xylopargus **9**: 1176
Xyrichthys novacula **21**: 2991
Xysticus cristatus **5**: 581

Y

Yabbie *see* Crayfish
Yak **3**: 305, *305*; **21**: 3000–3002
  Domestic (dzo) **21**: 3000, 3001, *3001*, 3002
  Northern (*extinct*) **21**: 3002
  Wild **21**: 3000, 3001, *3002*
Yapok *see* Opossum, Water

yellow fever **12**: 1664
Yellow jacket (*hornet*) **9**: 1233
Yellowjacket (*wasp*) *see* Wasp, Vespula
Yellowlegs
  Greater **16**: 2233
  Lesser **16**: 2233
Yeti **17**: 2324
Y-organs **17**: 2355
young
  identical in armadillos **1**: 87
  killing of *see* infanticide
  precocial **1**: 32
Ypsolophus pometellus **19**: 2687
Yuhina **1**: 114

Z

Zaedyus pichiy **1**: 85
Zaglossus **18**: 2486
  bruijni **18**: 2487
Zalembius rosaceus **18**: 2573
Zalophus californianus **16**: 2282, 2283, 2285
Zanclus cornutus **12**: 1653
Zanobatus **8**: 1103
Zapteryx **8**: 1103
Zebra **21**: 3009–3011
  Grevy's **21**: 3010, *3010*, 3011
  Mountain (Cape mountain zebra; Hartmann's mountain zebra) **21**: 3009–3010, 3011
  Plains (Chapman's zebra; Grant's zebra; Burchell's zebra—*extinct*; quagga—*extinct*) **8**: 1040; **16**: 2216; **21**: 3009, 3011
Zebrafish *see* Lionfish, Red
Zebra fish (zebra danio) **21**: 3014–3015
Zebra mouse (striped grass mouse) **21**: 3016–3017
Zebrasoma flavescens **18**: 2575
Zebu (Brahman cattle) **21**: 3018–3019
Zenkerella insignis **16**: 2251
Zeus faber **10**: 1328
Zeuzera pyrina **7**: 991
Ziphius cavirostris **2**: 170, *171*
Zirfaea crispata **14**: 1939, 1940
Zoarces viviparus **2**: 238
zoea larvae **4**: 573; **6**: 742; **9**: 1178; **16**: 2186; **17**: 2355, *2355*; **18**: 2475, 2506
zooplankton **4**: 475; **13**: 1770, 1771; **14**: 1980
zoo trade, orangutans **13**: 1799
zooxanthellae **4**: 463, 529; **16**: 2300; **17**: 2440
Zorilla (zorille; striped polecat) **21**: 3020–3021
Zorille *see* Zorilla
Zorro
  Crab-eating *see* Fox, Crab-eating
  Gray *see* Fox, South American gray
  Short-eared *see* Dog, Short-eared
Zosterops **21**: 2930, *2930*, 2931, *2931*
Zygaena **3**: 321–322, *321*
Zygogeomys **14**: 1996, 1997